Springer Proceedings in Earth and Environmental Sciences

Series Editors

The series Springer Proceedings in Earth and Environmental Sciences publishes proceedings from scholarly meetings and workshops on all topics related to Environmental and Earth Sciences and related sciences. This series constitutes a comprehensive up-to-date source of reference on a field or subfield of relevance in Earth and Environmental Sciences. In addition to an overall evaluation of the interest, scientific quality, and timeliness of each proposal at the hands of the publisher, individual contributions are all refereed to the high quality standards of leading journals in the field. Thus, this series provides the research community with well-edited, authoritative reports on developments in the most exciting areas of environmental sciences, earth sciences and related fields.

Khadija Baba · Latifa Ouadif ·
Abderrahman Nounah · Mounir Bouassida
Editors

Advances in Research in Geosciences, Geotechnical Engineering, and Environmental Science

Proceedings of the Fourth Scientific Conference on Geosciences and Environmental Management (GeoME'4), Morocco 2023

Springer

Editors
Khadija Baba
Higher School of Technology of Sale
Mohammed V University
Salé, Morocco

Latifa Ouadif
Mohammadia Engineering School
Mohammed V University of Rabat
Rabat, Morocco

Abderrahman Nounah
Higher School of Technology of Sale
Mohammed V University
Salé, Morocco

Mounir Bouassida
National School of Engineering of Tunis
University of Tunis El Manar
Tunis, Tunisia

ISSN 2524-342X ISSN 2524-3438 (electronic)
Springer Proceedings in Earth and Environmental Sciences
ISBN 978-3-031-49344-7 ISBN 978-3-031-49345-4 (eBook)
https://doi.org/10.1007/978-3-031-49345-4

This Springer imprint is published by the registered company Springer Nature Switzerland AG
The registered company address is: Gewerbestrasse 11, 6330 Cham, Switzerland

Paper in this product is recyclable.

Contents

Environmental Engineering

Influence of Organic Household Waste and Olive Pomace Composts on Fenugreek Seeds Germination and Growth

Ayoub Doughmi[1] (✉), Ghizlane Elkafz[1], Fatima Benradi[1], Essediya Cherkaoui[1], Mohamed Khamar[1], Abderrahman Nounah[1], and Abdelmjid Zouahri[2]

[1] Civil and Environmental Engineering Laboratory (LGCE), Water and Environmental Materials Team, Higher School of Technology in Salé, Mohammed V University in Rabat, MA11060 Salé, Morocco
ayoub_doughmi@um5.ac.ma

[2] INRA, Regional Center for Agronomic Research of Rabat, Research Unit on the Environment and the Natural Resources Conservation, MA10112 Rabat, Morocco

Abstract. A blend of olive pomace and organic household waste was meticulously combined in varying ratios, varied proportions were used for the olive pomace (ranging from 15% to 50%) and the organic household waste (ranging from 85% to 50%). This mixture was then subjected to a composting process lasting 4 months. Subsequently, the physicochemical and microbiological characteristics of each combination were thoroughly examined and analyzed. The main aim of this study is to investigate the potential of composts produced from a combination of olive pomace and organic household waste as soil enhancers, along with their effects on plant cultivation. To evaluate the maturity of these composts, a series of phytotoxicity tests were carried out. Specifically, the germination test was carried out in a controlled laboratory environment within an incubator, maintaining optimal conditions for temperature, light, and humidity. Each Petri dish contained 15 Fenugreek seeds placed on filter paper and was regularly irrigated every 48 h. Notably, the GD2 compost (25%) exhibited an impressive germination rate of 94%. Comparatively, the germination rates for various compost concentrations differed significantly from the control, indicating that certain compost concentrations had a negative effect on root quality. Overall, the germination tests demonstrated that the compost's favorable nutrient content makes it a promising organic amendment. Indeed, the enhancement in Fenugreek yield was directly related to the dosage of compost applied.

Keywords: Phytotoxicity test · Fenugreek · Organic household waste · Olive pomace · Compost

1 Introduction

Agriculture holds significant economic, social, and environmental importance in Morocco [1]. Notably, the agricultural sector contributed approximately 13.92 billion USD in added value in 2017, accounting for 15% of the country's GDP during the period

K. Baba et al. (Eds.): GeoME 2023, SPEES, pp. 3–14, 2023.
https://doi.org/10.1007/978-3-031-49345-4_1

from 2008 to 2017, surpassing an annual 11.80 billion USD [2], also it contributes to exports up to 10.5% [1]. In addition, A substantial proportion of the population, accounting for 38% nationwide and approximately 74% in rural areas, relies on agriculture as their primary source of income [2]. The sector also has a positive and negative impact on the country's overall environment. Since it uses natural resources [1], in particular arable land which covers 8.8 million ha (i.e. 18% of the country's total surface area) of which 15% is irrigable, and the rest is Bour land [3].

Morocco has embraced the Green Morocco Plan, which seeks to transform agriculture into a significant driver of growth and modern, efficient development. This plan is focused on achieving better valorization and sustainable management of natural resources in the country [3]. One of the actions identified in this plan is the national strategy for agricultural training and research (SNFRA), which aims to strengthen the capacities and potential of Moroccan agriculture [4], as well as the FERTIMAP tool or fertility map of cultivated soils in Morocco, which develops a sustainable fertilization advisory system based on the analysis of soil fertility and crop nutrient requirements for the different regions of Morocco [5].

As the world's population grows, so does the production of waste, particularly organic household waste, which is becoming worthless and harmless to the environment. Rich in organic matter and minerals, this waste can make valuable contributions to the soil if it is composted. The utilization of this waste in its raw form presents various challenges and constraints in terms of its handling and application, primarily due to the potential existence of pathogenic bacteria and microorganisms [6].

Moroccan organic waste production also includes residues from the agri-food sector, in addition to the 5.8 million tonnes/year fraction corresponding to biodegradable municipal waste from urban households and sludge generated by wastewater treatment [7]. Consequently, Morocco generates an estimated 1.57 tonnes of organic waste per inhabitant per year, which is available for management and utilization, compared to the 0.17 tonnes per inhabitant per year limited to household waste [8]. Fermentation in the broadest sense of the term is the only processing method suitable for recycling organic waste in agriculture, as it produces a co-product for soil improvement and fertilization.

Studies conducted in both the United States and Europe have demonstrated that composting is a straightforward biological process suitable for recycling waste, with the condition that trace elements such as Cd, Pb, Cu, and Zn are regulated. Composting is a natural process that aids in the decomposition of organic matter, ultimately yielding a highly stable end product free from phytotoxicity and pathogenic microorganisms [9].

Compost is abundant in humic substances, which effectively enhance the physicochemical properties of the soil and promote bioactivity [10]. Additionally, compost contains essential plant nutrients [11].

To be utilized as a soil enhancer, compost must attain a state of maturity. The evaluation of compost maturity and stability typically relies on various parameters, including chemical, physical, biological, and spectroscopic analyses [12].

In regions where livestock farming has diminished, urban composts play a significant role as a valuable source of organic matter for agricultural use. The utilization of these composts in agriculture is subject to strict regulations, with mandatory standards such as NFU 44 095 for sludge composts and NFU 44051 for other organic soil improvers.

Assessing their agronomic value involves laboratory characterization methods, while long-term field trials measure the effects of repeated applications. The ultimate goal is to use the produced compost to enhance the yield of specific crops and maintain green spaces [13], including the germination of plants using this compost.

Our study also addresses the issue of compost recovery in the agricultural sector.

2 Materials and Methods

Experimental Setup

The method consists of applying a germination test to fenugreek seeds [14–16], in petri dishes [17, 18, 20] by irrigating them with prepared compost solutions. The method consists of studying six compost mixtures. These had already been prepared in a previous study [19]. A well-ventilated environment is created by blending olive pomace with organic household waste in 30-L barrels, the mixture comprises different proportions, ranging from 15% to 50% for olive pomace and from 85% to 50% for organic household waste. To ensure proper aeration, the barrels are equipped with holes and strategically placed in a sunny location. It took 120 days for the composting process to reach its full maturity [19].

From each mixture, a stock solution and dilutions are prepared; each dilution is intended to irrigate three boxes of fenugreek seeds in repetition [20].

Previously, the proportions of the mixtures and the physicochemical and microbiological properties of the diverse composts were established [19].

Plant Material: Fenugreek

Fenugreek, or Trigonella Fenugreek (*Trigonella foenum-graceum*), grows in open fields on limestone-rich soils. This plant is distinguished by its tall stem, which can reach 50 cm. Its oval, toothed leaves and flowers oscillate between yellow and violet. It contains numerous yellow seeds resembling gravel [21].

The plant is versatile in its ability to thrive in various soil types and is typically cultivated in Bour without the need for supplementary irrigation in regions where rainfall ranges from 300 to 450 mm. Moreover, we chose it for the study because of its short germination time: 2–3 days and its ease of use [22].

The fenugreek seeds were subjected to one washing cycle using tap water, followed by two rinses with distilled water [14–16, 23]. And they were left to soak for 30 min in distilled water [20].

Seed Germination Conditions

Compost mixtures are dried at temperatures between 40 °C and 60 °C until a constant mass is obtained. After drying and mixing, solutions of (1/10) (m/V) were prepared. The solutions were mixed for 30 min. After decanting and filtration, stock solutions were obtained for all compost mixtures.

5 dilution percentages of the stock solutions were chosen to work with: 0%; 25%; 50%; 75%; 100% [17]. An average of 3 ml of irrigation per box under temperature conditions of 25 °C [20, 23] and 70–80% humidity [16].

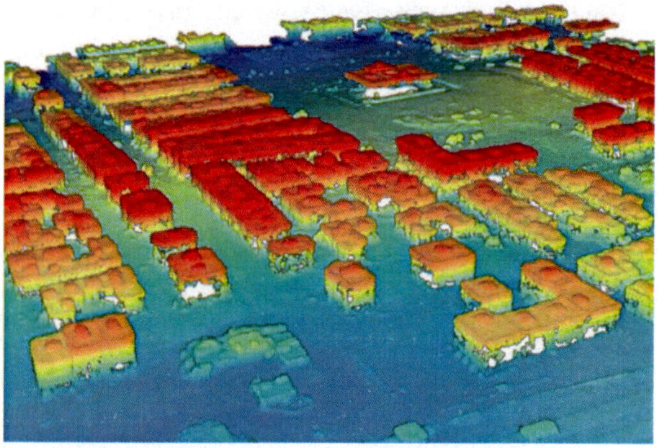

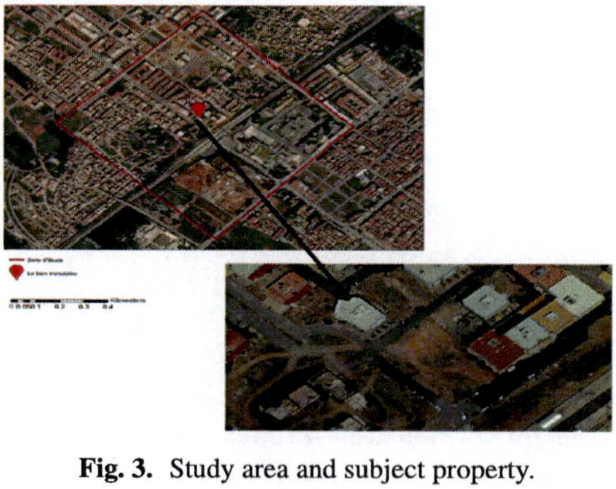

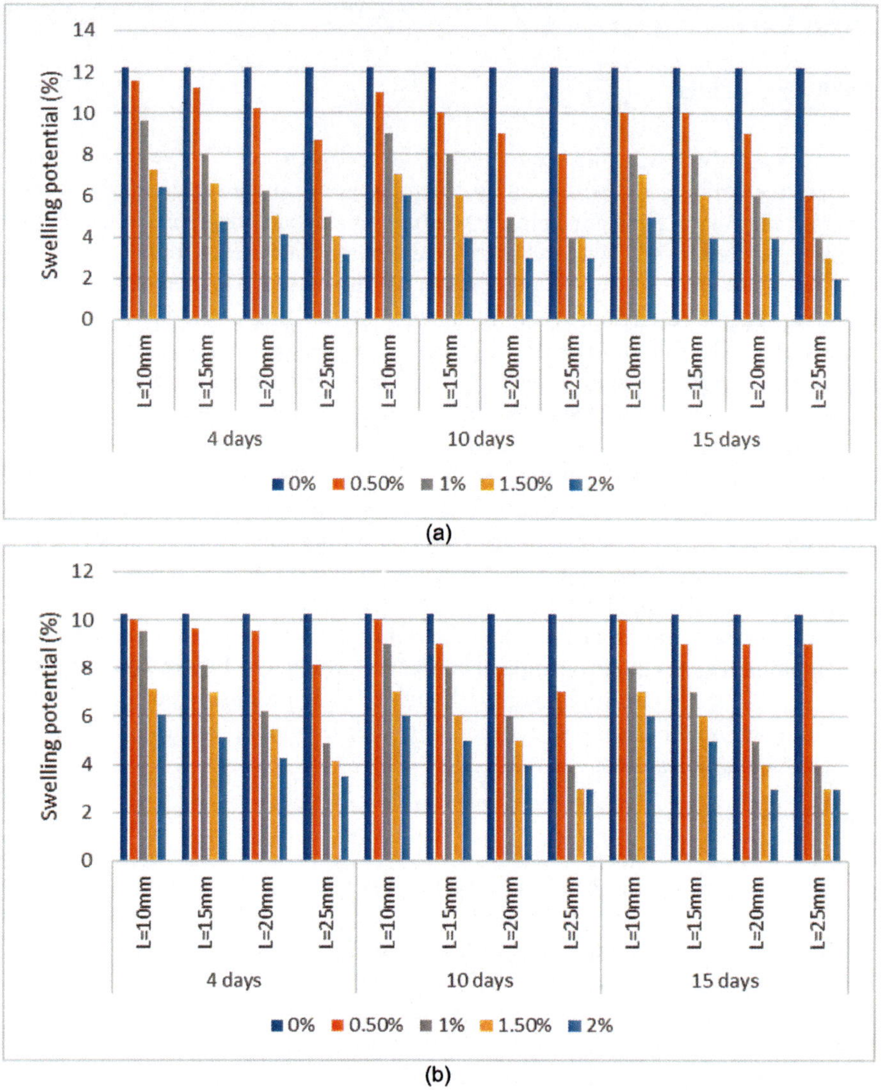

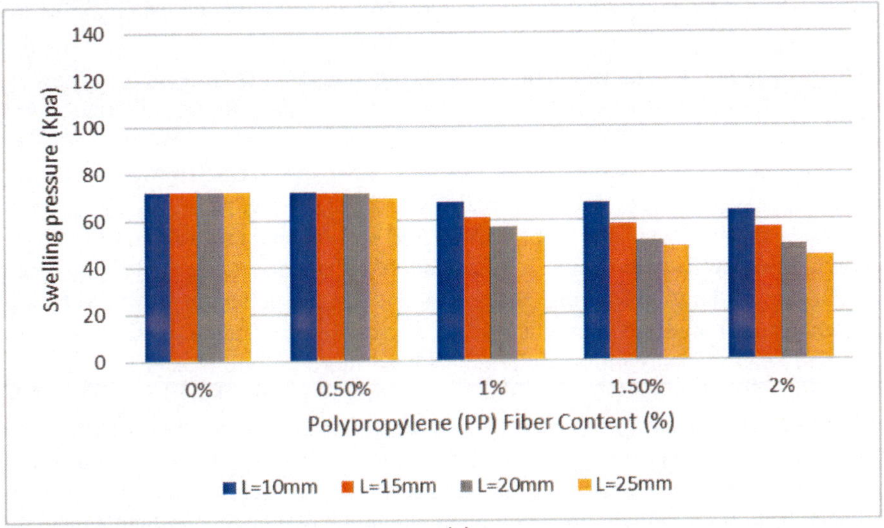

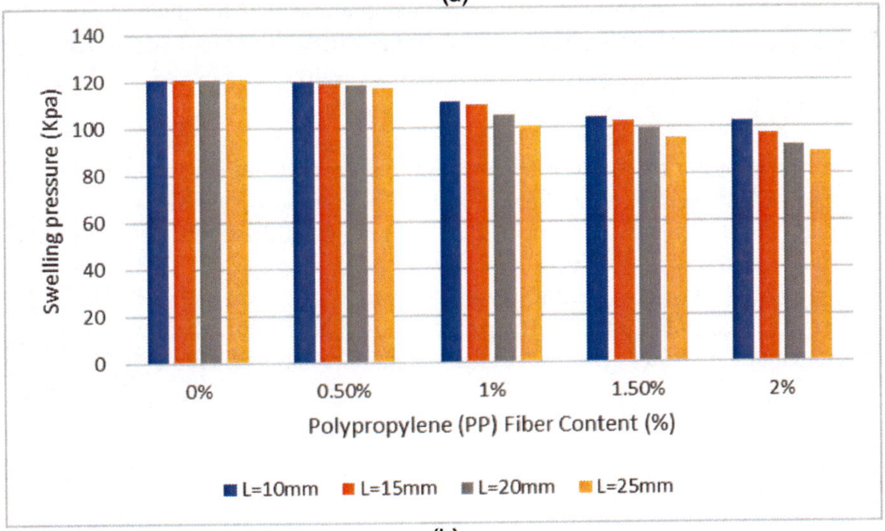

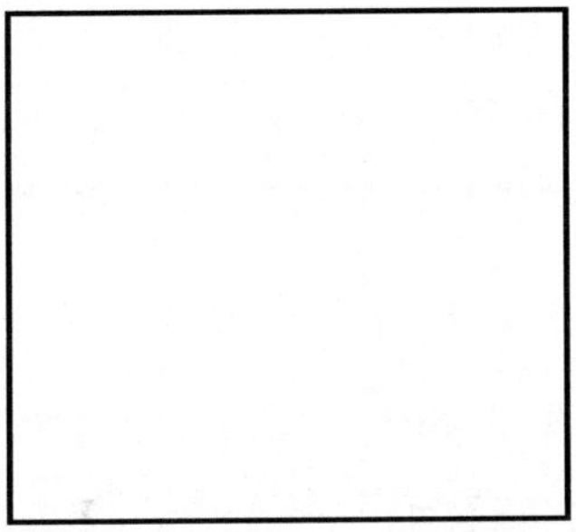

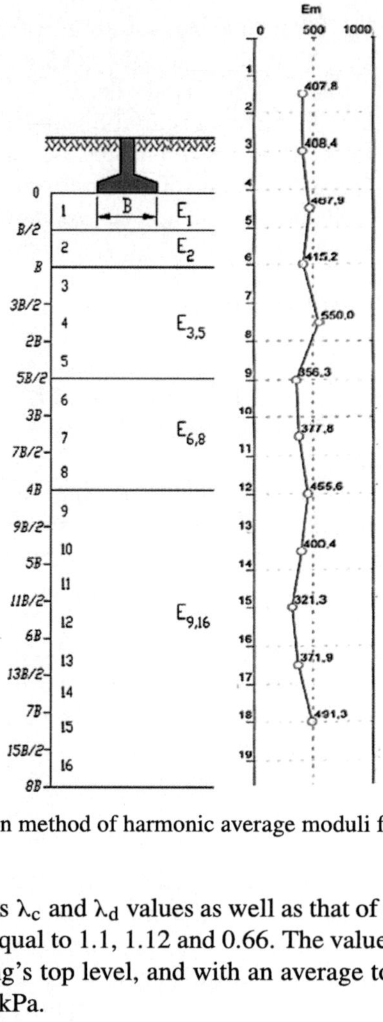

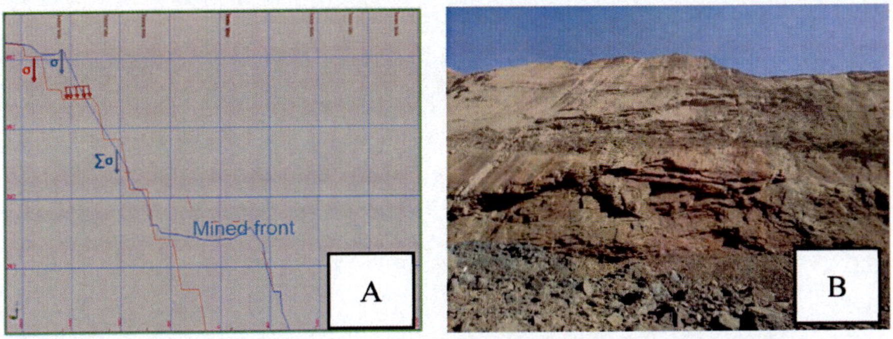

To determine the heavy vehicle traffic class, we will utilize the formula in Eq. 4, as shown below:

$$\text{NEE13T} = \text{C1} \times \text{C2} \times \text{C3} \times \text{C4} \times \text{N4} \times \text{N2} \times \text{F1} \tag{4}$$

where:

NEE13T represents the cumulative traffic expressed in the number of standard 13-ton vehicles.

C1 denotes the width coefficient.

C2 signifies the average aggressiveness coefficient.

C3 accounts for the increase in heavy vehicles.

C4 represents the growth of heavy vehicles.

N4 indicates the cumulative traffic.

N2 represents the updated traffic of heavy vehicles weighing more than 8 tons.

F represents the traffic distribution coefficient.

3 Results

3.1 Results of Reconnaissance Work

A field survey was carried out to collect essential information regarding the study sections. The survey involved conducting manual assessments that were strategically distributed across the designated areas. The primary goal was to gather precise data to facilitate the evaluation and analysis of the project. The manual surveys yielded the following findings (see Table 2).

Table 2. Type of pavement structure

Sounding depth	Type of material
[0 à – 0,1 m]	Embankment
[– 0,1 m à – 0,6 m]	Topsoil
[– 0,6 m à – 2,3 m]	Turf loam
[– 2,3 m à – 3 m]	Loamy tuff

The land's average configuration, based on the survey results, is described as follows:

The embankment has an average thickness of 10 cm.

This is followed by topsoil with an average thickness of 50 cm.

Next, there is turf loam with an average thickness of 1.7 cm.

In the coastal region, there is a presence of loamy tuff, which extends until the end of the soundings.

After 48 h of germination, the germinated seeds are counted and sized, and returned to the incubator for a period of 8 days, in the presence of light [24], to ensure photosynthesis, and under temperature conditions of 25 °C, and 70–80% humidity. Irrigated every 48 h with 3 ml of water (Table 1).

Table 1. Solutions prepared from the mature composts

Composts	25%	50%	75%	100%
Gr	25% Gr + 75% DW	50% Gr + 50% DW	75% Gr + 25% DW	100% Gr
D	25% D + 75% DW	50% D + 50% DW	75% D + 25% DW	100% D
GD1	25% GD1 + 75% DW	50% GD1 + 50% DW	75% GD1 + 25% DW	100% GD1
GD2	25% GD2 + 75% DW	50% GD2 + 50% DW	75% GD2 + 25% DW	100% GD2
GD3	25% GD3 + 75% DW	50% GD3 + 50% DW	75% GD3 + 25% DW	100% GD3
GD4	25% GD4 + 75% DW	50% GD4 + 50% DW	75% GD4 + 25% DW	100% GD4

DW: distilled water

Measured Parameters

Once 48 h have passed since germination, the number of seeds that have sprouted is tallied, and the lengths of their radicles are measured using a graduated ruler [15], such that a germinated seed has a radicle ≥ 5 mm [25]. In order to analyze the results obtained, a set of formulas is applied:

- **Germination rate %** [26, 27] = (germinated seeds number * 100)/tested seeds number.
- **Vigor index %** [14, 28] = (Root length + Stem length) * germination rate %.
- **Germination index %** [17, 29, 30] = the formula calculates the percentage of germinated seeds and their root lengths in the sample compared to the control group. It is represented as follows:

$$\begin{aligned}&(\text{Number of germinated seeds in the sample}/\\&\quad\text{Number of germinated seeds in the control})\times\\&(\text{Root lengths of germinated seeds in the sample}/\\&\quad\text{Root lengths of germinated seeds in the control})\times 100\end{aligned}$$

- **Dry matter**: The dry weight of the seedling's material, encompassing stems, leaves, and roots, was determined immediately after cultivation. This process was achieved by subjecting the seedlings to an oven at 105 °C until a consistent weight was reached, as specified in the provided Ref. [31].

Après les 48 h de germination, on compte le nombre de graines qui ont germées et on dimensionne les longueurs des radicules à l'aide d'une règle graduée [15], tel qu'une graine qui a germée est de radicule $\geq$ 5 mm [25].

Statistical Analysis

In this study, all the analyzed parameters were processed using SPSS software (Statistical Package for the Social Sciences, version 20). The results are presented as mean $\pm$ standard deviation and were analyzed using ANOVA variance analysis.

3 Results

3.1 Germination Rate

For olive pomace, the dilution that gave the best germination rate, exceeding 63.33% of the control, was the 75% dilution for olive pomace compost, with a value of 80%. The rate for the other dilutions remained close to the control, with a minimum of 60% corresponding to the 25% dilution.

For organic household waste, it is clear that the germination rate decreases with increasing concentrations of D, and always remains below the control germination rate. D's phytotoxicity is said to increase with this germination rate decrease.

We can clearly see the difference with the mixtures: GD1, GD2, GD3 and GD4, where all rates have evolved, most of them exceeding the control value, with germination rate of 78% in GD1, 94% in GD2, 80% in GD3 and 78% in GD4. From this, we can infer the significance of combining organic household waste with olive pomace to achieve an ideal C/N ratio and provide the essential nutrients required for optimal plant growth (Fig. 1).

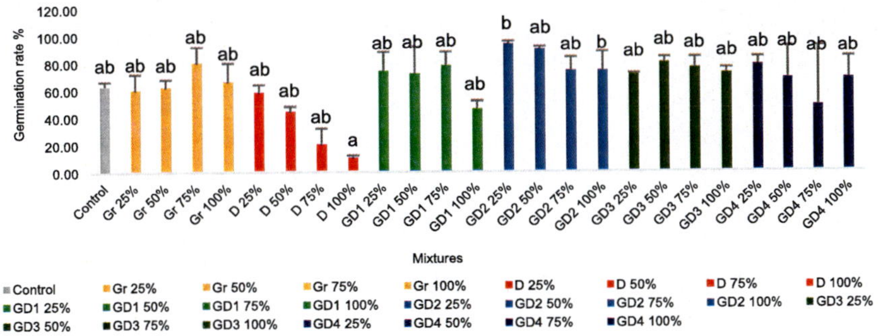

Fig. 1. The germination rate of fenugreek seedlings was assessed using various concentrations of composts. Distinct letters were used to denote values that exhibited significant differences ($p < 0.05$).

3.2 Germination Index

The obtained results indicate that the germination index of Fenugreek seeds exhibits an increase when compost mixtures are utilized. However, we observed a declining trend in the germination process as the dosage of organic household waste was elevated (Fig. 2).

Composts D 75% and D 100% show a germination index (25.47% and 6.14%) below 40%, indicating severe inhibition. In addition, the control showed a germination index (43.33%) between 40% and 60%, signifying strong inhibition. However, GD1 100% showed a germination index (61.12%) between 60% and 80%, indicating mild inhibition. On the other hand, all other composts show no inhibition (GI > 80%) with a maximum of (206.88%) in GD2 25% [17, 29, 30].

A 75% dose of organic extract from household waste compost gave a GI of 77% with lettuce seeds. A dose of 100% pure organic extract of household waste compost gave GI percentages of around 46% for lettuce [17].

According to Zucconi et al. [32], compost is considered non-toxic when its GI exceeds 50%. The findings indicate that the germination index varies depending on the dosage of compost extracts.

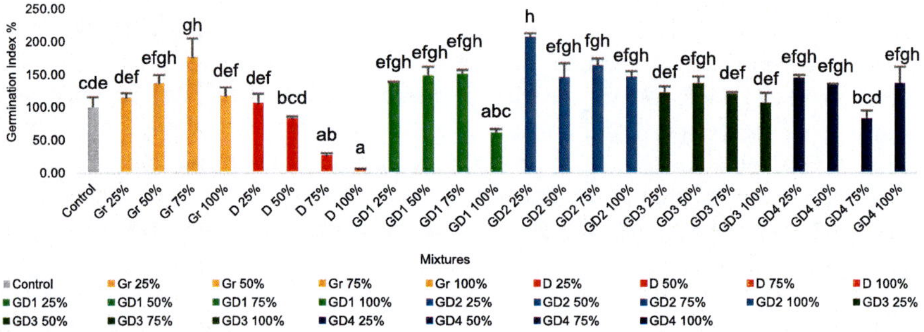

Fig. 2. The germination index of fenugreek seedlings was assessed using various concentrations of composts. Distinct letters were used to denote values that exhibited significant differences (p < 0.05).

3.3 Vigor Index

The vigor index demonstrates a beneficial impact resulting from the application of these composts, as evidenced by an increase when compared to the control across different compost percentages. Notably, seedlings cultivated in compost percentages of GD2 exhibited an optimal vigor index (355.22%), which was significantly distinct from both the other percentages and the control (Fig. 3).

Overall, the inclusion of organic amendments resulted in a significant enhancement of the vigor index in fenugreek seedlings across all tested percentages, in comparison to the control group. Seedling vigor is closely associated with the presence of various nutrients, with nitrogen playing a vital role in promoting growth, phosphorus stimulating root development, and potassium facilitating nutrient assimilation by plants [33]. These composts contain abundant quantities of these elements (NTK; P_2O_5; K_2O). This factor accounts for the notable vigor observed in plants fertilized with varying doses of compost, in contrast to the control.

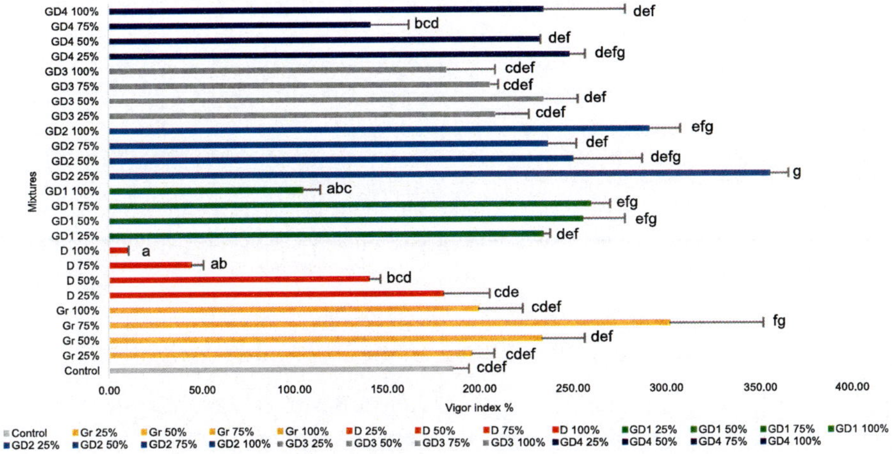

Fig. 3. The vigor index of fenugreek seedlings was assessed using various concentrations of composts. Distinct letters were used to denote values that exhibited significant differences ($p < 0.05$).

3.4 Composts Addition Effect on Root Seedlings Development

There were no notable differences in root length when comparing all compost concentrations to the control group ($P > 0.05$). However, across all concentrations except for D 100%, the root length showed a significant difference, being notably lower than the control. The relationship between root development and strength and the availability of soil phosphorus is widely recognized [34]. Hence, the increased concentration of phosphorus in the composts may explain the improved root development observed in plants treated with compost.

The same results were observed in the germination of Festuca and Italian ryegrass using six substrates obtained from the two pomace sources [35]. Nonetheless, the root development of Fenugreek seedlings showed a notable increase, particularly for the 5%, 10%, and 15% doses of olive pomace and cattle manure compost, in comparison to the control group [14] (Fig. 4).

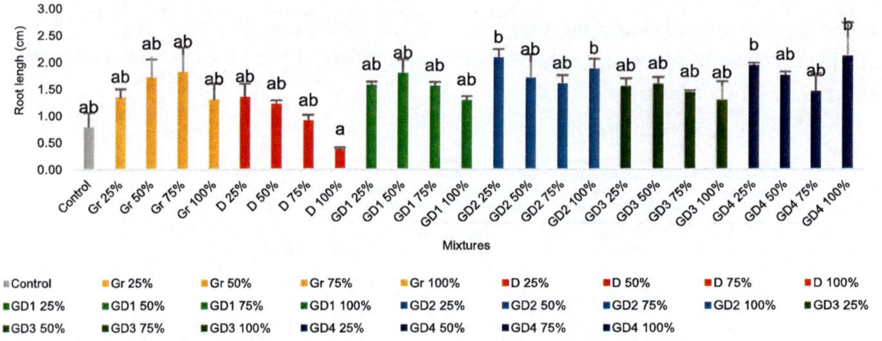

Fig. 4. The effects on fenugreek root seedlings development was assessed using various concentrations of composts. Distinct letters were used to denote values that exhibited significant differences ($p < 0.05$).

3.5 The Influence of Composts Addition on the Dry Matter Weight

The addition of composts at different percentages leads to a substantial increase in the dry matter weight of fenugreek seedlings, encompassing their stems, leaves, and roots, in comparison to the control group. Notably, this enhancement is statistically significant ($p < 0.05$) for compost concentrations GD1 50%, GD1 75%, GD2 25%, and GD2 50%. However, for all other compost concentrations, the disparity in dry matter weight is not statistically significant ($p > 0.05$) when compared to the control group. The observed increase in seedling dry matter weight can be attributed to the nutrient-rich composition of the composts, particularly their high nitrogen content and essential macroelements as sodium, calcium, magnesium, and potassium. These elements play a vital role in the development of plant tissues and together account for 99% of their total mass [36]. The relationship between the added compost concentrations and dry matter weight seems to follow an inverse proportionality, possibly due to the higher presence of phenols in the compost [35]. Comparable results were noted in prior studies that examined the utilization of olive pomace compost for the germination of Festuca and Italian ryegrass [35] (Fig. 5).

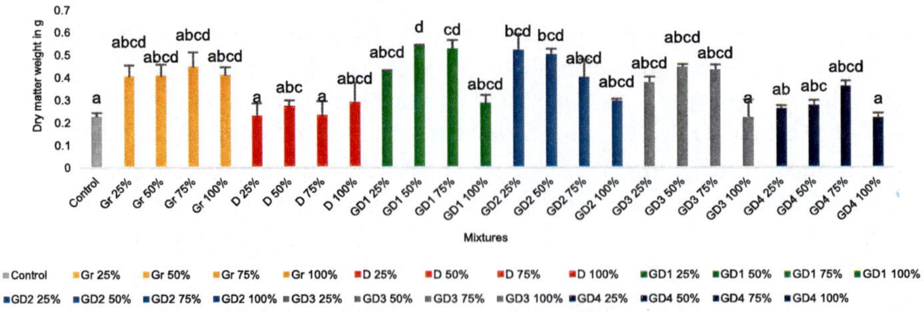

Fig. 5. The effects on fenugreek dry matter weight was assessed using various concentrations of composts. Distinct letters were used to denote values that exhibited significant differences ($p < 0.05$).

4 Discussion

Consequently, composts use made from olive pomace and organic household waste in different percentages can make a significant nutritional contribution to Fenugreek germination. Indeed, the inclusion of composts at various percentages resulted in the ideal circumstances that promote successful seed germination in comparison to the control group. Moreover, the addition of composts led to an increase in the dry matter weight of Fenugreek plants compared to the control group. These findings align with a previous study that reported similar results during the germination of Festuca and Italian ryegrass [35]. It was noted that ryegrass plants experienced a noteworthy surge in their dry matter weight following an 87-day amendment with olive pomace compost [37]. A research conducted showed that the incorporation of olive pomace compost and cattle manure resulted in a considerable improvement in seedling vigor and root development, especially at doses of 5%, 10%, 15%, and 20% [14]. This improvement can be attributed to the nutrient-rich composition of the composts, Comprising vital fertilizing elements (P_2O_5, NTK and K_2O) essential for the growth and development of the majority of plant species [38].

Our findings align with studies conducted in different countries and involving various plant species. As an example, the utilization of olive pomace compost on olive trees in Italy for an 8-year duration led to a substantial improvement in vegetative growth and fruit yield [39]. Following three consecutive years of amending olive trees with olive pomace compost, a notable rise in vegetative activity and productivity was observed [40]. This increase in yield can be attributed to elevated levels of total nitrogen, organic matter, exchangeable potassium and assimilable phosphorus within the compost [41]. Indeed, multiple studies have demonstrated that long-term soil amendment with pomace compost leads to the organic matter and nutrient levels augmentation into the soil, including total nitrogen, exchangeable potassium and assimilable phosphorus. Importantly, these amendments do not cause any significant alterations to soil pH or salinity [42–45]. The aforementioned studies did not observe any detrimental effects on olive oil quality. This reaffirms the notion that the use of pomace compost does not impact the overall quality of the fruit.

Following a 10-day incubation period of both maize and groundnuts crops, phytotoxicity tests were performed, and the results indicated that the inclusion of 25% compost in the growth medium led to a germination rate of 90% for maize, compared to 80% in the control group [13]. In the case of groundnuts, a slight improvement in germination rate was recorded, at 72% versus 70% for the control [13]. The findings presented align with those reported by Chennaoui et al. [17] in their phytotoxicity experiments involving wheat and tomato crops. In their research, they found that adding 25% organic household waste compost to the growing medium led to a germination rate of up to 85% for the wheat variety, as opposed to 70% in the control group.

In tomatoes case, a germination rate of around 67% was recorded, compared with 58% for the control. These results show that the profitability of compost use is excellent at the low rate of 25%, demonstrating that profitability does not depend on the rate applied.

5 Conclusions

The maturity of compost plays a crucial role in determining its agronomic quality, as immature compost can exhibit phytotoxic effects. To assess the phytotoxicity of the compost produced, a germination test was conducted. The test involved observing the number of seeds that successfully germinated during a cultivation period of 10 days. The results indicated that the compost did not exhibit any toxic effect, as all mixtures produced in this study demonstrated a germination rate exceeding 50% [46].

The germination index of fenugreek seeds in this study exceeded 50%, indicating compost maturity except for D 75% and D 100% (GI < 50%).

Our study corroborates the findings of previous authors and adds further support to the notion that the inclusion of a blend of olive pomace compost and organic household waste in the soil can present a sustainable alternative to costly and environmentally damaging chemical fertilizers.

References

1. Lahlimi Alami, A.: Prospective Maroc 2030 (agriculture 2030 quels avenirs pour le Maroc), 107 pp. (2008)
2. Ministère de l'Agriculture, de la Pêche Maritime, du Développement Rural et des Eaux et Forêts, Département de l'agriculture, Filières végétales (2020)
3. Harbouze, R., Pellissier, J.P., Rolland, J.P., Khechimi, W.: Rapport de synthèse sur l'agriculture au Maroc. Doctoral dissertation, CIHEAM-IAMM (2019)
4. Ministère de l'Agriculture, de la Pêche Maritime, du Développement Rural et des Eaux et Forêts, Département de l'agriculture, Filières végétales (2022)
5. Ministère de l'Agriculture, de la Pêche Maritime, du Développement Rural et des Eaux et Forêts, Département de l'agriculture (2016)
6. Aziable, E., et al.: Valorization of agro-industrial waste by bio-process aerobic "composting". J. Mater. Environ. Sci. (JMES) **8**(4), 1277–1283 (2017)
7. Ministère de la transition énergétique et du développement durable – département du développement durable: Stratégie nationale de réduction et de valorisation des déchets. Rapport de synthèse, 20 pp. (2019)
8. Belmakki, M., Bartali, E.H., Xiaoru, H., Anne-Belinda, B.: Identification and Characterization of Organic Waste in Morocco, an Important Step Towards the Valorization of Waste, pp. 37–45 (2015)
9. Bernal, M.P., Alburquerque, J.A., Moral, R.: Composting of animal manures and chemical criteria for compost maturity assessment. A review. Bioresour. Technol. **100**(22), 5444–5453 (2009)
10. Mustin, M.: Le compost: gestion de la matière organique Paris (1987)
11. Inkel, M., De Smet, P., Tersmette, T., Veldkamp, T.: Fabrication et Utilisation du compost, Wageningen, Pays-Bas, p. 37 (1990)
12. Bernai, M.P., Paredes, C., Sanchez-Monedero, M.A., Cegarra, J.: Maturity and stability parameters of composts prepared with a wide range of organic wastes. Bioresour. Technol. **63**(1), 91–99 (1998)
13. Dieng, M., Diedhiou, A.S., Sambe, F.M.: Valorisation par compostage des déchets solides fermentescibles collectés à l'Ecole Supérieure Polytechnique de l'Université Cheikh Anta Diop de Dakar: Etude de l'effet phytotoxique sur des plants de maïs et d'arachide. Int. J. Biol. Chem. Sci. **13**(3), 1693–1704 (2019)

14. Ameziane, H., Nounah, A., Khamar, M.: Olive pomace compost use for fenugreek germination. Agron. Res. **18**(3), 1933–1943 (2020)
15. Hojjat, S.S., Hojjat, H.: Effect of nano silver on seed germination and seedling growth in fenugreek seed. Int. J. Food Eng. **1**(2), 106–110 (2015)
16. Ivani, R., Sanaei Nejad, S.H., Ghahraman, B., Astaraei, A.R., Feizi, H.: Role of bulk and nanosized SiO_2 to overcome salt stress during fenugreek germination (*Trigonella foenum-graceum* L.). Plant Signal. Behav. **13**(7), e1044190 (2018)
17. Chennaoui, M., Salama, Y., Makan, A., Mountadar, M.: Valorisation agricole d'un compost produit à partir du compostage en cuve des déchets municipaux. Eur. Sci. J. **12**(35), 247 (2016)
18. Barral, M.T., Paradelo, R.: A review on the use of phytotoxicity as a compost quality indicator. Dyn. Soil Dyn. Plant **5**(2), 36–44 (2011)
19. Doughmi, A., Benradi, F., Cherkaoui, E., Khamar, M., Nounah, A., Zouahri, A.: Fertilizing power evaluation of different mixtures of organic household wastes and olive pomace. Agron. Res. **20**(1), 5–17 (2020)
20. Maurya, B.R., Sharma, P.K.: Studies on the germination and viability of *Parthenium hysterophorus* L. in its compost. Indian J. Weed Sci. **42**(3–4), 244–245 (2010)
21. Goetz, P.: Phytothérapie, la Santé par les plantes. Vidal Editeur - Plantes médicinales (2010)
22. Bernard, L.C.: Field Crop Production, References 2nd edn., 412 pp. (1999) (in French)
23. Moussa, O., Fatine, M., Driss, H., Houda, E., Lahcen, Z., Ahmed, D.: Effet du Chlorure de Sodium (NaCl) sur les Paramètres de Germination du blé Tendre (*Triticum aestivum* L.). Eur. J. Sci. Res. **127**, 298–310 (2014)
24. Crosaz, Y.: Le matériel végétal: un outil pour la protection des sols. Bull. rés. Éros. **5**, 449–460 (1995)
25. United States. Environmental Protection Agency. Office of Environmental Justice (OEJ) (1996)
26. Amani, S., Rajabi, M., Chaeechi, M.: Inhibitory effects of lavender, absinthium and walnut on germination and seedling growth of *Convolvulus arvensis*, *Portulaca oleracea* and *Triticum aestivum*. Pak. J. Weed Sci. Res. **21**(4), 575–591 (2015)
27. Anupama, S.: Content development for an agricultural expert system on organic vegetable cultivation. Doctoral dissertation, Department of Agricultural Extension, College of Agriculture, Vellayani (2014)
28. Abdul-Baki, A.A., Anderson, J.D.: Relationship between decarboxylation of glutamic acid and vigor in soybean seed 1. Crop Sci. **13**(2), 227–232 (1973)
29. Komilis, D.P., Karatzas, E., Halvadakis, C.P.: The effect of olive mill wastewater on seed germination after various pretreatment techniques. J. Environ. Manage. **74**(4), 339–348 (2005)
30. Pinho, I.A., Lopes, D.V., Martins, R.C., Quina, M.J.: Phytotoxicity assessment of olive mill solid wastes and the influence of phenolic compounds. Chemosphere **185**, 258–267 (2017)
31. Jacquemin, L.: Production of hemicelluloses from straw and wheat bran on a pilot scale. Study of the technical performance and environmental assessment of an agro-process. Doctoral thesis, National Polytechnic Institute of Toulouse, 345 pp. (2012) (in French)
32. Zucconi, F., Pera, A., Forte, M., de Bertoldi, M.: Evaluating toxicity of immature compost. Biocycle **22**, 54–57 (1981)
33. Delaire, M.: Variations in the mineral absorption capacity of the roots of young *Acer pseudoplatanus*, L. (Aceraceae) as a result of the recent and ancient nutritional history of the plant. In: Application to Off-Ground Cultivation of Woody Plants. Plant Biology, 180 pp. University of Angers (2005) (in French)
34. Plassard, C., et al.: Améliorer la biodisponibilité du phosphore: comment valoriser les compétences des plantes et les mécanismes biologiques du sol. Innov. Agron. **43**, 115–138 (2015)

35. Del Buono, D., Said-Pullicino, D., Proietti, P., Nasini, L., Gigliotti, G.: Utilization of olive husks as plant growing substrates: phytotoxicity and plant biochemical responses. Compost Sci. Util. **19**, 52–60 (2011)
36. Union of Fertilizer Industries. Fertilization, 7th edn., 78 pp. (1998) (in French)
37. Alburquerque, J.A., Gonzálvez, J., García, D., Cegarra, J.: Effects of a compost made from the solid by-product ("alperujo") of the two-phase centrifugation system for olive oil extraction and cotton gin waste on growth and nutrient content of ryegrass (*Lolium perenne* L.). Bioresour. Technol. **98**(4), 940–945 (2007)
38. Regni, L., Gigliotti, G., Nasini, L., Proietti, P.: Reuse of olive mill waste as soil amendment. In: Galanakis, C.M. (ed.) Olive Mill Waste: Recent Advances for Sustainable Management, pp. 97–117. Elsevier-Academic Press, Oxford, UK (2016)
39. Regni, L., et al.: Long term amendment with fresh and composted solid olive mill waste on olive grove affects carbon sequestration by prunings, fruits and soil. Front. Plant Sci. **7**, 20–42 (2017)
40. Proietti, P., et al.: Effects of amendment with oil mill waste and its derived-compost on soil chemical and microbiological characteristics and olive (*Olea europaea* L.) productivity. Agric. Ecosyst. Environ. **207**, 51–60 (2015)
41. Lopez-Pineiro, A., Albarran, A., Rato Nunes, J.M., Barreto, C.: Short and medium term effects of twophase olive mill waste application on olive grove production and soil properties under semiarid Mediterranean conditions. Bioresour. Technol. **99**, 7982–7987 (2008)
42. Ferarra, G., et al.: Effects of mulching materials on soil performance of cv. nero di troia grapevines in the Puglia region, southeastern Italy. Am. J. Enol. Vitic. **63**(2), 269–276 (2012)
43. Chartzoulakis, K., Psarras, G., Moutsopoulou, M., Stefanoudaki, E.: Application of olive mill wastewater to a Cretan olive orchard: effects on soil properties, plant performance and the environment. Agric. Ecosyst. Environ. **138**(3–4), 293–298 (2010)
44. Uygur, V., Karabatak, I.: The effect of organic amendments on mineral phosphate fractions in calcareous soils. J. Plant Nutr. Soil Sci. **172**, 336–345 (2009)
45. Montemurro, F., Convertini, G., Ferri, D.: Mill wastewater and olive pomace compost as amendments for rye-grass. Agronomie **24**, 481–486 (2004)
46. Chikae, M., Ikeda, R., Kerman, K., Morita, Y., Tamiya, E.: Estimation of maturity of compost from food wastes and agro-residues by multiple regression analysis. Bioresour. Technol. **97**(16), 1979–1985 (2006)

What Air Quality at the Hassan II Bridge? Case Of Rabat-Sale, Morocco

Soraya El Hamdouni(✉), Siham Bechar, Essediya Cherkaoui, Abderrahman Nounah, and Mohamed Khamar

Civil Engineering and Environment Laboratory (LGCE), Materials, Water and Environment Team, Higher School of Technology of Salé, Mohammed V University in Rabat, Rabat, Morocco
soraya_elhamdouni@um5.ac.ma

Abstract. Enjoying a crossroads situation, the concentration of administrations and higher education institutions, Rabat, the capital of Morocco, faces many challenges that directly or indirectly affect the quality of its air, such as migration and pendular migration. Air quality in Rabat is an emblematic subject when it comes to deciding whether the capital is a polluted city or not. If so, what is the source of its pollution? This regional and national influence made Rabat suffers from traffic jams. In this sense, we decided to study the impact of the car on its environment. To do this, we have set the measurement times in two phases: first period during peak hours (07:30–09:00 a.m.), second period during no peak hours (13:30–2:30 a.m.). The choice of the bridge HASSAN II measurement station is based on the fact that this station experiences the highest rate of cars during peak hours corresponding to the population living in Salé and working in RABAT. The gases chosen for this study are CO_2, NO_2, CH_4 and SO_2.

Keywords: Air quality · Cars · Pollution · Bridge Hassan II · Gases

1 Introduction

Cities are by nature concentrations of humans, materials and activities. They therefore exhibit both the highest levels of pollution and the largest targets of impact. Air pollution is, however enacted on all geographical and temporal scales, ranging from strictly "here and now" problems related to human health and material damage, over regional phenomena like acidification and forest die back with a time horizon of decades, to global phenomena, which over the next centuries can change the conditions for man and nature over the entire globe. In this respect the cities act as sources [1].

Cities are not equal regarding the impact of pollution emissions. Climate, topography, and town planning play a major role in the dispersion of pollutants. Anticyclonic conditions are the most favorable to the concentration of pollutants. Transport is today the main source of urban pollution. Undeniable technical progress has been made in terms of automobile pollution emissions. Between 1970 and 1993, the reduction was by a factor of 20–30 per vehicle and per year for unburned hydrocarbons and nitrogen oxides. However, at the same time, the French car fleet has doubled. Since 1980, we

K. Baba et al. (Eds.): GeoME 2023, SPEES, pp. 15–22, 2023.
https://doi.org/10.1007/978-3-031-49345-4_2

have observed an increase in the main pollutants emitted by transport: SO_2 (+ 23%), Nitrogen oxide (+ 27%), and dust [2].

Morocco is also affected by this phenomenon. To deal with it, it has launched numerous pilot projects as well as a legal framework to properly control and manage the situation. It is in this context that law 13-03 relating to the fight against air pollution, was born and defined the air as:

- Air: the gaseous envelope which surrounds the earth and whose modification of the physical or chemical characteristics can harm living beings, ecosystems, and the environment in general. This definition also includes the air in workplaces and that of enclosed and semi-enclosed public spaces [3].

Located in the northwest of Morocco, the twin cities of Rabat-Salé maintain a relationship of complementarity and exclusion. Being the capital of the Kingdom, Rabat enjoys a crossroads geographical location, the concentration of administrations, higher schools, etc. As for Salé, it has become a dormitory city of the agglomeration dominated by the weight and importance of the capital city [4].

The prefecture of Rabat and the prefecture of Salé are part of the Rabat-Salé-Kenitra region, bounded to the north by the province of Kenitra, to the south by the prefecture of Skhirat-Témara, to the east by the province of Khemisset and to the west by the Atlantic Ocean. With a population of 1546135 or 33% of the regional population [5].

Due to its position, the region of Rabat-Salé-Kénitra is distinguished by its semi-arid Mediterranean type climate. It is therefore under a double influence, namely maritime or continental. In other words, there prevails a mild, moderate and rainy climate in winter, and in summer it becomes humid and temperate with occasional waves of Chergui. The existence of a diversified relief means that the temperature is associated with significant variations ranging from 4 °C to 40 °C. In summer, the maximum temperature varies between 16 °C and 26 °C and exceptionally, it can reach between 38 °C and 40 °C in the continental part [6] (Fig. 1).

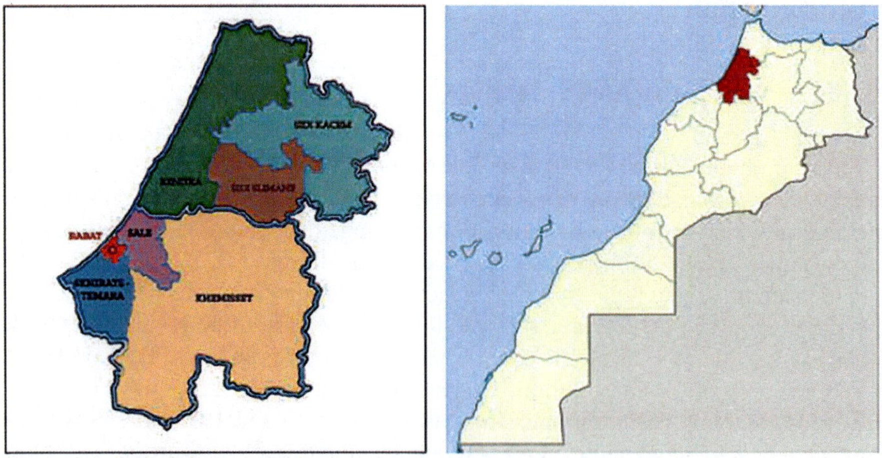

Fig. 1. Location of Rabat-Salé (RSK region)

As part of our thesis project on Rabat's transition to a sustainable and smart city, the study of air quality is one of the pillars for the success of this transition.

Very gradually, cities have become concerned with air quality by moving away from the paradigms on which a management more inspired by hygiene was based, to evolve towards a vision inspired by sustainable development. Indeed, according to the evolution of knowledge, the pollution that we deal with today is no longer the same as that of yesterday, it has interfered in all aspects of urban life since the interior of houses to agriculture [7].

Being a city that has no industrial activity, the only source of pollution for Rabat is the car. Is Rabat already a polluted city? Does the car really affect the quality of its air?

To answer these questions, fieldwork was carried out on all the entrances and exits of Rabat (peak hours and non-peak hours) with the aim of knowing the number of cars passing through each point as well as the rate gases emitted (CO_2, NO_2, CH_4, SO_2).

Rabat and Salé are linked by 5 bridges: the Hassan II bridge, the Al Fidaa bridge, the Moulay Youssef bridge, the Ribat Al Fath bridge, the Bouregreg bridge. The choice of the Hassan II bridge as a study station is based on the fact that it is the station that experiences the highest rate of cars during peak hours (Fig. 2).

Fig. 2. Location of the Hassan II bridge station [8]

2 Materials and Methods

To carry out this study, we divided the production process into 3 phases:

Phase 1: study preparation: to do this

- The choice of stations that represent the entrances and exits of Rabat being a city that suffers from pendular migration;
- Determination of measurement times using a sensor. It should be noted that the choice of gases to be studied was not random, they are gases emitted by vehicles;
- Creation of tables for taking measurements of air quality, temperature, humidity, and the number of cars;
- The purchase of the gas sensor.

Phase 2 data collection:

- Four days at each station (one day per gas) from 1:30 a.m. to 2:30 a.m. (non-peak hours) and from 7:30 a.m. to 9 a.m. (peak hours).

Phase 3 results: to do this

- Data entry for data processing;
- Data analysis by SPSS;
- Comparison of results with international air quality standards.

The objective of this study is to know if the air of Rabat is polluted, and what is the impact of the car on the quality of its air. Subsequent paragraphs, however, are indented (here insert the second paragraph).

3 Results

It should be noted that at the Pont Hassan II station, the average number of cars during peak hours from 07:30 a.m. to 09 a.m. is 5995 cars, the average temperature is 180 °C and the average humidity is 90%.

3.1 Descriptive Analysis

The main results of the descriptive analysis of the data collected in the field are summarized in Table 1. Thus:

- Average CO_2 during peak hours is 525 ppm, non-peak hours is 15.67 ppm. Regarding the standard deviation during peak hours 45.67 ppm, and during non-peak hours 15.46 ppm.
- Average CH_4 during peak hours is 1.26 ppm, and non-peak hours is 0.53 ppm. The standard deviation during peak hours is 0.45 ppm and during non-peak hours 0.51 ppm.
- Average SO_2 during peak hours is 0.20 ppm, and non-peak hours is 0 ppm. Regarding the standard deviation during peak hours 0.4 ppm, and during non-peak hours 0 ppm.

– Average NO_2 during peak hours is 0.01 ppm, and non-peak hours is 0.01 ppm. Regarding the standard deviation during peak hours 0 ppm, and during non-peak hours 0 ppm.

Table 1. Descriptive analysis

	ID	Moyenne	Ecart-type	N
CO_2	PH	525.210526	45.6771021	19
	NRH	563.846154	15.4641853	13
	Total	540.906250	40.9353222	32
CH_4	PH	1.263158	0.4524139	19
	NRH	0.538462	0.5188745	13
	Total	0.968750	0.5948367	32
SO_2	PH	0.021053	0.0418854	19
	NRH	0.000000	0.0000000	13
	Total	0.012500	0.0336011	32
NO_2	PH	0.014842	0.0083351	19
	NRH	0.011846	0.0043560	13
	Total	0.013625	0.0070654	32

3.2 Principal Component Analysis

In our study, we performed PCA and found that the first two principal axes together capture 73.6% of the total variance (Table 2). This indicates that these two axes account for a significant proportion of the variation present in our data.

This result is important because it suggests that the first two main axes are informative and contain much of the data structure. In other words, they provide us with a compact representation of the original variables, while preserving a substantial amount of information.

The principal component analysis (PCA) that we carried out (Fig. 3) made it possible to obtain significant results concerning the representation of individuals and variables. Two important findings emerged from our study:

First, we observed that the change in the amount of CO_2 is strongly influenced by the NBH phase. This suggests that variations in the amount of CO_2 are mainly associated with this specific phase of the process studied. This discovery highlights the importance of the NBH phase in understanding the factors influencing CO_2 levels.

Second, we identified that the change in the amount of NO_2, SO_2, the number of cars, and the average temperature largely depend on the PH phase. This observation suggests that these variables are closely related to the PH phase of the analyzed process. It is

Table 2. Variances explained by the PCA considered

Composante	Valeurs propres initiales			Extraction Sommes des carrés des facteurs retenus		
	Total	% de la variance	% cumulés	Total	% de la variance	% cumulés
1	3.803	54.325	54.325	3.803	54.325	54.325
2	1.347	19.243	73.568	1.347	19.243	73.568
3	1.125	16.065	89.633			
4	0.517	7.383	97.016			
5	0.209	2.984	100.000			
6	1.005E−013	1.064E−013	100.000			
7	− 1.000E−013	− 1.001E−013	100.000			

therefore essential to take this specific phase into account to understand the variations of these atmospheric pollutants and temperature.

These results highlight the importance of taking into consideration the different phases of the process studied to understand the variations of the different variables measured.

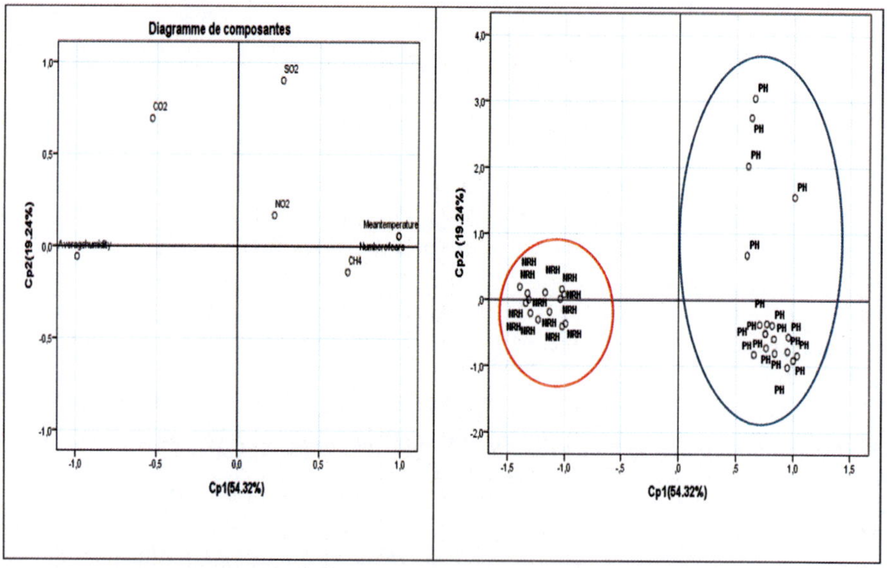

Fig. 3. Representation (x, y) of individuals and variables

3.3 Econometric Analysis

The results of the ANOVA analysis of variance (Table 3) are valid by displaying a coefficient of determination close to 1. They showed that the change from peak hours to non-peak hours impacts the quantity only for CO_2 and CH_4 for a margin of error of 5% (P-value less than 5%). On the other hand, SO_2, NO_2, number of cars and the average temperature are not influenced by this change.

Table 3. ANOVA regression results

Tests des effets inter-sujets						
Source	Variable dépendante	Somme des carrés de type III	ddl	Moyenne des carrés	D	Sig.
ID	CO_2	11521.869	1	11521.869	8.551	0.007
	CH_4	4.054	1	4.054	17.587	0.000
	SO_2	0.003	1	0.003	3.250	0.081
	NO_2	6.928E−005	1	6.928E−005	1.406	0.245

4 Discussion

The results of the analyzes confirm that the car has a significant impact on air quality with regard to CO_2. However, are we able to confirm that the air quality in Rabat is polluted?

In comparison with international standards in relation to CO_2, the Pont Hassan II station subject of this study 520 ppm during peak hours, which classifies it in the GOOD category (Fig. 4).

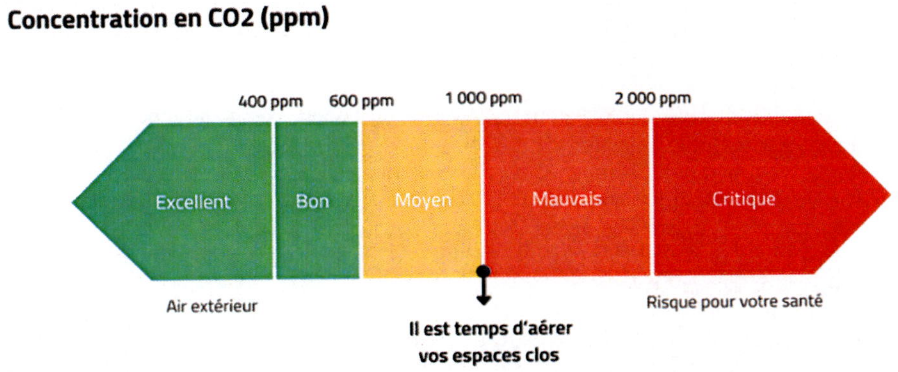

Fig. 4. Classification of air quality in relation to CO_2 [9]

According to the ATMO index, which is a daily air quality indicator calculated from the concentrations in the air of regulated pollutants [10], the rate of NO_2 and SO_2 is GOOD (Fig. 5).

		Bon	Moyen	Dégradé	Mauvais	Très mauvais	Extrêmement mauvais
Moyenne journalière	PM2.5	0-10	10-20	20-25	25-50	50-75	> 75
Moyenne journalière	PM10	0-20	20-40	40-50	50-100	100-150	> 150
Max Horaire Journalier	NO2	0-40	40-90	90-120	120-230	230-340	> 340
Max Horaire Journalier	O3	0-50	50-100	100-130	130-240	240-380	> 380
Max Horaire Journalier	SO2	0-100	100-200	200-350	350-500	500-750	> 750

Fig. 5. Classification of air quality in relation to NO_2 and SO_2

5 Conclusions

As a conclusion, the study on the gases emitted by cars affirms that at the Hassan II station does not suffer from air pollution. All the same, the fundamental question to ask is: Does the pendular migration have no direct impact on the air quality of Rabat or do the green spaces of Rabat as well as the Oued and the ocean help the latter to manage its air quality? The answer to this question will be the subject of an article that will deal with the question as a whole.

References

1. Fenger, J.: Urban air quality. Atmos. Environ. **33**(29) (1999)
2. Flament, G.: The Air and the City. In: Dab, W., Roussel, I. (eds.) (2004)
3. Dahir n° 1-03-61 of 10 rabii 1 1424 (May 12, 2003) promulgating law n° 13-03 relating to the fight against air pollution
4. Ali, O., El Kehal, A.: Salé citadine: governance of public spaces and urban structure. Rev. Geogr. Space Moroc. Soc. (41/42) (2020)
5. High Commission for Planning: General Population and Housing Census (2014)
6. Ministry of National Territory Planning, Town Planning, Housing, and Town Policy: Regional Monograph on the Housing Sector Region of Rabat-Sale-Kenitra (2019)
7. Roussel, I.: For a sustainable city that combines climate issues with those of air quality. Bull. Geogr. Soc. Liège (2017)
8. Google Maps
9. High Council for Public Health: TEQOYA, 19 rue chapon, 75003, Paris. https://www.teqoya.fr/capteur-co2/
10. ATMO Index: Ministry of Ecological Transition and Territorial Cohesion. https://www.airparif.asso.fr/indice-atmo

Diachronic Analysis of the Spatio-Temporal Evolution of an Estuary over Half a Century: The Case of Bouregreg Estuary in Morocco

Chaymae Najimi(✉), Essediya Cherkaoui, Mohamed Khamar, and Abderrahman Nounah

Civil Engineering and Environment Laboratory (LGCE), Materials, Water and Environment Team, Higher School of Technology of Salé, Mohammed V University, Rabat, Morocco
chaymae_najimi@um5.ac.ma

Abstract. The Bouregreg Estuary, located on the Atlantic coast of Morocco, has undergone significant morphological changes, as revealed by the findings of this study. The results of this study highlight concrete transformations in the intertidal zone of the Bouregreg Estuary and its banks, emphasizing the importance of our research to better understanding these changes. Satellite image analysis has revealed clear indications of morphological changes in the estuary before and after the construction of the SMBA dam in 1973 and the development project in 2006. The use of the unique dynamic index and transfer diagram allows for the identification of trends in different land use categories such as agriculture, urbanization, green spaces, and water bodies. These results underscore the importance of considering the temporal evolution of the Bouregreg Estuary for better long-term environmental management.

Keywords: Bouregreg Estuary · Morphological change · Management project · Diachronic analysis · Spatio-temporal evolution

1 Introduction

Estuaries represent transitional ecotones situated at the interface of terrestrial and marine environments, where fluvial freshwater mingles with oceanic saline waters. These dynamic ecosystems exhibit remarkable biodiversity, harboring diverse assemblages of flora and fauna, with particular importance attributed to their role as essential spawning and nursery habitats for numerous fish and crustacean species. Moreover, estuaries provide invaluable ecosystem services encompassing climatic regulation, erosion mitigation, nutrient and pollutant filtration, and substantial support to global economic sectors such as fisheries, tourism, and transportation. Nonetheless, these ecosystems face escalating anthropogenic pressures, imperiling their ecological integrity and hydrodynamic functionality.

Anthropogenic factors including urbanization, industrialization, tourism, port expansion, and land-use alterations can exert profound influences on estuarine morphological

K. Baba et al. (Eds.): GeoME 2023, SPEES, pp. 23–33, 2023.
https://doi.org/10.1007/978-3-031-49345-4_3

dynamics, yielding detrimental consequences for biotic communities, water quality, coastal stability, and human welfare. Hence, comprehending the intricate interplay of natural and human-induced processes governing estuarine morphological changes is imperative to promote sustainable management approaches and safeguard the intrinsic ecological and socio-economic values they encompass [1].

The Bouregreg Estuary, located on the Atlantic coast of Morocco, is an example of an estuarine ecosystem that has undergone significant morphological changes since the 20th century. This estuary has several peculiarities related to its natural setting, isolated on a long straight coastline, exposed to heavy Atlantic swells [2], but also to its disturbed state due to human activities. Indeed, the Bouregreg Estuary has experienced rapid urbanization, intensive port development, and increasing industrialization, which have instigated consequential modifications to its morphological structure, hydrodynamic patterns, and ecological dynamics. These alterations have engendered perturbations to the ecological equilibrium of the ecosystem, engendering multifaceted challenges for integrated management strategies.

Consequently, understanding the spatio-temporal evolution of estuaries is essential for the long-term management and conservation of these valuable ecosystems. It is in this context that our study aims to identify all the changes that have taken place along the banks of the Bouregreg Estuary over the past half-century. Understanding these changes is crucial for developing effective conservation and management strategies for this ecosystem, which profoundly influences biodiversity and sustains the socio-economic fabric. Through meticulous analysis and comprehensive documentation, we can gain a better understanding of how the banks of the estuary have evolved and how to better preserve it for future generations.

2 Materials and Methods

The Bouregreg Estuary, which is the subject of this study, represents a prominent fluvial system within the Moroccan hydrographic network. Located on the Atlantic coast of Morocco, this marine stretch separates two major urban agglomerations, Rabat and Salé (34° 02′ 09″ North, 06° 50′ 07″ West), bounded upstream by the Sidi Mohammed Ben Abdellah (SMBA) dam and downstream by the sandy beaches of Rabat and Salé [3].

With a length of 23 km and an average width of 150 m, it exhibits an overall orientation perpendicular to the coast in a SE-NW direction. The tidal regime is semi-diurnal and of a mesotidal type, with a tidal range of 2.3–2.8 m during high tides and 1.1–1.2 m during low tides [4].

From a historical perspective, the estuary has persistently served as a pivotal element fueling the economic, social, and cultural dynamics of the region. It has fostered maritime commerce, fisheries, agriculture, as well as leisure and tourism activities. Boasting an extraordinary expanse, the estuary stands out for its inherent natural and ecological opulence (wetlands, forests, shores, beaches, plateaus, slopes, etc.), thereby encompassing captivating scenic panoramas and historical legacies [5] (Fig. 1).

To investigate the morphological dynamics of the Bouregreg Estuary and interpret its environmental status, a diachronic methodology was employed spanning a half-century period. This study encompasses the period characterized by significant anthropogenic interventions, namely the establishment of the Sidi Mohammed Ben Abdellah

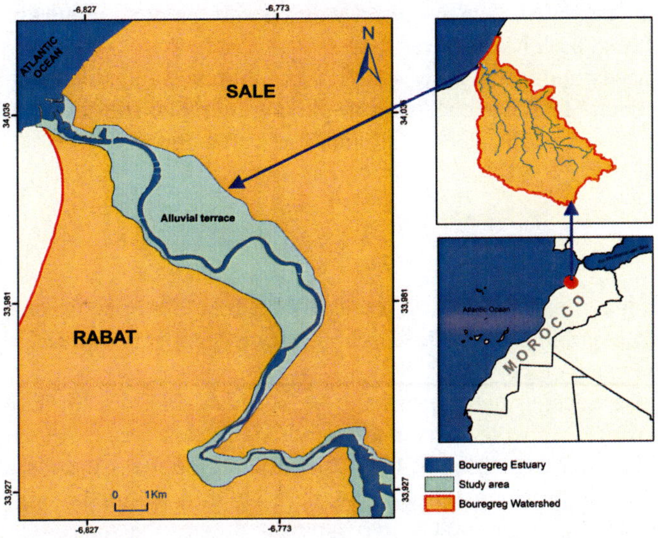

Fig. 1. Geographical location of the study area - the Bouregreg Estuary

(SMBA) dam in 1974 and a development project initiated in 2006. Historical documentary descriptions were meticulously analyzed to examine the pre- and post-intervention changes.

Thus, studies of this nature are reliant on the cartographic precision of different eras. Consequently, this study commenced in 1969 prior to the construction of the Sidi Mohammed Ben Abdellah dam, as it is the era with the most accurate available aerial photography allowing for precise or quantifiable information extraction. To evaluate the impact of the development project and the pollution remediation project in the Bouregreg Valley and coastline, satellite images from the years 2000 and 2022 were utilized.

Most of the documents were available in digital format, with the exception of the 1969 orthophotography, which required digitization and georeferencing through spatial alignment within a Geographic Information System (GIS) to enable digitization work.

The analyzed documents underwent the following processing steps: Firstly, the vectorization of the different studied classes was conducted. Due to the black and white format of the orthophotographs and the presence of potentially blurry areas, automated vectorization proved unsuitable. Thus, a manual approach was employed to ensure meticulous and accurate delineation of the various classes.

The processed documents underwent analysis to determine the percentages of each land cover class during the three study periods, identify the dynamic degree of the environment, and construct transfer matrices for each period. These analytical approaches allowed for the detection of emerging trends, strong/weak dynamic activity, and relative rates of change among classes, providing valuable insights into the spatio-temporal changes within the estuarine ecosystem.

Change detection analysis involves the identification and quantitative assessment of disparities between images captured at various time intervals, depicting the same geographic location. This analysis was conducted through the use of the unique dynamic index and transfer matrix, which collectively provided a comprehensive and rigorous assessment of land use changes over time.

The unique dynamic index served as a quantitative indicator to characterize the velocity of regional land use changes [6], playing a crucial role in the examination of regional differences and the analysis of land use change trends.

The formula is:

$$K = \frac{u_b - u_a}{u_a} \times \frac{1}{T} \times 100\% \tag{1}$$

In the formula presented, u_b, u_a represent the area of a specific land use category at the end and beginning of the research period, respectively, while T represents the duration of the research period [7].

The land use transfer matrix proves to be an invaluable tool for understanding the land use structure at a specific time and quantitatively describing the dynamic process of interconversion among different land categories throughout the research period. It provides insights into the transferred and exchanged information among each land category, thereby offering a comprehensive understanding of the spatio-temporal changes in land use [7].

3 Results

The landscapes and functioning of the Bouregreg Estuary bear the imprints of an ancient history that has progressively and profoundly transformed the territory in response to the practices and usages of the riparian societies.

The diachronic analysis of the Bouregreg Estuary's shores over a half-century period enabled the establishment of the spatial and temporal distribution of different land cover classes.

Spatial distribution maps depicting land use patterns at different time periods were generated (Fig. 2), along with corresponding statistical tables capturing land cover dynamics and the percentage changes in land cover between the three periods (Table 1), and a class transfer diagram (Fig. 3). These outputs were produced to facilitate quantitative data analysis and spatial characterization of land use transformations.

3.1 Assessment of Land Use Patterns and Spatial Attributes

A total of 14 land cover classes were detected (Fig. 2) along the estuarine region during the study period. The prominent land categories included agricultural land, prairies, and unused land, with agricultural land encompassing the largest extent. Notably, agricultural land exhibited a gradual decline, diminishing with a rate of 7% in 2000 to 9% in 2023. The agricultural land parcels were primarily concentrated within the alluvial terrace, covering more than 51% of the total study area.

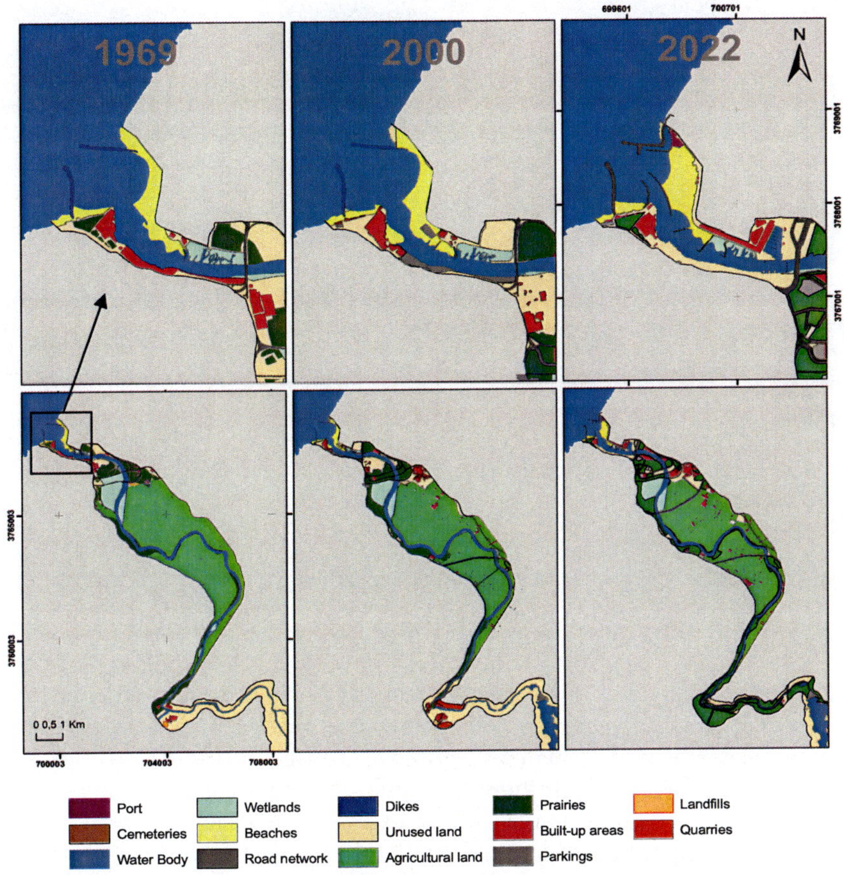

Fig. 2. Temporal analysis of land use spatial distribution along the banks of the Bouregreg Estuary in 1969, 2000, and 2022

3.2 Investigation of Land Use Transfer Characteristics in the Bouregreg Estuary

The Sankey diagram, generated using the transfer matrix, visually represents the transitions between various land cover classes and provides insights into the temporal dynamics of land use changes within the study region for the periods of 1969–2000 and 2000–2022 (Fig. 3).

3.3 Change Dynamic of Land Use Features of the Bouregreg Estuary

Using formula (1), the unique dynamic degree for each land use class was calculated for the periods of 1969–2000, 2000–2022, and 1969–2022. The results of these calculations are presented in the associated table, which also provides information on the total area and fraction of land cover in 1969, 2000, and 2022.

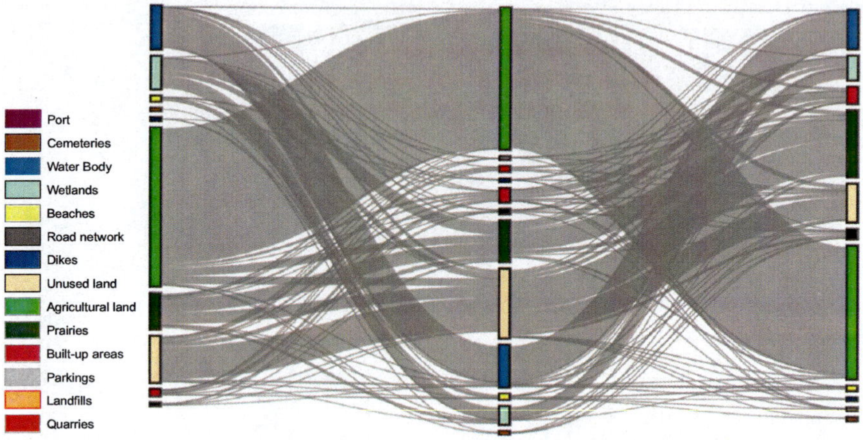

Fig. 3. Sankey diagram illustrating the complex interplay of land cover transformations from 1969 to 2000 and 2022

4 Discussion

4.1 Assessment of Land Use Patterns and Spatial Attributes

The Bouregreg Estuary is a complex and dynamic system that exhibits different land use categories and spatial characteristics along its course. Upstream, the estuary showcases a gorge-like valley flanked by high plateaus. Downstream, the gorge dissipates, and the estuary traverses an alluvial plain known as Oulja. This plain is replaced four kilometers from the estuary mouth by peri-urban and urban areas. The terminal parts of the estuary are formed by the sandy beaches of Rabat and Salé, protected from direct oceanic action by two converging arched seawalls [4].

In analyzing the land use categories and spatial characteristics within the Bouregreg Estuary, it is noteworthy that the land area devoted to built-up structures is comparatively smaller when compared to agricultural and prairie spaces.

Although the land area devoted to built-up structures is smaller compared to agricultural and prairies spaces, its relative dynamics have exhibited notable intensification, exhibiting an upward trajectory from 1969 to 2022. Spatially, these built-up areas have primarily concentrated at the estuarine mouth, manifesting as the zone of rapid development in contrast to other regions.

Quantitatively, the current built-up extent encompasses 110.8 ha, constituting nearly 5% of the total study area. However, these built-up areas were not as prevalent 50 years ago, with a mere 2% land coverage reflecting the intense urbanization witnessed over the past five decades.

The area between the estuarine mouth and the ONCF bridge emerges as a focal point for major investments, exemplified by noteworthy projects such as the 45-storey Mohammed VI Tower on the right bank and the grand theater on the left bank, both being flagship initiatives embedded within the comprehensive Bouregreg Valley development plan.

Table 1. Total area and fraction of land cover in 1969, 2000, and 2022.

	1969		2000		2022		1969–2000	2000–2022	1969–2022
	Area (ha)	%	Area (ha)	%	Area (ha)	%	Unique dynamic degree (%)		
Water body	314.57	13.51	309.64	12.776	283.32	12.02	− 0.21	− 1.11	− 1.32
Wetland	153.45	6.59	98.16	4.050	95.055	4.033	− 2.33	− 0.13	− 2.46
Built-up areas	38.98	1.6	83.44	3.443	110.74	4.698	1.88	1.15	3.03
Dikes	4.15	0.18	4.282	0.177	7.447	0.316	0.01	0.13	0.14
Road network	9.33	0.40	21.85	0.901	53.607	2.274	0.53	1.34	1.87
Parking	0	0	2.530	0.104	6.762	0.287	0.11	0.18	0.29
Port	0	0	0	0	1.000	0.042	0	0.04	0.04
Agricultural land	1194.2	51.29	1084.7	44.758	1006.6	42.708	− 4.62	− 3.29	− 7.91
Prairies	238.17	10.23	281.35	11.609	487.75	20.692	1.82	8.71	10.53
Unused land	343.81	14.77	475.35	19.613	278.64	11.821	5.55	− 8.30	− 2.75
Beaches	24.81	1.07	28.10	1.159	26.038	1.105	0.14	− 0.09	0.05
Landfills	7.01	0.30	7.013	0.289	0	0	0	− 0.30	− 0.30
Quarries	0	0	27.13	1.119	0	0	1.14	− 1.14	0
Cemeteries	0	0	0	0	0.102	0.004	0	0.0043	0.0043

The columns on the right indicate the unique dynamic degree percentage of land cover change between the three periods.

This zone now boasts a plethora of high-quality services and infrastructure, including a marina in Salé, a promenade in Rabat providing dining, entertainment, and leisure facilities, as well as residential projects and luxury hotels that enhance the allure of the estuary's riverbanks, contributing to its cultural and economic dynamism.

From a qualitative perspective, prior to the commencement of the development project in 2006, the habitat in the vicinity of the Akreuch and Oulja rivers was characterized by unsanitary conditions. These areas consisted of unauthorized and unplanned shantytowns, associated with the former unregulated Akreuch and Oulja dumps, which have undergone subsequent closure and rehabilitation within the overarching framework of the Bouregreg Valley development project. Notably, the Oulja industrial zone, accommodating polluting activities, has endured without significant alterations and persists as an extant feature.

4.2 Investigation of Land Use Transfer Characteristics in the Bouregreg Estuary

As depicted in the Sankey diagram in Fig. 3, the primary land category transferred from the agricultural zone between 1969 and 2000 was built-up land, which accounted for 23.41 ha, representing approximately 2% of the agricultural land area in 1969. Conversely, the agricultural zone saw an influx of water courses, spanning an area of 11.80 ha. No significant changes were observed in the transfer of land categories to or from other land cover classes, although the transferred areas varied among classes.

Furthermore, the diagram illustrates that urban and tourist development has resulted in an increase in built-up land area, road networks, and parking lots, at the expense of agricultural and prairie areas, which represent the most actively utilized categories. As various land categories experienced changes, the majority of them were exchanged with these two predominant types of land.

However, the conversion of natural areas into built-up land may have negative effects on water quality. Artificialization can lead to increased runoff of rainwater and the transfer of pollutants into water courses, which can adversely impact aquatic biodiversity and water uses [8].

Conversely, natural or semi-natural land covers such as wetlands, prairies, and forests can play a positive role in safeguarding water quality through their filtration capacity, flow regulation, and preservation of natural habitats.

In terms of quality, agricultural practices in the alluvial terrace have experienced changes primarily attributed to the construction of the Sidi Mohammed Ben Abdellah dam in 1974. The valley once thrived with family-owned agriculture, featuring orchards of various fruit trees, benefiting from fertile soil maintenance and access to freshwater resources through the permanent flow of the Bouregreg River and flood events. However, the accumulation of salt in the soil over time rendered it unsuitable for agriculture. This shift in cultivation patterns was driven by the salinization of shallow groundwater in the region. Traditional wells are no longer viable for irrigation throughout the plain, except for a small area around Ounk Jmel [9].

Regarding wetlands, a significant regression was also observed between 1969 and 2022. Wetland area decreased from 87.14 ha in 1969 to 78.06 ha in 2000, and further declined to 49.83 ha in 2022, representing a 42.8% decrease over the period. The main factors contributing to this regression were the conversion of wetlands into prairies and unused land, accounting for 49.83 ha and 30.92 ha, respectively, in outgoing transfers in 2022.

Wetland areas have exhibited relative instability and have demonstrated increased interactions with other land categories. Changes in water bodies were closely linked to the construction of calibration dikes at the estuarine mouth and beach, primarily resulting from the development projects, notably improvements in channel navigation and the construction of the Sidi Mohamed Ben Abdellah dam in 1974. Numerous authors argue that population growth and certain exploitation practices are responsible for land degradation, leading to disruptions in ecological balances [10], particularly in wetland areas that are highly vulnerable to climate variability [11].

The construction of the SMBA dam has altered the magnitude and frequency of river floods and changed the hydrodynamics of the river, resulting in seawater intrusion influenced by ocean tides. The natural interaction between river and marine waters no longer occurs. The Bouregreg Estuary has become an ocean inlet, solely dependent on releases from the reservoir and occasional high floods. Consequently, salinity levels have increased, and hydrodynamic processes have been attenuated [12, 13]. This regression is also due to the development of recreational infrastructure, namely the marina and the port on the right bank of the river and the development of both banks [14].

4.3 Change Dynamic of Land Use Features of the Bouregreg Estuary

As illustrated in Table 1, within the study area, the prairies exhibited the highest dynamic degree during the periods 2000–2022 and 1969–2022, reaching 8.71% and 10.53% respectively, indicating a substantial increase in green spaces during these two periods, attributed to ecological restoration and environmental protection policies implemented by the Bouregreg Valley Development Agency (AABV) as part of the Bouregreg Valley development project. This is evident in the aesthetically and functionally appealing landscapes that contribute to erosion control [9]. Although urbanization has fragmented these prairies, the presence of corridors and a network of parks and gardens has helped mitigate this fragmentation [15].

Overall, from 1969 to 2022, the prairies exhibited the highest dynamic degree followed by built-up areas at 3.03%. Both categories showed consistent growth, indicating a steady increase in green spaces and built-up areas with the deepening urbanization and ecological construction within the study area.

The changing trend of the unique dynamic degree of land use demonstrated significant variations among land categories in different periods.

In contrast, agricultural lands and water bodies experienced a negative dynamic degree, with their areas gradually decreasing over time, highlighting the ongoing challenges faced by these land categories.

While wetlands showed relatively low dynamic degrees, road networks experienced significant changes, similar to the dynamic degree of built-up areas in all periods. This can be primarily attributed to the construction of new bridges, two tramway lines, and a new highway across the estuary due to urban development and transportation demands.

The historical narrative of the Bouregreg River highlights the use of boats by individuals to cross the river and avoid traffic on the Moulay El Hassan Bridge [16]. Moreover, the construction of new bridges, such as the Fida Bridge in 2008, the Hassan II

Bridge in 2013, the renewal of the Moulay Youssef Bridge (Meknes Bridge) in 2014, the Mohammed VI Bridge in 2016, and the Ribat Fath Bridge in 2020, has significantly improved regional connectivity and mobility. These infrastructure developments have undoubtedly facilitated economic growth, social life, and tourism in the area, but they have also helped reduce the anthropogenic impact manifested by vehicular exhaust emissions into the air. These gases dissolve in the water and can alter estuarine life, particularly near the Mohamed V Bridge, where road traffic is often dense.

However, it is important to acknowledge the potential environmental consequences associated with bridge construction, particularly in terms of water quality degradation and alterations to hydrodynamic patterns. The construction of multiple bridges may have negative impacts by altering the relief and the original terrain, increasing the risk of soil erosion by reducing vegetation along the shoreline, disrupts the original ecosystem, contaminates groundwater, and even disconnects surface lithology and soil [17, 18]. The hydrodynamics of the estuary can be modified by creating flow obstacles, reducing the width of the channel, or altering water depth. These changes can affect flow velocity, direction, turbulence, as well as sediment and nutrient transport.

5 Conclusions

The diachronic analysis conducted on the Bouregreg Estuary's shorelines over a half-century has unveiled significant transformations within the Bouregreg Estuary. The findings have elucidated a discernible pattern of urban expansion and an escalating presence of built-up areas along the estuarine corridor, particularly from the river mouth extending to the Fida Bridge.

Agricultural lands have exhibited a progressive decline over the years, declining with a rate of 7% in 2000 to 9% in 2023. This discernible trend can be attributed to the salinization of underlying groundwater sources and the subsequent accumulation of saline deposits within the soil matrix, rendering the agricultural terrains progressively less amenable to cultivation. Furthermore, the wetland ecosystems have experienced a substantial regression, predominantly driven by land use conversions into prairies and unused lands.

The comprehensive development initiatives undertaken within the Bouregreg Estuary, including the construction of the Sidi Mohammed Ben Abdellah Dam and the calibration dikes, have elicited discernible morphological alterations within the estuary, fundamentally impacting the hydraulic and geomorphic characteristics of the watercourse.

These transformations have also been intimately associated with the enhancement of navigation efficiency and flood mitigation measures. Nonetheless, these changes have had ensuing ramifications on ecological equilibria and the intricate fabric of estuarine biodiversity which warrant careful consideration.

References

1. Cai, H., et al.: A novel approach for the assessment of morphological evolution based on observed water levels in tide-dominated estuaries. Hydrol. Earth Syst. Sci. **24**, 1871–1889 (2020). https://doi.org/10.5194/hess-24-1871-2020

2. Khamar, M., Cherkaoui, E., Nounah, A.: Bioaccumulation of heavy metals by the flora and benthic macrofauna of the Bouregreg estuary wetland. MATEC Web Conf. **149**, 02054 (2018). https://doi.org/10.1051/matecconf/201814902054
3. El Harim, A., Cherkaoui, E., Khamar, M., Nounah, A.: The impact of the depollution project on the quality of the estuarine ecosystem of Bouregreg (Morocco Atlantic coast) (2019)
4. Cherkaoui, E., Bayed, A.: Structure and distribution of intertidal benthic communities in a North African estuary: the Bou Regreg Estuary (in French). Mar. Life **15**(1–2), 29–41 (2005)
5. Alillouch, R., Mansour, M., Radoine, H.: The development project of the Bouregreg Valley. Context and methods for designing and implementing an urban project for a vulnerable site (in French). Afr. Mediterr. J. Archit. Urban. 60–78 (2019). https://doi.org/10.48399/IMIST.PRSM/AMJAU-V1I2.18885
6. Redo, D., Aide, T.M., Clark, M., Andrade Núñez, M.: Impacts of internal and external policies on land change in Uruguay, 2001–2009. Environ. Conserv. **39** (2012). https://doi.org/10.1017/S0376892911000658
7. Li, Y., Liu, G., Huang, C.: Dynamic changes analysis and hotspots detection of land use in the central core functional area of Jing-Jin-Ji from 2000 to 2015 based on remote sensing data. Math. Probl. Eng. **2017**, 1–16 (2017). https://doi.org/10.1155/2017/2183585
8. Issaka, S., Ashraf, M.A.: Impact of soil erosion and degradation on water quality: a review. Geol. Ecol. Landsc. **1**, 1–11 (2017). https://doi.org/10.1080/24749508.2017.1301053
9. Patrick, D., Bekkar, Y.: Producing quality food and landscapes by inventing sustainable urban agriculture – the case of the Bouregreg Valley (Rabat – Salé), Kingdom of Morocco (in French) (2016)
10. Chillasse, L., Dakki, M.: Potential and conservation status of wetlands in the Middle Atlas (Morocco), with reference to drought influences (in French) (2004)
11. Ichen, A., Messaoudi, C., El Malki, M., El Mderssa, M.: A diachronic study of the land uses of the Dayet Aoua wetland in the Middle Atlas in Morocco (in French). ecmed **47**, 107–115 (2021). https://doi.org/10.3406/ecmed.2021.2123
12. Laouina, A.: The coastline of the Rabat-Salé city, a threatened space. Evolution of activities and settlements, environmental degradation, and development options (in French). In: Camarda, D., Grassini, L. (eds.) Coastal Zone Management in the Mediterranean Region, pp. 137–142. CIHEAM, Bari (2002)
13. Cherkaoui, E., Bayed, A., Hily, C.: Spatial organization of subtidal macrozoobenthic communities in an estuary of the Moroccan Atlantic coast: the Bou Regreg Estuary (in French). Cah. Biol. Mar. **44**, 339–352 (2003)
14. Cherkaoui, E.: Structure and organization of macrozoobenthic communities in the Bou Regreg Estuary after the construction of the dam (in French) (2006)
15. Bennani, M.: City Landscapes of Morocco: Rabat, Marrakech, Meknes, Fez, Casablanca (in French). Carré, Paris (2017)
16. Chastel, R.: Rabat-Salé: vingt siècles de l'Oued Bou Regreg. Editions La porte, S. l. (1994)
17. Irish, L.B., Lesso, W.G., Barrett, M.E., Malina, J.F., Charbeneau, R.J., Ward, G.H.: Evaluation of the factors affecting the quality of highway runoff in the Austin, Texas area (1995)
18. Li, Q., Qian, R., Gao, J., Huang, J.: Environmental impacts and risks of bridges and tunnels across lakes: an overview. J. Environ. Manage. **319**, 115684 (2022). https://doi.org/10.1016/j.jenvman.2022.115684

Performance Evaluation of Trombe Wall with Multi-fold Glazing

Hasna Oukmi(✉), Meryem El Alaoui, Ouadia Mouhat, and Mohammed Rougui

Civil Engineering and Environment Laboratory (LGCE), Mohammadia Engineering School, Mohammed V University, Rabat, Morocco
hasna.oukmi@research.emi.ac.ma

Abstract. The Trombe wall system proves to be effective in the winter, but can lead to overheating in the summer. Its concept aims to use the maximum amount of solar radiation to regulate a comfortable temperature inside the local. The objective of this study is to demonstrate the effectiveness of a new Trombe wall conception, in reducing building energy consumption in both winter and summer periods. According to previous studies, the Trombe wall system has undergone certain adjustments to increase its effectiveness, resulting in a 30% reduce in building's energy usage by 30%. This paper is designed to construct a Trombe wall with multi fold glazing that can be opened or closed, depending on the season. During winter, the Trombe wall glazing will be closed to maximize heat retention, while in summer it will be opened to prevent overheating. Due to the system's orientation on the south side of the building, the greenhouse effect can be used to optimize the benefits of solar radiation. DesignBuilder software was used in this study in order to calculate heating and cooling energy consumption, as well as heat transfer and solar heat gains for Three Scenarios: Reference Configuration, Classic Trombe Wall, and Trombe Wall with Multi-Fold Glazing which is the new conception of this study. The findings reveal that when compared to the reference configuration, the implementation of this technology results in a remarkable 56% reduction in energy consumption. This reduction is achieved through a substantial 95% decrease in heating energy consumption and a notable 29% decrease in cooling energy consumption. Moreover, in comparison to the classic Trombe Wall, the Trombe Wall with Multi-Fold Glazing demonstrates a 58% reduction in energy consumption. The present paper demonstrates that the Trombe wall system effectively lowers the energy required for heating and cooling. The study's findings demonstrate a considerable decrease in energy use, demonstrating the effectiveness of this method in promoting sustainability and energy efficiency.

Keywords: Classic Trombe wall · Trombe wall with multi-fold glazing · Solar gains · Heating and cooling energy consumption · Energy efficiency

K. Baba et al. (Eds.): GeoME 2023, SPEES, pp. 34–41, 2023.
https://doi.org/10.1007/978-3-031-49345-4_4

1 Introduction

Energy saving is crucial for the transition to a decarbonized society. Renewable sources are essential for achieving energy efficiency, especially in the building sector, which accounts for a significant portion of energy consumption. To address this, there is growing interest in innovative building designs that reduce thermal exchanges with the outdoors, ensuring occupant comfort and utilizing renewable energy sources. Passive solar systems, like solar walls, are gaining attention as they effectively harness solar radiation to minimize energy needs [1].

Within the realm of passive solutions, one notable option is the Trombe wall, initially developed by Edward Morse in 1881, was later revived by the French inventor Felix Trombe (Duffie and Beckman 2006) [2]. This system is designed to effectively absorb heat by facilitating convection between the wall and the glazing, allowing the heated air to circulate through an opening [2].

Implementing the Trombe wall system can lead to a significant reduction of up to 30% in a building's energy consumption [3]. One of the advantages of this system is its ability to create a substantial temperature difference between the interior and exterior, ensuring thermal comfort for occupants and adjacent spaces. Additionally, it helps mitigate moisture and humidity issues in humid regions. However, Trombe walls do have some drawbacks, including low thermal resistance, susceptibility to an inverse thermo-siphon effect, and an aesthetic appearance that may not be appealing. Moreover, the amount of heat gained can be unpredictable due to fluctuations in solar intensity.

Over time, there have been advancements made to Trombe walls to enhance their efficiency. These walls are categorized into two types: heating-based and cooling-based, based on their primary utilization functions.

Extensive research has been conducted to explore various ideas and innovations to improve the performance of Trombe wall systems. These include the development of composite Trombe walls or Trombe–Michel walls, water Trombe walls, zigzag Trombe walls, solar trans-walls, fluidized Trombe walls, and photovoltaic Trombe walls. Additionally, to enhance their cooling capabilities, novel approaches such as ceramic evaporative cooling walls, adapting classic Trombe walls and photovoltaic Trombe walls for cooling purposes, and the integration of a solar chimney with Trombe walls have been investigated [4].

Research findings indicate that Trombe walls have a greater potential for reducing energy consumption in buildings compared to traditional walls [5]. These innovative variations aim to enhance the cooling efficiency and overall performance of the Trombe wall system.

The Trombe wall is well-known for its efficiency in passive heating during winter, but it can lead to overheating in summer. To address this issue, researchers have studied the impact of Venetian blinds on the performance of Trombe walls during the summer season.

Hu et al. conducted a modification of a Trombe wall by integrating a water blind flow channel. This modification serves multiple purposes, including providing shading, ventilation, and hot water supply, thereby enhancing occupant comfort during both summer and winter seasons. The system achieved a remarkable increase in performance, with a 42.6% improvement compared to walls without a Trombe wall system [6].

However, none of these studies have specifically explored the concept of fully opening the glass during the summer season. In the present work, our focus is on developing a Trombe wall design that can be adaptable to both summer and winter conditions. Our main objective is to address the issue of overheating during the summer months by incorporating multi-fold glazing that can be opened to facilitate natural ventilation using fans. In contrast, during winter, the glazing doors will be closed to harness the greenhouse effect, thereby enhancing the heating capabilities of the house.

2 Materials and Methods

2.1 Building Description

The present study works on a residential building located in Ifrane region, a town in the Middle Atlas Mountains of Morocco, consisting of two levels, where each level consists of four apartments. Each apartment is designed with two bedrooms, a bathroom, and a living room. The building features two apartments facing the north direction and two facing the south direction (Fig. 1). Furthermore, the building is surrounded by neighboring structures on both the eastern and western sides.

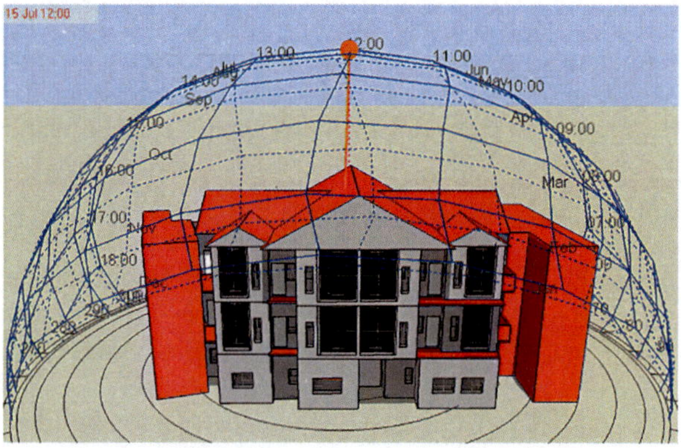

Fig. 1. 3D plan of the building.

According to Meteonorm software the average temperature in Ifrane is 15.6 °C, with the highest temperature reaching 41.2 °C and the lowest reaching – 3.3 °C. Additionally, during the summer months, temperatures exceeding 26 °C occur for approximately 11% of the year. Furthermore, Ifrane receives a total of 1,987,251 kWh/m^2 of global horizontal radiation per year.

The analysis will primarily focus on evaluating the influence of the Trombe wall on the south-facing wall of the building.

2.2 Trombe Wall Properties

The Trombe wall system was implemented in two rooms for each apartment. One of the rooms had dimensions of 9.6 m^2, which does not include the area occupied by the

Trombe wall. The external walls in each room measured 2.5 m in width and 3.2 m in height. The Trombe wall itself utilized only 64% of the total wall area in each room, resulting in dimensions of 1.6 m in width and 3.2 m in height. In addition to the Trombe wall, a window measuring 1.6 m × 0.4 m was integrated into the design of each room to ensure natural ventilation and provide daylighting.

The project will involve using multi-fold glazing (Fig. 2), which is a type of window or door system that consists of multiple glass panels that are hinged together and can be folded or stacked to both sides to create a wide opening.

We had a glazing width of 1.6 m, to create a folding glazing door that could be opened during summer, we selected 4 glass panels with a width of 0.4 m (Fig. 2).

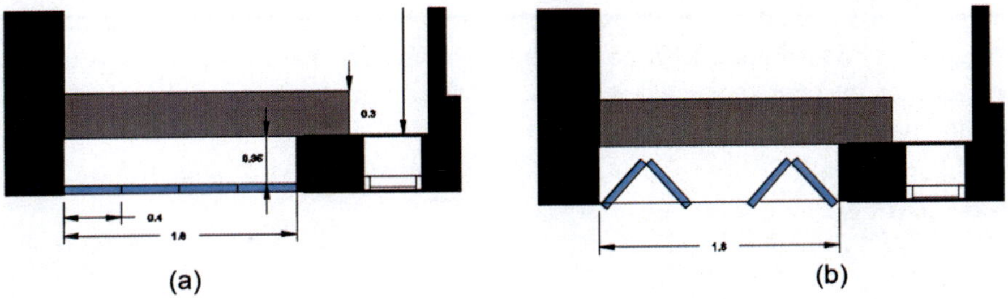

Fig. 2. (a) Plan displaying the dimensions of the Trombe wall. (b) Plan depicting the opening of the multi-fold glazing.

In order to utilize the heat generated by the greenhouse effect between the glazing and the Trombe wall, it is crucial to decrease heat loss through the glazing and enhance solar gain.

The massive wall serves two important functions: preventing the loss of thermal heat from the house and storing heat effectively, known as thermal mass. According to the findings of Agrawal and Tiwari [7], the ideal thickness for the massive wall falls within the range of 30–40 cm, and to enhance the absorption rate, the outer surface of the wall is coated with a black color.

In the present study, we have designed the massive wall to comprise two layers: a concrete layer measuring 30 cm in thickness and an insulation layer measuring 3 cm in thickness. As a result, the wall achieves a total U value of 0.795 W/m^2 K.

Considering a wall height of 3.2 m, researchers have determined that the optimal ratio between the depth and height of the cavity is approximately 1/10 [8]. Accordingly, a depth of around 0.32 m would be appropriate. In this study, a depth of 0.35 m was chosen, adhering closely to the suggested optimal ratio.

The research conducted by Stazi et al. in Italy highlights the positive impact of double glazing on enhancing the performance of Trombe walls [9]. In the present study, we investigated the specifications of the double-glazing unit, revealing a solar transmittance value of 0.849 and a thickness of 4 mm for each glass pane. The window frame is made of aluminum with a width of 57 mm, and its U-factor is 5.881 W/m^2 K. When combined, the U-value of the entire window system is 2.725 W/m^2 K, which Respect the limitations imposed by the Thermal Regulation of Construction in Morocco (RTCM) [10]. With an

SHGC of 0.779, the window system permits approximately 77.9% of the solar radiation to pass through the window and enter the cavity.

2.3 Numerical Simulation

In this numerical simulation, we used DesignBuilder software version v7 along with EnergyPlus. The indoor air temperature set-points for heating and cooling were kept at 20 °C and 26 °C, respectively, while a ventilation rate of 0.5 air renovations per hour was assumed for all scenarios. Our focus was on the zones oriented towards the south, and the average outdoor temperature in Ifrane was approximately 8 °C in winter and 23 °C in summer.

The purpose of the simulation analysis was to assess the effectiveness of the newly proposed Trombe wall in this study, focusing specifically on its impact on heating and cooling energy consumption. We performed a thorough comparison between the standard configuration and our suggested solution, which entailed opening the glass during the summer months.

3 Results

A parametric analysis will be conducted to analyze the results and assess the influence of the Trombe Wall on heating and cooling demands. The analysis will involve considering the following parameters: Reference building, Classical Trombe wall and Trombe wall with multi-fold glazing.

Initially, the study focuses on examining a single room to gather in-depth insights for analysis. Subsequently, it will be expanded to evaluate the energy performance implications of each configuration on the entire apartment.

This approach ensures a thorough understanding of the effects and enables comprehensive comparisons among the different scenarios.

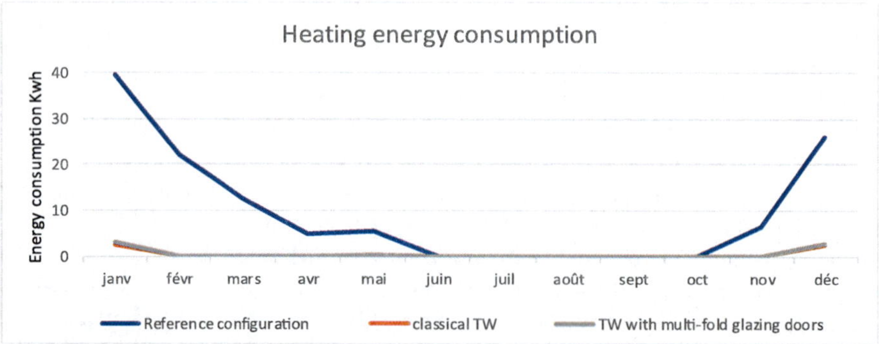

Fig. 3. Graph depicting the heating energy consumption for three scenarios: reference configuration, classic Trombe wall, and Trombe wall with multi-fold glazing.

Figure 3 illustrates the total heating energy consumption profiles, revealing a significant difference in heating energy consumption between the reference configuration (without Trombe wall) and the two scenarios of Trombe wall (classic TW and TW with multi-fold glazing doors). The reference configuration has a heating energy consumption of 116.17 kWh, while the classic Trombe wall exhibits a substantially lower value of 5.09 kWh. Similarly, the Trombe wall with multi-fold glazing doors shows a heating energy consumption of 5.8 kWh.

In January, the study observed that the classic Trombe wall exhibited a higher heat gain within the room from the partition wall when compared to the Trombe wall with multi-fold glazing. The difference in heat gain between the two was found to be 2.214 kWh.

Further analysis revealed that the incorporation of dividers in the glazing of the Trombe wall with multi-fold doors led to a reduction in solar gain. This reduction in solar gain can be advantageous in terms of minimizing heat loss. However, it is important to note that this improvement in heat loss is accompanied by a decrease in solar gains.

These details emphasize the influence of dividers in the glazing system on solar gains, which in turn impact the heating energy consumption of the Trombe wall.

To provide a comprehensive assessment of the Trombe wall's impact on energy consumption, it is imperative to analyze not only its effect on heating energy but also its influence on cooling energy consumption.

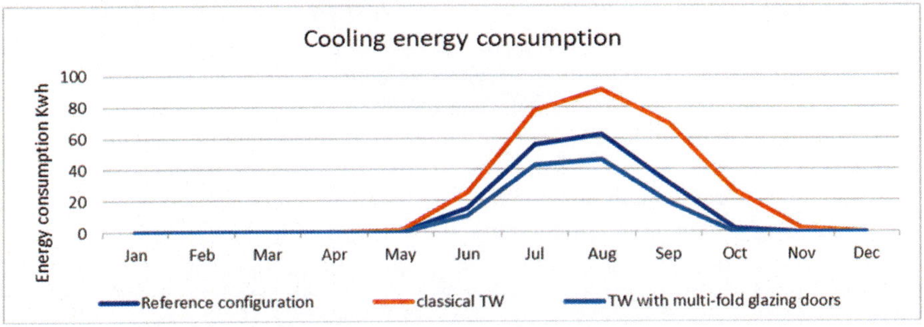

Fig. 4. Graph depicting the cooling energy consumption for three scenarios: reference configuration, classic Trombe wall, and Trombe wall with multi-fold glazing.

In this context, Fig. 4 provides an illustrative representation of the variation in cooling energy consumption throughout the year. We conducted a comparison of cooling energy consumption among the three previously mentioned scenarios.

Our findings indicate that the reference configuration experiences its peak cooling energy consumption in August, reaching 90.5 kWh and totaling 166.93 kWh annually. When comparing this profile to the classic Trombe Wall configuration, we observed that the classic Trombe Wall consumes 124.11 kWh, higher than the reference configuration. This significant difference can be attributed to the increased overheating generated by the classic Trombe Wall.

In contrast, the third profile representing the Trombe wall with multi-fold glazing doors (considering the opening of the multi-fold glazing during summer) exhibits lower cooling energy consumption compared to the classic Trombe wall. This reduction amounts to 172.44 kWh, representing a significant 59% decrease. Additionally, when compared to the reference configuration, this profile exhibits a 29% reduction. This reduction can be attributed to the integration of the Trombe wall, which reduces the room's surface area, as well as the reduction in window surface area.

These findings emphasize the significance of utilizing the Trombe wall's effect during winter, while also considering the removal of glazing during summer to mitigate overheating issues.

After examining the impact of the Trombe wall (TW) with a multifold glazing door on energy consumption in a single room and comparing it to the classic TW, it is now essential to assess the overall energy consumption for an entire apartment (Fig. 5).

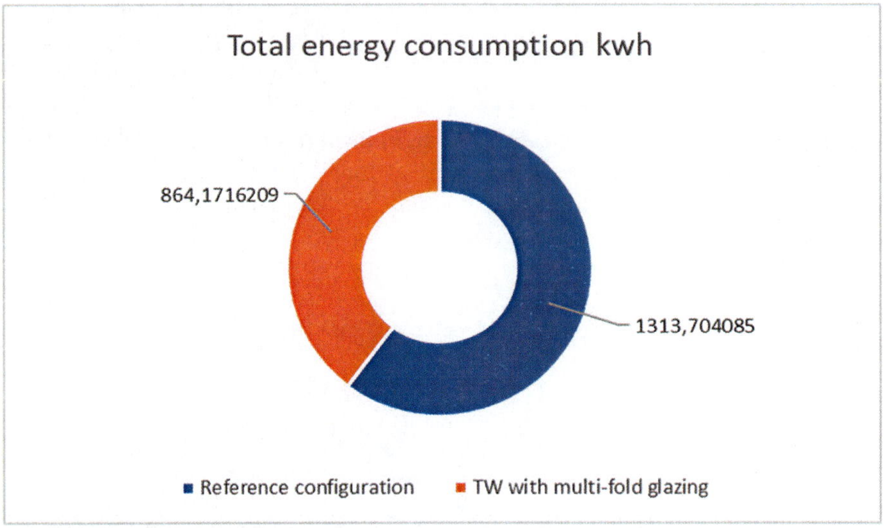

Fig. 5. Graph depicting the total energy consumption.

As previously stated, each apartment comprises two rooms and a living room. The application of the Trombe wall (TW) is restricted to one of the rooms and the living room. The findings reveal that the presence of the TW exerts a noteworthy impact on the overall energy consumption, resulting in a considerable reduction of 34%.

4 Conclusions

Trombe walls hold significant importance in the realm of energy-efficient building design. They offer a promising solution for reducing energy consumption, enhancing thermal comfort, and utilizing renewable energy sources. With their ability to effectively harness solar radiation, Trombe walls can contribute to substantial energy savings, potentially reducing energy needs in buildings by up to 30%.

Furthermore, the adaptability of Trombe walls to both summer and winter conditions is a valuable feature. By incorporating multi-fold glazing that can be opened during summer and closed during winter to harness the greenhouse effect, Trombe walls offer a versatile solution for maintaining thermal comfort throughout the year.

Additionally, it has been observed that Trombe wall with multi fold glazing can contribute to a substantial reduction of 56% in energy consumption for each zone. This highlights the significant potential of Trombe walls in achieving energy efficiency and enhancing occupant comfort.

Further studies should include an economic analysis to evaluate the cost-effectiveness of implementing Trombe wall systems in various building projects. This analysis would help assess the long-term financial viability and return on investment.

Moreover, expanding the scope of this project to include other cold regions around the world would be valuable. Different climatic conditions and architectural contexts may require adaptations and optimizations of the Trombe wall design. By studying and implementing this project in diverse cold areas, we can gather valuable insights and refine the technology to suit a broader range of environments.

References

1. Bevilacqua, P., Bruno, R., Szyszka, J., Cirone, D., Rollo, A.: Summer and winter performance of an innovative concept of Trombe wall for residential buildings (2022)
2. Saadatian, O., Sopian, K., Lim, C.H., Asim, N., Sulaiman, M.Y.: Trombe walls: a review of opportunities and challenges in research and development (2012)
3. Elsaid, A.M., Hashem, F.A., Mohamed, H.A., Ahmed, M.S.: The energy savings achieved by various Trombe solar wall enhancement techniques for heating and cooling applications: a detailed review (2023)
4. Hu, Z., He, W., Ji, J., Zhang, S.: A review on the application of Trombe wall system in buildings (2017)
5. Ruiz-Pardo, Á., Álvarez Domínguez, S., Antonio Sanz Fernández, J.: Revision of the Trombe wall calculation method proposed by UNE-EN ISO 13790 (2010)
6. Rabani, M., Kalantar, V., Dehghan, A.A., Faghih, K.A.: Empirical investigation of the cooling performance of a new designed Trombe wall in combination with solar chimney and water spraying system. Energy Build. (2015)
7. Agrawal, B., Tiwari, G.N.: Building integrated photovoltaic thermal systems: for sustainable developments (2010)
8. Liu, Y., Wang, D., Ma, C., Liu, J.: A numerical and experimental analysis of the air vent management and heat storage characteristics of a Trombe wall (2013)
9. Stazi, F., Mastrucci, A., di Perna, C.: The behaviour of solar walls in residential buildings with different insulation levels: an experimental and numerical study. Energy Build. (2011)
10. Règlement Thermique de Construction au Maroc (RTCM) (2014)

Greening of Oil Mills: Challenges and Constraints of Adaptation (Case of the Taounate Region in Morocco)

Brahim Idbendriss(✉) and Debbagh Bouchra

Faculty of Law, Economics and Social Sciences, University Sidi Mohamed Ben Abdellah University, Fes, Morocco
brahim.bendris@yahoo.fr

Abstract. Few research studies have focused on techniques for greening the olive industry in Morocco, where the olive tree is the main cultivated fruit species. In Taounate, a province known for the quality of its olive oil, there are only 40 modern units compared to about 3,000 traditional oil mills, which generate a large output of margines, waste that is harmful to the environment. Therefore, it is urgent for the actors in the sector to adopt appropriate technologies for the greening of their olive oil production activities in accordance with the environmental laws and the hygiene and quality standards in force. But even if the solutions exist, operators are still reticent to engage in this greening process. Thus, the objective of this article is to understand the constraints and production problems that greening efforts are facing and that hinder their success, and to propose some useful ideas to improve the efficiency of this sustainable development model. The results of our study show that the constraints are inherent to the dominant local consumption culture in Morocco and the investment costs of restructuring the milling activity.

Keywords: Olive oil · Liquid waste · Environment · Sustainable · Greening

1 Introduction

Morocco occupies the sixth place in the world after Spain, Italy, Tunisia, Turkey, and Greece in terms of olive production [1]. Taounate, a province of the Fez-Meknes region renowned for the quality of its olive oil, has enormous potential in the field of olive cultivation, which plays a fundamental socio-economic role through its multiple functions in the creation of income-generating activities and the fixation of populations in marginal areas [19]. For olive oil extraction, there are about 40 modern units compared to about 3000 maâsras (traditional olive crushing units), using three-phase centrifugal extraction facilities and thus generating a significant production of environmentally harmful margines [26].

In Taounate, as everywhere in Morocco, the olive industry, besides its basic production, which is olive oil, generates two sub-products: solids called pomace and liquids called margines (Balaid et al. 2002), the latter of which are hardly biodegradable and

K. Baba et al. (Eds.): GeoME 2023, SPEES, pp. 42–56, 2023.
https://doi.org/10.1007/978-3-031-49345-4_5

often discharged without prior treatment. Thus, olive oil extraction generates between 500 and 600 kg of highly toxic margins per tonne of olives [27]. This has a negative impact on the environment: pollution of surface water resources and all the species living there due to their high organic load of non-biodegradable polluting organic matter of high salinity and acidity.

Although the negative impact of margines discharges on surface water resources has been widely discussed [25, 32], few studies have been conducted on good olive oil production practices that preserve the environment. Moreover, even if solutions exist to ensure a harmonious development of the olive sector and control measures for professionals have been implemented by the competent authorities to reduce the effects of margines on the environment, the fact remains that the majority of traditional operators are still reticent to engage in this process of modernization by adopting appropriate technologies for the extraction of quality oil and/or treatment of margines in accordance with the environmental laws in force in Morocco. So what are the constraints that explain the disengagement of traditional olive growers from projects to green their olive activity?

To answer this question, we have opted to approach, in the first part, the theoretical and conceptual background to the ecological problems caused by margines and the measures implemented for their treatment or elimination, and to describe, in the second part, the research survey area, the data collection and analysis methods used, as well as the major results of the field study.

2 Theoretical and Conceptual Framework

2.1 Environmental Impact of the Olive Industry in Morocco

The national olive oil production is composed of 65% trituration and 25% conservation. The milling is done by a modern and semi-modern sector composed of 948 units. But also the traditional sector, made up of approximately 11,000 maâsras (traditional olive crushing units). The olive oil production process in Morocco is thus dominated by traditional crushing systems using extraction processes by centrifugation in three phases (oil, margines, and pomace) (Fig. 1).

In Morocco, the volumes of margines generated annually causes recurrent problems at each milling season in the principal olive oil production areas (Fez-Meknes, Marrakech-Safi …). For each ton of olives, the traditional units generate between 0.6 to 0.7 tons of margines and 300 to 500 kg of pomace. With an average annual production of 750,000 tonnes, recorded between 2014 and 2018, the average annual production of margines is between 450,000 and 500,000 tonnes [12].

It is easy to imagine the impact that the activity of traditional oil mills can have on the quality of our environment. The discharge of margines, which are very rich in nitrogenous elements, without any prior control, can cause the pollution of rivers, barrages and groundwater located in the area or close to the spreading site and contaminate the quality of drinking water (Fig. 2).

Moreover, these margines are not degradable because of the phytotoxic and antimicrobial substances (phenols, volatile oily acids, insecticides, etc. …) they contain [6]. In Morocco, water resources are limited and their preservation, both quantitatively and qualitatively, is crucial [16]. In addition to the visual inconvenience and bad odours, the

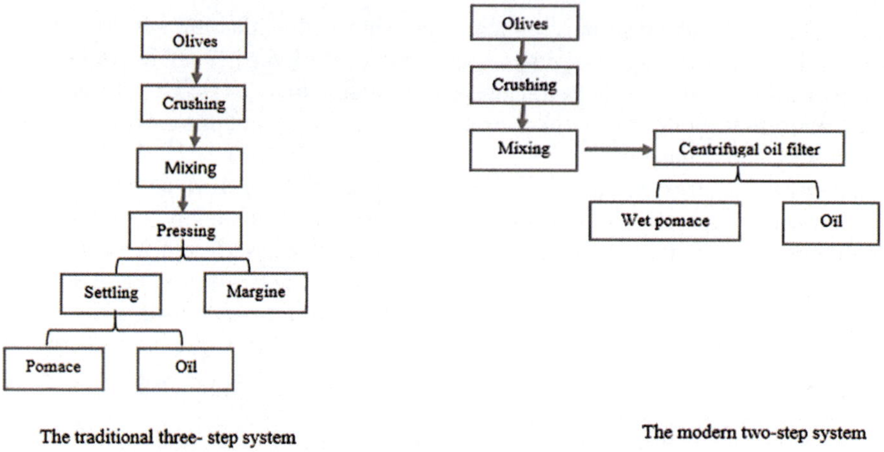

Fig. 1. Olive oil milling systems and sub-products.

Fig. 2. Ecological impact of margines

high organic load of the margines totally destroys the aquatic fauna and flora due to the deterioration of the water quality [28]. This type of pollution also causes other problems including, among others, difficulty in watering livestock, and the extinction of certain animal and plant species.

2.2 Greening Techniques of Oil Mills for a Harmonious Development of the Olive Sector

By greening practices, we mean any cognitive and normative reframing initiative aimed at an environmental inflection of current social practices in the field under consideration (agriculture, industry, sports and nature leisure activities, forestry, etc.). This greening can be based on environmental norms, generally supported by institutional actors, or more informally by a wide variety of actors (companies, associations, but also users, citizens, etc.) (Ginelli 2017).

In Morocco, the large quantities of margines and olive pomace often dumped in nature without treatment led the competent authorities to reflect on how to ensure better valorisation of these products, which are known to be very salty and acidic.

Thus, to ensure a harmonious development of the olive sector, it has become urgent to restructure and modernise olive production and use appropriate technologies for oil extraction. Such technologies would help to produce quality oil at a lower cost [3] (Fig. 3).

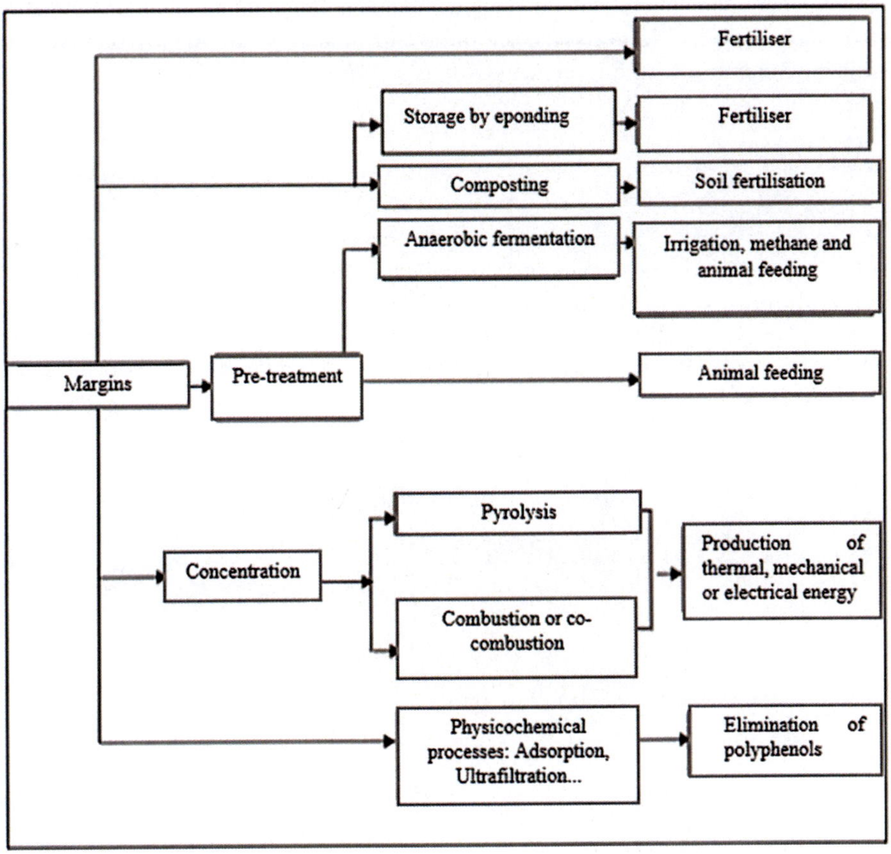

Fig. 3. Different processes for the valorisation of margines

The solutions proposed range from simple disposal by spreading in basins to more or less complicated recovery processes [10]. In this context, a series of technical and preventive measures aimed at limiting the ecological damage caused by margins are proposed, including:

- The introduction of continuous two-phase milling systems, which do not generate margines but only wet pomace, allows the separation of oil and pomace rich in vegetation water from the olives and reduces the costs and duration of olive processing, resulting in lower acidity in olive production [31];
- Spread storage in open-air assembly and evaporation basins is the cheapest and easiest solution [14], as well as a platform for storage and drying of pomace. However, evaporation is difficult because of the oily surface layer that forms;
- The use of margines as soil and crop fertilizer is a common practice that partially solves the problem of disposal of these liquid effluents [9]. Through this process, margines can be revalued into treated irrigation water, biogas, and protein-rich biomass for use as animal feed [23].

All ways are suitable to reduce the impact of margines on the environment while preserving the interests of farmers and owners of trituration units [7].

3 Methodology

3.1 Study Area

The Province of Taounate is a subdivision of the Fez-Meknes region located in the centre north of Morocco where the country's largest water reserve is located. It is a predominantly rural province delimited by the provinces of Al Hoceima and Chefchaouen to the north, Fez to the south, the province of Taza to the east and the province of Sidi Kacem to the west (Fig. 4).

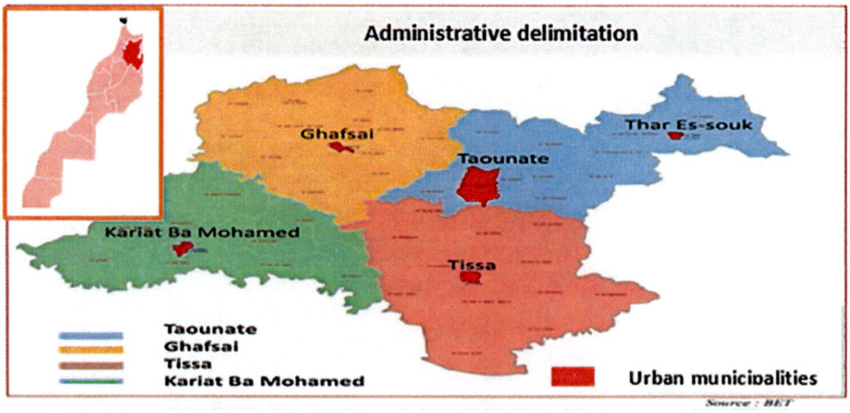

Fig. 4. Location of the study area (Taounat province, Morocco)

Taounate is endowed with natural resources, including a diversified forest cover, large and medium-sized barrages, notably Al Wahda, the largest barrage in the country, as well as an important network of rivers and valleys (Ouargha, Sebbou, Inawan, Bouadel valley, etc.). The olive tree is the main cultivated fruit species, with about 83% of the surface area of fruit trees in the province and 20% of the surface area of olive trees at the national level. Olive production, mainly for the national and local markets, is estimated at 200,000 metric tonnes on average. The province also has 3,000 traditional units, 27 semi-modern units, and 40 modern units (Fig. 5).

Fig. 5. Oliveries in the province of Taounate

Although these units contribute significantly to the production, conversion, and commercialisation of natural olive oil, with positive impacts on the region from a socio-economic development point of view in terms of employment and income for the local populations, they are the source of general contamination of the land, fields in general, and water resources in particular due to the olive margines. This poses a serious ecological problem when discharged untreated into natural environments such as surface water courses. Compared to the continuous system, the traditional discontinuous oil extraction process is commonly practised by maâsras located in mountainous areas and especially in remote regions where there is no infrastructure (water and electricity).

Whether in Fez, Taza, or Taounate, with the important number of oil mills implanted in the region, the environmental problems linked to the anarchic evacuation of margines in the natural environment have recently increased. Thus, in accordance with the environmental laws applicable in Morocco, the mixed regional committee in charge of giving its opinion on the environmental acceptability of the said projects and the control and follow-up of the olive oil extraction units of the region within the framework of the efforts aiming at fighting against the negative effects of the margine on the water resources carries out field visits, during which he alerts and sensitises the owners of the oil mills that resort to the anarchic discharge of margines in the rivers and dams of the region on the need to comply with the regulations and environmental standards in force. Measures have also been taken that have led to the closure of offending production units with regard to environmental standards and the terms of the required specifications (crushing techniques, hygiene conditions, construction of collection and evaporation tanks for margines, platforms for the storage of pomace and the drying of sludge, etc.).

3.2 Methods and Materials

Our empirical research is based on a qualitative methodology: in addition to a documentary study and observations of events and interactions (Sofaer 1999), we conducted semi-structured interviews with local olive operators, both face-to-face and by telephone. The selection of these interviewees was based on a broad documentary review of the actors in the region. We interviewed the owners or managers of traditional oil mills units in the communes of the province of Taounate, which were visited during the period between April and October 2022.

The objective is to identify the characteristics of these units, analyse the attitude towards the greening of olive oil production, and understand the constraints that explain the reluctance of olive growers to engage in greening. Our interest in these structures derives from two main considerations: On the one hand, these units are very popular among consumers, and on the other hand, they use the traditional technique, which generates a large quantity of margin residues and poor production performance. The sample was constituted according to snowball sampling for the choice of target interviewees. This involved asking each respondent to point out other people who could provide answers and insights to our questions [22] (Table 1).

Table 1. Characteristics of the exploitations and managers interviewed

Codification	Name of the exploitation	Locality	Interviewee's function
C111	Hadj Azouz Press	Oued Amlil	Owner
C112	Saiss Press	Commune Hamriya	Manager
C113	ELBARAKA Press	Commune Ras al-Wad	Owner
C114	Olive press Akhribish	Commune Otabouban	Manager
C115	Press l'or Vert	Karait Ba Mohamed	Manager
C116	Press Olive press Sebaa	Commune Beni Ammar	Owner
C117	Press Wadi El Darader	Commune Ghouazi	Responsible
C118 C119	Olive press Tamimi	Commune Galaz	Responsible Manager
C120 C121	Olive press Al-Atari	Commune Beni Yezzou	Manager Owner
C122	Bab Zriba Press	Commune Ourtzagh	Manager Owner

Source: Made by the author

The themes of the interview guides concern:

- The environmental impacts of the milling activities - the attitude of the olive oil producers towards the greening of their activities;
- The attitude of the olive oil producers towards the greening of their activities;
- The particular difficulties faced by traditional olive oil producers-the constraints that explain their reluctance to take part in greening processes (Table 2).

Table 2. Socio-demographic characteristics of the sample

Variables		%
Gender	Male	95%
	Female	5%
Age	Under 30 years	40%
	30–50 years	35%
	Over 50 years	25%
Formation	Less than Bac	68%
	Bac +2	20%
	Licence	10%

For the analysis of the data, we used the content analysis method (Andreani and Conchon 2005). This method allows us to follow a three-step process:

- Transcribing qualitative data: this involves taking stock of the information collected and transforming it into writing (Auerbach and Silverstein 2003);
- Coding the collected information: is a process that explores the interview texts and categorises the data according to units of meaning for descriptive information (Berg 2003);
- Processing the data: this is done in two ways, either semantically or statistically. The first consists of manual processing of data through empirical analysis of ideas, words, and their meanings, while the second uses computer processing for statistical analysis of words and sentences.

After transcribing, coding, and processing the information collected, they arrived at the interpretation stage of the findings. This is a step-by-step diagnosis of the solutions, starting from the faithful description of the interviews or observations and determining the consequences from the point of view of strategic choices or theoretical concepts (El Bakkouri et al. 2016) (Fig. 6).

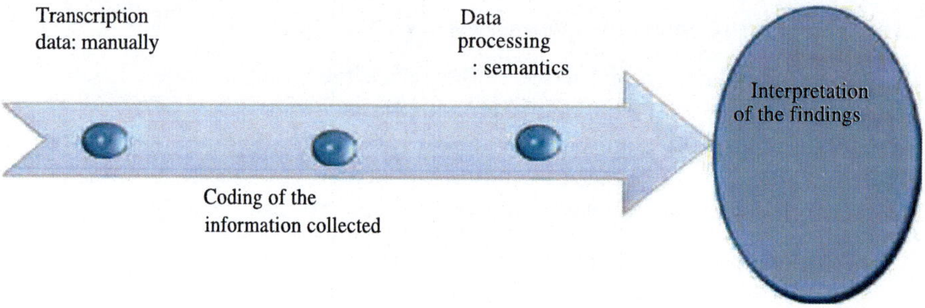

Fig. 6. Our process of content analysis and interpretation

4 Results

The main results of our survey show that:

- Concerning the efforts of depollution, a minority of units have or are considering investing in evaporation basins for margins. Indeed, the olive growers interested in this technique are those who have too many resources or who are ready to structure themselves into producer groups. One owner of an olive production unit stated:

  "A number of olive oil producers do not respect the environmental law and dispose of olive residues in the valley's sewers and streams."

Thus, 33% adopt a 'defensive' attitude, compared to 45% who have a non-compliant production. Instead of a 'conformist' attitude to environmental protection requirements (Fig. 7).

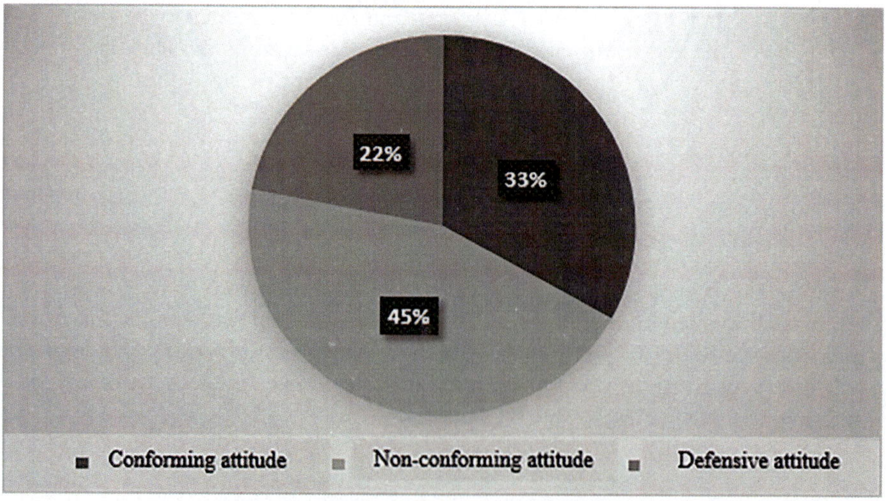

Fig. 7. Olive growers' attitudes to greening

- According to the interviews, the attitude of oil olive producers depends on several stated constraints, namely: the additional costs generated by the installation of evaporation ponds; the maintenance costs; the problem of low evaporation; and the lack of space to build the pond (Fig. 8).

Fig. 8. Olive growers' constraints (word cloud)

According to one manager:

> "Units that invest in facilities to control margins have high costs. Units that invest in facilities to control margins have high costs. And the owners resort to disposing of the residues and remains of the olives through the sewage channels and into the waterways of the valleys".

- The dynamics of greening through the use of the preventive system seem to evolve in modest measure. Indeed, the two-phase trituration system is particularly adapted to olive oil producers who have the necessary means to acquire modern trituration equipment, which remains expensive compared to the traditional trituration system.
- The reticence to engage in greening processes is also explained by socio-cultural factors. After a discussion, a unit manager expressed his recognition of the negative environmental impacts, but raised the problem that:

> « Moroccan households are going to source a large quantity for annual consumption from a traditional local producer, perceived as a trusted supplier, and continue to consume locally sourced olive oil from traditional milling that is estimated to be of better quality ».

Thus, the professionals point out the need to strengthen the training and awareness of farmers on good practices in olive harvesting, transport, and extraction, in addition to the marketing of this oil at the national and international level:

« The government and foreign partners have spent millions to set up modern and environmentally compatible production units, but they are left to be mismanaged and abandoned due to a lack of demand ». (Fig. 9)

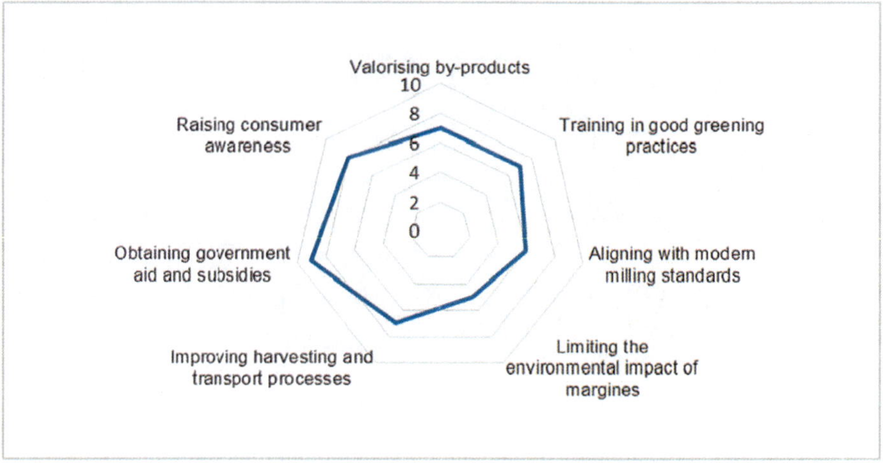

Fig. 9. Priorities according to olive oil producers

5 Discussion

The constraints that block olive oil producers from engaging in the greening process as imposed by the State for the protection of the environment are multiple and varied but can be grouped into three main categories, namely:

- Economic constraints: related to the additional costs of installing evaporation basins;
- Social constraints: olive oil coming from modern units is considered by consumers as an unnatural industrialised product of lower quality and reserved for "wealthy" social categories because of the prices charged;
- Cultural constraints: the large presence of traditional oil milling units is sustained by traditional buying behaviour (prevalence of informal sector) and a demand from consumers who consider the olive oil produced there as synonymous with an authentic product of better quality.

Satisfied with national consumption, the olive oil producers reveal that they have no interest in adopting ecological behaviour, which means increased costs for the installation and management of margines and does not adapt to the culture of the local consumer.

The results of our research are concordant with that of Benazzouz et al. [7] according to which the attitude towards margines is positively correlated with production costs. They are contradictory to the study of Klassen and McLaughlin [24], for whom there is a positive link between environmental management and financial profitability, and that of Boiral [8], according to whom such objectives can be the result of efforts to increase productivity through several components, including quality improvement, operations management, and the reduction of certain sources of waste. However, the analysis of the link between environmental actions and productivity remains very controversial: some research tends to validate Porter's 'win-win' hypothesis [4] (Azzone and Bertèle 1994), while others confirm, on the contrary, the main postulates of the classical model of the negative link between environmental actions and company competitiveness [11].

The attitude of producers in reaction to the objective of green milling practises refers to the theory of diffusion of innovation, which is "a process of spreading an innovation that brings new ideas and behaviours, and that is communicated through certain channels, over time, among members of a community or social system" (Rogers 1995). The probability of adoption of an innovation/behaviour depends on the target, the socio-cultural barriers, and the communication system of the innovation. Some people and groups in society are more ready than others to take up new ideas or "innovations" (Fig. 10).

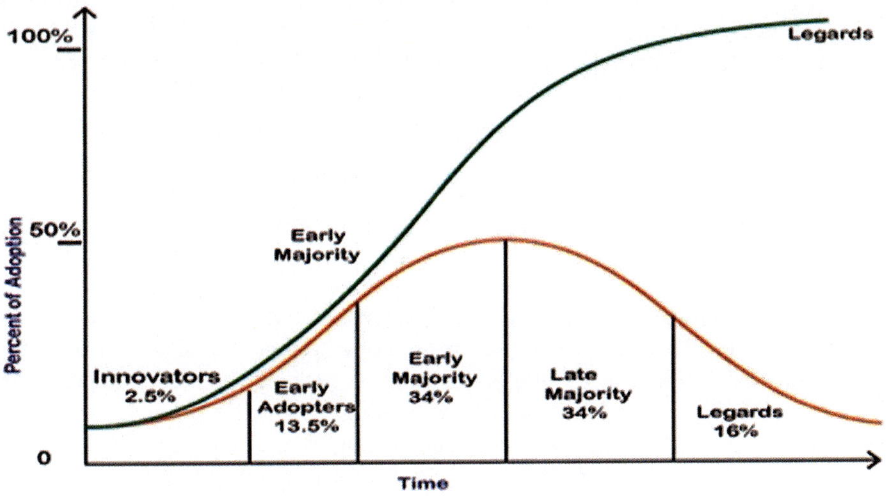

Fig. 10. Innovation diffusion curve (according to Rogers 1995)

Thus, in order to support the additional costs of greening deplored by the olive growers, aid from the departments concerned and in coordination with the various actors in the sector should be provided to encourage them to engage in the collection and collective treatment of margines, if necessary. Finally, only the implementation of strict regulations and consumer awareness, combined with awareness-raising on the field, can make a difference. Beyond its environmental impact, this process of modernisation and conversion would allow the region to comply with international standards of trituration, which would stimulate the capital of investors, often foreign, to invest in the region whose potential could transform it into a production and export platform on a national and international scale.

The professionals also emphasise the need to reinforce the training and awareness of consumers on responsible consumption and on good practices of harvesting, transport and extraction of olives by the farmers. However, although our study provides insight into the obstacles and productive problems hindering greening efforts, the small sample size, the small number of factors studied, and the lack of scientific rigour for generalisation are traditional criticisms of qualitative studies (Yin 2009) whose results are embedded in a particular context, but this is precisely what makes it possible to give them meaning, to appreciate their contribution and relevance, and to offer them a contextual force linked to culture, history, and the initial conditions of emergence of a phenomenon that mutually transform over time, so that they do not (re)occur in the same way in time and space (Proulx 2019).

6 Conclusion

The results of our research show that the margines resulting from the trituration in the traditional oil mills in the province of Taounate do not generally receive any prior treatment and are often discharged into nature. This has a negative impact on the environment, causing pollution of surface and groundwater. Thus, the treatment of the residues and by-products of this system is significantly delayed in Morocco despite the advantages of the two-phase technique and the control efforts of the competent authorities.

Satisfied with national demand and therefore little interested in exporting, the professionals have no motivation to invest in projects for the greening of their oil mills. Only the implementation of strict regulations and consumer awareness, combined with awareness raising in the field, can guarantee the harmonious development of the local olive oil industry.

Also, subsidies from the departments concerned should be provided to olive oil producers to encourage them to engage in the collection and treatment of margines, and aid should be provided to the poorest for the construction of open-air assembly and evaporation basins. Finally, training and awareness sessions, as well as control visits, must be regularly organised in coordination with the other stakeholders, to ensure that the units concerned comply with the requirements in terms of environmental protection and to sanction offenders. Nevertheless, repressive measures in the form of closing units for non-respect of the environmental standards in force are not enough to solve the problem. Only awareness on the part of the "responsible" consumer, combined with coordination with the various actors in the sector, particularly the traditional oil mills, to encourage them to commit to the collection and treatment of margines, can guarantee the harmonious development of the olive oil industry throughout Morocco.

References

1. Achak, M., Ouazzani, N., Mandi, L.: Élimination des polluants organiques des effluents de l'industrie oléicole par combinaison d'un filtre à sable et un lit planté. Rev. Sci. L'Eau/J. Water Sci. **24**(1), 35–51 (2011)
2. Agoumi, A., Debbarh, A.: Ressources en eau et bassins versants du Maroc: 50 ans de développement. Report Prepared Within the Framework of the "Water: Management of Scarcity" Organized by the Association of Moroccan Engineers of Bridges and Roads (2005)

3. Akika, F., Bouchefra, A., M'hamdioua, N., Roula, M.E.: Variété Sigoise de l'olivier: qualité des olives, de l'huile d'olive et étude des effets nutritionnels de l'huile d'olive chez le rat Wistar. Doctoral dissertation, Université de Jijel (2009)
4. Amara, H.A.: La transition de l'agriculture algérienne. Cah. Options Méditerr. **36**, 127–137 (1999)
5. Bajoub, A., Hurtado-Fernández, E., Fernández-Gutiérrez, A., Carrasco-Pancorbo, A., Ouazzani, N.: Quality and chemical profiles of monovarietal north Moroccan olive oils from "Picholine Marocaine" cultivar: registration database development and geographical discrimination. Food Chem. **179**, 127–136 (2015)
6. Belaid, C., Kallel, M., Elleuch, B.: Identification de nouveaux composés phénoliques présents dans les rejets liquides d'huileries d'olive (margines). Déchets sci. techn. **27**, 30–34 (2002)
7. Benazzouz, L., El Jaafari, S., Zahid, F., Mokhtari, F.: Écologisation des huileries de la Province de Meknès (Maroc): analyse technico-socio-économique. VertigO la rev. electron. sci. l'environ. **16**(2) (2016)
8. Boiral, O.: La gestion environnementale et la norme ISO 14 001 par Corinne Gendron, 347 pp. Presses de l'Université de Montréal, Montréal (2004). ISBN 2-7606-1809-9. Relat. Ind./Ind. Relat. **60**(2), 392–394 (2005)
9. Boudabia, H., Dahou, K.: Valorisation de la fraction organique des résidus agricoles et sous-produits oléicoles par la Co méthanisation. Doctoral dissertation (2019)
10. Bouamoud, M., Kebier, H.: Valorisation Des Déchets Organique D'origine Agricole Par Digestion Anaérobie. Doctoral dissertation, University of Ghardaia (2021)
11. Boyd, G.A., McClelland, J.D.: The impact of environmental constraints on productivity improvement in integrated paper plants. J. Environ. Econ. Manag. **38**(2), 121–142 (1999)
12. Bouzitoune, S.: Caractérisation phytochimique de margines monovariétal collectés dans la région East de l'Algérie (2018)
13. Btissam: Etude écotoxicologue des effluents des huileries de la ville de Marrakech: impact sur les milieux récepteurs et detoxication (2001)
14. Chouchene, A.: Etude expérimentale et théorique de procédés de valorisation de sous-produtis oléicoles par voies thermique et physico-chimique. Doctoral dissertation, Université de Haute Alsace-Mulhouse (2010)
15. El Haji, M.: Expérience de la coopération dans le cadre de l'AUF avec le Canada et l'action intégrée avec l'Europe. In: Congrès français de mécanique. AFM, Maison de la Mécanique, 39/41 rue Louis Blanc, 92400 Courbevoie, France (FR) (2011)
16. El Mouhtadi, I., Agouzzal, M., Guy, F.: L'olivier au Maroc. OCL **21**(2), D203 (2014)
17. Aktas, E.S., Imre, S., Ersoy, L.: Characterization and lime treatment of olive mill wastewater. Techn. Note Water Res. **35**(9), 2336–2340 (2001)
18. Essahale, A., Karrouch, L.: Contribution à l'étude de l'impact des huileries de la province d'El Hajeb sur l'environnement. Arch. Mal. Prof. Environ. **76**(4), 355–365 (2015)
19. Gharbi, I., Issaoui, M., Mehri, S., Hammami, M.: Assurance qualité des huileries tunisiennes. OCL **22**(4), A401 (2015)
20. Chahine, H.: Management of olive mills wastes in the Palestinian territories. In: Third Congress on the Environment Management (2005)
21. Hamdi, M.: Nouvelle conception d'un procédé de dépollution biologique des margines, effluents liquides de l'extraction de l'huile d'olive. Ph.D. theses, University of Provence (1991)
22. Johnston, L.G., Sabin, K.: Échantillonnage déterminé selon les répondants pour les populations difficiles à joindre. Methodol. Innov. Online **5**(2), 38–48 (2010)
23. Kapellakis, I.E., Tsagarakis, K.P., Crowther, J.C.: Olive oil history, production and by-product management. Rev. Environ. Sci. Bio/Technol. **7**(1), 1–26 (2008)
24. Klassen, R.D., McLaughlin, C.P.: The impact of environmental management on firm performance. Manage. Sci. **42**(8), 1199–1214 (1996)

25. Larif, M., Soulaymani, A., Elmidaoui, A.: Evaluation spatio-temporelle du degré de la pollution industrielle oléicole sur les cours d'eaux de l'oued Boufekrane dans la région de Meknès. J. Mater. Environ. Sci. **4**(3), 432–441 (2013)
26. M'Sadak, Y., Makhlouf, M.: Qualitative characterization of olive biomass resources and aptitude at valorization of energy in the delegation of Kalâa Kébira (Tunisia). J. Water Sci. Environ. Technol. **2**(1) (2017)
27. Nadjet, M.: Evaluation du compostage des sous produits d'huile d'olive sur le rendement de quelques espèces a intérêt agroalimentaire. Doctoral dissertation, Université Mohamed Boudiaf des sciences et de la technologi (2016)
28. Ouabou, E., Anouar, A., Hilali, S.: Élimination des polluants organiques présents dans la margine d'huile d'olive par filtration sur colonne d'argile et sciure de bois d'eucalyptus. J. Appl. Biosci. **75**, 6232–6238 (2014)
29. Rais, Z., et al.: Margines: traitement, valorisation dans la germination des graines de tomate et dans la filière de compostage. Rev. Sci. L'Eau/J. Water Sci. **30**(1), 57–62 (2017)
30. Salem, H.: Integrated waste management for the olive oil pressing industries in Lebanon, Syria and Jordan pressing industries in Lebanon, Syria and Jordan (project IMOOPW). In: National Workshop on Environmental Policy Integration & SMAPP III, Damascus, 13–14 Dec 2005
31. Saoudi, S.: Contribution à l'étude des sous-produits oléicoles générés par les huileries dans la région de M'chedallah. Doctoral dissertation, Université de Bouira (2017)
32. Zghari, B., Benyoucef, F., Boukir, A.: Impact environnemental des margines sur les eaux d'oued oussefrou: caracterisation physico-chimique et evaluation par chromatographie gazeuse couplee a la spectrometrie de masse (CPG-SM). Am. J. Innov. Res. Appl. Sci. **2429**, 5396 (2018)

Evaluation of Window-Blind's Effectiveness on Reducing Total Energy Consumption of a Library Building for Different Slat Angles

Meryem El Alaoui(✉), Hasna Oukmi, Ouadia Mouhat, and Mohammed Rougui

LGCE, Civil Engineering and Environment Laboratory, High School of Technology (EST) - Salé, Mohammed V University, P.O. Box 227, Rabat, Salé, Morocco
Meryem.elalaoui0@gmail.com

Abstract. Modern buildings have a high percentage of glass surfaces, which increases the amount of solar radiation that enters the structure and increases energy demand and discomfort for occupants. Consequently, the use of window shading devices is becoming more important than ever. This paper evaluates the effectiveness of blinds applied on a library building located in Errachidia region. The building is simulated in Designbuilder simulation programs during summer period (starting from 1st May to 31st August), with and without shading devices, for different blind slat angles varying from 10° to 170° with a step of 10°. The calculated total energy consumption concerns cooling and lighting. The comparison of total energy loads with and without blinds demonstrates that the use of window blinds optimized the total energy consumption by 17% for angles 10 and 170, and 10% as a minimum optimization percentage (angles from 70 to 120). This optimization leads to an economy of about 5800 DH during just four months (considering 0.951 DH/kWh). It can be concluded that the adoption of shading devices in buildings with a high percentage of glass surfaces (25% in our case study), is very useful for energy consumption management and energy conservation, especially in the 6th thermal Moroccan zone, where our building is located.

Keywords: Shading devices · Blinds · Slat angle · Designbuilder

1 Introduction

The use of shading devices is of utmost importance in buildings characterized by a high window-to-wall ratio. These devices play a critical role in regulating solar heat gain and reducing the ingress of sunlight into the interior spaces. By effectively managing solar heat gain, shading devices help minimize the need for excessive air conditioning, resulting in decreased cooling loads and lower energy consumption. Moreover, shading devices effectively address the problem of glare, which can cause discomfort and impair visibility. They either diffuse or block intense sunlight, creating a more comfortable and visually pleasing environment for occupants. Thus, the incorporation of shading devices is essential for optimizing energy efficiency, ensuring indoor comfort, and enhancing the

K. Baba et al. (Eds.): GeoME 2023, SPEES, pp. 57–64, 2023.
https://doi.org/10.1007/978-3-031-49345-4_6

functionality of buildings with a substantial expanse of glass surfaces. Numerous studies aimed to investigate the efficacy of shading devices in promoting energy conservation: In their study, Bellia et al. [1] explored the impact of external solar shading devices on energy consumption in a representative air-conditioned office building situated in Italian climates. The research specifically targeted a prevalent office building design prevalent in Europe. By utilizing a building energy simulation software, the researchers assessed the energy needs of heating, cooling, and lighting systems, while also quantifying the energy conservation benefits resulting from the adoption of solar shading devices. Results demonstrate that solar shading devices offer substantial energy efficiency benefits in warm summer climates, with potential annual energy savings ranging from 8% in Milan to 20% in Palermo. In another work, Kirimtat et al. [2] introduced novel design alternatives for energy-efficient shading devices with amorphous panels. To overcome the intricacies of design, the study implemented a methodology that relied on parametric modeling and performance-based optimization. The optimization model aims to minimize total energy consumption and maximize useful daylight illuminance using evolutionary multi-objective optimization algorithms. Numerical results demonstrate considerable energy savings of up to 14% while maintaining daylight availability above 50%. The role of shading devices in enhancing energy efficiency in buildings situated in hot, dusty, and dry tropical regions was investigated by Samanta et al. [3]. The researchers utilized a case study of a guest house that implemented a specific design approach aimed at reducing thermal loads. By inputting the building's plan, measurements, and occupancy schedules into a simulation program, a thermal model of the building was created. The study highlighted the impact of employing external shading, demonstrating that the use of movable shading devices on windows led to a reduction of approximately 1.5 °C in temperature for each thermal zone. Comparing the base case to a control case with movable shading devices, the simulations demonstrate an estimated 8% reduction in thermal energy loads. Experimental tests and simulation validations were conducted by Ye et al. [4] to investigate the possibility of replacing external shading devices with internal ones, using high reflectivity materials. The research findings suggest that an internal shading system with suitable materials can be an equally effective alternative to external shading. This substitution has the potential to reduce the overall cost of shading systems and offer more design flexibility for building facades. Additionally, the researchers performed a grey relational analysis to identify the key factors that influence the performance of internal shading devices, thereby aiding in their optimization. In another work, De Luca et al. [5] examined the effectiveness of exterior static shading devices in two classrooms in Tallinn, Estonia. In the study, a multi-objective optimization workflow was employed to evaluate the effects on glare reduction, daylight availability, view quality, and energy consumption. Through the use of simulations and spatial glare assessment techniques, the research revealed that static shading devices resulted in a significant reduction of visual discomfort, up to 89.8%, as well as a decrease in primary energy usage by up to 29.1%. The research highlights the potential benefits of properly designed static shading devices in enhancing occupants' well-being and improving building performance. In their study, Alsharif et al. [6] focused on rectifying the suboptimal arrangement of shading devices in buildings, which often results in increased solar heat gain and energy consumption. The research introduces enhancements to an existing framework that aims to optimize shading devices for multiple objectives such as reducing thermal discomfort, minimizing heating and cooling loads, and optimizing the surface area of the shading devices. The proposed improvements include the amorphous optimization of fixed shading devices, utilizing orthogonally structured training datasets for machine learning models, and incorporating the Multi-Objective Manta-Ray Foraging Optimizer. Results show that amorphous shading devices can significantly reduce heating and cooling loads by up to 12.7% and thermal discomfort hours by up to 2.8%. Koç and Kalfa [7] evaluated the effectiveness of fixed shading devices (SDs) in architectural design, specifically in hot climate regions. By simultaneously considering various parameters, energy-efficient scenarios for office buildings were sought to be identified. A comprehensive set of 1485 scenarios were done, encompassing different types of fixed shading devices (SDs). These scenarios took into account various factors such as direction, SD depth, glazing type, window-to-wall ratio, and slope. The research findings revealed that the most effective scenarios, which utilized high-performance glazing types, led to remarkable reductions in cooling energy consumption, ranging from 37% to 49% when compared to the scenario without any shading. Shahdan et al. [8] evaluated the effectiveness of external shading devices on a building's energy efficiency by reducing solar radiation. It highlighted that while some shading devices are designed for aesthetic purposes, their potential to reduce solar radiation and glare is often overlooked. The study emphasized the need for early consideration of shading devices during the building design phase. Using Building Information Modeling (BIM) and Autodesk Revit software, the researchers conducted computer simulations to analyze the impact of different shading device configurations on energy consumption. The results indicate that shading devices, significantly affect energy consumption.

In this study, the effectiveness of blinds on a library building in the Errachidia region is assessed. The building is simulated using the Designbuilder simulation program during the summer period (May 1st to August 31st) with and without shading devices. The study examines various blind slat angles ranging from 10° to 170° in increments of 10°.

2 Materials and Methods

2.1 Building Description

The building under investigation is a component of the Faculty of Medicine and Pharmacy library. It comprises two stories and has a conditioned area of 1300.52 m^2, with windows accounting for approximately 25% of the total wall area (Fig. 1). The project focused on studying this building situated in Errachidia city, which falls within the sixth Moroccan climate zone. The objective was to analyze the specific influence of solar radiation on the building. Table 1 provides a summary of the building's location, geometric features, and climatic parameters for the case study.

2.2 Shading Devices Presentation

Shading devices, such as venetian blinds, play a crucial role in enhancing energy efficiency and optimizing building performance. These devices are specifically engineered

Fig. 1. Project global view

Table 1. General information

Site	
Location	Errachidia City, Morocco
Latitude	31.93° N
Longitude	– 4.4° W
Dimensions	
Area	2089.81 m^2
Conditioned area	1300.52 m^2
Climatic parameters	
HDD (balance temperature 18 °C)	955 °C
CDD (balance temperature 21 °C)	3718 °C

to efficiently manage the influx of solar radiation into a space, consequently decreasing the necessity for excessive cooling during warmer seasons. Shading devices influence the transmittance of the system and the absorptance of the glass layer for both short-wave and long-wave (thermal) radiation [9]. The positioning of the shade (interior, exterior, or between-glass), its transmittance, and the extent of inter-reflection between the shading device and the glazing all have an impact on the end result [10]. The amount of radiation that the shade device absorbs is also interesting.

Venetian blinds are particularly notable for their versatility and ability to adjust the angle of their slats, allowing for precise control over daylight and solar heat gain. Furthermore, the adjustability of venetian blinds allows occupants to modulate natural lighting levels according to their preferences and needs. By altering the angle of the slats, users can regulate the amount of daylight entering a space, balancing natural illumination with the control of glare. This flexibility in daylight management not only enhances visual comfort but also contributes to energy conservation by reducing the reliance on artificial lighting.

In this paper, the studied building was equipped with venetian blinds featuring high reflectivity materials in their slats.

2.3 Designbuilder Simulation

The building under investigation was simulated in Designbuilder software to analyze the effect of venetian blinds on reducing energy consumption. The simulations were conducted by comparing scenarios with and without venetian blinds, while varying the slat angles of the blinds from 10° to 170° in 10° increments. By employing this approach, a thorough evaluation of the energy consumption reduction resulting from these shading devices was made possible.

Figure 2 represents the building simulated in designbuilder software.

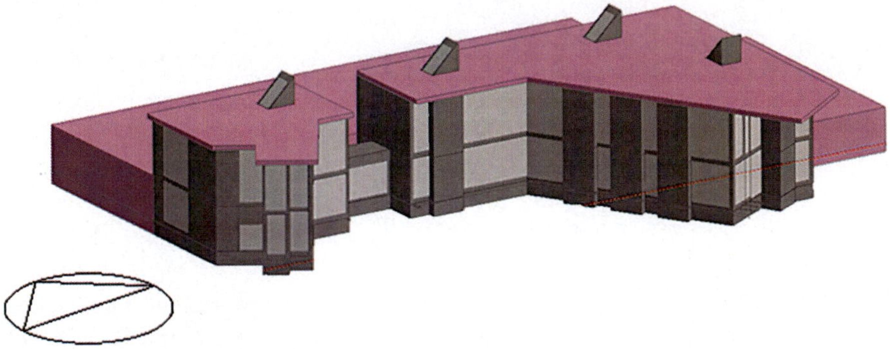

Fig. 2. Building designbuilder simulation

3 Results and Discussion

Building energy consumption for cooling and lighting without venetian blinds, during summer period (starting from 1st May to 31st August) is of about 36512 kWh.

The building was equipped with venetian blinds during the same period and for different slat angles (from 10° to 170° with 10° step). Table 2 summarized building energy consumption for each blinds slat angle.

To provide a more in-depth analysis of the impact of incorporating venetian blinds into the building, Table 3 presents the percentage of energy consumption optimization achieved for various slat angles. This data allows for a comprehensive understanding of the potential energy-saving benefits associated with different configurations of venetian blinds.

The results presented in Table 3 highlight the considerable energy consumption optimization achieved through the integration of venetian blinds in the studied building.

Table 2. Building energy consumption according to different slat angles

Slat angle (°)	Energy consumption (kWh)
10	30415.61
20	30916.50
30	31392.36
40	31838.42
50	32232.12
60	32507.72
70	32721.69
80	32857.59
90	32917.69
100	32891.49
110	32875.48
120	32752.33
130	32525.17
140	32114.65
150	31556.55
160	30916.07
170	30385.09

Notably, for all slat angle configurations, a minimum energy percentage optimization of 10% is observed. This demonstrates the consistent energy-saving benefits of venetian blinds across different angles. Significantly, the greatest optimization of 17% is achieved when the venetian blinds are positioned at slat angles of 10° and 170°. This wide range of slat angles exemplifies the remarkable effectiveness of venetian blinds in minimizing solar heat gain and reducing cooling loads, leading to substantial energy savings. To put the energy savings into perspective, if we consider a cost of 1.05 DH/kWh, the energy cost optimization amounts to approximately 5800 DH. This considerable cost reduction highlights the significant financial benefits that can be realized by incorporating venetian blinds into the building. It emphasizes the importance of implementing energy-efficient measures like venetian blinds to not only conserve energy but also optimize financial resources. These findings underscore the importance of carefully selecting and adjusting the slat angles of venetian blinds to maximize their energy efficiency potential. By optimizing the angle configuration, building owners and designers can effectively reduce energy consumption and contribute to a more sustainable and cost-effective operation of the building.

Table 3. Building energy consumption optimization using venetian blinds, according to different slat angles

Slat angle (°)	% Optimization
10	17
20	15
30	14
40	13
50	12
60	11
70	10
80	10
90	10
100	10
110	10
120	10
130	11
140	12
150	14
160	15
170	17

4 Conclusions

This study examined the effectiveness of venetian blinds as shading devices for energy consumption optimization in a library building located in the Errachidia region. Through simulations conducted in Designbuilder software, the building was evaluated with and without venetian blinds at various slat angles ranging from 10° to 170°.

The results demonstrated that the incorporation of venetian blinds led to significant energy savings. The total energy consumption was optimized by 17% for slat angles of 10° and 170°, while a minimum optimization percentage of 10% was achieved for slat angles between 70° and 120°. These energy optimizations resulted in substantial cost savings, estimated to be around 5800 DH over a four-month period.

The findings highlight the importance of shading devices, such as venetian blinds, in managing energy consumption and conserving energy in buildings with a high proportion of glass surfaces. Particularly in the Errachidia region, situated in the 6th thermal Moroccan zone, the adoption of shading devices can effectively reduce solar heat gain and enhance energy efficiency. Overall, the study emphasizes the value of incorporating venetian blinds as an energy-saving strategy for buildings, offering both environmental and economic benefits. By optimizing slat angles and utilizing shading devices, building owners and designers can achieve significant energy consumption reductions and contribute to sustainable building practices.

References

1. Bellia, L., De Falco, F., Minichiello, F.: Effects of solar shading devices on energy requirements of standalone office buildings for Italian climates. Appl. Therm. Eng. **54**(1), 190–201 (2013)
2. Kirimtat, A., Krejcar, O., Ekici, B., Tasgetiren, M.F.: Multi-objective energy and daylight optimization of amorphous shading devices in buildings. Sol. Energy **185**, 100–111 (2019)
3. Samanta, A., Saha, S., Biswas, J., Dutta, A.: Evaluation of impact of shading devices on energy consumption of buildings in tropical regions. J. Energy Resour. Technol. **136**(2) (2014)
4. Ye, Y., Xu, P., Mao, J., Ji, Y.: Experimental study on the effectiveness of internal shading devices. Energy Build. **111**, 154–163 (2016)
5. De Luca, F., Sepúlveda, A., Varjas, T.: Multi-performance optimization of static shading devices for glare, daylight, view and energy consideration. Build. Environ. **217**, 109110 (2022)
6. Alsharif, R., Arashpour, M., Golafshani, E., Rashidi, A., Li, H.: Multi-objective optimization of shading devices using ensemble machine learning and orthogonal design of experiments. Energy Build. **283**, 112840 (2023)
7. Koç, S.G., Kalfa, S.M.: The effects of shading devices on office building energy performance in Mediterranean climate regions. J. Build. Eng. **44**, 102653 (2021)
8. Shahdan, M.S., Ahmad, S.S., Hussin, M.A.: External shading devices for energy efficient building. IOP Conf. Ser. Earth Environ. Sci. 012034 (2018)
9. Winkelmann, F.C.: Modeling Windows in EnergyPlus 11 (n.d.)
10. A Review of Optical Properties of Shading Devices. Routledge (2013). https://doi.org/10.4324/9781849770378-7

Development of a BIM-BEM Approach for Modelling and Simulation of Indoor Thermal Comfort Factors Relating to Property Value: The Case of Residential Building

Hind Khana[1](✉), Rafika Hajji[2], and Moha Cherkaoui[3]

[1] Laboratory of Applied Geophysics, Geotechnics, Engineering Geology and Environment (L3GIE), Mohammadia School of Engineers, Mohammed V University of Rabat, Rabat, Morocco
khana.hind@gmail.com

[2] Department of Geometric Sciences and Surveying Engineering, Hassan 2nd Institute of Agronomy and Veterinary Medicine, Rabat, Morocco

[3] Department of Renewable Energy and Energy Efficiency, Higher National School of Mines, Rabat, Morocco

Abstract. In recent years, Environmental, Social and Governance (ESG) standards have become an important business consideration worldwide. The real estate sector, as an important investment, is gearing market decisions towards quality and sustainable outcomes, which will generate financial and economic benefits for the sector and investors, particularly over the medium to long term. These changes in macroeconomic scale are leading to transformations in the hierarchy of real estate values, with the level of appreciation of thermal comfort becoming a priority in the appraisal process. Thermal comfort, defined as the state of satisfaction of an occupant in relation to his or her thermal preferences and the ambient climate, involves measures of sensitivity to various behavioral and environmental parameters. This paper aims to integrate the capabilities of Building Energy Modeling (BEM) and Building Information Models (BIM) into an automated workflow based on Green Building XML (gbXML) to model the indoor thermal comfort factors of a property's value.

Keywords: BIM · BEM · Real estate appraisal · Thermal comfort · gbXML · Fanger

1 Introduction

Compared to other sectors, the real estate market has traditionally been regarded as a conservative one, in which innovative tools and operational approaches have often struggled to gain traction [1]. In the Construction 4.0 era, however, the convergence of digital and physical technologies (digital ecosystem and cyber-physical systems) has reshaped the way-built environments are operated and managed [2]. This dynamic has prompted economists and business leaders to take a quantum leap towards digital real estate operations, while forcing the real estate sector into its fourth phase [3].

K. Baba et al. (Eds.): GeoME 2023, SPEES, pp. 65–74, 2023.
https://doi.org/10.1007/978-3-031-49345-4_7

Property administration, an important segment of the real estate market, is at the forefront of this change [4]. Indeed, as the real estate industry adapts to digital realities dominated by cloud computing, analytics software and artificial intelligence (AI), the amount of data available on property is increasing exponentially [2]. This availability of data, reports and analytics is changing the way agents conduct transactions: from promoting properties to acquiring new assignments, to defining a property's fair market value [5]. In addition, data analysis promotes a more analytical and informed decision-making process that feeds into the investment process [6]. In this context, property price optimization remains the relevant element to consider, as it benefits all operators involved and, more generally, the financial stability of the real estate market [7]. Property valuation bases its efficiency on market data, represented analytically by the appraisal purpose. An econometric process that attributes the notion of "differentiated product" to the price/value of a property [8]. This value is the combination of the property's extrinsic, intrinsic, technical and economic features [9].

Extrinsic features are the three-dimensional (3D) elements associated to the property's immediate environment; they refer to the site's own dynamics in terms of urban fabric, accessibility, public facilities, services, physical environment, etc. [9].

Intrinsic features refer to the property's 3D parameters; they include structural and livability variables such as density, geometry, comfort, maintenance. Technical features concern the structural soundness of the building and the degree of finish of its envelope in terms of quality of materials, technical equipment, insulation and general design. Finally, economic characteristics involve the macro- and microeconomic parameters that influence the real estate cycle [9, 10].

Each of these features affects the value of a property to a different extent, depending on the level of appreciation that the market reflects for each of them [1]. However, they commonly require the implementation of information technology (IT) to represent and manage the third dimension of their parameters during the appraiser process. In particular, the literature indicates a growing trend towards the use of Building Information Models (BIM) to accurately manage the complexity of a building's 3D structure, including its intrinsic characteristics, while Geographic Information Systems (GIS) help to shape and manage the built environment through the accurate modeling of spatial data [11, 12].

The year 2020 marked an inflection point for real estate valuation [13]. Environmental, social and governance considerations are becoming increasingly important when purchasing residential real estate. This structural market shift, partly accelerated by the global pandemic, has altered the peculiarities of the property features [14]. According to The Royal Institution of Chartered Surveyors (RICS) Valuation - Global Standards, quality of life and affordability play an important role in the choice of location [15]. Thermal, acoustic and lighting comfort, well-being, safety and security, as well as reduced energy consumption and operating costs, are all factors in interior quality that benefit occupants. These benefits for the investor translate into improved satisfaction and a greater ability to invest in the space [15]. It is therefore essential to understand the performance of an energy building when making an investment decision, and this understanding needs to be integrated into a microeconomic and macroeconomic market context.

From this perspective, the integration of building energy modeling techniques holds great promise for thermal comfort analysis [11]. They make it possible to perform dynamic simulations of building 3D variables and personal parameters at an advanced stage. Although energy modeling and 3D spatial analysis are relevant in the context of real estate appraisal, their integration is still limited.

The overall aim of this study is to integrate BIM and BEM functionalities for a post-processing workflow, in order to model the impact of thermal comfort on the value of a residential property. More specifically, the idea is to adapt the sixth dimension of building information modeling to predict the comfort of building occupants, taking into account their health and productivity. The proposed method is that developed by Fanger, as a developed extension of the international standard ISO 7730, whose combines thermoregulation, the comfort equation and equilibrium theory with two indices, namely PMV (predicted mean) and PPD (Predicted Percentage of Dissatisfied occupants) [16]. Thermal comfort through those indices is measured using the 7-level thermal sensation scale (ambient environment and subjective personal sensation).

2 Methodology

To address research issue, we propose a BIM/Green Building XML model for a comparative study of comfort levels between three different living units. The process of this analysis comprises three main steps (Fig. 1): 1) modeling a 3D digital mock-up of the residential building with a degree of detail LOD 400; 2) mapping the building's architectural information through gbXML; 3) analyzing and interpreting the data to guide a better decision-making process.

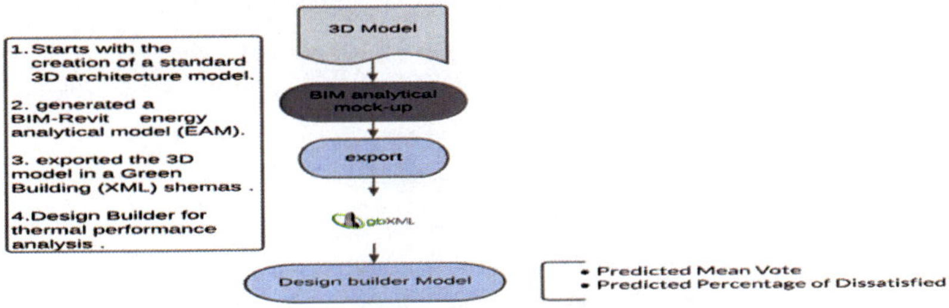

Fig. 1. BIM-BEM methodological workflow

- Case study selection:

The reference building adopted in this case study is an existent property, designated as condominium with a height of 13.80 m above street level. In terms of climatic conditions, the building is located in Témara, so meteorological data of this Moroccan thermal zone defined in the Moroccan Thermal Regulation of Construction (RTCM), are used in this investigation.

Zone (ref. Agadir), predominantly Mediterranean climate, with mild, rainy winters and hot, sunny summers [17].

The choice of this case study area is justified by: The level of complexity of the model comprising the issues studied, as this property has 3 different facades, each with its own view, orientation, surface and therefore its own degree of comfort, which designates its own market value, and which in turn serves the main objective of the study; the rectangular shape of the structure supports the transparent transfer of geometric data using the gbXML format [18, 19], and this makes gbXML more comprehensive in terms of exchanging data related to energy simulation [20] ultimately will help to obtain an energy performance test, which is one of the objectives of the study; the availability of Lidar data relating to the property's 3D built environment from a previous study carried out and processed in the IAV laboratory (Figs. 2 and 3). This will support one of the objectives of the study, which is to obtain more realistic results for validation.

Fig. 2. Lidar point clouds describing the study area.

Fig. 3. Study area and subject property.

- BIM modeling process:

On the basis of a 2D AutoCAD plan (Fig. 4), we modelled a 3D digital model of the building using Autodesk Revit 2022, the most accurate commercial BIM software, enabling a construction-scale representation of the interior and exterior (Fig. 5) [21].

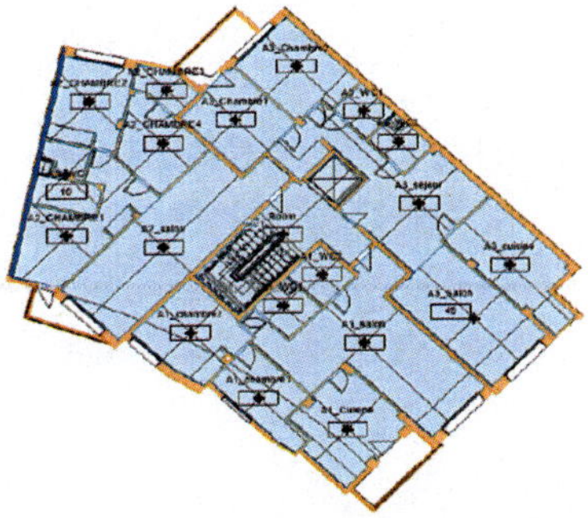

Fig. 4. AutoCAD plan.

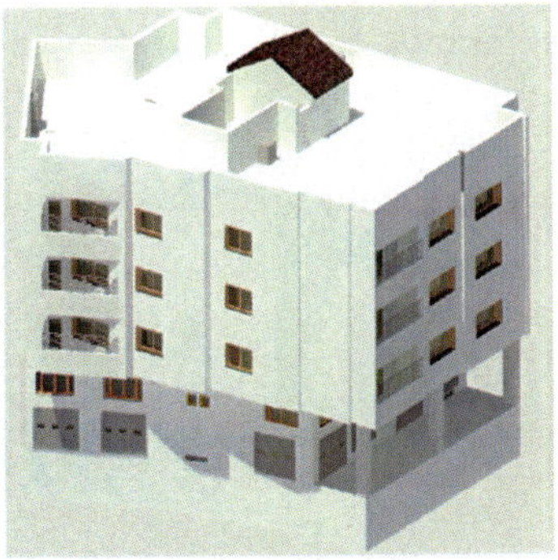

Fig. 5. BIM 3D mock-up.

- BIM model export:

To set up a dynamic energy simulation model, the information flow from the BIM building model is transmitted to a computerized database via interoperability languages represented by Green Building XML (gbXML), proven to be more suitable for energy modeling [22]. To do this, first generate an energy analysis model (EAM) and enter the energy parameters (location, fabrics, HVAC systems, operating schedule and internal loads) (Fig. 6). The resulting gbXML file contains the model's energy information in accordance with the gbXML schema.

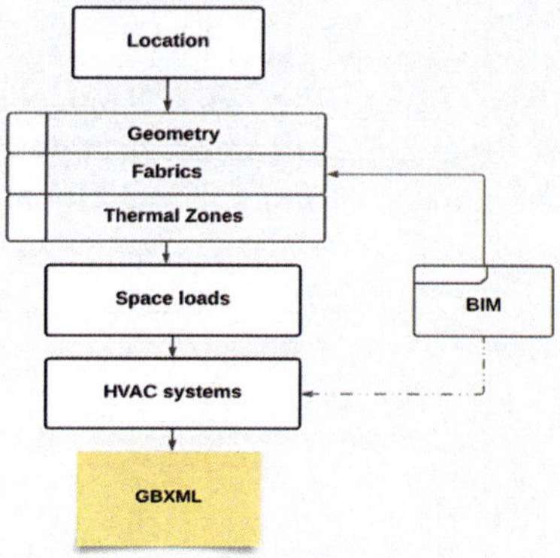

Fig. 6. Energy analysis model generation

- BEM thermal simulation:

The aim of this step is to measure occupant sensitivity to different parameters (environmental and behavioral), for 3 apartments in the same vertical building model thus comply with ASHRAE 55. For that purpose, a dynamic analysis of PPD and PMV indices is carried using Dynamic Thermal Simulation (DTS) on DesignBuilder v5.5 software based on the Energy Plus V8.1 calculation engine (Fig. 7). DesignBuilder, known as a global professional tool dedicated to building simulation and pre-processing software, offers interoperability with BIM models thanks to its gbXML import capability [23].

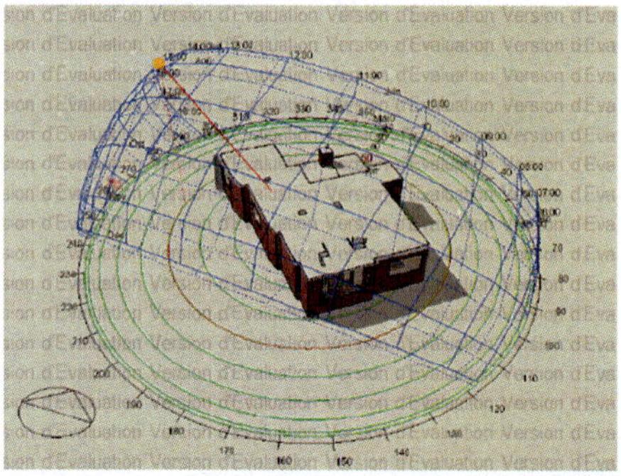

Fig. 7. Design analysis for an apartment.

The simulations are conducted using typical data files, databases for the US Department of Energy, based on the average value for the year 2021.

Also, a number of parameters and assumptions are also set out in the structure of the DesignBuilder program.

The prototype building is modeled with initial inputs, such as set-point temperatures: we have chosen values taken from Moroccan regulations, i.e., 16 °C in winter and 26 °C in summer, both of which provide a good compromise between comfort and energy consumption.

During the measurement phase, the simulation model is assumed to be occupied 7 days a week, 24 h a day.

Occupant density, was taken to be equal to average household per apartment surface area.

Clothing (1 clo in winter and 0.5 clo in summer).

The corresponding meteorological statistics show data for one year.

3 Results

For better 3D visualization of the results, we can exploit the various features of 3D GIS. For example, we can visualize the occupant dissatisfaction rate (PPD) for each housing unit, as well as their classification (Fig. 8).

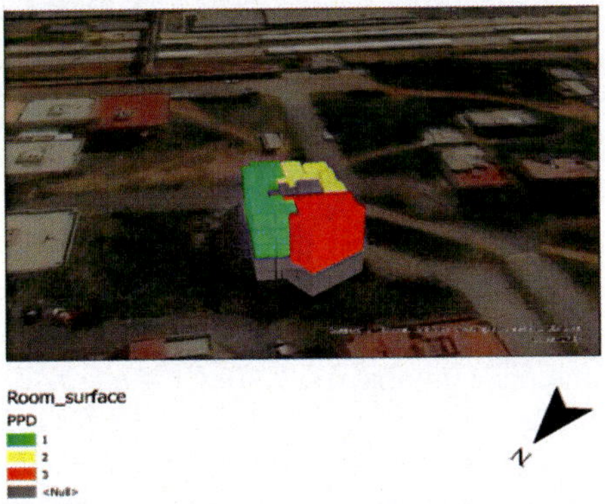

Fig. 8. Visualization of thermal comfort results on ArcGIS Pro

Figure 9 displays the simulation results of the PPD and PMV index for the apartment 1. The dotted line shows the period of the year when the PMV is between − 0.7 and + 0.7 (the recommended thermal limit for an existing building) and the PPD below 20%. This is the period when occupants feel most comfortable.

We calculate the thermal comfort parameters for the simulated period for the 3 apartments. The results are shown in Table 1.

Although the apartments are located on the same floor, have the same construction model and the same residential activity, they present very different PPD and PMV values. Apartment 3 is characterized by a low dissatisfaction rate, with only 2438 h of discomfort, making it the most comfortable apartment. Which ultimately affect its market value.

4 Discussion

For the simulation of intrinsic factors (thermal comfort indices), the 3D BIM model was enhanced with precise data (geometry, materials and deviations). The metrics chosen were deemed appropriate for a comparative analysis between housing units. Although

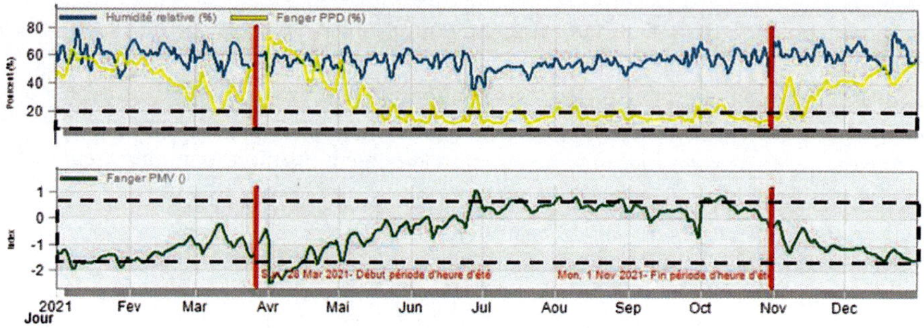

Fig. 9. Thermal comfort simulation results for apartment 1.

Table 1. Thermal comfort parameters for the simulated period for the 3 apartments.

	Apartment		
	1	2	3
Fanger PPD (%)	31.19	37.36	17.96
Fanger PMV	– 0.52	– 0.72	– 0.36
Hours of discomfort (h)	2617.45	2766.26	2437.72

the three flats are located on the same floor, have the same construction model, the same weather data and the same residential activity, they have very different comfort indices (Fig. 10). On the basis of these results, it can be estimated that the asset value of property with a quality indoor environment is higher than that of standard property. Hence the need to integrate property performance analysis into property valuation.

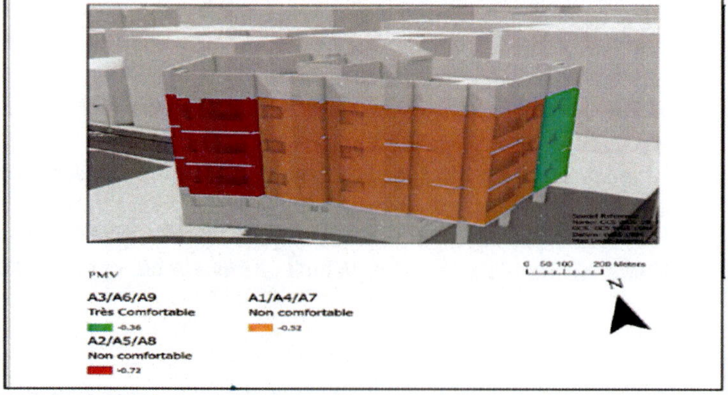

Fig. 10. Visualizing result of comfort criteria PMV on ArcGIS pro interface.

5 Conclusions

This article gives a new insight into the economic value of a property. In effect, in order to value real estate, it is necessary to determine the better use of the property, which is the basis of the market value. The optimal use is estimated following a complex and correct analysis procedure, on the basis of which a credible opinion can be issued regarding the market value.

The practical implementation of a BIM-BEM solution has demonstrated its contribution to the extraction of 3D factors for thermal comfort analysis. Indeed, the proposed approach enables comfort indices to be quantified in a precise manner, which is likely to improve the evaluator's subjective judgment.

References

1. Arcuri, N., De Reggiero, M., Salvo, F., Zinmo, R.: Automated valuation methods through the cost approach in a BIM and GIS integration framework for smart city appraisals. Sustainability **12**(18), 7546 (2020)
2. Tengxiang, S., Haijiang, L., An, Y.: A BIM and machine learning integration framework for automated property valuation. J. Build. Eng. **44**, 102636 (2021)
3. Siniak, N., Shavrov, S., Marina, N., Krajco, K.: Examining the feasibility of industry 4.0 for the real estate sector with a lens of value and job creation. In: Proceeding of Scientific Papers from the International Scientific Conference the Impact of Industry 4.0 on Job Creation, Trenčianske Teplice, Slovak Republic, 22 Nov 2018, pp. 179–185. Publishing House Alexander Dubček University in Trenčín (2019)
4. Siniak, N., Kauko, T., Shavrov, S., Marina, N., Krajco, K.: The impact of PropTech on real estate industry growth. In: Proceeding of Scientific Papers from the International Scientific Conference Materials Science and Engineering, vol. 869, p. 062041 (2020)
5. Zhang, X., Zhang, Y., Lin, Z.: Online advertising and real estate sales: evidence from the housing market. Electron. Commer. Res. **23**, 605–622 (2023)
6. RICS: Red Book – Normes professionnelles de la RICS. Edition internationale (2020)
7. Fuerst, F., McAllister, P., Nanda, A., Wyatt, P.: Does energy efficiency matter to home-buyers? An investigation of EPC ratings and transaction prices in England. Energy Econ. **48**, 145–156 (2015)
8. Conway Viriato, J.: AI and machine learning in real estate investment. J. Portf. Manag. Spec. Real Estate Issue **45**(7), 43–54 (2019)
9. Pagourtzi, E., Assimakopoulos, V., Hatzichristos, T., French, N.: Real estate appraisal: a review of valuation methods. J. Prop. Invest. Finance **21**(4), 383–401 (2013)
10. El Yamani, S., Hajji, R., Ettarid, M.: Building information modeling potential for an enhanced real estate valuation approach based on the hedonic method. WIT Trans. Built Environ. **192** (2019)
11. Polignac, B., Monceau, J.P., Cussac, X., Lesieur, P.: Real Estate Expertise, Practical Guide, 7th edn. (2019)
12. Oguzhan Mete, M., Guler, D., Yomralioglu, T.: Towards a 3D real estate valuation model using BIM and GIS. In: Innovations in Smart Cities Applications, vol. 5 (2021)
13. El Yamani, S., Hajji, R.: BIM and 3D GIS integration for real estate valuation. In: Hajji, R. (ed.) Building Information Modeling for a Smart and Sustainable Urban Space. ISTE Ltd. (2022)

14. Kara, A., Oosterom, V., Çağdaş, V., Işıkdağ, U., Lemmen, C.: 3-Dimensional data research for property valuation in the context of the LADM Valuation Information Model. Land Use Policy **98**, 104179 (2020)
15. Kempeneer, S., Peeters, M., Compernolle, T.: Bringing the user back in the building: an analysis of ESG in real estate and a behavioral framework to guide future research. Sustainability **13**(6), 3239 (2021)
16. Renigier-Bitozor, M., Zrabek, S., Walacik, M., Janowski, A.: Hybridization of valuation procedures as a medicine supporting the real estate market and sustainable land use development during the covid-19 pandemic and afterwards. Land Use Policy **99**, 105070 (2020)
17. Cheung, T., Schiavon, S., Parkinson, T., Li, O., Brager, G.: Analysis of the accuracy on PMV – PPD model using the ASHRAE Global Thermal Comfort Database II. Build. Environ. **153**, 205–2017 (2019)
18. Thermal Building Regulations in Morocco (2014), the Agence Nationale pour le Développement des Énergies Renouvelables et de l'Efficacité Énergétique (ADEREE)
19. Gourlis, G., Kovacic, I.: Building information modelling for analysis of energy efficient industrial buildings - a case study. Renew. Sustain. Energy Rev. **68**(1) (2017)
20. Gao, H., Christian, K., Yupeng, W.: Building information modelling based building energy modelling: a review. Appl. Energy **238**, 320–343 (2019)
21. D'Oca, S., Hong, T.: A data-mining approach to discover patterns of window opening and closing behavior in offices. Build. Environ. **82**, 726–739 (2014)
22. Ferrandiz, J., Banawi, A., Peña, E.: Evaluating the benefits of introducing "BIM" based on Revit in construction courses, without changing the course schedule. Univ. Access Inf. Soc. **17**, 491–501 (2018)
23. Elnabaw, M.H.: Building information modeling-based building energy modeling: investigation of interoperability and simulation results. Front. Built Environ. **6** (2020)

The Energy Rehabilitation of a Riad's Building Located in the Mediterranean Climate

Najoua Eraza[1], Najma Laaroussi[1](✉), Amine Hajji[1], Latifa El Farissi[2], and Mohammed Garoum[1]

[1] Material, Energy and Acoustics Team (MEAT), Higher School of Technology Salé, Mohammed V University in Rabat, Avenue Prince Héritier, B.P: 227, Salé Médina, Morocco
Najma.laaroussi@est.um5.ac.ma

[2] Complexe des Centres Techniques Industriels Marocains (CTPC), 20280 Sidi Maârouf, Casablanca, Morocco

Abstract. The Riads represent traditional Moroccan houses that testify to the glorious past of the country. These buildings are located in the old medinas of the imperial cities of Morocco. The studied Riad is situated at Rabat city in the proximity of the Atlantic Ocean and is subjected to maritime influence. Therefore, it has a semi-arid Mediterranean type of climate which is mild, moderate, and rainy in winter; and humid and temperate in summer with Chergui days. Several materials are used in the construction of the studied Riad, the zellige, the plaster, the wood, and the metal. The Riad is composed of a ground floor and a first floor (R + 1) and consists of two living rooms on the ground floor, but the second floor consists of four rooms and an open patio. The thermal transmittance coefficients U (W/m^2 K) for each wall were calculated, considering the materials composing the walls of the Riad. The dynamic thermal simulation of the Riad was carried out using DesignBuilder software [13]; to evaluate a building's thermal performance without energy consumption for heating and air-conditioning.

Keywords: Energy efficiency · Building materials · Insulation · Thermal simulation · Comfort

1 Introduction

Although construction is a key sector of the Moroccan economy, it is one of the country's most energy-consuming sectors. This consumption is mainly due to the sharp increase in energy demand for air conditioning and heating in Morocco. To reduce its energy bill, Morocco has adopted a national energy efficiency strategy aimed at saving energy. In order to meet its energy challenges and promote the building sector in terms of energy economics and sustainability, the integration of energy efficiency and renewable energy resources seems the most appropriate solution. The choice of building materials is a fundamental step in the construction. Unfired clay is a common building material in rural Morocco, especially in areas with hot arid weather. These bricks are chosen for their low environmental impact and cost-effectiveness as a building material. Several research

K. Baba et al. (Eds.): GeoME 2023, SPEES, pp. 75–86, 2023.
https://doi.org/10.1007/978-3-031-49345-4_8

studies have investigated the performances in traditional buildings built in unfired clay. An energy analysis [1] was carried out on a residential building in the new town of Lakhyayta near Casablanca using DesignBuilder software [13], using alternative wall compositions as opposed to standard walls built with conventional materials in this new urban hub. Based on the simulations carried out, a significant 30% reduction in overall energy consumption for heating and cooling was achieved, compared with the standard scenario. This further indicates that insulation factors have the most significant impact on energy requirements for heating, with a potential reduction of up to 72% when using 6 cm of external wall insulation. In another study [2], three types of unfired clay were compared to determine their thermal performance in different regions. This was then simulated to determine the best ecological material with the best thermal properties. According to a dynamic thermal simulation of the three unfired clay envelopes, The performance observed in the studied buildings is nearly identical. El Azhary et al. [3] studied the thermal characteristics of unfired clay bricks and their impact on providing optimum thermal conditions in buildings during colder climates. It confirms that optimized envelope design can improve thermal performance through passive solar techniques based on the profound role played by earth materials such as unfired bricks. El Azhary et al. [4] was devoted to achieving thermal comfort in the studied house through the use of clay as a natural raw material in the building materials. By this, the orientation and the use of sustainable building materials seem to have a significant impact on the creation of a better climate. The aim of the study [5] is to assess the impact of building orientation and vernacular architecture on interior thermal quality and determine, using simulation software, the optimum orientation for achieving thermal comfort. The proposed methodology studies the thermal performance of traditional buildings under the Rissani's climate to manage, naturally, good thermal comfort in summer and winter. The objective of the study [6] was to determine the thermal behavior of Ksourian vernacular architecture and the thermal performances affecting the interior temperature of buildings. Characterization of the thermophysical properties of Rissani earth was carried out using well-defined experimental devices. Simulations have made it possible to determine the optimum orientation, showing that taking into account the criterion of building orientation helps to achieve good thermal comfort and a more thermally efficient design throughout the year. El Harrouni et al. [7] analyzes the thermal performance of two pilot buildings located in two extreme climatic zones in Morocco. A series of bioclimatic building design methods and techniques were used to determine the right strategy for ensuring comfortable indoor conditions and limiting energy consumption. Diz-Mellado et al. [8] studied the possibility of developing process modeling to optimize the architectural design of buildings with courtyards. The results of the study, obtained using TRNSYS simulation software, suggest the optimum configurations for improving the energy performance of courtyards; the square shape being compatible with cold climates (48% reduction in heating demand), the deepest and narrowest courtyard is recommended for arid climates (10% reduction in cooling demand). For temperate climates, the narrowest, medium-depth shape guarantees the lowest energy consumption in all seasons (around 58% energy saving). M'Saouri El Bat et al. [9] investigated the influence of interior courses on cooling energy demand in the construction field. To this end, an experimental and numerical data approach was applied to examine the influence of this geometric factor. The results obtained showed a reduction in cooling demand of between 8% and 18%, depending on the geometry of the courtyard. The analysis also demonstrated the effect of the orientation of adjacent zones. The study thus confirms that the geometric design of the courtyard and its presence in the building are significant factors in the building's energy performance. The study [10] evaluates the performance of an internally insulated historic building in a cold climate, focusing on the effects of the insulation on the building's hygrothermal behavior. The authors [10] conducted simulations and measurements to assess the building's energy performance, indoor air quality, and risk of moisture damage. The results of the study can provide insights into the potential benefits and risks of internal insulation of historic buildings in cold climates. Akkurt et al. [11] debate the challenges and possible solutions in conducting dynamic thermal and hygrometric simulations of historical buildings. They emphasize the importance of accurate input data, such as the material properties and climatic conditions, and the need to consider the complex interactions between the building envelope, the indoor environment, and the outdoor climate. The work also highlights the limitations and uncertainties in the simulation results, as well as the need for validation and calibration with experimental data.

This research study is being conducted within the framework of established norms, aiming to unveil a novel architectural style that integrates both thermal comfort and energy efficiency properties. The Riad buildings, blending Andalusian and Moroccan architectural influences, have captivated the admiration of both Moroccans and foreigners. These traditional houses were meticulously designed to provide inhabitants with a comfortable environment, complete with gardens adorned with fruit trees, shrubs, and fountains. The distinctive feature of this construction style lies in the organization of public and private rooms around a central open courtyard, a hallmark of Islamic architecture that has remained remarkably consistent throughout the Arab world. The Riad's layout is carefully crafted to maximize privacy for families, shielding them from the outside world. Typically, rooms feature windows and balconies that overlook the central courtyard, drawing attention to it as the focal point for all activities within the house.

As part of this project, numerous energy simulations were conducted for the studied Riad in Rabat. This Riad, being a traditional urban house in Morocco with a central patio, adheres to specific design and architectural requirements. DesignBuilder software [13] was employed to model the building as a case study, incorporating various wall compositions using standard materials suitable for the climate zone. This study assesses the cooling and heating cost savings that can be made in a Riad based on the microclimate of the courtyard.

2 Case Study: Riad Building

2.1 Climate Analysis of Rabat City

The Riad under investigation is situated in Rabat city, close to the Atlantic Ocean, which exposes it to maritime influences. Consequently, it experiences a semi-arid Mediterranean climate characterized by mild to moderate rainfall in winter and humid, temperate summer with occasional Chergui days. The Rabat climate is characterized by an apparent variability, the minimum temperature is 4 °C and the maximum is 40 °C. In the summer, the atmosphere heats up significantly, with maximum temperatures in July varying between 16 °C and 26 °C. As for rainfall, it varies between 900 mm and 300 mm.

Fig. 1. Geographical location of studied Riad.

Located in the old Medina of Rabat which is known for its lively atmosphere, more precisely in the Souk Sebbate street (Fig. 1) which is under its new aspect: ceiling made of rare wood, paved floor, and diffused natural light filtered by moucharabieh give this leather Souk a functional environment faithful to traditions.

2.2 Riad Building Materials

Riads, generally one to five stories high, are entirely closed in, insulated by high walls with a minimum of openings to protect them from the heat and noise of the street. Riads are organized around a central patio (Fig. 2a) as the basis for an architectural structure in the form of a shaft with balconies facing inwards (Fig. 2b), inspired by traditional Arab-Andalusian housing, Persian heritage, and Roman heritage. The rooms are in most Riads on the floor. Depending on the season, the level of light, heat, and weather, the top of the patio can be opened, or closed by a canvas or a retractable canopy. The roof is arranged as a terrace with an open sky, overlooking the other roofs and the urban environment.

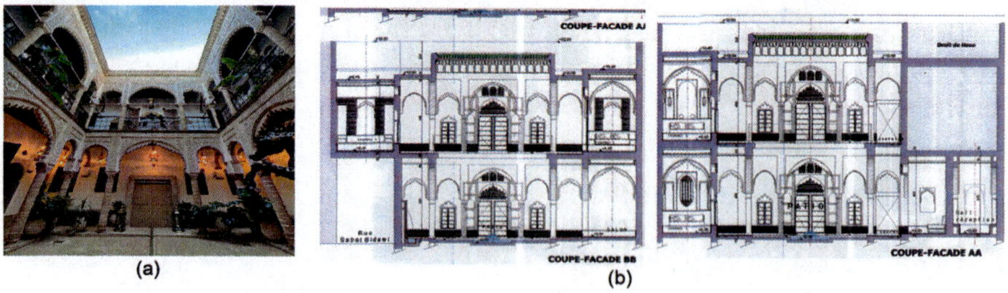

Fig. 2. a) Interior view of Riad, b) building facades of Riad.

Several materials are used in the construction of Riads, the zellige (Fig. 3), which is a surface coating in the form of small colored tiles assembled against each other dressing the floor and walls of mosaics. The plaster is a material that allows the realization of very floral decoration. Wood is an insulating material used for doors (Fig. 4), roofs (Fig. 5), and windows and gives an inimitable charm. Metal, mainly wrought iron, is used on a variety of occasions, but mainly on entrance doors (Fig. 4), for decoration in the form of nails, knockers, hinges, or closures.

Fig. 3. The zellige.

Fig. 4. The door of Riad.

The studied Riad is composed of a first floor and a single floor (R + 1). Its first floor consists of two rooms called The Mauresque and the Bouhemienne and two living rooms and an open patio. The second floor consists of 4 rooms: Amazigh, Andalusian, Berber, and Nomad. The Riad also contains a roof terrace with a breathtaking view of the old Medina and its mosques.

The naturally cool and air-conditioned patio (Fig. 2a) is inspired by the oasis, the Islamic garden, and the Persian garden. It is usually planted with trees, and ornamental plants and has refreshing pools and fountains. The open living and dining rooms face the patio and allow you to enjoy its coolness.

Fig. 5. The roof of Riad.

2.3 Wall Compositions

To calculate the coefficient of thermal transmission (W/m^2 K) for each wall of the Riad, it is imperative to conduct an inventory of the materials comprising the walls.

Table 1. Composition of internal and external walls

Walls	Internal wall (U = 1.16 W/m^2 K) int → ext			External wall (U = 1.35 W/m^2 K) Uref (RTCM) ≤ 1.20 W/m^2 K int → ext		
Materials	Tadelakt	Adobe	Tadelakt	Tadelakt	Adobe	Coating-straw
Thickness, e (cm)	3	30	3	3	44	3
Density (kg/m^3)	1800	2000	1800	1800	2000	2000
Thermal conductivity (W/m K)	0.65	0.7	0.65	0.65	0.7	0.9

The Riad employs single glazing for its windows (U_w = 5.20 W/m^2 K), with an overall rate of glazed windows (τ = 13%) determined by the ratio of the total area of glazed windows to the total area of all external walls in contact with the conditioned external space. Situated in the Z_1 climatic zone, as defined by the Moroccan Building Thermal Regulation (RTCM) [12], the Riad adheres to the guideline of having a global rate of glazed windows below 15% with 103 m^2 of occupied surface area.

Tables 1, 2, and 3 provide the thermophysical properties of the construction materials used, along with the thermal transmission coefficients U (W/m^2 K) for each wall. Upon comparison with the reference values mandated by the RTCM, it is evident that the performance of the external wall does not meet the requirements according to the

Table 2. Composition of the external roof

External roof (U = 0.18 W/m^2 K) Uref (RTCM) ≤ 0.75 W/m^2 K	int → ext						
Materials	Wooden beams	Floor wood	Rock wool	Solid slab	Bitumen	Slope shape	Zellige
Thickness, e (cm)	13	5	15	15	1.5	5	2
Density (kg/m^3)	1000	120	100	2400	1050	600	2400
Thermal conductivity (W/m K)	0.5	0.036	0.044	2	0.23	0.56	0.23

Table 3. Composition of intermediate floor

Intermediate floor (U = 0.17 W/m^2 K)	int → ext					
Materials	Wooden beams	Floor wood	Polystyrene (EPS)	Solid slab	Slope shape	Smooth concrete tiles
Thickness, e (cm)	13	5	15	15	5	3
Density (kg/m^3)	1000	120	35	2400	600	1800
Thermal conductivity (W/m K)	0.5	0.036	0.039	2	0.56	0.65

prescriptive approach of the Moroccan Building Thermal Regulation [12]. To comply with the RTCM regulation, an optimized solution incorporating a 4 cm thick interior insulation layer made of polystyrene foam is proposed, resulting in a reduced thermal transmission coefficient of U = 0.53 W/m^2 K for the external walls.

As for the flooring, it consists of concrete tiles placed on a concrete slab for both durability and insulation purposes. The thermal transmittance coefficient of the flooring is U = 0.23 W/m^2 K, and there are no specific regulatory requirements for Zone 1.

3 Results and Discussion

DesignBuilder [13] uses the EnergyPlus dynamic simulation engine to simulate the thermal performance of the building including a series of environmental performance parameters. The Riad (Figs. 6 and 7) is an establishment representing a guest house receiving several guests for several nights. The Riad can also be a meeting place for a musical evening or a dinner for the participants of a scientific event.

An occupancy density of 0.15 per/m^2 was considered. Several solutions have been proposed to achieve thermal and hygrometric comfort within the Riad without heating or cooling energy consumption.

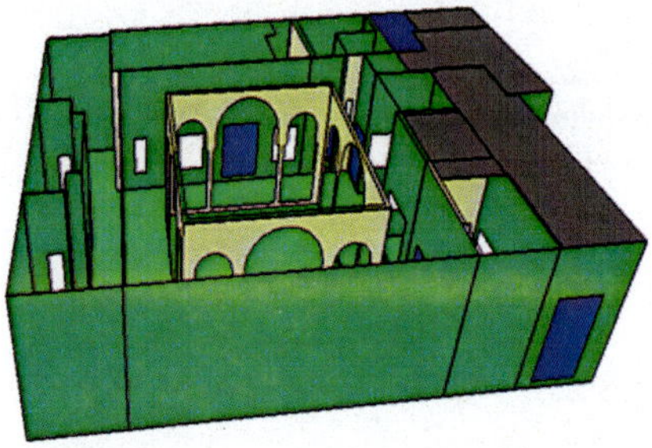

Fig. 6. An interior view of de Riad.

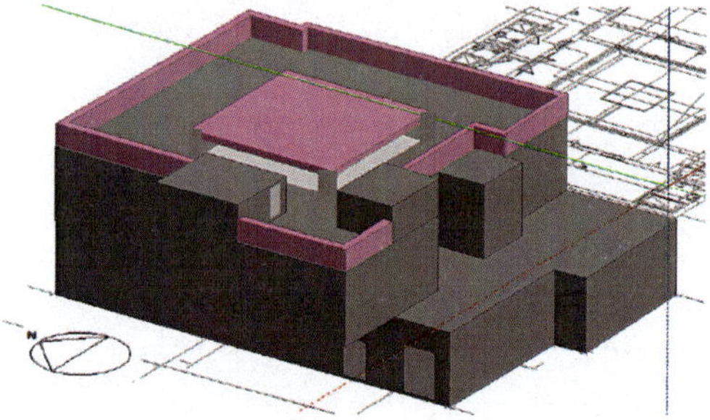

Fig. 7. A geometric 3D model of the Riad with a covered patio.

The exterior wall of the Riad has been effectively insulated with a 4 cm thick layer of expanded polystyrene, providing thermal insulation and enhancing energy efficiency. Additionally, the space above the patio has been partially enclosed using glazed walls, which incorporate openings to facilitate natural ventilation. To further ensure optimal air circulation, a mechanical ventilation system has been implemented as well with an air renewal rate of approximately 0.3 air changes per hour (0.3 V/h). These measures combine to create a comfortable and well-regulated indoor environment within the Riad. The reference case (Fig. 8) represents the construction of a Riad with an open patio, where the exterior walls are not insulated, and mechanical ventilation systems are not applied. This serves as the baseline scenario against which the case of a building with improvement can be compared. It helps us assess the impact and benefits of implementing insulation, mechanical ventilation, and other measures in terms of energy efficiency, hygro-thermal comfort, and overall performance.

As a result of the improvements made to the building, the indoor temperature profiles (Fig. 8) exhibit a smoothed-out pattern, minimizing the impact of fluctuations in the outdoor ambient temperature. These enhancements have led to more comfortable indoor conditions throughout the year. In winter, the indoor temperature ranges between 18 and

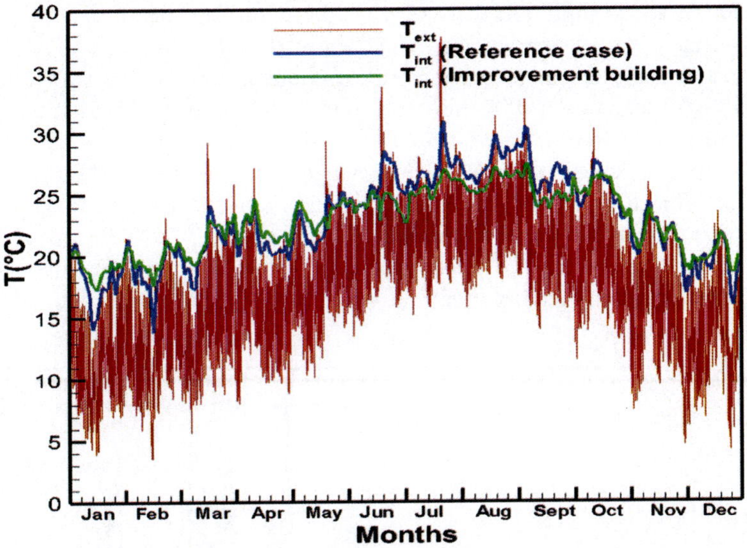

Fig. 8. The hourly variation of temperatures according to the seasons.

21 °C, providing a warmer environment. In contrast, during summer, the indoor temperature ranges between 20 and 24 °C, creating a cooler and more pleasant atmosphere. These temperature ranges demonstrate the effectiveness of the building modifications in maintaining a stable and desirable indoor climate, contributing to the overall comfort and well-being of the occupants.

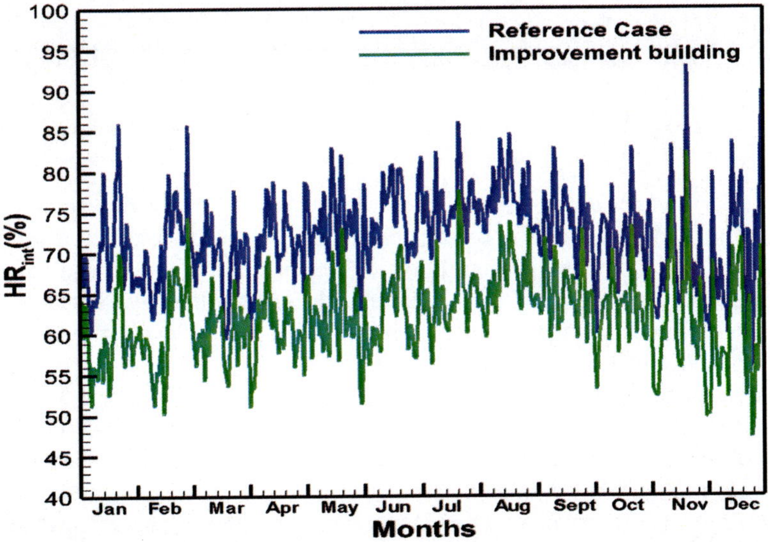

Fig. 9. The hourly variation of internal humidity according to the seasons

It is important to highlight that covering the patio can lead to a reduction in the ventilation rate (Fig. 9), particularly when coupled with a low overall glazing ratio. This decrease in airflow increases the risk of condensation on the walls because the indoor humidity can reach 95% in some rooms. To mitigate this risk and maintain a healthy indoor environment, we have implemented mechanical ventilation systems. These systems are designed to effectively evacuate excess humidity ($HR_{min} = 50\%$ and $H_{max} = 70\%$), ensuring proper ventilation and minimizing the potential for condensation-related issues within the Riad. This recommended rate (0.3 V/h) ensures an adequate exchange of fresh air within the enclosed spaces of the Riad, promoting indoor air quality and a comfortable living environment.

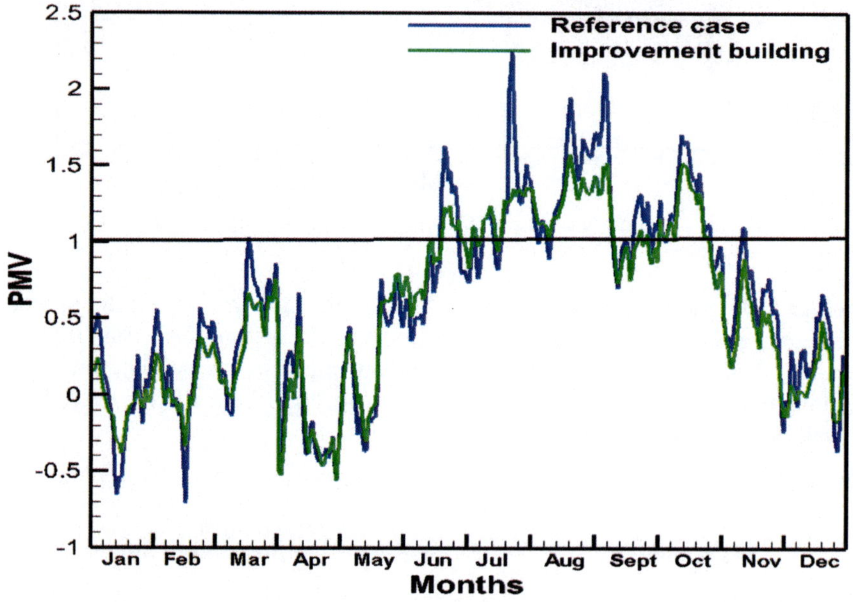

Fig. 10. The hourly variation of PMV according to the seasons.

The hourly variation of the Predicted Mean Vote (PMV) varies according to the seasons (Fig. 10). The PMV is a measure used to assess thermal comfort based on factors such as air temperature, humidity, air velocity, and clothing insulation. During the different seasons, the PMV values fluctuate throughout the day. In winter, the PMV tends to be within the interval [– 1, 1], indicating a higher level of thermal comfort, particularly during the colder hours.

In summer, the PMV values tend to be within the interval [– 0.5, 1.5], indicating a certain level of thermal comfort for the improved building within the interval [– 0.5, 2.5] for the reference case due to warmer temperatures, reflecting the increased discomfort experienced in higher temperatures.

Overall, the hourly variation of PMV according to the seasons reflects the dynamic nature of hygro-thermal comfort, with adjustments needed to maintain an optimal indoor environment as external conditions change throughout the day.

4 Conclusions

By insulating the roof terrace and external walls, and partially covering the patio, we have successfully achieved compliance with Moroccan thermal regulations for Zone 1. These measures ensure that the Riad meets the required standards for thermal performance and energy efficiency. The insulation helps to minimize heat transfer and maintain a high-quality indoor environment while reducing energy consumption. The partial covering of the patio and by applying mechanical ventilation further contributes to temperature and humidity control and enhances the overall hygro-thermal efficiency of the Riad, aligning it with the prescribed regulations for Zone 1. By comparing the reference case with the upgraded building, we have a better understanding of the benefits and improvements achieved through insulation and mechanical ventilation interventions in Riad construction.

Acknowledgments. We extend our sincere gratitude and deep appreciation to the following individuals, whose invaluable contributions and unwavering support have played a pivotal role in the resounding success of our work.

Mr. Emmanuel Muguet: Owner of Riad Al BAHI.

Mr. Mohammed Taibi Mouline: Director of MMT architectures and author of the Riad's refurbishment.

References

1. El Azhary, K., Laaroussi, N., Garoum, M., El Harrouni, K., Mansour, M.: A dynamic thermal simulation in new residential housing of Lakhiayta City in Morocco. In: Proceedings of 3rd International Sustainable Buildings Symposium 2017, vol. 2–3, pp. 26–35. Springer (2018)
2. El Azhary, K., Raefat, S., Laaroussi, N., Garoum, M.: Energy performance and thermal proprieties of three types of unfired clay bricks. Energy Procedia **147**, 495–502 (2018)
3. El Azhary, K., et al.: The improving energy efficiency using unfired clay envelope of housing construction in the south Morocco. J. Mater. Environ. Sci. **8**(10), 3771–3776 (2017)
4. El Azhary, K., Chihab, Y., Mansour, M., Laaroussi, N., Garoum, M.: Energy efficiency and thermal properties of the composite material clay-straw. Energy Procedia **141**, 160–164 (2017)
5. El Azhary, K., Ouakarrouch, M., Laaroussi, N., Garoum, M.: Energy efficiency of a vernacular building design and materials in hot arid climate: experimental and numerical approach. Int. J. Renew. Energy Dev. **10**(3), 481 (2021)
6. El Azhary, K., Ouakarrouch, M., Laaroussi, N., Garoum, M., Mansour, M.: Impact of traditional architecture on the thermal performances of building in south Morocco. In: Green Buildings and Renewable Energy, Med Green Forum 2019, Part of World Renewable Energy Congress and Network, pp. 339–347. Springer (2020)
7. El Harrouni, K., Filali, M., Kharmich, H., Mansour, M., Laaroussi, N., Garoum, M.: Energy efficient houses meeting both bioclimatic architecture principles and Moroccan thermal regulation. In: 6th International Renewable and Sustainable Energy Conference (IRSEC), pp. 1–8. IEEE (2018)
8. Diz-Mellado, E., Ruiz-Pardo, Á., Rivera-Gómez, C., Sanchez de la Flor, F.J., Galán-Marín, C.: Unravelling the impact of courtyard geometry on cooling energy consumption in buildings. Build. Environ. **237** (2023)

9. M'Saouri El Bat, A., Romani, Z., Bozonnet, E., Draoui, A., Allard, F.: Optimizing urban courtyard form through the coupling of outdoor zonal approach and building energy modeling. Energy **264** (2023)
10. Blumberga, A., Freimanis, R., Biseniece, E., Kamenders, A.: Hygrothermal performance evaluation of internally insulated historic stone building in a cold climate. Energies **16**(2), 866 (2023)
11. Akkurt, G.G., et al.: Dynamic thermal and hygrometric simulation of historical buildings. Critical factors and possible solutions. Renew. Sustain. Energy Rev. **118** (2020)
12. Règlement Thermique de Construction au Maroc RTCM. The Moroccan Agency for Energy Efficiency AMEE (2012)
13. DesignBuilder Software, Version 5

Analysis of Wind Energy Production in Five Cities in the Southern Region of Morocco

Youssef El Baqqal(✉), Mohammed Ferfra, and Abdessamade Bouaddi

Department of Electrical Engineering, Mohammadia School of Engineers, Mohammed V University in Rabat, Rabat, Morocco
Youssef.elbaqqal@um5r.ac.ma

Abstract. Renewable energy adoption is being acknowledged worldwide as an imperative approach for countries to mitigate their environmental impact and combat the effects of climate change. Additionally, there has been a significant increase in renewable energy development globally, with a particular focus on wind energy due to technological advancements and a reduction in installation costs. As a result, Wind energy has emerged as one of the most promising options for generating electricity on a large scale. As part of its drive towards sustainable development, Morocco has prioritized decreasing its reliance on fossil fuels by cultivating alternative energy sources. Impressive progress has been made in the utilization of solar and wind energy, with this later showing immense promise, especially in the southern region. In particular, Morocco's southern region accounted for about 52.6% of the country's total wind power capacity by 2021. This article provides an overview of the current and future state of wind energy in Morocco, including the country's operating and planned wind projects. The primary purpose of this paper is to examine wind energy production, along with the simulation of the capacity factor and the annual electrical energy produced by the two selected wind turbines installed at a hub height of 50 m. The study focuses on five different locations in the southern region of Morocco: TanTan, Tarfaya, Laayoune, Boujdour, and Dakhla. A predictive model of wind turbines from the literature was used to estimate the generated electrical energy. Wind speed data was obtained from a TMY file containing typical meteorological year data, which was directly downloaded from the System Advisor Model (SAM) Software. The findings of this study demonstrate the economic viability of implementing wind farms across all five locations. Boujdour has the highest capacity factor at 49.9%, while TanTan has a lower potential for wind turbine installation with a capacity factor of 31.4%. However, the chosen wind turbines had a good capacity factor in these locations, indicating that they could produce a high amount of electricity over the course of a year. In summary, the study revealed that the selected five locations in the southern regions of Morocco possess considerable wind potential, in addition to solar potential. Therefore, by combining wind and solar energy systems in these sites, renewable energy can be generated even when the amount of solar radiation is low. This will lead to a more stable and reliable energy supply, which can help decrease dependence on conventional energy sources.

Keywords: Renewable energy · Wind energy · Capacity factor · Wind speed · Wind turbine

K. Baba et al. (Eds.): GeoME 2023, SPEES, pp. 87–97, 2023.
https://doi.org/10.1007/978-3-031-49345-4_9

1 Introduction

Wind energy has developed to be one of the most economical sources of electricity in terms of electricity production costs. Furthermore, significant reductions in technology installation costs and advancements in wind energy technology have been observed (Olczak and Surma, 2023). These advancements have been so significant that the onshore wind projects commissioned had electricity costs that were lower than the cheapest fossil fuel-fired option International Renewable Energy Agency (International Renewable Energy Agency (IRENA), 2022a). Driven by an ambitious energy strategy, Morocco is currently facing challenges in implementing its renewable energy promotion plan (Šimelytė et al., 2016). Azeroual et al. (2018) provide a comprehensive overview of the current state and future prospects of renewable energy in Morocco. It examines the existing challenges that hinder the growth of renewable energy, as well as the national strategy for addressing these challenges and ensuring energy security. Morocco has been making efforts to tap into its wind energy potential. The country has established a goal to reach a 52% share of renewable energy in its energy mix by 2030 as part of its Integrated Energy Strategy, accomplished through investments in solar and wind power projects (The Moroccan Ministry of Energy Transition and Sustainable Development, 2023). As a result, renewable energy projects, including those focusing on wind energy, have seen substantial investments. To reach these goals, several renewable energy projects are being implemented in several locations in Morocco, especially wind and solar projects. As shown in Fig. 1, the actual (International Renewable Energy Agency (IRENA), 2022a) renewable penetration is 3950 MW, with 19% coming from solar (CSP/PV), 36% from wind, and 45% from hydroelectricity (Chauhan and Saini, 2014), these capacities have variable energy production, especially for solar and wind power and it's depending in particular on weather conditions.

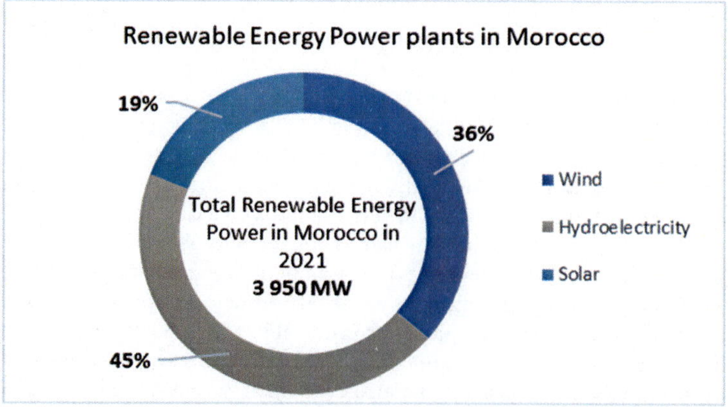

Fig. 1. Observed Hydroelectricity, Wind, and Solar (CSP/PV) capacity in Morocco in 2021

In the literature, numerous studies focus on wind energy in Morocco. The authors (Haidi et al., 2021) present the evolution of the installed wind capacity in Morocco between 2000 and 2019. Four hypothetical locations in Morocco were the subject of a study to determine their wind potential (Daoudi et al., 2022). Additionally, a thorough analysis was carried out to evaluate and compare the cost of energy production at nine wind farms located in Morocco (Tizgui et al., 2018). The authors in reference (El-Hadri et al., 2019), identified potential locations that exhibit favorable conditions for the expansion of wind energy in the coming years. They also established the characteristics of the wind power density distribution in Morocco across the period from 2021 to 2050. A study (Elmahmoudi et al., 2020) identified areas with high wind potential, resulting in the creation of a map that identifies the best wind farm locations in Morocco. The objective of the study described in article (Benazzouz et al., 2021) was to Assess the wind energy potential along the Moroccan continental shelf in specific places. The assessment was conducted over a duration of ten years, from 2008 to 2017. Finally, a summary of Morocco's current renewable energy situation and prospects is provided (Tizgui et al., 2018). It includes a forecast of Morocco's wind and solar power potential until the year 2030, providing insights into the long-term renewable energy outlook for the country.

2 Status of Wind Installed Capacity in Morocco

According to a report International Renewable Energy Agency (International Renewable Energy Agency (IRENA), 2022b) from the International Renewable Energy Agency (IRENA), between 2012 and 2021, Morocco's total installed capacity increased from 255 to 1435 MW. Notably, the share of wind energy in the overall renewable energy mix has significantly reached 41% in 2021 against only 16% in 2012. Figure 2 shows the global installed capacity in Morocco between 2012 and 2021.

Wind farms are operating in more than 10 cities, and more than 2.7 GW are currently under construction or in the planning stages, particularly in the southern region (The Moroccan Ministry of Energy Transition and Sustainable Development, 2023; Power and Database, 2023). It is worth mentioning that by 2021, The southern regions of Morocco account for approximately 52.6% of the country's total installed wind capacity. Table 1 presents an overview and a summary of the wind farms that have been installed, While Table 2 resumes those which are programmed or on construction and development.

The wind conditions in the southern regions of Morocco have shown great promise (Daoudi et al., 2022; El-Khchine and Sriti, 2021) leading to a significant emphasis on wind energy development in these areas.

3 Materials and Methods

The research focuses on choosing a model, existing in the literature, to accurately estimate the energy generated by wind turbines. Key factors considered in this estimation include wind speed, hub height, and characteristics of the wind turbine. To calculate wind energy production, we rely on parameters such as the annual energy generation and the annual capacity factor.

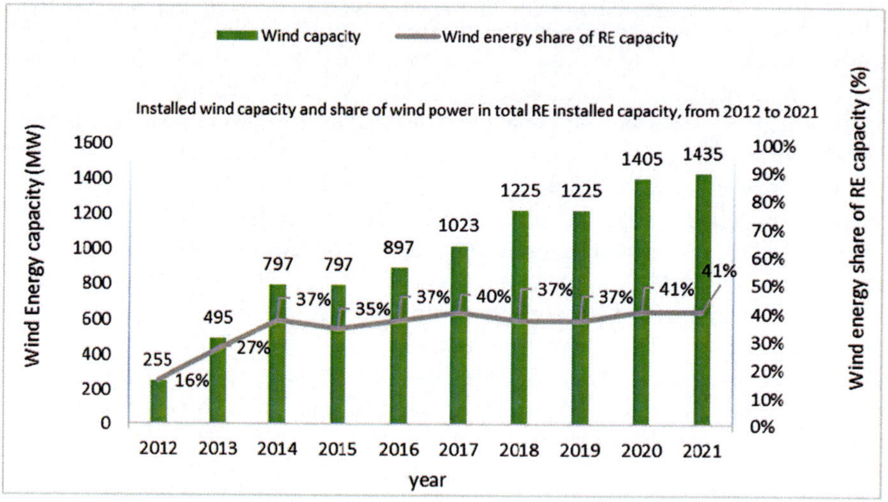

Fig. 2. The development of wind capacity installed in Morocco.

Table 1. Operational wind projects in Morocco by the end of 2021.

Power plant name	City	Power MW	Production start date	Number of turbines
Abdelkhalek Torres	Tetouan	54	2000	90
Midelt	Midelt	210	2020	50
Lafarge	Tetouan	32.2	2010	33
Tanger 1 (Dhar Saadane)	Tanger	140	2011	165
Haouma	Fahs Anjra	50.6	2014	22
Jbel Sendouk/Khalladi	Ksar sghir	120	2021	40
Amougdoul	Essaouira	60	2007	71
CIMAR	Laayoune	5	2011	5
Aftissat 1	Boujdour	200	2019	67
Akhfennir 1	Tarfaya	102	2014	61
Akhfennir 2	Tarfaya	102	2017	61
Foum El Oued	Laayoune	50.6	2014	22
Tarfaya	Tarfaya	300	2014	131
Oualidia 1	Safi	18	2021	–

Table 2. Programmed wind projects in Morocco by the end of 2021

Power plant name	City	Power MW	Planned commissioning date	Expected generation (GWh/an)
Oualidia 2	Safi	18	2021	–
Aferkat	Guelmim	80	2023	320
Akhfennir 3	Tarfaya	50	2023	–
AM Wind	Dakhla	100	2023	320
Continuity of the AM wind project	Dakhla	800	–	–
Cap Cantin	Marrakech-Safi	108	2024	–
Dakhla	Dakhla	40	2024	–
Aftissat 2	Boujdour	200	2022	1000
Tanger 2	Tanger	70	2024	290
Biranzarane	Dakhla	200	2024	750
Taza Phase 1	TAZA	87	2022	–
Taza Phase 2	TAZA	63	2025	–
Jbel Lahdid	Essaouira	270	2023	600
Koudiat Baida 1	Tétouan	120	2022	–
Koudiat Baida 2	Tétouan	80	2024	
Tiskrad	Laâyoune	100	2024	325
Ghrad Jrad	Laâyoune	80	2023	300

3.1 Research Object

As previously mentioned, more than half of the planned wind farms in Morocco are intended to be located in the southern region of the country. In this context, this work focuses on five different locations in the southern region of Morocco: TanTan, Tarfaya, Laayoune, Boujdour, and Dakhla. The geographical characteristics are presented in Table 3. Furthermore, the calculation was performed using MATLAB, and the results were obtained from the typical metrological year (TMY) data over a period of 15 years. TMY data are widely used for designing and testing solar and wind energy systems. The wind speed Data are provided from the European Commission PVGIS weather database (European Commission, n.d.).

Table 3. The geographical characteristics of the selected sites

Site	Latitude °	Longitude °
Tan tan	28.26	– 11.60
Tarfaya	27.56	– 12.55
Laayoune	27.02	– 13.38
Boujdour	26.07	– 14.28
Dakhla	23.41	– 15.57

3.2 Wind Energy Calculation Method

The relationship between the power output of wind turbines and wind speed is directly proportional, and can be expressed in the following way (Olczak and Surma, 2023):

$$P_{WT}(V_h) = \begin{cases} 0 & \text{if } V_h \langle V_{in} \text{ or } V_h \rangle V_{out} \\ \frac{1}{2}\rho \cdot A \cdot C_p \cdot V_h^3 & \text{if } V_{in} \leq V_h < V_{rated} \\ P_{rated} & \text{if } V_h \geq V_{rated} \end{cases} \quad (1)$$

where is the rated power, A is the rotor Area, V_{in} is the cut-in wind speed, V_{rated} is the rated wind speed, and V_{out} is the cut-off wind speed. Using reference (Besseau et al., 2020), we have set the air density, ρ, value to 1225 kg/m^3 while also utilizing a C_p of 0.44 for our calculations.

Calculating the energy output from wind turbines requires determining the wind speed at the hub height of the turbine. Typically, wind speed measurements are taken at a height of 10 m (Liu et al., 2022). To accomplish this, an extrapolation method is utilized by applying Eq. (2) to estimate the wind speed V_h, at the hub height h. This equation takes into account the wind speed, V_0, at a specific reference height, h_0, and the Wind Shear, α (Marih et al., 2020).

$$V_h = V_0\left(\frac{h}{h_0}\right)^a \quad (2)$$

The wind shear which is a dimensionless constant used to characterize the wind speed profile in the atmosphere, it's defined by Khchine et al. (2019):

$$\alpha = 0.37 - 0.088 \cdot \ln V_0 \quad (3)$$

3.3 Evaluating Wind Energy Generation

The selected power model given in Eq. (1) is employed to calculate the hourly electrical energy generated by each wind turbine. In this study, the calculation of hourly electrical energy production is determined by the hourly power value of the wind turbine. The calculation is performed in one-hour intervals, and the hourly electrical energy produced can be obtained by multiplying the hourly power by the duration of one hour using Eq. (4).

$$E_h(V_h) = P_{WT}(V_h) \cdot 1h \quad (4)$$

The capacity factor is utilized to assess and evaluate the performance of a wind turbine (Wang et al., 2016). This capacity factor serves as an indicator of the turbine's efficiency (Sohoni et al., 2016; Akpinar and Akpinar, 2005). In addition, the estimation of the annual electrical energy, E_{out}, produced by each wind turbine model, as well as the calculation of the capacity factor, C_f, are determined, using Eqs. (5) and (6).

$$E_{out} = \sum_{1}^{h=8760} E_h \tag{5}$$

$$C_f = \frac{E_{out}}{P_{rated} \cdot 8760} \times 100 \tag{6}$$

4 Results and Discussion

4.1 Meteorological Data

By analyzing the wind speed data provided and presented in Fig. 3, we can see that wind speed varies significantly across the five sites in southern Morocco. Throughout the year, Boujdour and Laayoune demonstrate the highest average wind speeds, which vary between 5.10 and 8.99 m/s. while Tarfaya maintains an intermediate average wind speed, varying between 4.61 and 7.98 m/s. TanTan and Dakhla have the lowest average wind speeds ranging from 3.56 to 7.42 m/s.

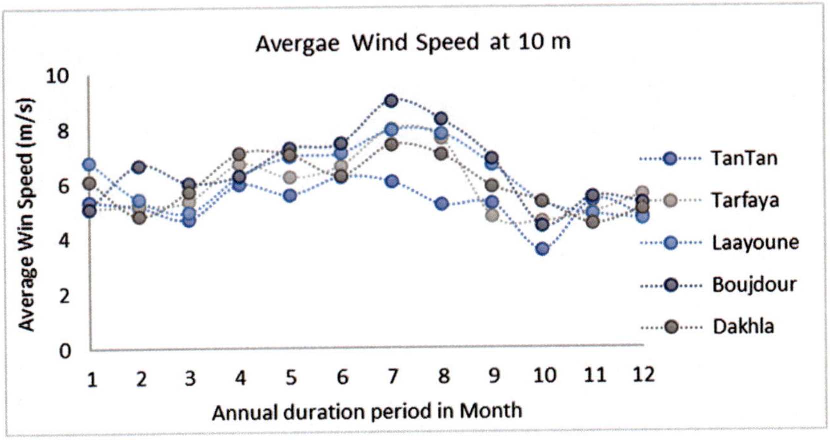

Fig. 3. Wind speed at 10 m

4.2 Wind Turbines

The technical characteristics of the wind turbines that were chosen for this study are presented in Table 4. These characteristics were obtained from the wind power database

(Power and Database, 2023). Moreover, Fig. 4 illustrates a representation of the power curves of the chosen wind turbines. The nominal powers of these turbines are 750 and 1000 kW. The power curves displayed in Fig. 4 were also obtained from the wind power database (Power and Database, 2023).

Table 4. Comparison of Technical specifications for Mitsubishi MWT62–1.0 and Unison U50 wind turbines

Wind turbine	V_{in} (m/s)	V_{rated} (m/s)	V_{out} (m/s)	P_{rated} (KW)	Rotor diameter (m)
Mitsubishi MWT62–1.0	5	13.5	25	1000	61.4
Unison U50	2.5	12.5	25	750	50

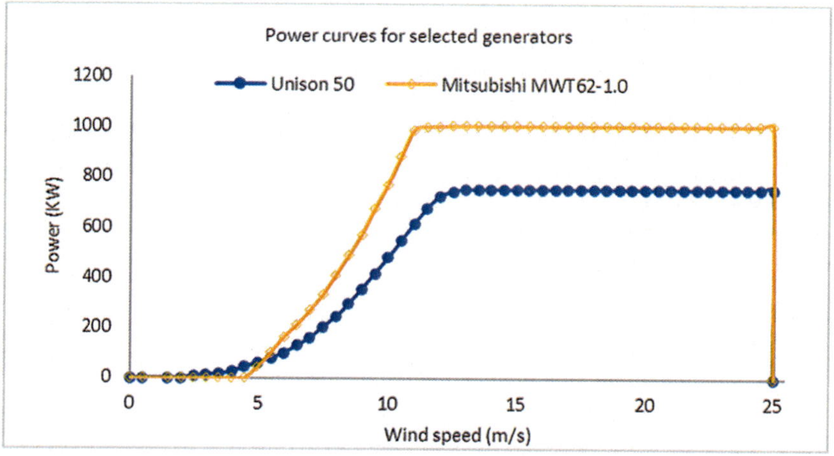

Fig. 4. Comparison of Power Curves for Mitsubishi MWT62–1.0 and Unison U50 Wind Turbines

4.3 Energy Production of Wind Turbines

The various factors such as the characteristics of the turbines, wind speed, and location influence the annual electrical energy generated by the wind turbines at each site. Table 5 provides an overview of the annual electrical energy output produced by the wind turbines at each site. To ensure a profitable investment in wind energy, it is advisable to select turbines with a minimum capacity factor of 25%. This recommendation is widely supported by several sources in the field, including references (Daoudi et al., 2022) and (Chauhan and Saini, 2014). Based on this recommendation, our analysis indicates that both the MWT62/1.0 and Unison U50 wind turbine models are suitable for grid integration at all five sites, as their capacity factors exceed 25%, making them economically feasible for wind energy production.

When examining the results between the five sites, it becomes evident that there are notable differences in the capacity factor and annual energy production values for both turbines. The results indicate that Boujdour achieves the highest capacity factor and annual electrical energy production for both turbines, whereas Tarfaya demonstrates the lowest values. This indicates that the overall performance of a wind turbine can be significantly influenced by the selection of an appropriate turbine model for a specific location.

According to the data presented in Table 5, the capacity factor depending on the site and turbine models. Boujdour stands out as the location with the highest capacity factor values for both turbines, ranging from 46.3 to 49.9%. Moreover, Boujdour also boasts the highest annual electrical energy production values, with 4.37 GWh for the Mitsubishi MWT62-1.0 turbine and 3.04 GWh for the Unison U50 turbine. However, the maximum energy production at Laayoune is 4.08 GWh, while Tarfaya exhibits a production range between 2.58 and 3.74 GWh. In contrast, Laayoune achieves a maximum energy production of 4.08 GWh generated by the Mitsubishi MWT62-1.0 turbine, whereas Tarfaya demonstrates a production range between 2.58 and 3.74 GWh.

TanTan produces the lower annual energy production of 3.89 GWh for Mitsubishi MWT62/1.0 and 2.06 GWh for Unison U50. In terms of wind energy production, Dakhla can generate ~ 2.06–3.76 GWh. This range is achieved using the Unison U50 and Mitsubishi MWT62/1.0 turbines, respectively.

Comparing the results between the two wind turbines, we can see that the Unison U50 generally has lower capacity factors and annual energy production values compared to the Mitsubishi MWT62/1.0 for all five sites. For example, at Tan Tan, the Mitsubishi turbine exhibits a capacity factor of 44.4%, resulting in an annual electrical energy generation of 3.89 GWh. In comparison, the Unison turbine achieves a lower capacity factor of 31.4%, leading to an annual electrical energy production of 2.06 GWh. Similarly, at Boujdour, the Mitsubishi turbine demonstrates a capacity factor of 49.9%, enabling it to generate an annual energy production of 4.37 GWh. On the other hand, the Unison turbine demonstrates a slightly lower capacity factor of 46.3% and an annual energy production of 3.04 GWh.

Table 5 presents the recorded capacity factor of the wind turbine for each specific location. Boujdour has the highest capacity factor at 49.9%, while TanTan has a lower potential for wind turbine installation with a capacity factor of 31.4%. However, the chosen wind turbine had a good capacity factor in these five locations, indicating that it could produce a high amount of electricity over the course of a year.

5 Conclusions

The southern region of Morocco has emerged as a promising hub for wind energy generation, accounting for a significant portion of the country's total wind power capacity. The study revealed that the selected locations within this region exhibited substantial wind potential, making them suitable for the implementation of wind turbines. Based on the obtained results, both MWT62-1.0 and Unison U50 wind turbine models are suitable for grid integration at all five sites, as their capacity factors exceed 25%, making them economically feasible for wind energy production. Furthermore, for all selected

Table 5. Analysis of capacity factor and annual electrical energy generation of Mitsubishi MWT62/1.0 and Unison U50 wind turbines

Site		Mitsubishi MWT62/1.0	Unison U50
Tan tan	C_f (%)	44.4%	31.4%
	E_{out} (GWh)	3.89	2.06
Tarfaya	C_f (%)	42.6%	39.3%
	E_{out} (GWh)	3.74	2.58
Laayoune	C_f (%)	46.6%	42.9%
	E_{out} (GWh)	4.08	2.82
Boujdour	C_f (%)	49.9%	46.3%
	E_{out} (GWh)	4.37	3.04
Dakhla	C_f (%)	42.9%	31.4%
	E_{out} (GWh)	3.76	2.06

wind turbines, the highest energy production is calculated at Boujdour, while the lowest energy production is at TanTan. Among the selected sites, Boujdour displayed the highest capacity factor at 49.9%, indicating its exceptional energy production capabilities. On the other hand, TanTan showcased a lower potential for wind turbine installation with a capacity factor of 31.4%.

In conclusion, the research findings indicate that the selected locations, situated in the southern region of Morocco, have significant wind potential in addition to solar energy potential, indicating the importance of continued investment and development in renewable energy infrastructure in this region. Therefore, by integrating both wind and solar energy systems at these sites, it becomes possible to generate renewable energy consistently, even during periods of limited solar radiation. This will lead to a more stable and reliable energy supply, which can help to reduce the reliance on non-renewable energy sources. Future research will focus on evaluating and optimizing a hybrid system that combines solar and wind energy systems. This analysis will assess the feasibility, advantages, and potential optimizations of integrating both energy sources, to further enhance renewable energy generation in the southern region of Morocco.

References

1. Akpinar, E.K., Akpinar, S.: An assessment on seasonal analysis of wind energy characteristics and wind turbine characteristics. Energy Convers. Manage. **46**(11–12), 1848–1867 (2005). https://doi.org/10.1016/j.enconman.2004.08.012
2. Azeroual, M., El-Makrini, A., El-Moussaoui, H., El-Markhi, H.: Renewable energy potential and available capacity for wind and solar power in Morocco towards 2030. J. Eng. Sci. Technol. Rev. **11**(1), 189–198 (2018). https://doi.org/10.25103/jestr.111.23
3. Benazzouz, A., Mabchour, H., El-Had, K., Zourarah, B., Mordane, S.: Offshore wind energy resource in the Kingdom of Morocco: Assessment of the seasonal potential variability based on satellite data (2021)

4. Besseau, R., et al.: A parametric model for wind turbine power curves incorporating environmental conditions. Renew. Energy **157**, 754–768 (2020). https://doi.org/10.1016/j.renene.2020.04.123
5. Chauhan, A., Saini, R.P.: Statistical analysis of wind speed data using Weibull distribution parameters. Iconce **14**, 160–163 (2014)
6. Daoudi, M., Ait, A., Mou, S., Naceur, L.A.: Evaluation of wind energy potential at four provinces in Morocco using two-parameter Weibull distribution function. Int. J. Power Electr. Drive Syst. **13**(2), 1209–1216 (2022). https://doi.org/10.11591/ijpeds.v13.i2.pp1209-1216
7. El-Hadri, Y., Khokhlov, V., Slizhe, M., Sernytska, K.: Wind energy land distribution in Morocco in 2021–2050 according to RCM simulation of CORDEX-Africa project (2019)
8. El-Khchine, Y., Sriti, M.: Performance evaluation of wind turbines for energy production in Morocco's coastal regions. Results Eng. **10**, 100215 (2021). https://doi.org/10.1016/j.rineng.2021.100215
9. Elmahmoudi, F., Bahatti, L., Abra, K.: Elaboration of a wind energy potential map in Morocco using GIS and analytic hierarchy process. Eng. Technol. Appl. Sci. Res. **10**(4), 6068–6075 (2020)
10. European Commission PVGIS: European Commission PVGIS. https://re.jrc.ec.europa.eu/pvg_tools/fr/#TMY (n.d.)
11. Haidi, T., Cheddadi, B., El Mariami, F., El Idrissi, Z., Tarrak, A.: Wind energy development in Morocco: evolution and impacts. Int. J. Electr. Comput. Eng. **11**(4), 2811–2819 (2021). https://doi.org/10.11591/ijece.v11i4.pp2811-2819
12. International Renewable Energy Agency (IRENA): Renewable Power Generation Costs in 2021. https://www.irena.org (2022a)
13. International Renewable Energy Agency (IRENA): Renewable Energy Statistics 2022. https://www.irena.org/ (2022b)
14. Khchine, Y., El-Sriti, M., Eddine, N., Kadri, E.: Evaluation of wind energy potential and trends in Morocco. Heliyon **6**, e01830 (2019). https://doi.org/10.1016/j.heliyon.2019.e01830
15. Liu, X., Kang, Y., Chen, H., Lu, H.: Comparison of surface wind speed and wind speed pro fi les in the Taklimakan Desert. PeerJ **10**, 1–18 (2022). https://doi.org/10.7717/peerj.13001
16. Marih, S., Ghomri, L., Bekkouche, B.: Evaluation of the wind potential and optimal design of a wind farm in the Arzew industrial zone in Western Algeria. Int. J. Renew. Energy Develop. **9**(2), 177–187 (2020). https://doi.org/10.14710/ijred.9.2.177-187
17. Olczak, P., Surma, T.: Energy productivity potential of offshore wind in Poland and cooperation with onshore wind farm. Appl. Sci. **13**, 4258 (2023). https://doi.org/10.3390/app13074258
18. Šimelytė, A., Ševčenko, G., Monni, S.: Promotion of renewable energy in Morocco. Energy Transf. Towards Sustain. **4**, 249–287 (2016). https://doi.org/10.9770/jesi.2016.3.4(2)CITATIONS
19. Sohoni, V., Gupta, S. C., Nema, R. K.: A Critical Review on Wind Turbine Power Curve Modelling Techniques and Their Applications in Wind Based Energy Systems (2016)
20. The Wind Power Database: The Wind Power Database. https://www.thewindpower.net/ (2023)
21. The Moroccan Ministry of Energy Transition and Sustainable Development: The Moroccan Ministry of Energy Transition and Sustainable Development. https://www.mem.gov.ma (2023)
22. Tizgui, I., El-Guezar, F., Bouzahir, H., Vargas, A.N.: Estimation and analysis of wind electricity production cost in Morocco. Int. J. Energy Econ. Policy **8**(3), 58–66 (2018)
23. Wang, J., Hu, J., Ma, K.: Wind speed probability distribution estimation and wind energy assessment. Renew. Sustain. Energy Rev. **60**, 881–899 (2016). https://doi.org/10.1016/j.rser.2016.01.057

Sound Absorption and Transmission of Homogeneous and Heterogeneous Micro-Perforated Plates

Brahim El Kharras(✉), Mohammed Garoum, Abdelmajid Bybi, and Najma Laaroussi

Mohammed V University in Rabat, Higher School of Technology in Salé, Material, Energy and Acoustics Team (MEAT), Salé, Morocco
brahim_elkharras@um5.ac.ma

Abstract. The use of glass in both exterior and interior buildings has become more and more popular over the past few years. Multi-layered micro-perforated plate (MPP), on the other hand, has long been recognized as a replacement for traditional porous materials due to its ease of installation, extended durability, ecological friendliness, and appealing look. By perforating the conventional glass structure currently available, improvement of the acoustic performance could be achieved. Extensive research on MPP absorbers has been published, to improve their absorption frequency bandwidth. This study presents the absorption and transmission characteristics of multilayered MPPs with homogeneous and inhomogeneous perforation whose facings are excited by a plane wave. The transfer matrix method was employed to calculate the transmission and absorption coefficient of the MPPs. The theoretical findings were compared to available results in the literature. Key parametric research on the effect of inhomogeneous MPP parameters on the MPP performance is also carried out. These parameters are the plate's micro-perforation, the hole size, and the cavity thickness. It is demonstrated that the addition of inhomogeneous perforation results in similar absorption performance compared to the homogenous one. It was also proved that the transmission properties of the multilayered MPPs deteriorate when the perforation increases.

Keywords: Microperforated plates · Absorption · Heterogeneous perforation · Transfer matrix method

1 Introduction

The research on the micro-perforated plate (MPP) absorbers is still ongoing. The MPP absorber, proposed by Maa [1], has been used in a variety of acoustics and noise control techniques [2]. In comparison to standard mineral and synthetic fibrous materials, they are appropriate when light visibility and fiber-free qualities are also desired.

However, the MPP's absorption frequency bandwidth is small, which restricts its practical applicability, especially in the wideband sound field typical of room acoustics. As a result, some study has suggested expanding its frequency range using structural modifications or increasing its perforation qualities. By creating several peaks at different

K. Baba et al. (Eds.): GeoME 2023, SPEES, pp. 98–107, 2023.
https://doi.org/10.1007/978-3-031-49345-4_10

frequencies, the broadband absorption characteristic may be recovered. Consequently, an array of MPPs will be used in the MPP absorber system. This arrangement may be achieved by placing the MPPs in series [3–5] and parallel [6–8]. The group of overlapping peaks can constitute a broad absorption bandwidth, and experimental work has confirmed such techniques.

However, there have been relatively few studies on the sound-insulating efficacy of MPPs. To achieve excellent sound insulation effectiveness in the mid-frequency range on MPP structures, an air-cavity-subdivision method is used [9], owing to mass-air-mass resonance, MPPs were further employed to increase the insulation capabilities of multi-layer structures [10].

In this paper, We propose a systematic and comprehensive examination of the absorption and transmission characteristics of multi-layered MPPs with homogeneous and inhomogeneous perforation in the whole frequency spectrum for future implementation in glass structures. A prediction model based on the transfer matrix method is developed for the computation of the sound absorption coefficient (SAC) and the sound transmission loss (STL) of multi-layered MPPs.

2 Theoretical Formulations

2.1 Homogeneous Perforation (MPP in Series)

As illustrated in Fig. 1, we investigate N-layer MPPs subjected to an incident plane wave. The depth of the cavity between the two MPPs is L_n.

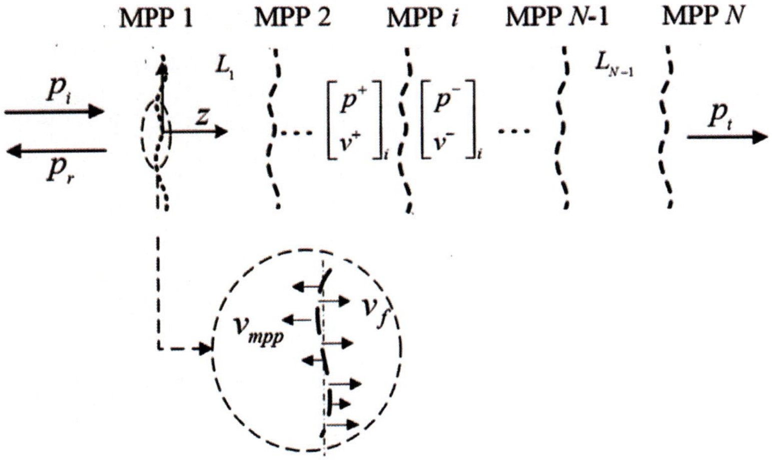

Fig. 1. Propagation of acoustic wave in N-layer MPPs subjected to normal incident wave.

After omitting $e^{i\omega t}$, as a function of displacement, the equation that governs the nth MPP is presented.

$$D_n \nabla^4 w_n - M_n \omega^2 w_n = \Delta p_n \tag{1}$$

where M_n and D_n are the mass per unit area and the flexural rigidity of the nth MPP respectively, D_n may be expressed as:

$$D_n = E_n h_n^3 / 12(1 - v_n^2) \tag{2}$$

The solution of Eq. (1) is obtained using the following equation

$$w_n = \overline{w_n} e^{-ikx \sin\theta} \tag{3}$$

Equation (3) is replaced with Eq. (1) to produce the following relationship:

$$v_{n,p} = i\omega w_n = \frac{\Delta p_n}{Z_{n,p}} \tag{4}$$

where $Z_{n,p}$ is the infinite plate impedance.

$$Z_{n,p} = \frac{(D_n k^4 \sin^4\theta - M_n \omega^2)}{i\omega} \tag{5}$$

As observed in Fig. 1, The MPP's average velocity is linked to both the plate's velocity v_p and the fluid velocity v_f.

$$\overline{v_n} = v_{n,p}(1 - \sigma_n) + v_{f_n}\sigma_n \tag{6}$$

where σ_i is the perforation ratio.

As previously demonstrated in the study [1], the MPP's impedance and the pressure difference are related as follows:

$$Z_{resist,n}(v_{f,n} - v_{p,n}) + Z_{react,n} v_{f,n} = \Delta p_n \tag{7}$$

The impedance of the hole, $Z_n = Z_{resist,n} + Z_{react,n}$, is given as

$$Z_{resist,n} = \frac{8\eta_0 h}{(d_i/2)^2}\left(\sqrt{1 + \frac{X^2}{32}} + \frac{\sqrt{2}dX}{32h}\right) \text{ and}$$

$$Z_{react,n} = j\rho\omega h\left(1 + \frac{1}{\sqrt{9 + X^2/2}} + \frac{8d_i}{3\pi h}\right) \tag{8}$$

where η_0 is the air viscosity, and $X = \left(d/2\right)\sqrt{\rho\omega/\eta}$.

When we eliminate $v_{j,f}$ from Eqs. (3) and (4), we get the following relation:

$$\overline{v_n} = \gamma v_{n,p} + \frac{\sigma_n \Delta p_n}{Z_n} \tag{9}$$

In which $\gamma = 1 - \sigma\left(\frac{Z_{react}}{Z}\right)$.

Replacing Eq. (4) for Eq. (9) yields:

$$\overline{v} = \left(\frac{\gamma}{Z_{n,p}} + \frac{\sigma}{Z}\right)\Delta p_n = \frac{1}{Z_{mpp,n}}\Delta p_n \tag{10}$$

The velocity and acoustic pressure on the sides of the nth layer of the MPPs may be represented as:

$$\begin{bmatrix} p^+ \\ v^+ \end{bmatrix} = \begin{bmatrix} 1 & Z_{Mpp,n} \\ 0 & 1 \end{bmatrix} \begin{bmatrix} p^- \\ v^- \end{bmatrix} = [A_n] \begin{bmatrix} p^- \\ v^- \end{bmatrix} \tag{11}$$

v^+ and v^- represent the vibration velocity on both sides of the MPP. $[A_n]$ is The transfer matrix of the nth MPP.

2.2 Transmission and Absorption Properties

The cavity transfer matrix between two successive MPPs is:

$$[B_n] = \begin{bmatrix} \cos(kL_i) & jZ_0 \sin(kL_i) \\ j\frac{1}{Z_0} \sin(kL_i) & \cos(kL_i) \end{bmatrix} \tag{12}$$

where Z_0 is the air's characteristic impedance.

The total transfer matrix for N-layer MPPs, denoted $[T]$, is as follows:

$$[T] = [A_1][B_1]....[A_n][B_n] = \begin{bmatrix} T_{11} & T_{12} \\ T_{21} & T_{22} \end{bmatrix} \tag{13}$$

The transmission coefficient τ and reflection coefficient R are:

$$\tau = \frac{2}{T_{11} + \frac{T_{12}}{Z_0} + Z_0 T_{21} + T_{22}} \quad \text{and}$$
$$R = \frac{T_{11}Z_0 + T_{12} - Z_0(Z_0T_{21} + T_{22})}{T_{11}Z_0 + T_{12} + Z_0(Z_0T_{21} + T_{22})} \tag{14}$$

The transmission loss of the N-layer MPPs and the absorption coefficient are:

$$TL = -20 \log 10(|\tau|) \quad \text{and}$$
$$\alpha = 1 - |R|^2 - |\tau|^2 \tag{15}$$

2.3 Inhomogeneous Perforation (MPP in Parallel).

The schematic illustration of the suggested inhomogeneous MPP configuration is represented in Fig. 2, which consists of a finite plate comprising two regions, each with distinct hole diameter and perforation ratios with rigidly partitioned cavities.

The Parallel Transfer Matrix Method is utilized to compute the absorption performance of the entire inhomogeneous MPPs, which is viewed as an arrangement of parallel acoustic components. Transfer matrices were established for every region and connected using the technique outlined below. we assume normal incident plane wave and each area reacting locally. Moreover, the finite size impedance of the plate is used

$$\begin{bmatrix} p_i^+ \\ v_i^+ \end{bmatrix} = [T_i] \begin{bmatrix} p_i^- \\ v_i^- \end{bmatrix} = \begin{bmatrix} T_{i,11} & T_{i,12} \\ T_{i,21} & T_{i,22} \end{bmatrix} \begin{bmatrix} p_i^- \\ v_i^- \end{bmatrix} \tag{16}$$

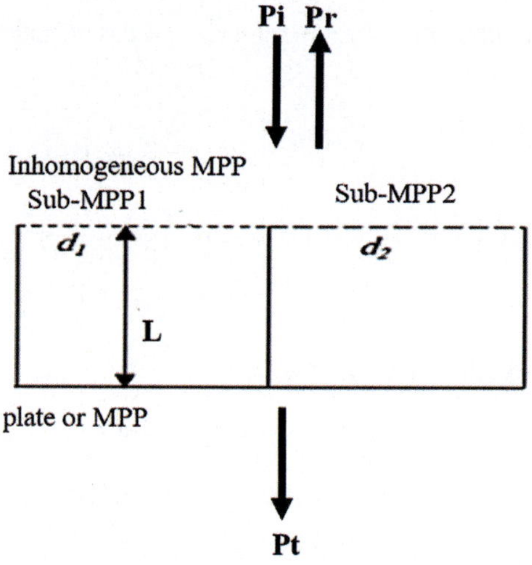

Fig. 2. Sound propagation of 2-layer MPPs with inhomogeneous perforation under normal incidence conditions

Given that the MPPs are parallel, admittance matrices should be established for each MPP as

$$[Y_i] = \begin{bmatrix} Y_{i,11} & Y_{i,12} \\ Y_{i,21} & Y_{i,22} \end{bmatrix} = \frac{1}{T_{i,12}} \begin{bmatrix} T_{i,22} & T_{i,21}T_{i,12} - T_{i,22}T_{i,11} \\ 1 & -T_{i,11} \end{bmatrix} \tag{17}$$

Subsequently, the transfer matrix of the inhomogeneous MPPs derived from Eqs. (16) and (17) is as follows:

$$\begin{bmatrix} p^+ \\ v^+ \end{bmatrix} = [T] \begin{bmatrix} p^- \\ v^- \end{bmatrix} = \frac{-1}{\sum r_i Y_{i,21}} \begin{bmatrix} \sum r_i Y_{i,22} & -1 \\ \sum r_i Y_{i,22} \sum r_i Y_{i,11} - \sum r_i Y_{i,12} \sum r_i Y_{i,21} & -\sum r_i Y_{i,11} \end{bmatrix} \begin{bmatrix} p^- \\ v^- \end{bmatrix} \tag{18}$$

where r_i is the surface ratio of sub-MPP 'i'. The acoustic properties of the inhomogenemous MPPs can be estimated from this global matrix using Eqs. (14) and (15).

3 Results and Discussions

In this part, the model will be compared to available results in the literature and a parametric analysis will be carried out, following the theoretical model and solution technique described in earlier sections of the paper.

3.1 Model Validation

This section would provide comparisons to demonstrate the reliability and accuracy of the present approach. We have compared the findings with those that have already been documented in the literature to validate the suggested model. We have chosen the work provided by [2] in particular that considers the computation of the absorption coefficient of multi-layered MPPs.

In this analysis, the MPPs have the chosen material properties: Young modulus $E = 7 \times 10^{10} Pa$, Poisson ratio $v = 0.22$, and density $\rho_p = 2500\ kg/m^3$. The properties of air are $\rho_0 = 1.2\ kg/m^3$ and $c_0 = 343\ m/s$, and the viscosity is $\eta_0 = 17.9 \times 10^{-6}\ kg/ms$. The cavity thickness (Li) is 33 mm. The thickness of the MPPs is 0.002 m. The perforation ratio is: $\sigma_1 = 5.1\%$, $\sigma_2 = 2.1\%$, $\sigma_3 = 0.8\%$, $\sigma_4 = 0\%$.

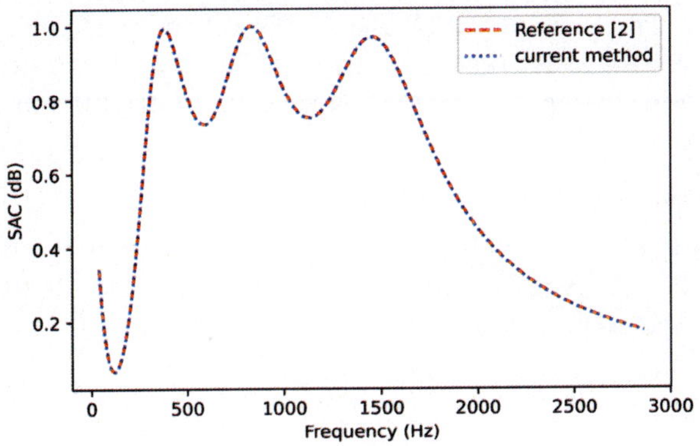

Fig. 3. SAC of multilayered MPPs using the current method and method of reference [2]

To confirm the suggested model's accuracy in computing the STL of multi-layered MPPs. The outcomes were compared with those that had already been reported in the literature. We have specifically chosen the work that [7] provided.

From Figs. 3 and 4, the current method shows excellent agreement with the results published in the literature.

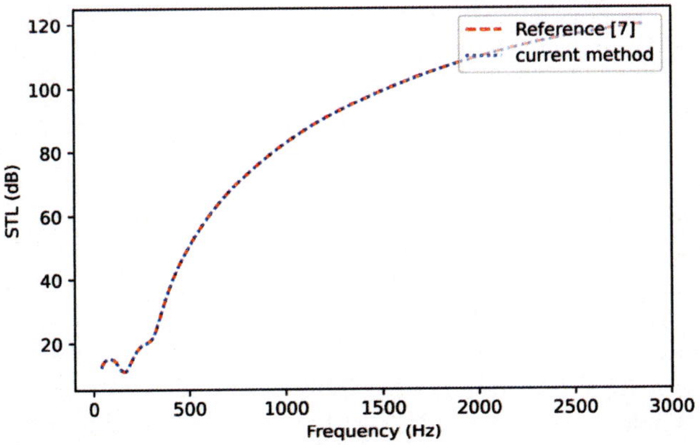

Fig. 4. STL of multilayered MPPs using the current method and method of reference [11]

3.2 Parametric Study

The effects of many important system factors on the acoustic performance of multi-layered MPP structures for both cases are examined using the proposed model. In the simulations, the properties are given as follows. The properties of air are $\rho = 1.2\ kg/m^3$ and $c = 343\ m/s^3$, and the viscosity is $\eta_0 = 17.9 \times 10^{-6}\ kg/ms$. The material properties of the MPPs are Young modulus $E = 7 \times 10^{10} Pa$, Poisson ratio $v = 0.22$, and density $\rho_p = 2500\ kg/m^3$. The MPPs have a thickness of 0.002 m with pores of 0.5 mm in diameter.

The dimensions of the two sub-MPPs are $L_x \times L_y = 0.5 \times 0.35$. The first MPP is under acoustic excitation azimuth angle $\theta = 0$. Several important system variables, such as, micro-perforation ratio, hole size, and cavity thickness are examined using numerical analysis to determine their effects on the SAC and TL of the different configurations.

3.2.1 Influence of Micro-Perforation Ratio

Figure 5 depicts the STL of triple MPPs for various perforation ratios, with a cavity thickness of q = 33 mm. The 0.5 mm hole diameter is the same for all three MPPs. The third MPP's perforation ratio is zero. The MPPs are infinite.

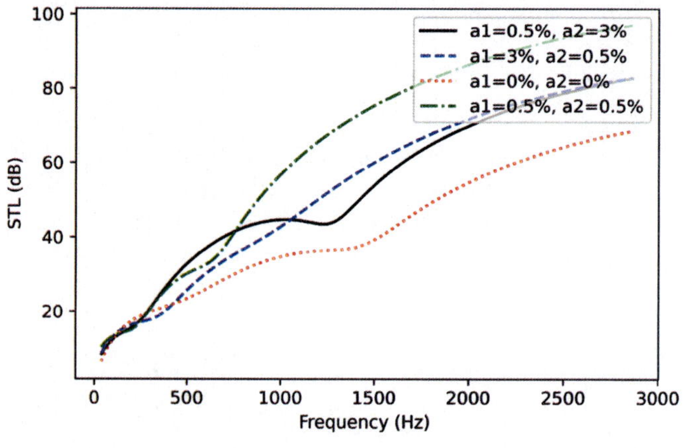

Fig. 5. The STL of triple-MPPs with homogeneous perforation for different perforation ratio values.

The absorption coefficient for the case of homogenous perforation of multi-layered MPPs is compared in Fig. 6(a) for various perforation ratios. The cavity depths and other MPP characteristics remain unchanged. The third MPP's perforation ratio is left at zero.

The absorption capabilities of MPP in parallel (inhomogeneous MPP) with various perforation ratios are shown in Fig. 6(b). This is carried out under the scenario when each sub-MPPs hole diameter has a distinct hole diameter. Maintaining one sub-MPP's perforation ratio while adjusting the perforation ratio of the other sub-MPP simulates the impact of perforation on sound absorption. The third MPP's perforation ratio is left at zero.

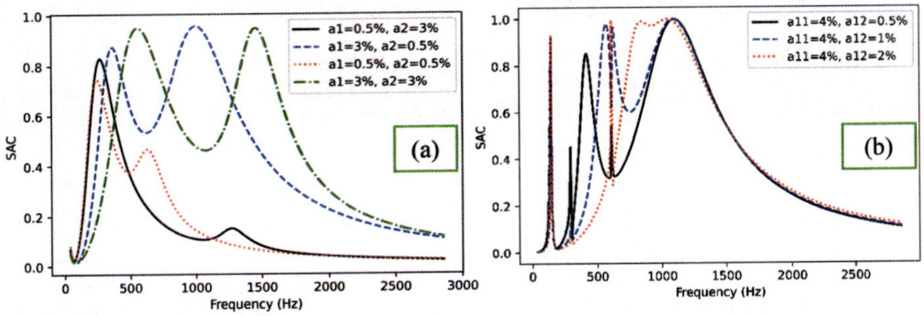

Fig. 6. Effect of perforation on the SAC: (a) triple-MPPs with homogeneous perforation and (b) double-MPP with inhomogeneous perforation ($d_1 = 0.4\ mm, d_2 = 0.8\ mm$)

3.2.2 Influence of Hole Size

In this part, the pore size is changed but the value of perforation ratio remains constant on the MPPs. The perforation ratios are $\sigma_1 = 3\%$ and $\sigma_2 = 0.5\%$. The third MPP's perforation ratio is zero.

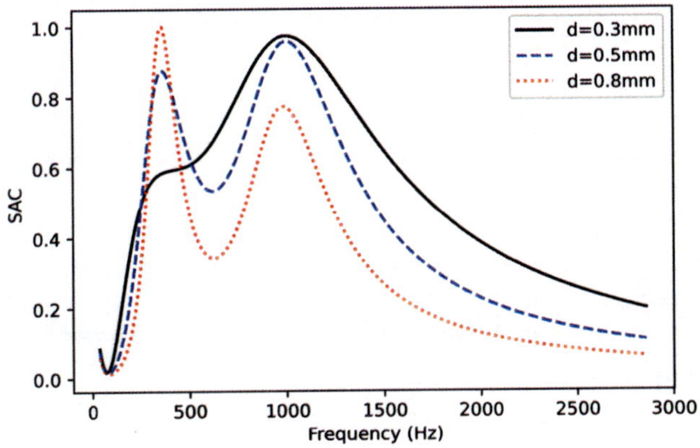

Fig. 7. Influence of pore diameter on the SAC of triple-MPPs with homogeneous perforation

3.2.3 Influence of Air Gap Thickness

Figure 8(a) depicts the influence of changing the cavity thickness in the case of multi-layered MPP with homogeneous perforation. The parameters of the MPPs are: $\sigma_1 = 3\%$, $\sigma_2 = 0.5\%$, and $\sigma_3 = 0\%$ with 0.5 mm hole diameter.

Figure 8(b) depicts the influence of changing the air gap thickness for the MPP with inhomogeneous perforation. The parameters are: $\sigma_{11} = 4\%$, $\sigma_{12} = 0.5\%$ and $\sigma_2 = 0\%$ with $d_1 = 0.4\ mm$, $d_2 = 0.8\ mm$.

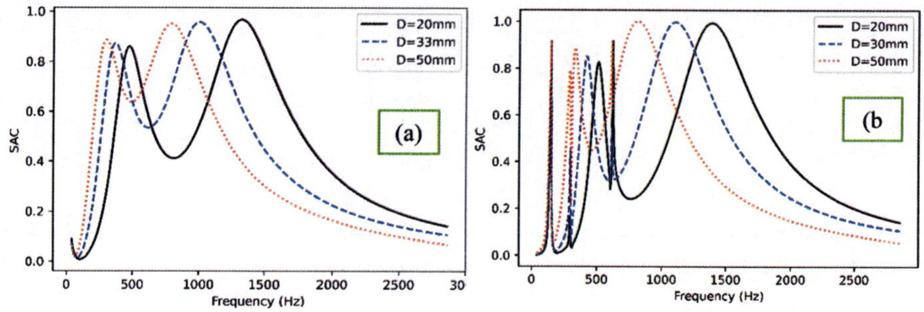

Fig. 8. Effect of air gap thickness on the absorption coefficient: (a) triple-MPPs with homogeneous perforation ($L_1 = L_2 = D$) and (b) double-MPP with inhomogeneous perforation ($d_1 = 0.4\ mm$, $d_2 = 0.8\ mm$) and ($L = D$).

4 Discussions

The perforation ratio $\sigma_1 = \sigma_2 = 3\%$, and $\sigma_1 = 0.5\%$, $\sigma_2 = 3\%$ provides a greater average sound absorption value coefficient than the other two cases as shown in Fig. 6(a). However, it is observed In Fig. 5, that as σ_1 and σ_2 is increased, the STL is degraded at mid and high-frequency ranges. The effect of perforation on absorption coefficient of inhomogeneous MPP is calculated in Fig. 6(b) by keeping one sub-MPP's perforation ratio constant while modifying the other sub-MPP's perforation ratio. The findings show general tendencies, where the wideband frequency may be calculated from the difference in perforation ratios between the sub-MPP. It is shown that inhomogeneous MPP gives wideband absorption.

Figure 7 depicts the absorption coefficient findings obtained by altering the hole diameter. The lower resonance frequency's peak (420 Hz) decreases as the hole width increases. However, there is an improvement in sound absorption. The performance improvement is only observed for a restricted frequency range around the corresponding resonance frequency.

The effect of air gap change on the absorption coefficient for the two examples is depicted in Fig. 8. As L is raised, the peaks move towards a lower frequency. Controlling the difference in cavity depth allows you to construct an adequate absorption bandwidth.

5 Conclusion

In terms of practicality, the MPP series configuration is a solid choice for on-site deployments. However, because of the use of many air cavities that separates MPP layers, this technique necessitates a huge amount of area. It was demonstrated that a collection of distinct MPPs may be stacked in parallel, with a one-panel surface containing numerous sub-MPPs and all sub-MPPs sharing the same cavity depth providing similar acoustic performance (SAC). As a result, this configuration does not need additional installation space. Regarding the STL, the increase in the perforation ratio degrades the insulating performance of the multi-layered MPPs with homogeneous perforation. A similar trend is expected for the case of inhomogeneous perforations. Hence to optimize the performance of inhomogeneous MPPs it is necessary to evaluate both acoustic indicators: SAC and STL.

References

1. Maa, D.: Theory and design of microperforated panel sound-absorbing constructions. Sci. Sin. **18**(1), 55–71 (1975)
2. Asdrubali, F., Pispola, G.: Properties of transparent sound-absorbing panels for use in noise barriers. J. Acoust. Soc. Am. **121**(1), 214–221 (2007)
3. Kang, J., Fuchs, H.V.: Predicting the absorption of open weave textiles and microperforated membranes backed by an air space. J. Sound Vib. **220**(5), 905–920 (1999)
4. Sakagami, K., Morimoto, M., Koike, W.: A numerical study of double-leaf microperforated panel absorbers. Appl. Acoust. **67**(7), 609–619 (2006)
5. Sakagami, K., Matsutani, K., Morimoto, M.: Sound absorption of a double-leaf micro-perforated panel with an air-back cavity and a rigid-back wall: detailed analysis with a Helmholtz-Kirchhoff integral formulation. Appl. Acoust. **71**(5), 411–417 (2010)
6. Sakagami, K., Nagayama, Y., Morimoto, M., Yairi, M.: Pilot study on wideband sound absorber obtained by combination of two different microperforated panel (MPP) absorbers. Acoust. Sci. Technol. **30**(2), 154–156 (2009)
7. Wang, C., Huang, L., Zhang, Y.: Oblique incidence sound absorption of parallel arrangement of multiple micro-perforated panel absorbers in a periodic pattern. J. Sound Vib. **333**(25), 6828–6842 (2014)
8. Wang, C., Huang, L.: On the acoustic properties of parallel arrangement of multiple micro-perforated panel absorbers with different cavity depths. J. Acoust. Soc. Am. **130**(1), 208–218 (2011)
9. Toyoda, M., Takahashi, D.: Sound transmission through a micro-perforated-panel structure with subdivided air cavities. J. Acoust. Soc. Am. **124**(6), 3594–3603 (2008)
10. Mu, R.L., Toyoda, M., Takahashi, D.: Improvement of sound insulation performance of multilayer windows by using micro-perforated panel. Acoust. Sci. Technol. **32**(2), 79–81 (2011)
11. Kim, H.-S., Kim, S.-R., Kim, B.-K., et al.: Sound transmission loss of multilayered infinite microperforated plates. J. Acoust. Soc. Am. **147**(1), 508–515 (2020)

Optimization of Building Envelopes Types for Maximizing the Energy Saving

Yousra M'Hamdi(✉), Khadija Baba, and Mohammed Tajayouti

Civil Engineering and Environment Laboratory (LGCE), Mohammadia Engineering School, Mohammed V University, Rabat, Morocco
yousra.mhamdi.d@gmail.com

Abstract. The pressing issue of global warming has become a major concern for scientists and researchers worldwide. To address this, various solutions are being explored to enhance energy efficiency, especially in the large energy consumer that is the building industry. However, merely implementing energy-saving measures in buildings is not enough; an optimization strategy is crucial to implement these solutions in a successful manner. This study's goal is to create an optimization technique that uses a genetic algorithm to enhance the energy saving in various building envelopes. This algorithm facilitates determining the most suitable values for decision variables that align with the defined objective functions. In this case, the primary goals are to minimize energy consumption and reduce costs associated with energy usage. The study examines various building configurations, each with distinct envelope material compositions, and includes an economic analysis to ascertain the financial viability of the proposed setups. The results demonstrate that optimal building envelopes can significantly enhance energy efficiency, resulting in improved thermal comfort for occupants and substantial energy savings within the designated time frame. To visualize the trade-offs between different variables, a Pareto Diagram is used after conducting multiple iterations, highlighting the correlations among design choices. Furthermore, a comprehensive cost analysis of energy-saving expenses offers valuable guidance to professionals in selecting appropriate construction solutions for building envelopes. This analysis aids in making informed decisions to balance energy efficiency with economic considerations.

Keywords: Envelope materials · Energy saving · Optimization · Cost analysis · Buildings

1 Introduction

Global warming is an urgent issue that poses significant risks to our planet and its inhabitants. The increase in global temperatures is mostly attributed to the increase of greenhouse gas emissions, especially carbon dioxide (CO_2), resulting from human activities, including the burning of fossil fuels. One sector that contributes significantly to these emissions is the construction industry. The International Energy Agency reports that the building industry accounts for 40% of all primary energy consumption and produces the most CO_2 emissions [1].

K. Baba et al. (Eds.): GeoME 2023, SPEES, pp. 108–116, 2023.
https://doi.org/10.1007/978-3-031-49345-4_11

The construction sector is accountable for a substantial portion of energy consumption and greenhouse gas emissions worldwide. This includes the energy required for building operations, such as heating, cooling, lighting, and the use of appliances and equipment. Additionally, the production and transportation of construction materials add to the industry's carbon footprint. To address the challenge of global warming, optimizing the energy used in the constructions is of utmost importance. By reducing and optimizing energy consumption in buildings, we can achieve environmental, economic, and social benefits. It helps lower greenhouse gas emissions, promotes energy efficiency, saves money, enhances energy security, and improves comfort and health.

In this context, numerous scientists are actively seeking renewable energy solutions to replace conventional energy sources, aiming to mitigate the repercussions of this alarming situation, particularly in building construction. These solutions focus on reducing the cooling and heating requirements of occupants by employing intelligent construction materials capable of daytime storing of solar energy and nighttime release, similar to phase change materials (PCM). These materials undergo a storage/release cycle has a phase transition that goes from a liquid state when energy is received to a solid state when energy is released [2, 3].

Several researchers have explored the thermal benefits of incorporating phase change materials (PCM) in different building materials to assess their impact on energy reduction. M'hamdi et al. [4] conducted an energy analysis of buildings integrated with PCM in different climates. They carried out an economic and environmental study to determine the most effective solution. Additionally, M'hamdi et al. [5] assessed the impact of using PCM mortar in building envelopes through a parametric analysis, comparing different configurations in two climates in Morocco. The results demonstrated a significant reduction in energy consumption, estimated at approximately 11%, with the use of PCM.

These studies are complemented by other investigations focused on optimizing energy performance. Researchers have employed two types of optimization methods: Studies on single- and multi-objective optimization aimed at improving the energy consumption of building envelopes [6].

Saffari et al. [7] conducted an optimization using simulation method to optimize the PCM melting temperature aimed at enhancing the energy performance of buildings in various climates. They explored three different scenarios with distinct objective functions. The study concluded that PCM melting at 26 °C is more effective for achieving higher energy savings in cooling-dominant climates, while melting at 20 °C is more suitable for heating-dominant climates.

In a similar vein, Wu et al. [8] conducted a multi-objective optimization of both the energy system and building envelope in a typical residential building. Their approach involved simultaneous optimization of building retrofit and energy systems to improve overall energy performance.

Reviewing the previous up to date works, it becomes evident that various researches have explored energy performance optimization in buildings using different approaches. However, no study has systematically evaluated the energy performance of different envelope materials through a Pareto analysis to determine the best-performing material type.

In our paper, we applied an optimization method to enhance energy savings in buildings, considering three different scenarios of bi-objective functions. This study's primary goal is to offer an optimization strategy using genetic algorithms and cost analysis, while considering different scenarios, to offer valuable insights for professionals in the construction industry.

2 Materials and Methods

- Building model data

In this paper, a residential house is the subject of the case study. It's located in Morocco's Benguerir City's Green Energy Park. The residence was built in 2019 as part of the Solar Decathlon of Africa. Figure 1 displays the house's plan view.

In this study, we will explore various envelope types, namely concrete block, cast concrete, brick, and earth. Each envelope type will be examined under four different cases: without PCM, with PCM interior layer, and with PCM exterior layer, with two different thicknesses (15 and 40 mm) respectively.

The climatic conditions of Benguerir city are characterized by a semi-arid climate with warm winters.

Table 1 presents the building case study's characteristics, while Table 2 provides information on the PCM used in the study.

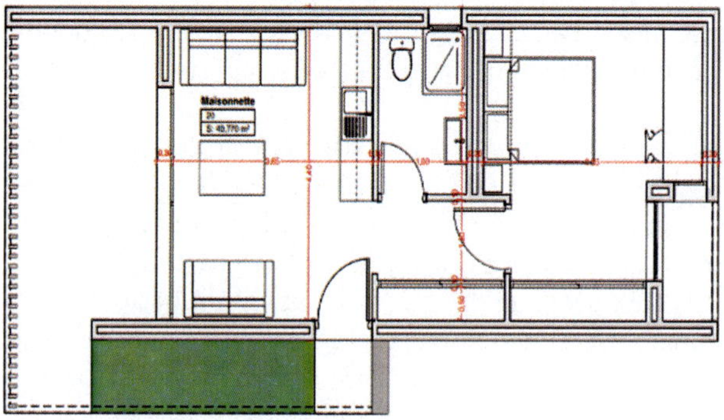

Fig. 1. Floorplan building

- Optimization method

Genetic Algorithms (GAs) are computational algorithms inspired by the principles of natural selection and genetics. They efficiently explore various design alternatives and identify the best possible solutions.

To employ the traditional genetic algorithm for calculations, it is crucial to carry out operator selection, crossover, and mutation, creating an appropriate initial population and generating a new population. Through iterative evaluations, the strengths and weaknesses

Table 1. Building data

Designation	Building
Case study type	Residential house
Floor area m^2	77.8 m^2
Number of floors	1
Windows	15 mm (double layered glass of 6 mm with air gap)

Table 2. PCM characteristics

Property	PCM
Nature	Cement + microencapsulated PCM
Thermal conductivity (W/m·k)	0.37
Specific heat capacity (J/kg·k)	1255
Density (kg/m^3)	1248
Melting area °C	26

of the objective function are assessed, ultimately leading to the selection of the optimal outcome [9].

For our optimization method, we utilized the Design Builder software, which employs Genetic Algorithms (GA) to search for optimal design solutions. To implement this optimization, we considered three distinct scenarios as outlined in Table 3. Each scenario had its unique objective function, but all shared the same design variable, which was the selection of the exterior wall material.

Table 3. Optimization scenarios

Scenario	Objective functions
1	Minimize cooling load Minimize heating load
2	Minimize cooling load Minimize discomfort hours
3	Minimize heating load Minimize discomfort hours

3 Results

3.1 Multi-objective Optimization

Regarding the initial optimization scenario, aimed at minimizing both heating and cooling loads simultaneously, the Pareto diagram displayed in Fig. 2 reveals the outcomes.

The results indicate that two types of envelopes represent the optimal solutions: An envelope consisting of a concrete block with an inside side layer of 40 mm PCM, resulting in 2131.6 kWh of heating load and 37.37 kWh of cooling load. The same envelope type with the outside side having the PCM layer, leading to ~ 2055.03 kWh of heating load and 65.67 kWh of cooling load (Fig. 3).

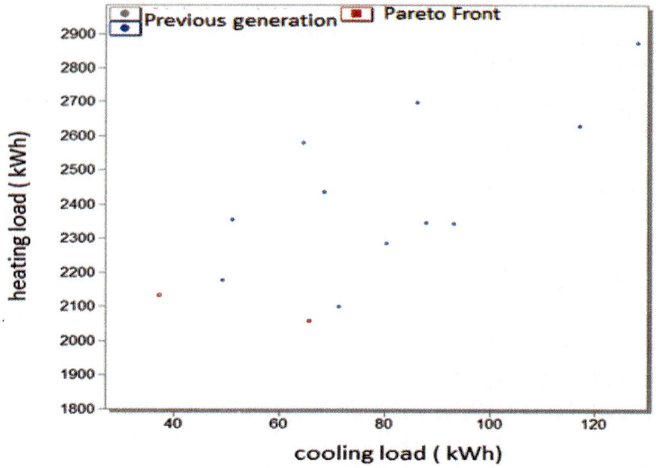

Fig. 2. Optimization analysis results (scenario 1)

In the second optimization scenario, which focuses on reducing both cooling load and discomfort hours, the research revealed that the optimal design involves envelopes with a concrete block and a 40 mm PCM layer. This configuration not only demonstrated the lowest cooling load, as previously explained in the first scenario but also resulted in the lowest total discomfort hours.

The total number of uncomfortable hours was roughly 5588 when the PCM layer was applied to the inside side. The overall number of discomfort hours decreased even further to roughly 5485.8 h when the PCM layer was located on the outer side.

In the third scenario, the pareto front is different and represent a unique optimum design variable: concrete block envelope with PCM layer of 40 mm placed in the interior side which represent a total discomfort hours of 5588 h and a heating load of 2131.6 kWh (Fig. 4).

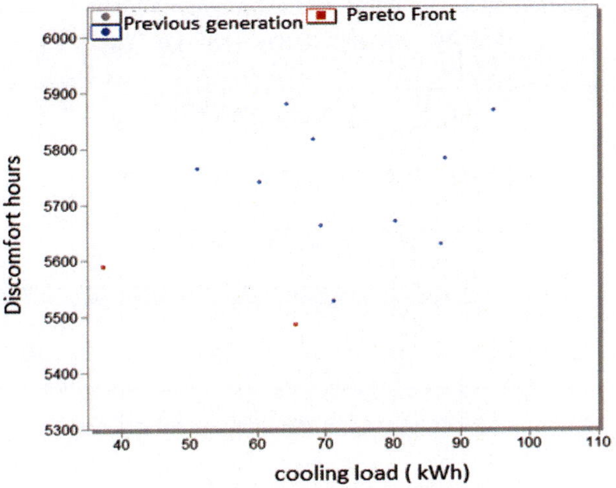

Fig. 3. Optimization analysis results (scenario 2)

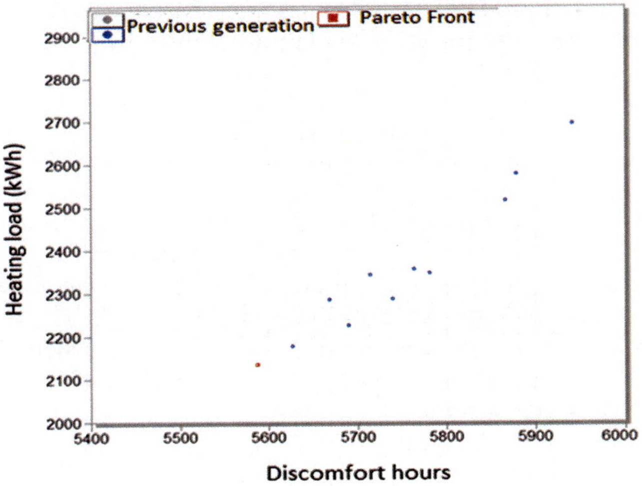

Fig. 4. Optimization analysis results (scenario 3)

According to the optimum solutions obtained for each envelope type, we conducted an analysis of energy saving load per year for each envelope considering the without PCM configuration as a baseline and the PCM of 40 mm thickness placed in interior side to conclude on the energy load saving.

Figure 5 shows that the concrete block envelope realize the best energy saving successed by reinforced concrete one comparing to the other configurations.

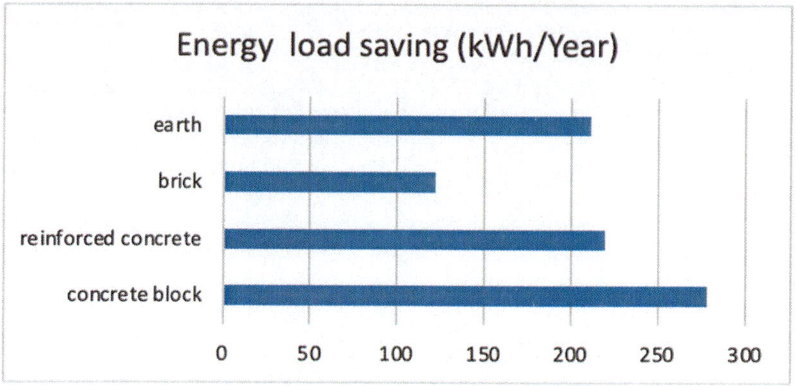

Fig. 5. Energy saving load (kWh/year)

3.2 Cost Analysis

To conduct a cost analysis of the energy-saving potential achieved due to the building envelope's integration of PCM, our focus lies in calculating the Energy Cost Saving (ECS) using Eq. 1 provided below. In this equation, ES stands for energy savings, and CE is the price of electricity.

$$ECS = ES * CE \tag{1}$$

The electricity price estimation in Morocco stands at 1.09 MAD/kWh. To assess the energy-saving benefits, the research team calculated the difference between the total energy load of the optimum solution (concrete block envelope with PCM) and the baseline (concrete block envelope without PCM). The results, as shown in Table 4, indicate an estimated total energy saving of 278.22 kWh per year.

Regarding cost savings, the calculation reveals a yearly reduction of 300 dh if the internal side of the concrete block building is covered with a 40 mm of PCM layer, equivalent to 12% in cost savings.

Table 4. Energy load and energy saving

	Heating load (kWh)	Cooling load (kWh)	Total energy load (kWh)
Concrete block envelope (baseline)	2212.47	234.72	2447.19
Concrete block with PCM (40 mm) envelope (optimum solution)	2131.6	37.37	2168.97
Energy saving (kWh)	80.87	197.35	278.22

3.3 Environmental Analysis

We are concentrating on the CO_2 emissions rate (CO_2-eq) associated with coal as a source of energy production in order to undertake an environmental analysis of our study. Coal has a carbon intensity of around 1 kg/kWh.

Based on the energy savings found in the previous section, an environmental analysis was conducted. As a maximum value achieved with the best possible envelope solution utilized in the analyzed building, we discovered that the usage of PCM on our building can cut CO_2 emissions by up to 278 kg/year.

4 Discussion

The multi-objective optimization results clearly demonstrate that the envelope composed by the concrete block with a 40 mm PCM layer outperforms the other configurations in achieving the three set objectives: minimizing cooling load, heating load, and total discomfort hours.

These favorable outcomes can be attributed to the interconnected relationship between these factors. Specifically, the need for energy to maintain a comfortable temperature is directly influenced by both heating and cooling loads, and the optimal solutions capable of addressing these loads also lead to the reduction of discomfort hours.

Interestingly, these findings align with a study conducted by Shen et al. [10], which similarly supports the promising use of PCM in combination with concrete block for enhancing energy efficiency.

Overall, the research highlights the effectiveness of the selected concrete block envelope with PCM layer in attaining superior performance across all specified objectives, making it a favorable choice for achieving energy efficiency and improved comfort in building designs.

The finding of cost analysis are in alignment with the results found by Cunha et al. [11], which reported an 11% reduction in energy costs through PCM usage. Similarly, Yun et al. [12] demonstrated that employing PCM as a decorative element can lead to a maximum energy cost reduction of 7.48%.

In summary, the results highlight the considerable energy and cost-saving potential of incorporating PCM into building envelopes, corroborating previous research in this domain.

5 Conclusions

This research aimed to assess the energy efficiency of buildings through an optimization method and cost analysis. The methodology employed a genetic algorithm optimization approach using Design Builder Software.

The study's findings allow for the following main inferences:

- The most favorable building envelope solution, considering minimum cooling load, heating load, and discomfort hours in the specific climatic conditions, consists of a concrete block envelope with a 40 mm phase change material (PCM) layer.

- The overall energy burden is greatly reduced when phase change materials are used in building envelopes. Implementing the optimal building envelope solution can lead to a 12% cost saving for users over one year.

These findings hold important implications for professionals and policymakers in the construction sector. The research paper's results offer valuable recommendations and insights, especially with regard to the application of materials for the envelopes that are energy-efficient during the design process. This study provides practical directions for material selection, thereby ensuring energy efficiency in construction projects.

References

1. Al-Yasiri, Q.: Incorporation of phase change materials into building envelope for thermal comfort and energy saving: a comprehensive analysis. J. Build. Eng. **36**, 102122 (2021). https://doi.org/10.1016/j.jobe.2020.102122
2. Soares, N., Costa, J.J., Gaspar, A.R., et al.: Review of passive PCM latent heat thermal energy storage systems towards buildings' energy efficiency. Energy Build. **59**, 82–103 (2013). https://doi.org/10.1016/j.enbuild.2012.12.042
3. Zhang, Y., Zhou, G., Lin, K., Zhang, Q., et al.: Application of latent heat thermal energy storage in buildings: state-of-the-art and outlook. Build. Environ. **42**(6), 2197–2209 (2007). https://doi.org/10.1016/j.buildenv.2006.07.023
4. M'hamdi, Y., Baba, K., Tajayouti, M., et al.: Energy, environmental, and economic analysis of different buildings envelope integrated with phase change materials in different climates. Solar Energy **243**, 91–102 (2022). https://doi.org/10.1016/j.solener.2022.07.031
5. M'hamdi, K.B., Tajayouti, M., et al.: Parametric analysis of the use of PCM in building energy performance. In: Springer Proceeding: EMCEI Conference (2016)
6. Roberti, F., Oberegger, U.F., Lucchi, E., et al.: Energy retrofit and conservation of a historic building using multi-objective optimization and an analytic hierarchy process. Energy Build. **138**, 1–10 (2017). https://doi.org/10.1016/j.enbuild.2016.12.028
7. Saffari, M., de Gracia, A., Fernández, C., et al.: Simulation-based optimization of PCM melting temperature to improve the energy performance in buildings. Appl. Energy **202**, 420–434 (2017). https://doi.org/10.1016/j.apenergy.2017.05.107
8. Wu, R., Mavromatidis, G., Orehounig, K., Carmeliet, J.: Multiobjective optimisation of energy systems and building envelope retrofit in a residential community. Appl. Energy **190**, 634–649 (2017). https://doi.org/10.1016/j.apenergy.2016.12.161
9. Wang, D., Ding, F., Chu, Y.: Data filtering based recursive least squares algorithm for Hammerstein systems using the key-term separation principle. Inform. Sci. **222**, 203–212 (2013). https://doi.org/10.1016/j.ins.2012.07.064
10. Shen, Y., et al.: Experimental thermal study of a new PCM-concrete thermal storage block (PCM-CTSB). Constr. Build. Mater. **293**, 123540 (2021). https://doi.org/10.1016/j.conbuildmat.2021.123540
11. Cunha, S., Aguiar, J.B., Tadeu, A.: Thermal performance and cost analysis of mortars made with PCM and different binders. Constr. Build. Mater. **122**, 637–648 (2016). https://doi.org/10.1016/j.conbuildmat.2016.06.114
12. Yun, B.Y., Park, J.H., Yang, S., Wi, S., Kim, S.: Integrated analysis of the energy and economic efficiency of PCM as an indoor decoration element: application to an apartment building. Sol. Energy **196**, 437–447 (2020). https://doi.org/10.1016/j.solener.2019.12.006

Liquid–Liquid Extraction Studies of Heavy Metals (Cd, Cr and Zn) from Phosphoric Acid Solutions Using $C_{11}H_{18}N_2O$ as a Synthetic Agent

Kaoutar Berkalou[1,3,4](✉), Abderrahman Nounah[1], Mohamed Khamar[1], Essediya Cherkaoui[1], Nisrine Boughou[1], and Ratiba Boussen[2]

[1] Civil Engineering and Environment Laboratory (LGCE), Materials, Water and Environment Team, Higher School of Technology of Salé, Mohammadia School of Engineering, Mohammed V University of Rabat, Rabat, Morocco
Kaoutar.berkalou@lecnam.net

[2] Laboratory of Nanotechnology Materials and Environment, Faculty of Sciences of Rabat, Mohammed V University of Rabat, Rabat, Morocco

[3] CNAM, 75003 Paris, France

[4] Paris-Saclay University, INRAE, AgroParisTech, UMR SayFood, 91300 Massy, France

Abstract. Many researches have investigated the removal of cadmium as an impurity from phosphoric acid using various commercial and synthetic extracting agents. In this paper, the reaction mechanism of liquid-liquid extraction process of heavy metals like cadmium from the commercial phosphoric acid was studied. The extraction is carried out using the synthetic extracting agent hexahydrotrimethylquinazoline-2one ($C_{11}H_{18}N_2O$) diluted in benzene. Afterwards, a feasibility test of this synthesized extracting agent has been done to extract other metals like zinc and chromium from industrial phosphoric acid produced by hydrometallurgical way (wet process). The results showed that the reaction of liquid-liquid extraction process of cadmium from phosphoric acid solution by using $C_{11}H_{18}N_2O$ as a extracting agent can be describe at equilibrium by: $Cd(H_2PO_4)_{2aq} + (HA)_{2org} \rightleftharpoons Cd(H_2PO_4)_2(HA)_{2org}$. The equilibrium constants were found to be $\log(K) = 1.38$. By using industrial phosphoric acid, the extracting agent revealed its ability to extract cadmium, chromium and zinc with the extraction percentages of 77.6, 66.5 and 42.7% respectively.

Keywords: Reaction mechanism · Liquid–liquid extraction · Phosphoric acid · Synthetic extracting agents

1 Introduction

Many research have examined extraction process of heavy metals from phosphoric acid, by using synthetic or commercial extractants [1–5]. The organic compound Kelex100 has been used as extracting agent to extract several heavy metals like cadmium (Cd), zinc (Zn) and chromium (Cr) from a concentrate solution of phosphoric acid [6]. In those trials, addition of 10% of n-decanol enhanced the metals extraction rate of 60% and

K. Baba et al. (Eds.): GeoME 2023, SPEES, pp. 117–125, 2023.
https://doi.org/10.1007/978-3-031-49345-4_12

the equilibrium time goes from 24 to 0.5 h. Tributylphosphate diluted in hydrocarbon liquid was used as extracting agent to extract the same metals (Zn, Cd and Cr) from phosphoric acid [7]. In our previous works [8–10], a modeling and optimization of cadmium extraction from a solution of phosphoric acid has been carried out in batch reactor. The synthetic extractant hexahydroquinazolin-2-one diluted in benzene (C_6H_6) was used. The percentage of cadmium extraction achieved is 98% in the following operating conditions: phosphoric acid concentration $[H_3PO_4] = 2.5M$, extracting agent concentration $[EA] = 1.10^{-2}$ M, pH = 3, O/A = 1/1, agitation speed 800 rpm and time of 90 min.

Persistence of this work consists first, to understand the reaction mechanism of liquid-liquid cadmium extraction and then, to test the extraction performance for cadmium and others metals like chromium and zinc in an industrial acid. The difficulty of industrial phosphoric acid is manifested by the presence of impurities that can affect the extraction process. The extraction mechanism of different extractants and the nature of the cadmium complexes have been investigated in several studies. D2EHPA was used as commercial extractant agent to determine the stoichiometric relation extraction of this metal [11]. The reaction obtained by D2EHPA was: $Cd^{2+}_{aq} + \frac{5}{4}(H_2A_2)_{org} \rightarrow CdA_2(HA)_{\frac{1}{2}\,org} + 2H^+_{aq}$ with an equilibrium constant equal to 3.50×10^{-4} mol$^{3/4}$ L$^{-3/4}$. In another study, 3-methyl-quinoxaline-2-thione (LH) was tested as extractant synthetic agent dilueted in toluene [12]. In general, this system of extraction that involved Cd–H_3PO_4, Cd–LH complexes, H^+ and OH^- exchange is a complicated process [13] and such examination has not been done before by using hexahydroquinazolin-2-one ($C_{11}H_{18}N_2O$) as an extractant.

This paper is divided in four sections: Sect. 2 introduces reagents and solutions used in this study and the operating conditions of extraction. In Sect. 3 is divided in two parts. The first one is dedicated to the study of the extraction reaction. The last part, shows the results of metal extraction (Cd, Zn and Cr) by the extracting synthesized agent from industrial phosphoric acid. The conclusion of this study is represented in the last section.

2 Materials and Methods

2.1 Reagents and Solutions

The aqueous phase used for the study of the reaction mechanism is constituted by a solution of commercial phosphoric acid of concentration 2.5 molar [VWR CHEMICALS, (85%)] which contains a cadmium concentration of 10^{-3} mol/l.

Four different concentrations of the extracting agent Hexahydrotrimethylquinazoline-2one ($C_{11}H_{18}N_2O$): 0.5×10^{-2} M, 1.10^{-2} M, 1.5×10^{-2} M and 2×10^{-2} M diluted in benzene (C_6H_6) (Riedel-dehën) were dissolved to prepare the organic phase.

Industrial phosphoric acid produced by hydrometallurgical process (density = 1.62 at room temperature), containing 17.7, 70.66, 27.87 mg/l of cadmium, chromium and zinc respectively.

2.2 Extraction Process

Trials were conducted in a reactor of 100 ml with the same volume of aqueous ans organic phases, under stirring. The mixture was separated after 120 min. Metals concentrations was mesured by using inductively coupled plasma spectrometer (ICP-OES). A small amount of NaOH is added to the solutions to change the pH to the desired value. A pH meter was used to measure the pH (model JENWAY 3520 pH Meter).

3 Results and Discussions

3.1 Reaction Mechanism

3.1.1 Identification of Extracted Metal–Organic Complexes Nature

Numerous investigations have suggested an extraction reaction mechanism [11, 14–16], considering the following presumptions: In the aqueous phase, extractant solubility is minimal, the same case for metal complex. Metals recovered are not related to one another, the organic phase contains the extractants as dimers $(HA)_2$. Metal (N) ion extraction can be represented by the reaction in Eq. (1):

$$N^{i+}_{aq} + \left(\frac{i+j}{2}\right)(HA)_{2org} \rightarrow NA_n(HA)_{porg} + iH^+_{aq} \tag{1}$$

i valence of the metal ion, j number of extractant molecules.

The reaction constant K, can be determined throughout the reaction in Eq. 2:

$$K = \frac{\left[NA_i(HA)_j\right]_{org}\left[H^+\right]^i_{aq}}{\left[N^{i+}\right]_{aq}[(HA)_2]^{\frac{i+j}{2}}_{org}} \tag{2}$$

The distribution factor D is defined in Eq. (3).

$$D = \frac{\left[NA_i(HA)_j\right]_{org}}{\left[N^{i+}\right]_{aq}} \tag{3}$$

By replacing Eqs. 2 in 3, distribution factor D becomes:

$$D = \frac{K[(HA)_2]^{\frac{i+j}{2}}_{org}}{\left[H^+\right]^i_{aq}} \tag{4}$$

Using logarithms in Eq. (4), we obtain:

$$\log D = \log K + \left(\frac{i+j}{2}\right)\log[(HA)_2]_{org} - i\log\left[H^+\right] \tag{5}$$

$$\log D = \log K + \left(\frac{i+j}{2}\right)\log[(HA)_2]_{org} + ipH \tag{6}$$

Additionally, graphing log D against $\log[(HA)_2]_{org}$ gives a slope of $\frac{i+j}{2}$.

$$\left[\frac{\partial \log D}{\partial pH}\right] = i \text{ constant } [(HA)_2]_{org} \tag{7}$$

j value can be obtained, when during the extraction process the equilibrium pH does not change.

Reaction stoichiometry is obtained by replacing i and j in Eq. (1). K may deduced from the Fig. 3.

Figures 1 and 2 show link among $\log D$ and equilibrium pH as function of extractant concentration. Equation (7) describe link among $\log D$ and pH. The equilibrium pH is between 1.0 ± 0.2 and 5.0 ± 0.2. $i = \frac{1}{5}$ independently of the extractant concentration. This value is considered zero, because according to the experimental measurements carried out no variation in pH was noted before and after extraction of the cadmium.

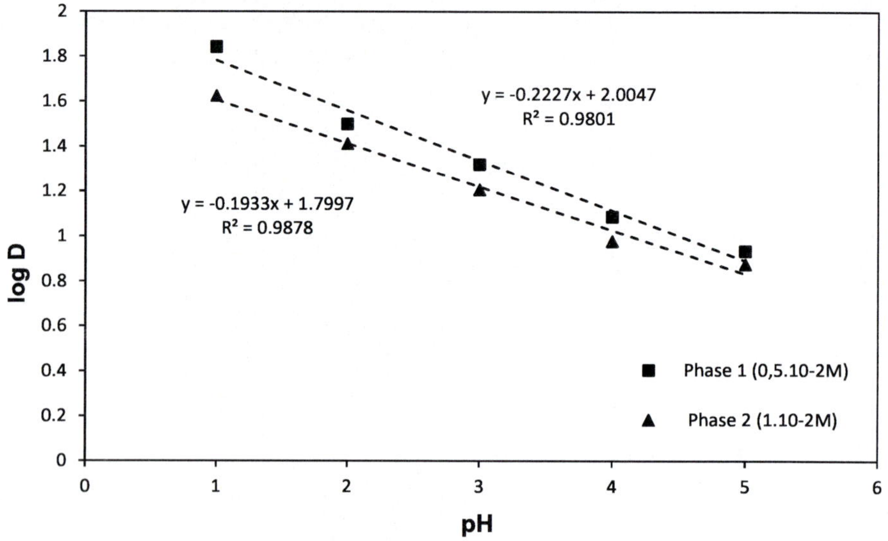

Fig. 1. pH effect on the distribution coefficient of cadmium, with $[H_3PO_4] = 2.5$ M, $[C_{11}H_{18}N_2O] = 0.5 \times 10^{-2}$ M and 10^{-2} M, $[cd^{2+}] = 10^{-3}$ M, O/A = 1/1 and T = 20 °C.

Based on the graph of $\log D$ against $\log[\mathit{extractant}]$, at $pH = 1.8 \pm 0.2$ (Fig. 3), $\frac{i+j}{2} = 1$ and the value of j become $1.8 \approx 2$.

Based on the results above the reaction process is written in Eq. (8):

$$Cd(H_2PO_4)_{2(aq)} + (HA)_{2org} \rightleftharpoons Cd(H_2PO_4)_2(HA)_{2org} \tag{8}$$

K deduced from Fig. 3 is equal to:

$$\log(K) = 1.38 \tag{9}$$

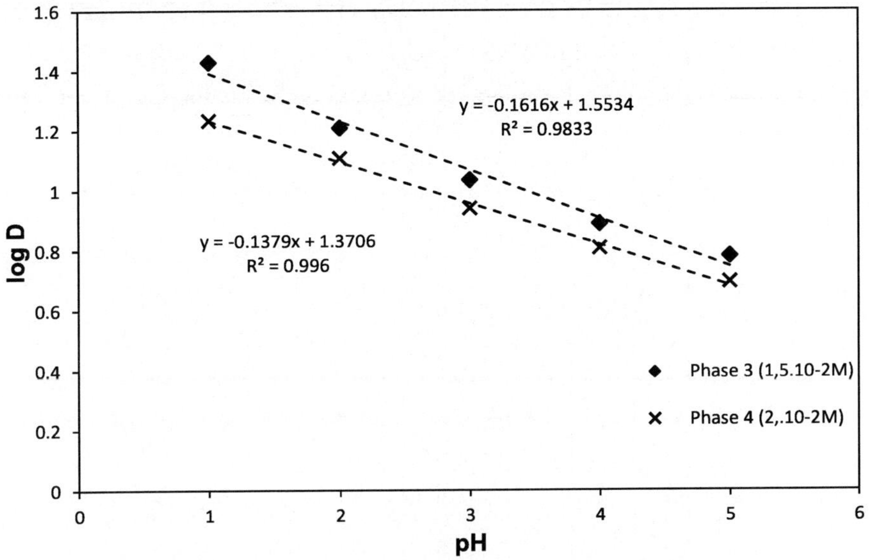

Fig. 2. pH effect on the distribution coefficient of cadmium, with $[H_3PO_4] = 2.5$ M, $[C_{11}H_{18}N_2O] = 1.5 \times 10^{-2}$ M 2×10^{-2} M, $[cd^{2+}] = 10^{-3}$ M, O/A = 1/1 and T = 20 °C.

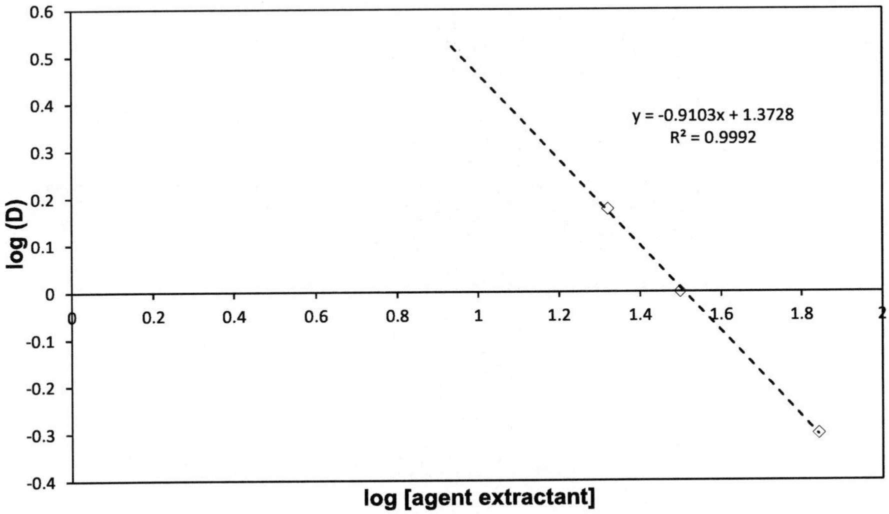

Fig. 3. Variation of $(C_{11}H_{18}N_2O)$ concentration on distribution factor, with $[H_3PO_4] = 2.5$ M, $[cd^{2+}] = 10^{-3}$ M, O/A = 1/1 and T = 20 °C.

3.2 Extraction of Cadmium (Cd), Chrome (Cr) and Zinc (Zn) of H_3PO_4 Industrial

The extractant ($C_{11}H_{18}N_2O$) approved its efficiency in extracting Cd of commercial H_3PO_4, next work is to test the feasibility of $C_{11}H_{18}N_2O$ to extract cadmium from industrial H_3PO_4 obtained through hydrometallurgical process, complexity of this medium manifests by the presence of other impurities that can influence the extractive efficiency of $C_{11}H_{18}N_2O$.

Extraction of Cd, Cr and Zn of industrial H_3PO_4 has been tested for different concentrations of extractant $C_{11}H_{18}N_2O$ vary from 0.5×10^{-2} M to 2.5×10^{-2} M. The extraction evolution is shown in the Figs. 4, 5 and 6.

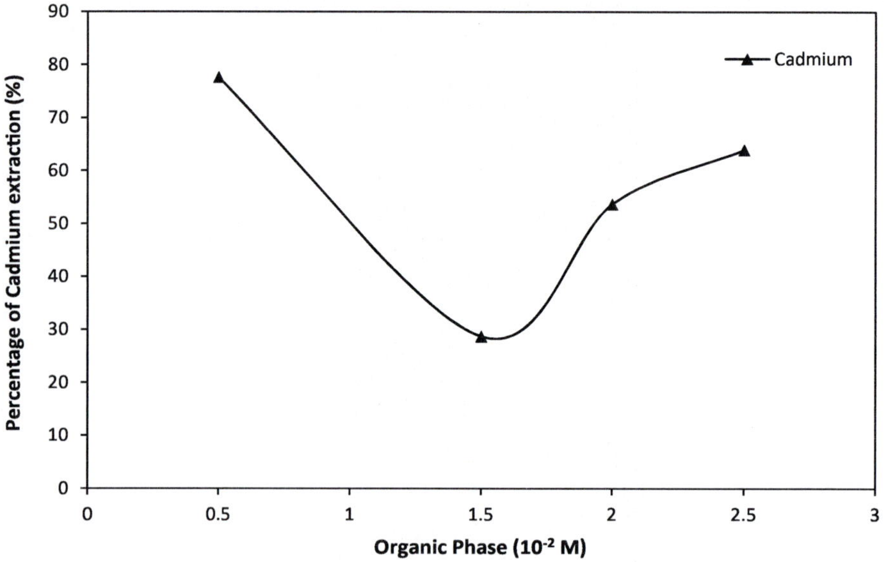

Fig. 4. Evolution of $C_{11}H_{18}N_2O$ concentration on Cd extraction of H_3PO_4 industrial, with O/A = 1/1 and T = 20 °C.

Based on Fig. 4, a non-uniform evolution of Cd extraction of H_3PO_4 industrial was found. This is due to the complexity of the industrial phosphoric acid solution which contains several elements such as chromium and zinc (Figs. 5 and 6) which can influence the process of extraction.

However, cadmium extraction reached 77.6% by using a concentration of 0.5×10^{-2} M of the extractant ($C_{11}H_{18}N_2O$). This percentage decreased to 28.7% at a concentration of 1.5×10^{-2} M of the extractant. Afterwards the percentage of extraction increased to 63.9% at 2.5×10^{-2} M of extractant.

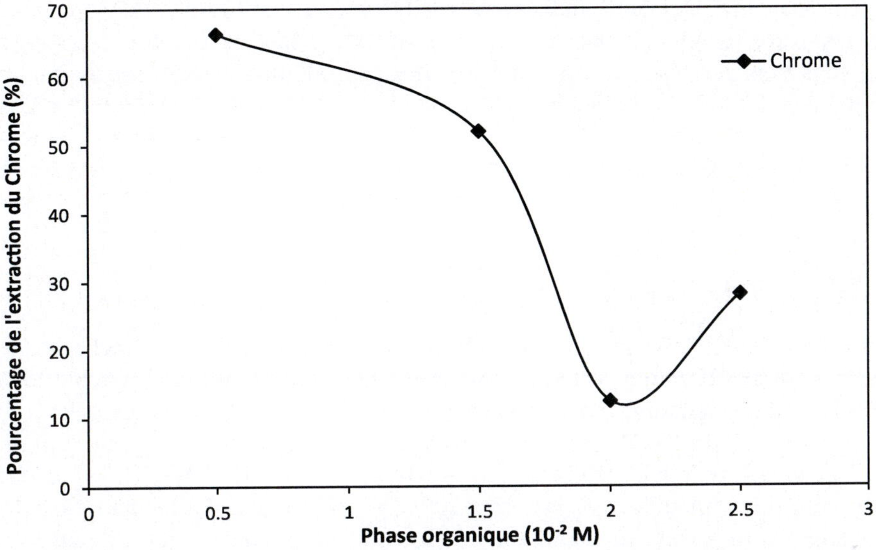

Fig. 5. Evolution of $C_{11}H_{18}N_2O$ concentration on Cr extraction of H_3PO_4 industrial, with O/A = 1/1 and T = 20 °C.

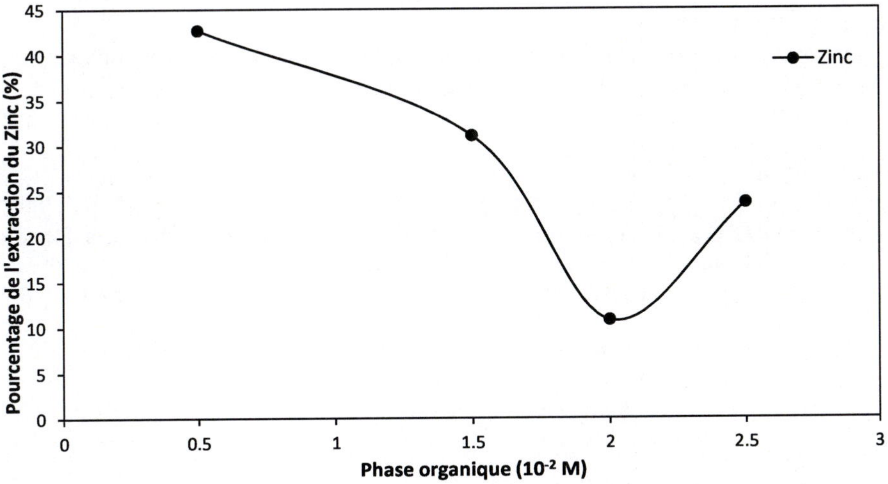

Fig. 6. Evolution of $C_{11}H_{18}N_2O$ concentration on Zn extraction of H_3PO_4 industrial, with O/A = 1/1 and T = 20 °C.

The extractant agent showed its efficiency to extract cadmium and its ability to extract two other metals like Chromium and Zinc (Figs. 5 and 6), with an extraction percentage of Chromium of 66.5, and 42.7% of Zinc under an organic/aqueous phases ratio of (1/1), room temperature, with an extracting agent concentration of 0.5×10^{-2} M stirred during 120 min.

However, in the paper of Mellah et al. [6] the use of Kelex100, allows to obtain a low rate compared to our study. 58% of Zn, 34% of Cr and 15% of Cd have been extracted at organic/aqueous phases ratio of 1/1 in 4 h [6]. In addition, to optimise their results, the organic phase has received a modification reagent (decanol). Therefore, extraction rate becomes 83% for Zn, 80% for Cr and 71% for Cd in half hour compared to 4 h. Modification of reagent will also be studied in future works to improve our extraction results.

4 Conclusion

Hexahydrotrimethylquinazoline-2one ($C_{11}H_{18}N_2O$) diluted in benzene has been used in this study as an extracting agent to extract cadmium, chrome and zinc from phosphoric acid. The conclusions are: extraction reaction of 10^{-3} mol/l of Cd from 2.5 M of H_3PO_4 by used the $C_{11}H_{18}N_2O$ diluted in benzene could be represented at equilibrium by the following reaction: $Cd(H_2PO_4)_{2(aq)} + (HA)_{2org} \rightleftharpoons Cd(H_2PO_4)_2(HA)_{2org}$. With a equilibrium constant of $\log(K) = 1.38$ at 20 °C, an organic/aqueous phases ratio of (1/1) with an extracting agent concentration of 0.5×10^{-2} M stirred during 120 min.

The same extractant was used this time to extract chromium and zinc in addition to cadmium from industrial phosphoric acid in the same conditions. The results of this second study show a relatively good extraction yield for cadmium and chromium with an extraction percentage of 77.6 and 66.5% respectively and a lower percentage for zinc extract of 42.7%.

In order to optimize all parameters that influence the percentage of extraction of these metals, the extraction kinetics will be studied in the next investigations.

References

1. Beltrami, D., Cote, G., Mokhtari, H., Courtaud, B., Chagnes, A.: Modeling of the extraction of Uranium(VI) from concentrated phosphoric acid by synergistic mixtures of bis-(2-ethylhexyl)-phosphoric acid and tri-n-octylphosphine oxide. Hydrometallurgy **15**, 129–130 (2012)
2. Bidari, E., Irannajad, M., Gharabaghi, M.: Investigation of the influence of acetate ions on cadmium extraction with D2EHPA. Hydrometallurgy **144–145**, 129–132 (2014)
3. Daryabor, M., Ahmadi, A., Zilouei, H.: Solvent extraction of cadmium and zinc from sulphate solutions: comparison of mechanical agitation and ultrasonic irradiation. Ultrason. Sonochem. **34**, 931–937 (2017)
4. Daryani, M., Jodeiri, N., Fatehifar, E., Shahbazi, J.: Optimization of operating conditions in purification of wet process phosphoric acid in a liquid-liquid extraction column. Chem. Eng. Commun. **209**, 1082–1095 (2022)
5. Ocio, A., Almela, A., Elizalde, M.P.: Cadmium(II) extraction from phosphoric media by bis(2,4,4-Trimethylpentyl)dithiophosphinic acid (CYANEX 301). Solvent Extr. Ion Exch. **22**, 961–977 (2004)
6. Mellah, A., Benachour, D.: Solvent extraction of heavy metals contained in phosphoric acid solutions by 7-(4-ethyl-1-methyloctyl)-8-hydroxyquinoline in kerosene diluent. Hydrometallurgy **81**, 100–103 (2006)

7. Mellah, A., Benachour, D.: The solvent extraction of zinc, cadmium and chromium from phosphoric acid solutions by tri-n butyl phosphate in kerosene diluent. Sep. Purif. Technol. **56**, 220–224 (2007)
8. Berkalou, K., Nounah, A., Khamar, M., Boussen, R., Cherkaoui, E.: Extraction of cadmium from phosphoric acid by a synthesized extracting agent. E3S Web Conf. **150**, 02007 (2020)
9. Berkalou, K., Nounah, A., Khamar, M., Cherkaoui, E., Boussen, R.: Removal of cadmium from phosphoric acid by the liquid-liquid extraction process using synthetic agent. Inter. J. Adv. Res. Eng. Tech. **11–16**, 28–35 (2020)
10. Berkalou, K., Nounah, A., Chaair, H., Khamar, M., Boussen, R., Driouich, A.: Modelling and optimizing the solvent extraction ofcadmium from phosphoric acid using experimental design. Asian J. Chem. **33**, 637–643 (2021)
11. Mellah, A., Benachour, D.: The solvent extraction of zinc and cadmium from phosphoric acid solution by di-2-ethyl hexyl phosphoric acid in kerosene diluent. Chem. Eng. Process. Process Intensif. **45**, 684–690 (2006)
12. Senhaji, S., Elyahyaoui, A., Boulassa, S., Essassi, E.M.: Liquid-liquid extraction of cadmium by 3-methyl-quinoxaline-2-thione from phosphoric medium. Inter. J. Sci. Res. **5**, 2277–8179 (2016)
13. Senhaji, S., Elyahyaoui, A., Boulassa, S., Essassi, E.M.: The solvent extraction of cadmium from phosphoric acid solution by3-methyl-quinoxaline-2-thione in toluene diluent. Orient. J. Chem. **32**, 3035–3041 (2016)
14. Mansur, M.B., Slater, M.J., Biscaia, E.C.: Kinetic analysis of the reactive liquid–liquid test system ZnSO4/D2EHPA/n-heptane. Hydrometallurgy **63**, 107–116 (2002)
15. Kumar, V., Kumar, M., Jha, M.K., Jeong, J., Lee, J.: Solvent extraction of cadmium from sulfate solution with di-(2-ethylhexyl) phosphoric acid diluted in kerosene. Hydrometallurgy **96**, 230–234 (2009)
16. Mansur, M.B., Slater, M.J., Biscaia, E.C.: Equilibrium analysis of the reactive liquid–liquid test system ZnSO4/D2EHPA/n-heptane. Hydrometallurgy **63**, 117–126 (2002)

Unveiling the Physicochemical Characteristics of Organic Materials for Composting: Unleashing Their Potential for Sustainable Waste Management in Morocco

Chadia Majdouline[1,2](✉), Mohamed Khamar[1,2], Mounaim Halim El Jalil[1,2], Essediya Cherkaoui[1,2], and Abdelmjid Zouahri[3]

[1] Civil Engineering and Environment Laboratory (LGCE), Materials Water and Environment Team, Mohammed V University in Rabat, High School of Technology (ESTS), MA11060 Sale, Morocco
chadia_majdouline@um5.ac.ma, chadiamajdouline@research.emi.ac.ma
[2] Doctoral Studies Center, Engineering Sciences and Techniques, Mohammadia School of Engineers (EMI), Mohammed V University in Rabat, Rabat, Morocco
[3] Research Unit On Environment and Conservation of Natural Resources, Regional Center of Rabat, National Institute of Agricultural Research (INRA), BP 6570, Rabat, Morocco

Abstract. This study explores the composting potential of organic materials, including cow and poultry manure, sawdust, garden grass, and household waste, with the primary objective of waste valorization. Thorough physicochemical analyses were performed to evaluate essential parameters of the raw materials. The composting process was closely monitored, and the resulting compost product underwent assessment for nutrient content and its suitability for agricultural and horticultural applications. The results demonstrate that all raw materials exhibit highly favorable characteristics for composting process. Composting these organic materials offers significant waste reduction, diverting substantial amounts from landfills and mitigating methane emissions, a potent greenhouse gas. Moreover, the composting process yields nutrient-rich compost while considerably reducing waste volume and weight. These compelling findings underscore the robust viability and sustainability of composting as a highly efficient waste management approach in Morocco, utilizing locally available organic waste to produce valuable soil amendments. The retrieval of nutrients and organic matter through composting profoundly improves soil health, delivering significant benefits to agriculture and the broader environment. This investigation underscores the importance of the proper characterization and selection of raw materials for composting to ensure their safe and effective use in agriculture and horticulture. By harnessing locally abundant organic resources, this study proposes a practical and sustainable solution to address environmental challenges associated with waste disposal and promote soil health. Ultimately, this research contributes to the advancement of sustainable agriculture practices in Morocco and beyond.

Keywords: Composting · Organic waste · Waste management · Physicochemical characteristics · Valorization

K. Baba et al. (Eds.): GeoME 2023, SPEES, pp. 126–136, 2023.
https://doi.org/10.1007/978-3-031-49345-4_13

1 Introduction

Waste management is a critical global issue, especially in developing countries, where factors such as rapid population growth, changing consumption patterns, and inefficient waste management systems contribute to the escalating volume of waste generated [1]. Projections indicate that by 2030, the annual generation of municipal solid waste is estimated at 2.59 billion tonnes, with a further increase of 70% to 3.081 billion tonnes by 2050 [2, 3]. In Morocco, waste management poses significant challenges. In 2019, the country generated 6.7 million tonnes of waste categorized as household and similar, with urban areas making up 79% of the total, in comparison to 4.9 million tonnes/year in 2017 [4]. Researchers in [5] examined the issue of household waste in Morocco and discussed the potential advantages of implementing a highly efficient waste management approach. They also emphasized the large amounts of organic waste that could be composted, benefitting **soil**, vegetable production, and personal health related to physical activity. On the other hand, animal husbandry, such as poultry and cows, significantly contributes to Morocco's agricultural output, representing ~ 25–30% [6]. Consequently, the management of the substantial amount of animal waste generated becomes crucial. Composting offers a promising solution for valorizing organic livestock wastes, along with other organic materials, such as organic household waste, sawdust and grass clippings. Grasslands, covering about 30–40% of the global land surface, have traditionally served as animal feed [7]. However, they can also be processed for waste management purposes, including composting, due to their composition suitable for material recovery. Sawdust, a lignocellulosic by-product from the furniture and wood industry, is readily available and rich in carbon content [8]. It can also help balance the C:N (carbon to nitrogen) ratio in the composting mix. However, the improper management of organic wastes in Morocco, with over 95% being either disposed of in the environment or used without proper pretreatment, poses significant risks to both human health and the environment [9]. Achieving sustainable services requires innovative, cost-effective and locally appropriate technologies and management approaches [5]; In fact, inadequate waste management leads to pollution of air, soil and water, contributing to climate change indirectly [10]. Effectively processed compost derived from animal waste and other organic refuse holds immense potential as a valuable nutrient resource for agricultural fields, simultaneously mitigating environmental impact. To ensure effective management and recovery of organic wastes, understanding their physicochemical properties is essential, enabling the implementation of appropriate treatment techniques [11]. In this investigation, a range of experimental methods were employed to evaluate key parameters, such as electrical conductivity (EC), pH, moisture content (MC), organic matter (OM), as well as the analysis of essential macronutrients and microelements. Intensive evaluations were undertaken to acquire a profound comprehension of the sample characteristics.

Composting has been identified as a potential treatment method for agricultural waste, offering the stabilization of horticultural wastes and enhancing their properties as organic fertilizers [12, 13]. Proper waste management through composting not only restores organic matter in depleted soils, but also contributes to reducing greenhouse gas emissions, promoting sustainable agricultural practices. This study explores the potential of composting as a viable solution for managing organic waste in Morocco. It offers valuable insights into the benefits of organic wastes and serves as a guide for developing effective and sustainable waste management strategies, aligning with the overarching goal of achieving efficient and environmentally friendly waste practices. Specifically, the focus of this research is to evaluate the physical, chemical, and characteristic properties of the most commonly encountered organic wastes in Morocco, namely cow and poultry manure, sawdust, garden grass, and household waste. These organic solid wastes pose challenges such as odor, land occupation, and environmental contamination, necessitating proper handling. Hence, it is crucial to develop a technology that not only effectively manages and treats these wastes, but also generates valuable products. This investigation highlights composting as a promising approach to mitigate the environmental impact and cost associated with waste management while simultaneously producing compost that enhances soil health and fertility. Moreover, this study not only conducts a thorough evaluation of composting's appropriateness for organic wastes but also tackles the escalating need for sustainable waste management methods, especially in urban regions. The results of this research hold significant value for policymakers, waste management professionals, and community members, equipping them with crucial insights to make well-informed decisions concerning sustainable waste management practices.

2 Materials and Methods

The research was conducted in collaboration between the Civil Engineering and Environmental Laboratory at the High School of Technology in Sale and the Environmental Laboratory of the National Agricultural Research Institute in Rabat. The objective of the study was to assess the composting potential of organic waste from different sources, including organic household waste from a housing estate in Sale city, cow and poultry manure collected from commercial holdings and a dairy farm in the suburbs of Sale city, sawdust from a sawmill near the studied residence, and garden grass from the garden of the higher school of technology in Sale city. The waste collected is presented in Fig. 1. To ensure the purity of the samples, the waste was sorted into an organic waste category and transported to the laboratory where it underwent further treatment to remove contaminants. These steps helped to achieve the most accurate and reliable results possible, providing valuable insights into the composting potential of these organic materials.

2.1 The Analytical Methods

In order to guarantee reliable results, we implemented a pretreatment process to assess the physico-chemical characteristics of the studied organic waste. Initially, the samples underwent a thorough mixing process, followed by drying in an oven at 65 °C for 24 h (hours). Subsequently, the samples were ground and sieved resulting in a particle size of 2 mm for the analysis of the remaining parameters and 0.2 mm specifically for determining total carbon, nitrogen, organic matter, macronutrients and microelements. For pH measurement, a pH meter was utilized, while a conductivity meter was employed to measure the electrical conductivity (EC); The aqueous extract was used to measure pH and EC using a ratio of 2:5 (w/v) and 1:5 (w/v), respectively. The moisture content was measured as a percentage by subjecting the sample to drying at 105 °C for 24 h until a constant mass was attained. Additionally, organic matter was calculated through a 3-h calcination process at 525 °C.The total nitrogen (NTK) was evaluated as a percentage using the method of Kjeldahl, while the total organic carbon (TOC) was evaluated using the method of volatile solids, employing loss on ignition at 550 °C for 2 h. The levels of microelements (Pb, Fe, Cd, Zn, Cr, Ni, Cu, Mn) in (mg kg^{-1}) were measured using atomic absorption spectrometry, while macronutrients (P_2O_5, Ca, Na, K_2O and Mg) in (g kg^{-1}) were measured in (g kg^{-1}) using a flame photometer.

3 Results and Discussion

The physicochemical characterization data of different organic materials, including cow manure, sawdust, poultry manure, grass clippings, and OHW (Organic household waste) are presented in Table 1.

The ratio of carbon to nitrogen (C:N) is a crucial indicator for assessing the decomposition of organic matter. Higher C:N ratios are associated with slower decomposition processes. In this study, various materials were analyzed, and poultry manure exhibited the lowest C:N ratio of 9, indicating advanced decomposition. On the other hand, sawdust had the highest C:N ratio of 140, suggesting a less decomposed state. Recent research demonstrated that the optimal C:N ratio for microbial activity is ~ 20–40 [14–17]. Consequently, materials with C:N ratios outside this range may not provide an ideal environment for microbial activity unless appropriately adjusted and balanced within the required range.

Moisture content plays a crucial role in determining the suitability of organic materials for composting. In particular, the moisture content of different raw materials can significantly impact the composting process and the overall quality of the final compost product. For instance, sawdust exhibits a considerably lower moisture content of 9.78%, while organic household waste (OHW) has the highest moisture content of 84.6%. These variations in moisture values among the raw materials can have implications for the composting process. High moisture content, as observed in OHW, can create a wet and anaerobic environment, leading to potential odor issues and a slower composting process. On the other hand, lower moisture content, like that of sawdust, may require additional moisture supplementation to ensure optimal microbial activity and decomposition. Thus, maintaining an appropriate moisture level is crucial for successful composting. The ideal moisture content level for biodegradation can significantly vary based on the compost mixture and composting duration. Some research suggests that the ideal moisture content typically ranges from about 50–70% on a wet basis [18, 19]. Nevertheless, surpassing a moisture level of 70% presents the risk of water movement, resulting in anaerobic conditions and undesirable odors. Additionally, elevated moisture levels can excessively reduce the available oxygen for worms, potentially leading to their demise. To meet the standards for initiating composting, additional compost feedstock, bulking agents, or co-substrates may be required due to specific factors such as high moisture levels and a low C:N ratio [5]. Considering the importance of sawdust as a carbon-rich material is essential for ensuring a successful composting process; Sawdust can help achieve the appropriate C:N ratio and absorb excess moisture present in other wet raw materials under investigation. As an optimal catalyst and foundation for initiating the composting process, sawdust plays a crucial role in maintaining the overall balance and effectiveness of the composting system. Sawdust, with its low moisture content and carbon-rich nature, can help regulate moisture levels and achieve the desired C:N ratio.

The presence of organic matter serves as a crucial carbon and nutrient source for microbial activity, which plays a vital role in decomposition and the production of compost. It is worth emphasizing that the organic matter content significantly impacts the composting process, and ultimately affects the quality of the final compost product. A higher organic matter content generally signifies a greater potential for the formation of nutrient-rich compost. In our study, the raw materials exhibited higher organic matter content, ranging between 39.87 and 62.23%. These findings underscore the substantial variation in organic matter content across different raw materials, highlighting its critical role in the composting process.

Electrical conductivity (EC) is a measure of soluble salt concentration in organic materials [20]. Higher EC values indicate a higher concentration of soluble salts. In this study, grass clippings exhibited the highest electrical conductivity (4 dS m^{-1}), indicating a high salt concentration, while sawdust showed the lowest electrical conductivity (1.68 dS m^{-1}). Studies such as [21] have found a direct relationship between EC and phytotoxicity, suggesting that elevated EC levels may have inhibitory effects on plants and soil. The pH values provided in the table indicate the acidity or alkalinity of the raw materials. Cow manure displayed a slightly alkaline pH of 7.67, sawdust had a neutral pH of 6, poultry manure had a pH of 6.21, grass clippings were more acidic with a pH of 4.91, and organic household waste had a pH of 5.40. Throughout the composting process, an acidic pH fosters the growth of bacteria and fungi, while a basic pH promotes the proliferation of actinomycetes and alkaline bacteria [22]. The optimal pH range for composting is typically reported to be between 5.5 and 7.5 [23, 24]. The variations in electrical conductivity and pH among the different raw materials emphasize the importance of considering their composition and properties when designing composting systems. Adjustments may be necessary to optimize composting conditions, such as regulating moisture levels or incorporating amendments to neutralize acidity or alkalinity.

The concentrations of microelements and macronutrients vary across the different organic materials. Cow manure, for example, exhibits relatively high levels of manganese and zinc, which can contribute to the nutrient content of the compost. Sawdust has elevated nickel concentrations, while poultry manure contains notable amounts of chromium. Grass clippings exhibit higher copper content, which can be beneficial for plant growth. Among the macronutrients, poultry manure stands out with higher levels of potassium, calcium and phosphorus. This makes poultry manure a valuable nutrient source for composting. Grass clippings, on the other hand, exhibit higher magnesium content. Understanding these variations in nutrient composition is important for designing composting systems that optimize nutrient levels and enhance compost effectiveness. By tailoring composting practices to specific nutrient profiles, it becomes possible to create compost that is more beneficial for plant growth and soil health. While microelements are required in smaller quantities, they still play crucial roles in physiological processes and can influence compost quality, albeit in limited amounts. Compost quality standards, such as those set by Canadian and American directives [25, 26], recognize the significance of micronutrients. However, data on microelement levels in composts are often scarce in many African countries, highlighting the need for more research in this field. It is important to note that excessive concentrations of microelements beyond quality standards can lead to pollution hazards and environmental concerns. In fact, understanding the nutrient composition of organic materials used in composting is crucial for designing effective composting systems. By considering the specific macro and micronutrient profiles, nutrient levels can be optimized, and compost can be tailored to enhance plant growth and soil health. It is important to adhere to compost quality standards to avoid excessive concentrations of micronutrients, which can lead to pollution concerns. Further research is needed, particularly in African countries, to gather more data on microelement levels in composts and ensure environmentally sound composting practices.

Fig. 1. Raw materials investigated

The physicochemical characterization of these organic materials can provide useful information for composting and soil improvement purposes. The differences in the parameters provide an insight into the material for a particular application. Table 2 provides valuable insights for composting and soil improvement purposes. The table displays the initial and final physicochemical properties of the organic mixture compost, offering a comprehensive view of its transformation during 90 days of the composting process in compost container with holes for drainage as shown in Fig. 2.

During the composting process, the temperature initially started at 18 °C and gradually rose to reach the ambient temperature of 25 °C. This increase in temperature served as evidence of ongoing microbial activity. The C:N ratio decreased from 23 in the preliminary analysis to 15.09 in the maturity analysis. A decrease in the C:N ratio suggests an increase in the decomposition and mineralization of organic matter. Moisture content decreased significantly from 61.21 to 17.87%. This reduction indicates that moisture was lost during composting, likely due to evaporation. The organic matter content decreased from 62.73 to 59.88% during the composting process. This decrease indicates the breakdown and decomposition of organic materials, which can contribute

Table 1. Physical-chemical characterization of the raw materials intended to be composted

Sample	Cow manure	Sawdust	Poultry manure	Grass clippings	OHW
C:N	38.1	140	9	23.6	20.2
Moisture (%)	45	9.78	68.56	58.8	84.6
Electrical conductivity (dS m^{-1})	3.12	1.68	2.9	4	3.16
pH	7.67	6	6.21	4.91	5.40
Organic matter (%)	39.87	47.08	36.34	40.4	62.23
Manganese (mg kg^{-1})	165.89	142.8	155.54	89	163.1
Zinc (mg kg^{-1})	189.4	133.98	21.98	4.57	177.9
Nickel (mg kg^{-1})	22.8	279.09	9.86	68.9	44.81
Cadmium (mg kg^{-1})	58	70.32	2	3.01	67.3
Chromium (mg kg^{-1})	144.1	175.09	19.43	2.60	14.9
Lead (mg kg^{-1})	84	148.76	1.92	2.5	34.72
Copper (mg kg^{-1})	48.98	59.77	33.13	69.84	27.2
Sodium (mg kg^{-1})	11.84	3.98	4.65	50.4	21.03
Nitrogen (g kg^{-1})	31.32	29.21	48.29	61.1	1
Potassium (g kg^{-1})	8.4	2.2	16.51	0.020	2.97
Calcium (g kg^{-1})	14.52	0.71	24	0.033	2.18
Magnesium (g kg^{-1})	0.1	0.02	9.22	0.01	2.45
Phosphorus (g kg^{-1})	7.92	1.12	20.3	17	28.8
Iron (mg kg^{-1})	12.31	1	7.86	2.2	7.26

to the eventual formation of humus over time. Electrical conductivity decreased considerably from 66.09 to 5.12 dS m^{-1}. Lower electrical conductivity values indicate a reduction in the concentration of soluble salts, which is beneficial for the quality of compost. The pH increased from 4.42 to 7.88. The rise in pH indicates a decrease in acidity and a movement towards neutrality, which is also favorable for compost quality. Among the heavy metals analyzed, only manganese (Mn) and zinc (Zn) showed slight changes. Manganese levels remained relatively stable, while zinc increased from 188 to 304 mg kg^{-1}. The other metals, such as nickel (Ni), cadmium (Cd), chromium (Cr), lead (Pb) and copper (Cu), did not show significant variations. Regarding the nutrient content, nitrogen (N) increased from 1.83 to 2.13 g kg^{-1}, indicating an enrichment of this essential nutrient during composting. Potassium (K) levels remained relatively stable, while calcium (Ca) increased from 6.28 to 9.68 g kg^{-1}. Magnesium (Mg) showed a significant increase from 3.78 to 30.78 g kg^{-1}. Phosphorus (P) content decreased slightly from 42.79 to 36.77 g kg^{-1}. Iron (Fe) content increased from 22.09 to 33.80 mg kg^{-1}. The changes observed in the physicochemical characteristics of the compost indicate the decomposition and transformation of organic matter, resulting in improved nutrient availability and reduced levels of certain heavy metals. These results provide valuable information for further investigations and can aid in the development of strategies to enhance compost quality and efficiency.

Fig. 2. Compost container with holes for drainage

Table 2. Initial and final Physical-chemical characterization of the organic mixture compost

Sample	Preliminary analysis	Maturity analysis
Temperature (°C)	18	25
C:N	23	15.09
Moisture (%)	61.21	17.87
Organic matter (%)	62.73	59.88
Electrical conductivity (dS m^{-1})	66.09	5.12
pH	4.42	7.88
Manganese (mg kg^{-1})	225.13	223
Zinc (mg kg^{-1})	188	304
Nickel (mg kg^{-1})	68.29	80
Cadmium (mg kg^{-1})	5.29	7.9
Chromium (mg kg^{-1})	34.24	67.23
Lead (mg kg^{-1})	132.06	161.2
Copper (mg kg^{-1})	53.91	70.98
Sodium (mg kg^{-1})	2.39	8.67
Nitrogen (g kg^{-1})	1.83	2.13
Potassium (g kg^{-1})	6.28	6.20
Calcium (g kg^{-1})	6.28	9.68
Magnesium (g kg^{-1})	3.78	30.78
Phosphorus (g kg^{-1})	42.79	36.77
Iron (mg kg^{-1})	22.09	33.80

4 Conclusion and Perspective

In conclusion, this study aimed to determine the physicochemical characteristics of organic wastes commonly generated in Morocco, with the objective of building a national database and evaluating their potential for safe and sustainable composting. The findings emphasize the importance of considering the characteristics of feedstock materials for composting, particularly the proportions of carbon and nitrogen. Nitrogen-rich materials, such as grass, fruit and vegetable peels in organic household waste (OHW), and manure, decompose quickly and have low carbon content but high nitrogen content. On the other hand, carbon-rich materials, like sawdust, decompose slowly and have high carbon content but low nitrogen content. Achieving an ideal carbon-to-nitrogen ratio of 25–50:1 is crucial for initiating an effective composting process, and the careful selection and combination of feedstock materials are essential for producing high-quality compost.

The findings indicate that the collected organic wastes are highly suitable for composting. The pH level of the materials was measured at 7.88, suggesting a neutral environment, while the electrical conductivity recorded a value of 5.12 ms cm^{-1}. Moreover, the organic matter content was observed to be within the range of 59.88%, highlighting its substantial contribution to the composting process. Furthermore, both macro and micro-elements present in the feedstock were found to fall within safe limits, confirming their suitability for agricultural applications.

From a perspective point of view, the organic wastes generated in Morocco present several important considerations. Firstly, immediate intervention is needed to treat and valorize these wastes rather than disposing of them in open-air environments. Secondly, the incineration of these wastes is not an energy-efficient solution. Thirdly, they hold great potential as soil fertilizers. Finally, exploring the potential of these organic wastes as bioresources can contribute to the production of green energy and organic amendments for Moroccan farms. Therefore, it is crucial to adopt a sustainable approach in managing these wastes and to further explore their potential for agricultural applications, benefiting both the environment and the economy.

References

1. Guerrero, L.A., Maas, G., Hogland, W.: Solid waste management challenges for cities in developing countries. Waste Manag. **33**(1), 220–232 (2013)
2. Kaza, S., Yao, L., Bhada-Tata, P., Van Woerden, F.: What a waste 2.0: a global snapshot of solid waste management to 2050. World Bank Publications (2018). https://doi.org/10.1596/978-1-4648-1329-0
3. World Bank: Global waste to grow by 70 percent by 2050 unless urgent action is taken: World Bank Report. https://www.worldbank.org/en/news/press-release/2018/09/20/global-waste-to-grow-by-70-percent-by-2050-unless-urgent-action-is-taken-world-bank-report (2018). Accessed 1 May 2023
4. SNRVD: National strategy for waste reduction and valorization. https://www.environnement.gov.ma/fr/133-a-la-une/2358-atelier-de-presentation-de-la-strategie-nationale-de-reduction-et-de-valorisation-des-dechets-snrvd (2019). Accessed 1 May 2023
5. Majdouline, C., Khamar, M., El Jalil, M.H., Cherkaoui, E., Zouahri, A.: Unlocking the potential of organic household waste: a comprehensive assessment of composting suitability in Salé City, Morocco. Int. J. Environ. Stud. **124**, 1–17 (2023). https://doi.org/10.1080/00207233.2023.2231274
6. Ducrotoy, M.J., Ammary, K., Ait Lbacha, H., Zouagui, Z., Mick, V., Prevost, L., et al.: Narrative overview of animal and human brucellosis in Morocco: intensification of livestock production as a driver for emergence? Infect. Dis. Poverty **4**, 1–21 (2015)
7. Blair, J., Nippert, J., Briggs, J.: Grassland ecology 14. Ecol. Environ. **389**, 389–423 (2014)
8. Kayalvizhi, K., Alhaji, N.M.I., Saravanakkumar, D., Mohamed, S.B., Kaviyarasu, K., Ayeshamariam, A., et al.: Adsorption of copper and nickel by using sawdust chitosan nanocomposite beads: a kinetic and thermodynamic study. Environ. Res. **203**, 111814 (2022)
9. Elasri, O., Afilal, M.E.: Study a risk of contamination Moroccan waters by chickens droppings. Int. J. Innov. Appl. Stud. **7**(2), 593 (2014)
10. Ayilara, M.S., Olanrewaju, O.S., Babalola, O.O., Odeyemi, O.: Waste management through composting: challenges and potentials. Sustainability **12**(11), 4456 (2020)
11. Chandrappa, R., Das, D.B., Chandrappa, R., Das, D.B.: Waste quantities and characteristics. Solid Waste Manag. Principles Pract. **56**, 47–63 (2012)
12. Gavilanes-Terán, I., Jara-Samaniego, J., Idrovo-Novillo, J., Bustamante, M.A., Moral, R., Paredes, C.: Windrow composting as horticultural waste management strategy: a case study in Ecuador. Waste Manag. **48**, 127–134 (2016)

13. Azim, K., Soudi, B., Boukhari, S., Perissol, C., Roussos, S., Thami Alami, I.: Composting parameters and compost quality: a literature review. Org. Agricult. **8**, 141–158 (2018)
14. El-Mrini, S., Aboutayeb, R., Zouhri, A.: Effect of initial C/N ratio and turning frequency on quality of final compost of turkey manure and olive pomace. J. Eng. Appl. Sci. **69**(1), 1–20 (2022). https://doi.org/10.1186/s44147-022-00092-6
15. Rynk, R., et al.: The composting handbook: a how-to and why manual for farm, municipal, institutional and commercial composters. Academic Press, New York (2021)
16. Thomas, C., Idler, C., Ammon, C., Amon, T.: Effects of the C/N ratio and moisture content on the survival of ESBL-producing *Escherichia coli* during chicken manure composting. Waste Manag. **105**, 110–118 (2020). https://doi.org/10.1016/j.wasman.2020.01.031
17. Nguyen, V.T., et al.: Effects of C/N ratios and turning frequencies on the composting process of food waste and dry leaves. Bioresour. Technol. Rep. **11**, 100527 (2020). https://doi.org/10.1016/j.biteb.2020.100527
18. Madejón, E., Díaz, M.J., López, R., Cabrera, F.: New approaches to establish optimum moisture content for compostable materials. Bioresour. Technol. **85**(1), 73–78 (2002). https://doi.org/10.1016/S0960-8524(02)00030-5
19. Richard, T.L., Hamelers, H.V.M., Veeken, A., Silva, T.: Moisture relationships in composting processes. Comp. Sci. Utiliz. **10**(4), 286–302 (2002). https://doi.org/10.1080/1065657X.2002.10702093
20. Zhang, Y., He, Y.: Co-composting solid swine manure with pine sawdust as organic substrate. Bioresour. Technol. **97**(16), 2024–2031 (2006)
21. Ait-El-Mokhtar, M., Fakhech, A., Ben-Laouane, R., Anli, M., Boutasknit, A., Ait-Rahou, Y., et al.: Compost as an eco-friendly alternative to mitigate salt-induced effects on growth, nutritional, physiological and biochemical responses of date palm. Int. J. Recycl. Org. Waste Agricult. **11**(1), 85–100 (2022)
22. Sánchez-Monedero, M.A., Roig, A., Paredes, C., Bernal, M.P.: Nitrogen transformation during organic waste composting by the Rutgers system and its effects on pH, EC and maturity of the composting mixtures. Bioresour. Technol. **78**(3), 301–308 (2001). https://doi.org/10.1016/S0960-8524(01)00031-1
23. Bernal, M.P., Alburquerque, J.A., Moral, R.: Composting of animal manures and chemical criteria for compost maturity assessment: a review. Bioresour. Technol. **100**(22), 5444–5453 (2009). https://doi.org/10.1016/j.biortech.2008.11.027
24. James, D.W., Topper, K.F.: Utah fertilizer guide (1993)
25. Moral, R., Moreno-Caselles, J., Perez-Murcia, M.D., Perez-Espinosa, A., Rufete, B., Paredes, C.: Characterisation of the organic matter pool in manures. Bioresour. Technol. **96**(2), 153–158 (2005). https://doi.org/10.1016/j.biortech.2004.05.003
26. Brinton, W.F.: Compost quality standards and guidelines: an international view. Woods End Research Laboratory Inc. (2000)

Evolution of Vegetation and Forests with Future Expectations of Changes in Lakhdar Sub-basin

Fatiha Ait El Haj(✉), Latifa Ouadif, and Ahmed Akhssas

L3GIE Laboratory, Mohammadia Engineering School, Mohammed V University in Rabat, Rabat, Morocco
aitelhaj@emi.ac.ma

Abstract. Land use practices and environmental factors such as climate change, erosion and other global changes. Have a negative effect on vegetation cover and well on agricultural land and forests. Using both a geographical and temporal examination of cells, the Ca-Markov model was employed. This study simulates and forecasts future patterns in land use and land cover for the years 2029 and 2039. Landsat satellite images and spatial data on the area were processed and classified by the Liklihood maximum method under GIS geographic information systems, and then a simulation of the future state of LULC was performed under IDRISI software. Over a 19-year period (from the year 2000–2019), the vegetation state was evaluated and based on the probability matrices of changes, maps for the year 2029 and 2039 were generated. The results revealed a decrease of forest land in the year 2039 to an area of 75 Km^2 instead of 250 Km^2 in the year 2000 and agricultural land will also undergo significant decreases and will be replaced by bare land and habitats (29% of the area). The validation of the results obtained was verified by the kappa precision coefficient, which shows a very high degree of agreement of about 86%. For a better management and protection of the area, strategic plans must be combined with the obtained results in order to propose adaptable plans to the Lakhdar basin in order to ensure the sustainability of these resources.

Keywords: Ca-Markov · Prediction · LULC · Landsat · Lakhdar

1 Introduction

Due to urbanization, deforestation, erosion, climate change and others, the world has recently seen immense changes in land usage. Monitoring and assessment of LULC, is becoming a necessity, for the protection and conservation of biodiversity and ecosystem [1, 2]. The understanding and analysis of LULC changes was established on the basis of satellite data, Landsat images were used for the years 2000, 2007, 2010 and 2019, and the images followed a supervised classification by the maximum Liklihood method [3], with a determination of accuracy by the kappa coefficient. The prediction of LULC changes using satellite and real field data is important to establish strategic plans for sustainable management and adapted to future changes. The stochastic model (Ca-Markov) use

K. Baba et al. (Eds.): GeoME 2023, SPEES, pp. 137–144, 2023.
https://doi.org/10.1007/978-3-031-49345-4_14

the current state and real data to make the future prediction. Based on a transition probability matrix of changes created by the Markov chain for two study periods, future changes are evaluated [1]. The model (Ca-Markov) is based on multi-spectral images for the prediction of different land cover categories [4, 5]. Influencing factors have been integrated into the model to provide a more realistic estimate of future land cover, such as: distance to rivers, distance to roads, elevation, slope [6, 17]. In Morocco, the changes of LULC are a relevant event influencing the economic activity of the country, which is mainly based on agricultural activity. Recent studies have indicated that the country has followed very rapid increases in terms of deforestation, urbanization and soil erosion phenomena, which affect land use in several regions and areas. For instance, Tairi et al. [7] demonstrated that soil erosion poses a high to very high danger to 48% of the territory surrounding Tifnout-Askaoun in southern Morocco. The Lakhdar sub-basin (Fig. 1), which has an area of around 3503 km^2 and is a part of the Oum Er-rbia hydraulic system, is situated in the province of Beni Melal's hilly region. This area is known by the agricultural activity as the first economic sector of this region (8). The sub-basin's plain is covered by a temperate winter climate of the desert type [8], while the foothills are covered by a chilly winter climate and just 3% of the sub-basin's area is covered by a wet, temperate winter climate.

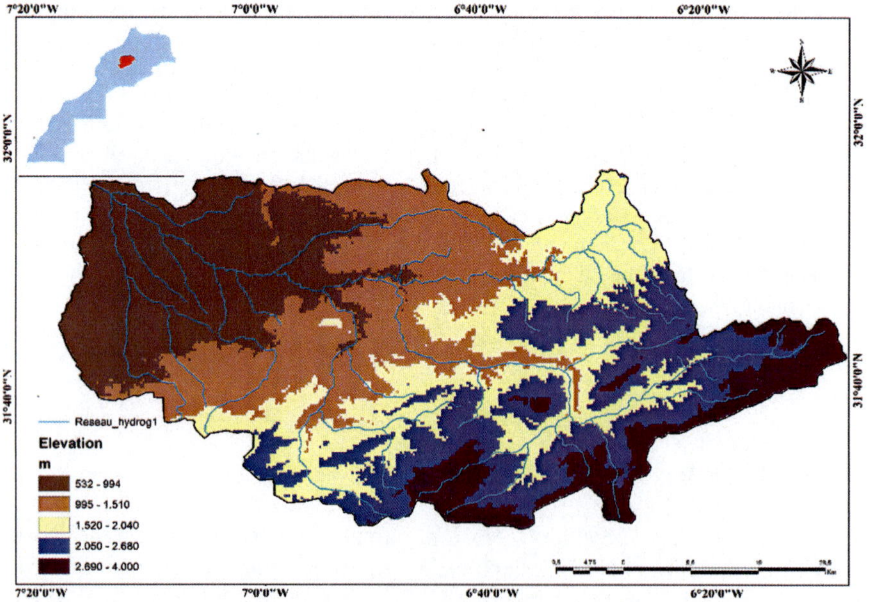

Fig. 1. Map of the Lakhdar sub-basin's location

2 Methods and Materials

2.1 Methodology Used in This Study Flowchart

Flowchart summarizes experimental protocol used (Fig. 2).

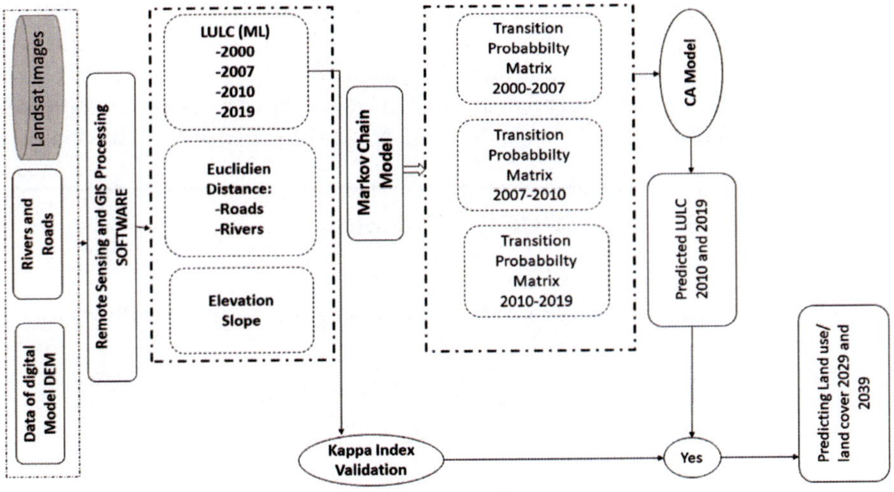

Fig. 2. Flowchart of method used

2.2 LULC Cover Classified Maps

The final results of the supervised classification for the years 2000, 2007, 2010, and 2019 [11, 13] are shown in Fig. 3 [11, 13]. The Landsat images used were chosen based on their availability for the same period of year and according to the clarity. The processing and classification is done in GIS, four classes were obtained: Water, Forest, Vegetation and Urban area with Bareland.

The satellite images were downloaded on Earth Explorer and the processing of the data set was done in GIS and Idrissi Selva. To verify the accuracy of the images classified the coefficient kappa is determined, and it gave good values that fit in the interval between 0.9 and 0.95 for all images.

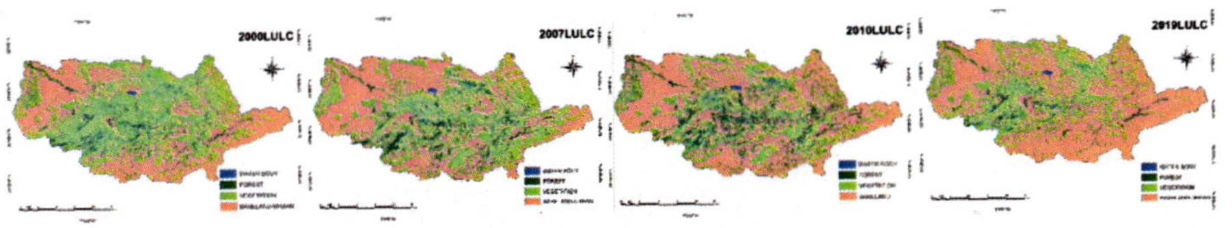

Fig. 3. LULC maps: years 2000, 2007, 2010, 2019

2.3 LULC Change Simulation Using the CA-Markov Chain

2.3.1 The Markov Chains

The Landsat imagery captured between 2000 and 2007 were used to create the LULC land cover maps (after processing, extraction and classification as shown in Fig. 2), were used to produce two transition matrices between the two periods (2000–2007), and a set of conditional probability images [10] under Idrisi Selva software.

The transition matrix gives for each class the probability of remaining in that class or transforming into another. Considering two dates t0 and t1, the matrix of changes between dates t0 and t1, corresponding to land cover classes i and j respectively; Based on the conditional probability formula, future changes in cover can be predicted (1) [9, 15, 16]:

$$S(t+1) = P_{ij} \times S(t), \tag{1}$$

S(t), S (t + 1): State of the system at the time of t or t + 1;

P_{ij}: The transition probability matrix in a state that is calculated as follows

$$P_{ij} = \begin{pmatrix} P11\ P12\ P1n \\ P21\ P22\ P2n \\ Pn1\ Pn2\ Pnn \end{pmatrix} \text{ with } \left(0 \le P_{ij} < 1 \text{ and } \sum_{j=1}^{n} Pij = 1, \quad (i, j = 1, 2, \ldots, n)\right) \tag{2}$$

2.3.2 Celullar Automaton (CA)

Cellular automata are a particular model of discrete dynamic systems. They are used to take into account the spatial dynamics, and also the probable spatial transition that occur in a particular area over time by combining it with the Markov chain. They present a set of similar cells belonging to a regular network [12, 14]:

$$S(t, t+1) = f(S(t), N) \tag{3}$$

S: Set of limited and discrete cellular states;

N: Cellular field.

t and t + 1: Rule for changing cellular states in local space at various periods.

The neighborhoods of each cell in the land cover class were defined using the suitability images and the conventional 5 5 pixel adjacency filter. Each cell's center is surrounded by a matrix space made up of 5 5 cells in order to significantly influence its center. The gain of a category happens close to where the category already existed thanks to the 5 5 spatial filter.

2.4 LULC Prediction Validation

A validation step using the kappa index is necessary to assess the level of agreement between the prediction maps and the actual maps. This step consists in comparing and calculating the kappa index under Idrisi Selva software between two maps, one simulated by the Ca-Markov model, and another reference map for the same date (LULC of Fig. 2). In this study and in the validation phase, kappa indices are generated to evaluate the accuracy of the models at two different dates. Several kappa index components have been provided by Pontius [9] with concise statistical formulations that are explained in [9]: Kstandard, Klocation, and Kquantity. They [9] specify "Kstandard as an index of agreement that attempts to account for the expected agreement due to random spatial reassignment of categories in the comparison map, given the proportions of categories in the comparison and reference maps, regardless of the size of the disagreement on quantity." Klocation is the agreement in their spatial arrangement, and Kquantity is a ratio of the quantitative differences between the categories on the comparison map and the reference map.

If the kappa value is > 0.75, it is deemed exceptional. If it is between 0.40 and 0.75, it is deemed ordinary to good. If it is lower than 0.40, it is deemed poor [10, 13].

3 Results and Discussion

3.1 Amount Change in Area Over 19 Years in Square Kilometer Graph

Figure 4 shows a graph of the total changes affected over the 19-year period (2000–2019). The results of the graph show the following changes:

- The area of water body has experienced a very small change, in 2000 the area was about 4.45 Km^2, a remarkable increase is noted in 2010 to 6.18 km^2 area. A rapid decrease observed to 4.76 Km^2 which makes the situation stable of water departure.
- Forest area: in 2000, the land occupied by forests reached 250.18 km^2. A considerable loss of area was recorded in 2019 (183.3 km^2). A slight increase was observed between 2007 and 2010 (from 309.54 to 481.61 Km^2).
- The area of agricultural land: between 2000 and 2010, the rate of agricultural land dramatically dropped. The amount of change reached a peak of 1027.76 km^2 in 2019;
- Area of buildings with bare soil: the area occupied by this category has undergone a rapid growth from 1754.73 km^2 in 2000 to 2422.4 km^2 in 2019.

3.2 Validation of Model Predicting Land Use Changes

Finding the kappa coefficient between the simulated map for 2010 and the LULC map that was produced from the supervised classification of the Landsat TM 5 picture constitutes the validation step. Before going on to the future projection of land cover and land use changes, the same approach was used to determine the kappa index for the year 2019. Ten iterations were accepted as the number of iterations [13]. The values of kappa coefficient obtained for the two dates are shown in Fig. 5 which was determined under Idrisi Selva. The two values obtained are acceptable and very good which indicates a good probability for the future simulation.

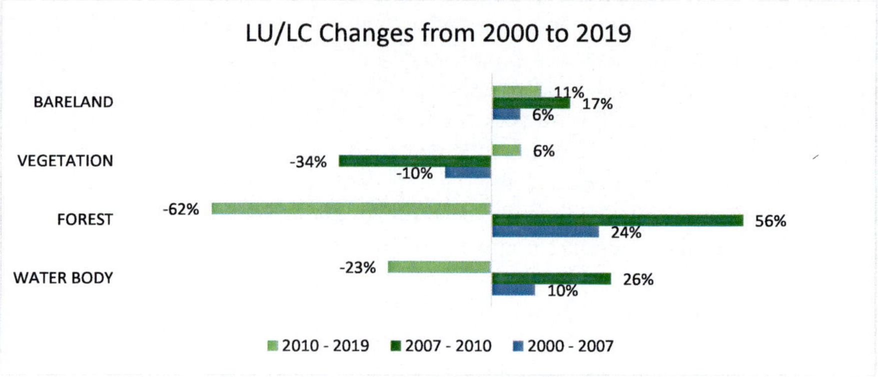

Fig. 4. Amount change in area of land cover over 19 years in Km^2.

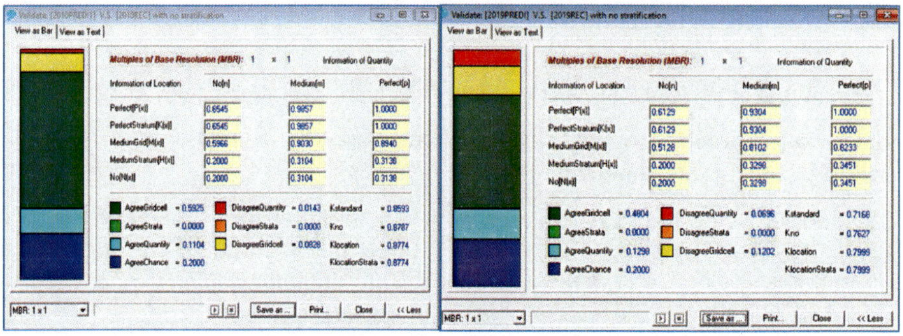

Fig. 5. Kappa index for 2010 and 2019.

3.3 Current and Predicted Maps of Land Use Change Combination

The transition matrices obtained have been validated by simulation (kappa index), which allows to adopt these matrices from the Markov chain for future predictions. The prediction maps for the years 2029 and 2039 were created by fusing the Markov chain with the cellular automata (CA) (Fig. 6). These have allowed to evaluate and observe the changes in land use, which shows that bare land with urbanization will undergo significant increases in the order of 37% from 2000 to 2039, and an intense decrease in green spaces (Vegetation and forest) that is − 29 and − 70% respectively. This means that urban growth in Lakhdar has a negative impact on the sustainability of green spaces in the study area. The water body will also decrease in the future years. The results obtained make clearer the problems that will arise in the future years, which gives a clear vision to the decision makers in terms of solutions to consider for the protection of this area. Ca-Markov model simulation produced accurate results in terms of both temporal changes and spatial distribution.

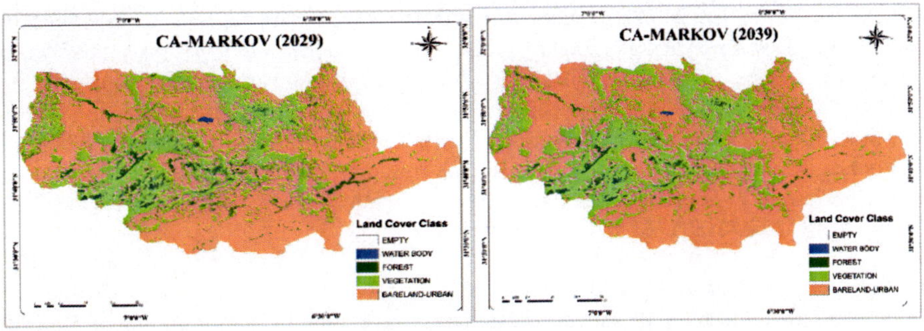

Fig. 6. Land use of predict maps in: (e) 2029, (f) 2039

4 Conclusion

In recent years, the world has experienced radical changes in the potential and natural resources available to man. These changes are mainly due to human activity that have caused the imbalance of these resources, and which has a negative effect on the environment. Therefore, the effects of changes on natural resources were identified. By incorporating the Ca Markov approach for the modeling and mapping of future land occupations, the present study was carried out with the aim of studying and evolving the key changes impacting the Lakhdar basin at the level of Oum Er-rbia. To do this, the results of the Markov chain of LULC images classified for the years 2000, 2007, 2010 and 2019, was combined with cellular automata to give the spatio-temporal distribution of LULC in the Lakhdar sub-basin. The resulting maps actually show the influence of human activity (urbanization, climate change effects, etc.) on the distribution of water bodies and green spaces. This should be of interest to decision makers and planners, in order to take measures to protect this area in the prospects of sustainable development of natural resources.

References

1. Hyandye, C., Martz, L.W.: A Markovian and cellular automata land-use change predictive model of the Usangu Catchment. Int. J. Remote Sens. **38**(1), 64–81 (2017)
2. Sleeter, B.M., Sohl, T.L., Loveland, T.R., et al.: Land-cover change in the conterminous United States from 1973 to 2000. Glob. Environ. Change **23**(4), 733–748 (2013)
3. Gallego, F.J.: Remote sensing and land cover area estimation. Int. J. Remote Sens. **25**(15), 3019–3047 (2004)
4. Zadbagher, E., Becek, K., Berberoglu, S.: Modeling land use/land cover change using remote sensing and geographic information systems: case study of the Seyhan Basin, Turkey. Environ. Monit. Assess. **190**(8), 494 (2018)
5. Roy, S., Farzana, K., Papia, M., Hasan, M.: Monitoring and prediction of land use/land cover change using the integration of Markov chain model and cellular automation in the southeastern tertiary hilly area of Bangladesh. Int. J. Sci. Basic Appl. Res. **24**, 125–148 (2015)
6. Sleeter, B.M., Sohl, T.L., Bouchard, M.A., et al.: Scenarios of land use and land cover change in the conterminous United States: utilizing the special report on emission scenarios at ecoregional scales. Glob. Environ. Change **22**(4), 896–914 (2012)

7. Tairi, A., Elmouden, A., Aboulouafa, M.: Soil erosion risk mapping using the analytical hierarchy process (AHP) and geographic information system in the Tifnout-Askaoun Watershed, Southern Morocco. Eur. Sci. J. ESJ **15**(30), 338 (2019). https://doi.org/10.19044/esj.2019.v15n30p338
8. AHT Group, F., Ag, A.H.T.G., Avril, R.: Diagnostic du sous-bassin de Lakhdar (2016)
9. Pontius, R.G., Millones, M.: Death to Kappa: birth of quantity disagreement and allocation disagreement for accuracy assessment. Int. J. Remote Sens. **32**, 4407–4429 (2011). https://doi.org/10.1080/01431161.2011.552923
10. Eastman, J.R.: IDRISI Selva help system. Clark Labs, Clark University, Worcester (2012)
11. Ait El Haj, F., Ouadif, L., Akhssas, A.: Monitoring land use and land cover changes using remote sensing techniques and the precipitation-vegetation indexes in Morocco. Ecol. Eng. Environ. Technol. **24**(1), 272–286 (2023). https://doi.org/10.12912/27197050/154937
12. Liping, C., Yujun, S., Saeed, S.: Monitoring and predicting land use and land cover changes using remote sensing and GIS techniques: a case study of a hilly area, Jiangle, China. PLoS ONE **13**(7), 1–23 (2018). https://doi.org/10.1371/journal.pone.0200493
13. Ait El Haj, F., Ouadif, L., Akhssas, A.: Simulating and predicting future land-use/land cover trends using CA-Markov and LCM models. Case Stud. Chem. Environ. Eng. **7**, 100342 (2023). https://doi.org/10.1016/j.cscee.2023.100342
14. Souidi, H., Ouadif, L., Bahi, L., Elhachmi, D., Edderkaoui, R.: Application of stochastic process and cellular automata integrated to GIS for land monitoring: Coastal Chaouia, Morocco. Int. J. Adv. Res. Eng. Technol. **11**(5), 245–252 (2020). https://doi.org/10.34218/IJARET.11.5.2020.026
15. Souidi, H., Ouadif, L., Bahi, L., Edderkaoui, R., Jaouda, I., Bahi, Y., Belhaj, S.: Research of suitability area of agriculture in Coastal CHAOUIA by integration the AHP, IDW, and MCDA to GIS. E3S Web Conf. **150**(20), 3009 (2020). https://doi.org/10.1051/e3sconf/202015003009
16. Souidi, H., Ouadif, L., Bahi, L., Habitou, N.: Spatiotemporal monitoring and prediction of land use integrating the Markov chain and cellular automata in the Coastal Chaouia. Int. J. Recent Technol. Eng. **8**(4), 1704–1711 (2019). https://doi.org/10.35940/ijrte.c5760.118419
17. Jaouda, I., Akhssas, A., Ouadif, L., Bahi, L., Elkasri, J., Souidi, H., Soussi, H.: Study of soil erosion risks using remote sensing in Ouergha River watershed (Morocco). E3S Web Conf. **150**, 3012 (2020). https://doi.org/10.1051/e3sconf/202015003012

First Step Towards Ecological Heating in Green School Procurement Relating to Operations and Maintenance: A Case Study

Taoufik Jebli(✉), Issam Aalil, and Mohammed Radouani

National Higher School of Engineering (ENSAM), Moulay Ismail University, Meknes, Morocco
taoufikjebli@gmail.com

Abstract. Among all Public Schools in Morocco, there are those located in areas with low temperatures. Hence the need for heating to improve learning conditions for students and teachers. However, the types of combustibles used to operate the furnaces are either coal, wood or fuel oil. These combustibles can cause asphyxiation and fainting. Hence the interest of using more ecological and more economical means. The overall purpose of this research is to show how to get rid of the classic scheme adopted by school maintenance managers who focus during maintenance work on the building per se and not on the building's performance in order to improve, among others, the thermal comfort and indoor air quality of educational spaces, and especially work relating to heating. The analysis is demonstrated in a case study conducted on school using anthracite as combustible. Switching to another combustible that is less polluting and easily mobilized and quickly usable, in this case Liquid Petroleum Gas (LPG), as a first step towards environmentally friendly combustibles has made it possible to (i) reduce the carbon impact and (ii) improve the climate inside the classrooms and (iii) meet the expectations of stakeholders based on the Value-Based Maintenance Management (VBMM) approach. This is a first and unique experience in schools at nationwide which has highlighted the principle of green purchasing.

Keywords: Public schools · Maintenance · Value-based maintenance management · Heating · Green public procurement

1 Introduction

Morocco has a climatic diversity due to the geographical position of each region, the presence or not of mountains, deserts, sea or oceans. The country is distinguished by four types of climate: humid, sub-humid, semi-arid and arid. In the interior of the country, the climate varies with altitude. Summer is warm and pleasant in the mountains. Winter is very cold and wet on the heights [1]. Therefore, the installation of heating devices, especially in schools located in mountainous areas, is essential in order to avoid absenteeism and school dropout of learners during these winter periods of the year.

K. Baba et al. (Eds.): GeoME 2023, SPEES, pp. 145–152, 2023.
https://doi.org/10.1007/978-3-031-49345-4_15

In order to ensure access to education for all students and to fight against school dropouts, Morocco spares no effort to generalize education through the construction of schools close to the population. The number of Provincial Directorates housing educational buildings located in areas with low temperatures represents one third (1/3) of the Provincial Directorates of Education in the Kingdom and required a budget of 39 million DH for heating in 2014. Schools' heating consumed nearly 4850 tons of anthracite and 934 tons of wood during the 2013–2014 school year [2]. These combustibles would have adverse effects on the environment.

Morocco is committed to reduce its greenhouse gas emissions by 42%, compared to projected emissions for the year 2030 [3] to reach "Net Zero" emissions around 2050 [4]. To do this and to be limited to the education sector, it is wise for managers to establish a management system for the maintenance of educational spaces (building, heating, air conditioning, plumbing, etc.) based on value and to promote sustainable purchases.

This paper studies heating in a Moroccan school building and how to bring together the energies of maintenance managers and stakeholders in a dynamic of co-construction through responsible purchases to heat schools aimed at improving reception conditions and school equipment. This contributes to improving students' concentration (Fig. 1) [5], to the motivation of learners and teachers, which reduces absenteeism [6], promotes learning and reduces the risk of school dropout.

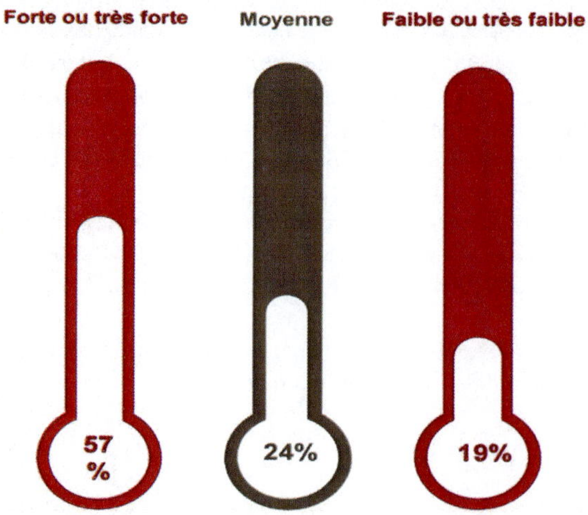

Fig. 1. Lack of concentration of students linked to thermal discomfort, estimated by the teachers interviewed

2 Study Area and Methodology

The site under consideration in this study is a middle school in Dayet Aoua, Ifrane, Morocco. An aerial view of the site is shown in Fig. 2. It comprises teaching classrooms (general, scientific, and specific), administration rooms, a boarding school dorm including canteen. It also comprises sports fields, sports locker rooms, restrooms, staff accommodations, and a wall fencing. This middle school was built in 2011 and was extended by adding a multipurpose block in 2022. The latter had the particularity of having the thermal insulation of its outdoor structural components. The buildings are one to two levels.

The middle school is located in a mountainous and snowy area, at an altitude of 1460 m. Temperatures vary between an average maximum of 21.1 °C and an average minimum of 4.8 °C [7]. January is the coldest month, while July and August are the hottest months [8].

Fig. 2. Satellite image

The middle school benefited from a dedicated budget, among other things, for the maintenance of its heating system. The upgrading needs were determined through an exhaustive diagnosis of all the buildings and facilities of the such school. Priority works' program was drawn up in consultation with the beneficiaries in order to best meet their expectations.

The study is decomposed into three phases: (i) diagnosis (ii) Heating's Maintenance Management user-focused (iii) Promoting green purchasing through public procurement.

3 Results and Discussion

3.1 Diagnosis

The existing heating system at the middle school was mixed. Indeed, a central coal boiler connected to aluminum radiators provided heating at the level of the boarding school dorm while coal stoves was installed in the classrooms (Fig. 3). Some administration's office was heated by individual electrical appliances. The amount of anthracite needed varied from 12T to 16T annually. Heating ignition mobilized a person for two hours. He filled the stoves in the evening after the departure of the learners and the teaching staff. In the morning, he started work to heat the classrooms before and during the arrival of the students. He had to feed the fire while controlling the supply of air. This was a complex operation that consumed time and energy and was sometimes doomed to failure. As for the boarding school dorm, the boiler is lit in the evening, well before the installation of the interns.

Fig. 3. Equipment and combustible for coal heating

Anthracite waste was difficult to evacuate following an eco-responsible approach given the absence of dedicated landfills with appropriate treatment.

When lighting the anthracite contained in the stoves, a lot of dust was settling inside the classrooms, fine particles bad for the quality of the air was released from the fire, which can cause cases of suffocation and fainting. This would have a detrimental effect on the health of learners and teaching staff [9]. Furthermore, this was contributing to the loss of school time considered as the central dimension in terms of pedagogical effectiveness [10]. The previous maintenance and rehabilitation activities undertaken concerned only minor works (painting, glazing, etc.) and did not seek to install a clean heating system that satisfies for the beneficiaries. Maintenance is oriented to the physical conditions of the buildings and cost driven.

3.2 Heating's Maintenance Management User-Focused

It considers maintenance as a value-generating action that could lead to more effective and more efficient maintenance strategies. Buildings only have value if they are useful to users and meet their needs and their expectations [11]. Unlike the existing practices maintenance management, which focuses on the building itself, this approach puts the beneficiary at the center of the Operation and Maintenance (O&M) system, ensuring their satisfaction by taking into account their expectations and priorities. Indeed, it is not the physical conditions themselves that matter for the beneficiaries but what is crucial for them is that the building is able to meet their expectations to carry out their activities inside and outside the building [12–14].

Through this approach promoting the active participation of the beneficiaries the need for heating using gas boilers has emerged. This new heating system comes out of what is usually practiced, namely coal stoves. In principle, in case there were maintenance and rehabilitation activities, the maintenance managers would have thought of a coal-fired heating system. They could also make this decision without consulting the users. While the organization of consultation meetings and sharing and concertation workshops with the final beneficiaries, at the preliminary stages of the studies and before the start of the works, made it possible to detect their concerns, to prioritize their needs and to involve them in making decisions by confirming their choices in writing. This choice reduced carbon dioxide (CO_2) emissions (Table 1) and saved two hours of school time and made classrooms conducive to student learning (Fig. 4). Students, during the oral interviews, expressed that they no longer felt fainting and that they felt more secure. The teaching and administrative staff of the middle school are also satisfied.

The promotion of levers allowing the reduction of the carbon footprint due to heating is initiated through Green Public Procurement to take into account sustainability during maintenance management.

Fig. 4. Equipment and combustible for gaz heating

3.3 Promoting Buying Green Using Public Procurement

Green Public Procurement (GPP) is defined as "a process by which public authorities seek to procure goods, services, and works with a reduced environmental impact throughout their life-cycle when compared to goods, services, and works with the same primary function that would otherwise be procured" in the European Commission's Communication Public procurement for a better environment [15]. Public Procurement is an important leverage and a crucial pillar to the successful Green purchasing. Public procurement is a key element and a powerful vector for the promotion and development of sustainable consumption and production [16]. In Morocco, planned public investment amounts to 300 billion dirhams in 2023, accounting for ~ 25% of the Morocco's Gross Domestic Product (GDP) [17]. More than 40,000 tenders are put out to tender each year throughout Morocco [18]. For the Province of Ifrane, the provincial directorate of

National Education had mobilized budget of one million three hundred thousand dirhams (1,300,000 MAD) for the 2022 budget year allocated to the heating of 75 schools in the province [19], i.e. 1% of the operating budget at the regional level [17].

Environmental and ecological dimensions as well as the objectives of sustainable development has been introduced into Moroccan public procurement codes since 2013 [20, 21]. Although the majority of public project owners are not sufficiently aware of sustainable purchases and do not yet take these dimensions into account [22]. This project has contributed not only to raising awareness among managers of the need for decarbonation of purchases, but also among stakeholders and civil society. Taking advantage of this experience, it was highlighted on the need, through public procurement, to strengthen the accountability of actors to reduce the carbon impact and encourage sustainable practices.

The CO_2 Emission Factors (ED) of Coal (anthracite) and Liquid Petroleum Gas (LPG) combustibles are given according to the Table 1.

Table 1. CO_2 emission factors (ED) of Anthracite/LPG combustibles according to the ADEME carbon base (France data—January 2015) [23].

Combustibles	FE (gCO_2eq/kWh)
Anthracite	345
LPG	233

This transition from an anthracite heating combustible to that of LPG allows a reduction in greenhouse gas emissions of around 30%, which is a first step towards decarbonation of consumption.

4 Conclusions and Recommendations

This Paper analyzed the concept of value in maintenance management which focuses on the user and his expectations, unlike the existing practices maintenance management, which focuses on the physical conditions of buildings and cost driven. It also addressed sustainable Public Procurement, a crucial aspect of the move towards schools heated with less carbon-intensive combustibles. All of these concepts were applied to a case study.

The study was decomposed into three phases: (i) Diagnosis (ii) Heating's Maintenance Management user-focused (iii) Promoting buying Green using Public Procurement.

It was observed that:

- 100% participation of the administrative and pedagogical body as well as the students and their parents throughout the process from the diagnosis, the choice of the gas boiler heating system to the completion of the work.
- Reduction of greenhouse gas emissions by 30%.
- The climate inside the classrooms is made healthy and safe, which promotes student learning.

- Middle school's stakeholders are 100% satisfied with the gas boiler heating system.
- Wrestling against the loss of school time. The latter can reach 25% per day of the time recovered.
- Rising the attendance of children at school, especially girls.
- Saving time when switching on the heating and saving the salaries of the people responsible for igniting the coal-fired heating systems.
- Promoting sustainable purchases and consumption through environmentally friendly purchases given that more than 40,000 calls for tenders are launched each year throughout Morocco and that mobilizes about 25% of the Moroccan Gross Domestic Product.
- 92% of purchased anthracite is directed to stoves, showing that the majority of schools in the province of Ifrane are heated via stoves. This is an important resource for maintenance focused on environmentally friendly purchases which would transform the educational space towards more low-carbon heating systems.

References

1. Direction Générale de la Météorologie: Maroc Etat de climat en 2022 (2022)
2. Chambre des Conseillers: Réponse de Mr le Ministre de l'Éducation Nationale Rachid Belmokhtar (2023)
3. Ministre de l'Energie, des Mines et de L'Environnement, STRATÉGIE NATIONALE DE L'EFFICACITÉ ÉNERGÉTIQUE À L'HORIZON 2030 (2020)
4. Ministère de la Transition Energétique et du Développement Durable Département du Développement Durable, Stratégie Bas Carbone à Long Terme MAROC 2050 (2021)
5. Giz and EDMITA (Energie durable dans les provinces de Midelt et les Tata) and Ministère de l'Energie, des Mines et du Développement Durable, Etude de chauffage par poêle à bois amélioré pour les écoles de la province de Midelt (2018)
6. Howden-Chapman, P., Pierse, N., Nicholls, S., et al.: Effects of improved home heating on asthma in community dwelling children: randomised controlled trial. Br. Med. J. **337**, a1411 (2008)
7. Sayad, A., Chakiri, S.: Impact de l'évolution du climat sur le niveau de Dayet Aoua dans le Moyen Atlas marocain. Sci. Chang. Planet. Secheresse **21**(4), 245–251 (2019)
8. Sayad, A., Chakiri, S., Chahlaoui, A., Bejjaji, Z.: Impact des stress hydriques sur les potentiels de dayet aoua (MOYEN ATLAS—MAROC). Sci. Edn. Mersenne **4**, 12060 (2012)
9. Ige, J., Pilkington, P., Orme, J., et al.: The relationship between buildings and health: a systematic review
10. Carroll, J.B.: A model of school learning. Teach. Coll. Rec. **64**(8), 723–733 (1963)
11. Wordsworth, P.: Lee's Building Maintenance Management, 4th edn. Blackwell Science Limited, Oxford (2001)
12. Chapman, K., Beck, M.: Recent experience of housing associations and others registered social landlords in commissioning stock condition surveys. COBRA 1998 Conference. RICS Res. **1**, 23–24 (1998)
13. Jones, K.: Sustainable buildings maintenance. In: Kelly, J., Morledge, R., Wilkinson, S. (eds.) Best Value in Construction. Blackwell Publishing, London (2002)
14. Jones, K., Sharp, M.: A new performance based process model for built asset maintenance. Facilities **25**(13/14), 525–535 (2007)

15. Commission of The European Communities: Public procurement for a better environment. EU Bruxelles. https://eur-lex.europa.eu/legal-content/EN/TXT/?uri=CELEX:52008DC0400 (2008). Accessed 25 Feb 2022
16. Iraldo, F., Melis, M., Testa, F.: L'attuale sviluppo del green public procurement. Economia delle Fonti di Energia e dell'Ambiente **1**, 5–19 (2007)
17. Ministère de l'Economie et des Finances (MEF) Marocain: Loi de Finance Marocaine (2023)
18. Dhouibi, W., Cerruti, C., Desponts, C.: Verdir l'économie marocaine en dématérialisant les marchés publics, Voix Arabes Banque Mondiale (2023)
19. 2M.ma, Vendredi (2022)
20. Décret no. 2-12-349 du 8 joumada 1 1434 (20 mars 2013) fixant les conditions et les formes de passation des marchés de l'Etat ainsi que certaines règles relatives à leur gestion et à leur contrôle (2013)
21. Décret no 2-22-431 du 15 chaabane 1444 (8 mars 2023) relatif aux marchés publics marocains (2023)
22. El-Haddadi, T., Mourabit, T., El-Haddadi, A.: Sustainable public procurement in Morocco: an investigative survey regarding tender preparation (2021)
23. ADEME (Agence de la transition écologique) (2020)
24. Idrus, A., Khamidi, M.F., Lateef, O.A.: A value-based maintenance management model for university buildings in Malaysia: a critical review (2009)
25. Finamore, M., Oltean-Dumbrava, C.: Green Public Procurement and the circularity of the built environment. IOP Conf. Ser. Earth Environ. Sci. **1122**, 012054 (2022)
26. Lateef, O.A.: Case for alternative approach to building maintenance management of public universities. J. Build. Appraisal **5**(3), 201–212 (2010)
27. Mong, S.G., Mohamed, S.F., Misnan, M.S.: Maintenance management model: an identification of key elements for value-based maintenance management by local authority. Int. J. Eng. Technol. **7**(3.25), 35–43 (2018)

Quality Assessment in Terms of Agricultural Water Supply and Macro Element Contents in Water of Çorlu Stream (Thrace Region, Türkiye)

Cem Tokatli[1](✉) and Memet Varol[2]

[1] Trakya University, Evrenos Gazi Campus, İpsala, Edirne, Türkiye
tokatlicem@gmail.com
[2] Malatya Turgut Özal University, Vocational School of Doğanşehir, Malatya, Türkiye

Abstract. Freshwater contamination by macro elements because of agrogenic—domestic discharges and threatening almost all the aquatic habitats is a significant global problem today. Çorlu Stream is one of the main components of the Ergene River Basin (sub-basin of Meriç River Watershed). It is also the fluvial ecosystem known to be exposed to the most anthropogenic pressure in the Thrace Region of Türkiye, where has a significant agricultural production potential. In this study, levels of Na, Mg, Ca and K in the water of upstream (Ç1), midstream (Ç2) and downstream (Ç3) of Çorlu Stream were investigated and water qualities were assessed in terms of agricultural use by using Magnesium Rate (MR), Sodium Absorption Rate (SAR), Kelly Index (KI) and Sodium Percentage (Na%). The results indicated that the mean accumulation order of investigated macro elements in water of Çorlu Stream were as follows: Na (183.53 ppm) > K (8.35 ppm) > Mg (6.99 ppm) > Ca (6.13 ppm). The results also showed that the order of locations in terms of irrigation water qualities were as follows: Ç1 > Ç2 > Ç3. The results of applied indices indicated that all of the locations selected on Çorlu Stream were recorded as not suitable for agricultural use.

Keywords: Çorlu stream · Macro elements · Agricultural water quality

1 Introduction

Freshwater pollution is a type of pollution that occurs when freshwater resources become inactive directly or indirectly. This type of pollution is a very serious problem in terms of managing the life of living things. Since < 1% of freshwater resources, which are necessary for life on earth, are accessible and drinkable, we cause this limited resource to become unusable by polluting it in different ways. Among the main causes of water pollution are many pollutants such as domestic and industrial wastewater, agricultural irrigation, oil spills, mining activities, chemicals discharged from factories. These pollutants pose serious threats to the aquatic life and to prevent water pollution, it is very important to keep water sources clean. In order to keep water resources clean, it is of

K. Baba et al. (Eds.): GeoME 2023, SPEES, pp. 153–160, 2023.
https://doi.org/10.1007/978-3-031-49345-4_16

great importance to manage wastewater correctly, to reduce the use of chemicals such as pesticides, to control chemicals discharged from factories, and to raise environmental awareness to society [1–6].

Irrigation water quality is of great importance for successful agriculture and has a significant importance on the quality of soils. It is known that good quality agricultural products may be provided by using good quality irrigation waters. Accordingly, evaluation of freshwater quality in terms of irrigation water supply is a significant topic for an effective water management. Irrigation water quality is a critical factor for the healthy growth and development of plants. Good quality irrigation water does not damage the roots of the plants and does not prevent them from getting nutrients. In addition, a clean water source protects plant health by preventing the spread of plant diseases and pests. Irrigation water quality affects the soil fertility. Irrigation waters with high salt content, pollution or harmful substances can disrupt the structure of the soil and reduce the nutrient uptake capacity of plants. Proper irrigation water quality preserves the nutrient and water holding capacity of the soil, balances soil acidity and supports the activity of microorganisms. Irrigation water quality is a significant factor that has a direct impact on plant productivity and product quality. Inadequate irrigation water quality can negatively affect the growth of plants and cause yield loss. Using good quality irrigation water ensures that the plants receive enough water and nutrients, resulting in higher yields and better quality. Irrigation water quality has also importance in terms of protecting water resources and the environment. Contaminated irrigation water can cause soil erosion, soil and water pollution. This can negatively affect natural life and make sustainable agricultural practices difficult. For these and similar reasons, irrigation water quality should be tested regularly and improved by taking appropriate measures. In order to ensure the efficiency of agricultural activities, the protection of water resources and the improvement of irrigation water quality are of great importance [7–12].

Çorlu Stream with an extreme agricultural, rural, and industrial pressure is a stream located in the Marmara Region of Türkiye. Located within the borders of Tekirdağ province, the creek has a length of approximately 65 km. Çorlu Stream is a significant water source for irrigation of agricultural lands and for industrial use. In addition to agricultural activities, the creek, which is used to provide irrigation water, is also used by the local people as a fishing and picnic area. However, there are significant pollution problems in the creek. Pollutants such as industrial wastes and sewage water adversely affect the water quality in the stream. Therefore, it is necessary to take measures to protect and improve the water quality in the stream. Çorlu Stream is also the most significant and polluted tributary of Ergene River that is the main sub-basin of Meriç River. Contamination in the Çorlu Stream cause very serious adverse effects on the entire watershed and it causes the Ergene River to become unusable in terms of agricultural water supply, which is ranked as the highest water usage in the region [12–22].

The main purpose of this scientific research was to investigate the macro element contents in water of the upper, middle and lower basin of Çorlu Stream and assess the water qualities in terms of their usability as irrigation water.

2 Materials and Methods

2.1 Collecting Surface Water Samples from Çorlu Stream

Three stations were selected in the Çorlu Stream considering the organic and inorganic contamination resources (upstream—Ç1, midstream—Ç2, downstream—Ç3). The map of Çorlu Stream watershed and sampling locations are given in Fig. 1.

The upstream station (Ç1), which was located quite close to the source area of the Çorlu Stream, is far from residential areas and any point sources of pollution. The midstream station (Ç2) was in the Çerkezköy District one of the Türkiye's largest industrial centres, where the industry is concentrated in textile, plastic, paint, metal, food, health, mining, wood, electronics, construction materials and automotive sectors. The downstream station (Ç3) was in the Çorlu District most crowded residential area of the Thrace Region (about 300,000 people) and one of the Turkey's most significant industrial areas.

Surface water samples were taken from about 0.3 m under the surface area into the 1 L bottles at the end of dry season of 2021 (summer) [23].

2.2 Analysis of Macro Elements

pH values of surface water samples (sws) (one liter) were set to 2 (with adding HNO3). Sws were filtered (with a filter of cellulose nitrate—0.45 μm). Capacities of the sws were made up to 50 ml (with ultrapure water). Then the levels of macro elements were read with an Agilent 7700 xx ICP—MS (TS EN/ISO IEC 17025) [23, 24].

2.3 Calculation of Applied Indices

The methods of calculation for the four widely used indices including SAR [25], Na% [26], MR [27] and KI [28] are given in Table 1. Macro element units are converted to meq/L in all calculations.

3 Results and Discussion

In this investigation, macro element accumulation levels in water of Çorlu Stream were determined and the water qualities of up–mid–downstream locations of Çorlu Stream were assessed in terms of its usability as irrigation water by using some widely used indices including SAR, Na%, MR and KI. The recorded macro element levels are given in Fig. 1 and the data of applied indices are shown in Fig. 2.

The mean macro element contents in water of the Çorlu Stream occurred in the following order: Na (183.53 ppm) > K (8.35 ppm) > Mg (6.99 ppm) > Ca (6.13 ppm). Also, the total macro element contents in water of investigated locations occurred in the following order: downstream (Ç3) > midstream (Ç2) > upstream (Ç1).

As a result of applied indices, all the sampling stations (up–mid–downstreams) were detected as not suitable for agricultural use in terms of applied MR, SAR, KI and Na%. All the locations selected on the Çorlu Stream were found to be as "Unsuitable (> 9)" in terms of SAR, "Not suitable (> 50)" in terms of MR, "Not suitable (> 1)" in terms of KL and "Not applicable (> 80)" in terms of Na%.

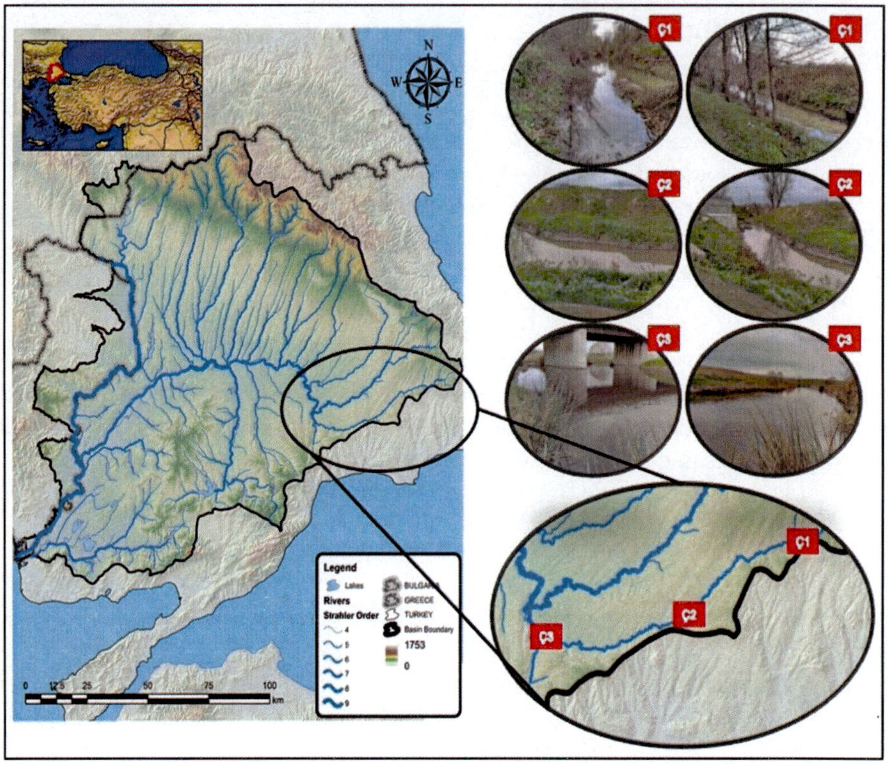

Fig. 1. Çorlu stream watershed and sampling locations

Table 1. Names, abbreviations, formulations and scales of the indices

Name of index	Formulas	Evalution scale
Sodium adsorption rate	$SAR = \frac{Na}{\sqrt{\frac{(Ca+Mg)}{2}}}$	< 6: Good 6–9: Doubtful > 9: Unsuitable
Sodium percentage	$Na\% = \left(\frac{Na+K}{Na+K+Mg+Ca}\right) * 100$	< 20 Excellent 20–40 Good 40–60 Permissible 60–80 Doubtful > 80 Not applicable
Magnesium rate	$MR = \left(\frac{Mg}{Mg+Ca}\right) * 100$	< 50 Suitable > 50 Not suitable
Kelly index	$KI = \frac{Na}{Mg+Ca}$	< 1 Suitable > 1 Not suitable

In general, a gradual increase was detected in macro element values and irrigation water index values from the upper basin to the lower basin in the Çorlu Stream. Çorlu Stream is known as a critically contaminated riverine ecosystem. Industrial and domestic waste discharges from the settlements (Çorlu and Çerkezköy) are known as the main pollution resources for the basin. In the present investigation, significant spatial differences in macro element levels of Çorlu Stream were recorded [29–34]. The recorded quite high macro element values and quite high coefficients of indices in the water of Çorlu Stream especially at the mid–downstream locations may be due to the domestic and industrial wastes from these settlements (Fig. 3).

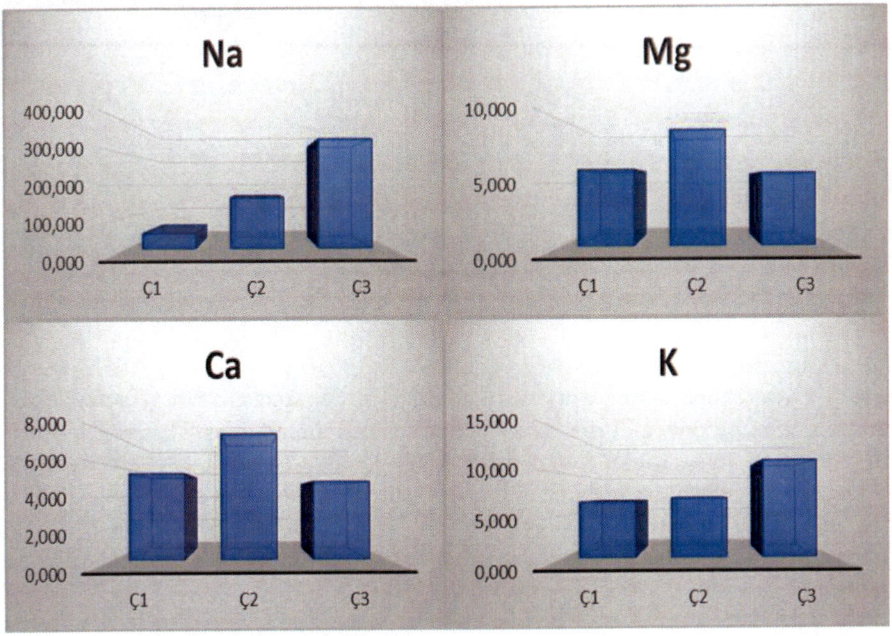

Fig. 2. Macro elements in water samples

4 Conclusions

In the current scientific research, surface water quality of Çorlu Stream was assessed by using some widely used irrigation water quality assessment indices. Water quality of Çorlu Stream is significantly decreasing from upstream to downstream in terms of macro element accumulations and the contamination is being started from almost just after the upstream, when the stream enters to the settlements. It has been determined that the use of Çorlu Stream as irrigation water is also very risky. Various measures are taken on many rivers, including Çorlu Stream, in order to protect and sustainably use the water resources of Türkiye. These measures include the establishment of wastewater treatment plants, the implementation of sustainable irrigation techniques in agricultural

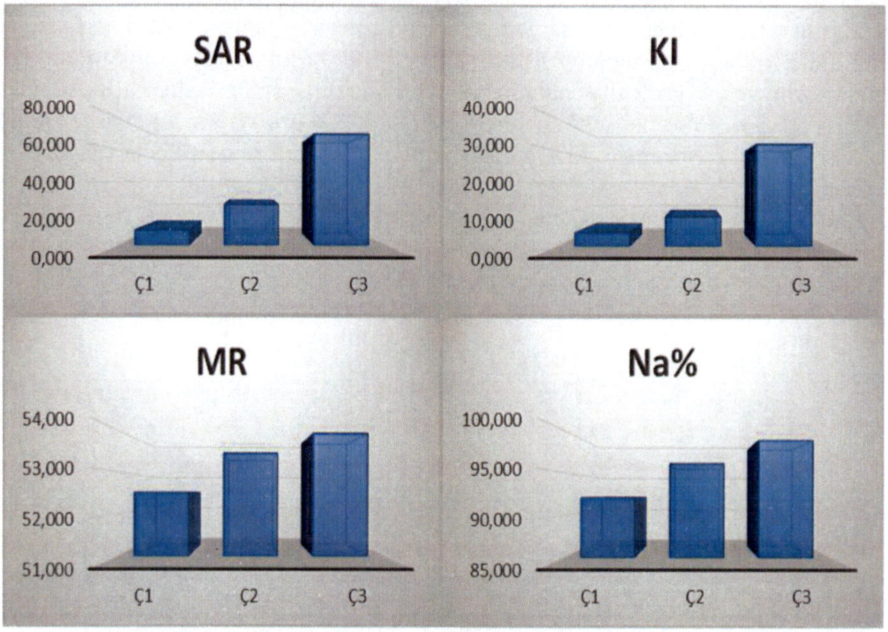

Fig. 3. Data of applied indices

areas and awareness raising activities. Industrialization and growth of population are increasing day by day, and this situation causes contamination of the water, soil and atmosphere. Serious problems may arise in the quality and amount of freshwater that can be used for irrigation, with the pollution of fluvial and lacustrine habitats, and other water resources. Therefore, in order to prevent possible environmental and health risks, it is necessary to constantly monitoring the water quality of Çorlu Stream, which is an important irrigation source in the region.

References

1. Arslan, N., Tokatlı, C., Çiçek, A., Köse, E.: Determination of some metal concentrations in water and sediment samples in Yedigöller Region (Kütahya). Rev. Hydrobiol. **4**(1), 17–28 (2011)
2. Arslan, N., Köse, E., Tokatlı, C., Emiroğlu, Ö., Çiçek, A.: Ecotoxicological effects of solid waste storage areas on aquatic systems: example of Yedigöller. Kütahya. Karaelmas Sci. Eng. J. **2**(1), 20–26 (2012)
3. Çiçek, A., Köse, E., Tokatlı, C.: Use of factor analysis to evaluate the sediment quality of a significant mining area: Seydisuyu Stream Basin (Turkey). Polish J. Environ. Stud. **28**(3), 2021–2025 (2019)
4. Tokatlı, C., Ustaoğlu, F.: Health risk assessment of toxicants in Meriç River Delta Wetland, Thrace Region. Turkey. Environ. Earth Sci. **79**, 426 (2020)
5. Köse, E., et al.: Assessment of ecologic quality in terms of heavy metal concentrations in sediment and fish on Sakarya River and Dam Lakes, Turkey. Soil Sedim. Contam. Int. J. **29**(3), 292–303 (2020)

6. Tokatlı, C., Varol, M.: Variations, health risks, pollution status and possible sources of dissolved toxic metal(loid)s in stagnant water bodies located in an intensive agricultural region of Turkey. Environ. Res. **201**, 111571 (2021)
7. Yüksel, B., Ustaoğlu, F., Tokatlı, C., İslam, S.: Ecotoxicological risk assessment for sediments of Çavuşlu Stream in Giresun, Turkey: association between garbage disposal facility and metallic accumulation. Environ. Sci. Pollut. Res. **29**, 17223–17240 (2022)
8. Jannat, J.N., et al.: Hydrochemical assessment of fluoride and nitrate in groundwater from east and west coasts of Bangladesh and India. J. Clean. Prod. **14**, 133675 (2022). https://doi.org/10.1016/j.jclepro.2022.133675
9. İslam, S., et al.: Potentially toxic elements in vegetable and rice species in Bangladesh and their exposure assessment. J. Food Comp. Anal. **106**, 104350 (2022)
10. Varol, M., Tokatlı, C.: Seasonal variations of toxic metal(loid)s in groundwater collected from an intensive agricultural area in Northwestern Turkey and associated health risk assessment. Environ. Res. **204**, 111922 (2022)
11. Haque, J., et al.: Effects of Covid-19 era on a subtropical river basin in Bangladesh: Heavy metal(loid)s distribution, sources and probable human health risks. Sci. Total. Environ. **857**, 159383 (2023). https://doi.org/10.1016/j.scitotenv.2022.159383
12. Mia, Y., et al.: Identifying factors affecting irrigation metrics in the Haor Basin using integrated Shannon's entropy, fuzzy logic and automatic linear model. Environ. Res. **226**, 115688 (2023)
13. http://www.cerkezkoy.gov.tr
14. http://www.corlu.gov.tr
15. Anonymous: Meriç-Ergene basin industrial wastewater management main plan study final report. T.R. Ministry of Environment and Forestry (2010)
16. Tokatlı, C.: Drinking water quality assessment in villages located in Meriç River Basin (Edirne, Turkey). Sigma J. Eng. Nat. Sci. **36**(3), 871–886 (2018)
17. Tokatlı, C.: Use of the potential ecological risk index for sediment quality assessment: a case study of dam lakes in the Thrace part of the Marmara Region. Aquat. Sci. Eng. **34**(3), 90–95 (2019)
18. Tokatlı, C.: Pesticide residues in water and sediment of Ergene River and tributaries in Turkey. Sigma J. Eng. Nat. Sci. **38**(1), 361–370 (2020)
19. Orak, E., Akkoyunlu, A., Can, Z.S.: Assessment of water quality classes using self-organizing map and fuzzy c-means clustering methods in Ergene River. Turkey. Environ. Monit. Assess. **192**, 638 (2020)
20. Tokatlı, C., Köse, E., Çiçek, A., Emiroğlu, Ö.: Pesticide accumulation in Turkey's Meriç River basinwater and sediment. Polish J. Environ. Stud. **29**(1), 1–6 (2020)
21. Emadian, S.M., Sefiloglu, F.O., Balcioglu, İA., Tezel, U.: Identification of core micropollutants of Ergene River and their categorization based on spatiotemporal distribution. Sci. Total. Environ. **758**, 143656 (2021)
22. Tokatlı, C., Islam, A.R.T.: Spatial-temporal distributions, probable health risks and source identification of organic pollutants in surface waters of an extremely hypoxic river basin in Türkiye. Environ. Monit. Assess. **195**, 435 (2023)
23. APHA (American Public Health Association): Standard Methods for the Examination of Water and Wastewater. American Public Health Association, Washington DC (2005)
24. EPA (Environmental Protection Agency), METHOD 200.7.: Determination of Metals and Trace Elements in Water and Wastes by Inductively Coupled Plasma-Atomic Emission Spectrometry (2001)
25. Richards, L.A.: Diagnosis and Improvement of Saline and Alkali Soils. United States Department of Agriculture, Washington, DC (1954)
26. Wilcox, L.V.: Classification and Use of Irrigation Waters, pp. 1–19. Unıted States Department of Agrıculture, Washington, DC (1955)

27. Raghunath, I.I.M.: Groundwater, 2nd edn. Wiley Eastern Ltd, New Delhi (1987)
28. Kelly, W.P.: Use of saline irrigation water. Soil Sci. **95**(6), 385–391 (1963)
29. Tokatlı, C., Baştatlı, Y., Elipek, B.: Water quality assessment of dam lakes located in Edirne Province (Turkey). Sigma J. Eng. Nat. Sci. **35**(4), 743–750 (2017)
30. Çingiroğlu, F.: Determination of river pollution sources using source apportionment method: Ergene River. Master's Thesis. Istanbul Technical University (2018)
31. Tokatlı, C.: Health risk assessment of toxic metals in surface and groundwater resources of a significant agriculture and industry zone in Turkey. Environ. Earth Sci. **80**, 156 (2021)
32. Tokatlı, C.: Invisible face of Covid-19 pandemic on the freshwater environment: an impact assessment on the sediment quality of a cross boundary river basin in Turkey. Int. J. Sedim. Res. **37**, 139–150 (2022)
33. Celen, M., Oruc, H.N., Adiller, A., Töre, G.Y., Engin, G.O.: Contribution for pollution sources and their assessment in urban and industrial sites of Ergene River Basin, Turkey. Int. J. Environ. Sci. Technol. **19**, 11789–11808 (2022). https://doi.org/10.1007/s13762-022-03919-0
34. Varol, M., Tokatlı, C.: Evaluation of the water quality of a highly polluted stream with water quality indices and health risk assessment methods. Chemosphere **311**, 137096 (2023)

Geosciences and Geotechnical Engineering

Thermophysical Properties of an Eco-friendly Mortar Incorporating Drinking Water Treatment Sludge

Oumaima Bourzik[1](✉), Khadija Baba[1], and Nacer Akkouri[2]

[1] Civil Engineering and Environment Laboratory (LGCE), Mohammadia Engineering School, Mohammed V University, Rabat, Morocco
oumaimabourzik27@gmail.com

[2] Moroccan Foundation for Advanced Science Innovation and Research (MAScIR), University Mohamed 6 Polytechnic (UM6P), Batteries and Smart Materials Center, Benguerir, Morocco

Abstract. The process of purifying water generates a significant amount of sludge from treating drinking water, which can lead to an ecological imbalance when it accumulates in the environment. Currently, about 73% of treated sludge is repurposed for agricultural use, where it can be transformed into compost or fertilizer for land application. The decision to repurpose the sludge into compost is generally made by local authorities. One potential approach for recycling this sludge involves partially replacing sand with it in cement mortars. While previous studies have investigated the mechanical properties of cement mortars that contain drinking water treatment sludge (DWTS), this study is novel in its exploration of the sludge's thermophysical properties, such as its thermal conductivity and thermal diffusivity. To conduct the study, a group of five mortar mixtures was prepared, with sand substitution levels ranging from 0% to 20%. Various tests were conducted to assess the mixtures' workability, porosity, density, and thermal properties. The results of these tests indicated a decrease in workability, density, and thermal conductivity, as well as an increase in porosity. Thus, the study demonstrated that replacing sand with DWTS in cement mortars leads to an improvement in thermal properties, relative to conventional mixtures, and suggests that DWTS can effectively replace sand in cement mortars. In addition to its advantages for the construction industry as a source of raw materials, the utilization of DWTS for cement mortars provides an eco-friendly solution for disposing of DWTS and preserving natural aggregate reservoirs.

Keywords: Drinking water treatment sludge · Eco-friendly mortar · Thermophysical properties · Materials recycling · Energy efficiency

1 Introduction

Water purification processes generate a substantial amount of sludge during the treatment of drinking water (DWTS) [1]. The accumulation of this sludge in the environment can pose significant ecological challenges [2]. Currently, approximately 73% of treated

K. Baba et al. (Eds.): GeoME 2023, SPEES, pp. 163–171, 2023.
https://doi.org/10.1007/978-3-031-49345-4_17

sludge is repurposed for agricultural use, where it is transformed into compost or fertilizer for land application [3]. However, there is a need to explore alternative methods for utilizing this sludge effectively to minimize its environmental impact and promote sustainable practices [4].

One potential approach for recycling drinking water treatment sludge (DWTS) involves its incorporation into cement mortars by partially replacing sand [5]. Cement mortars are widely used in construction for various applications, including plastering, masonry, and flooring [6]. The addition of DWTS to cement mortars not only offers a solution for sludge disposal but also presents an opportunity to enhance the properties of the resulting mortars [7].

Previous studies have primarily focused on evaluating the mechanical properties of cement mortars containing DWTS [8]. These investigations have demonstrated the feasibility of using DWTS as a substitute for sand, highlighting its potential for improving the strength, durability, and workability of mortars [6]. However, little attention has been given to exploring the thermophysical properties of DWTS-incorporated mortars, particularly their thermal conductivity and thermal diffusivity.

Thermophysical properties play a crucial role in the energy performance and thermal behavior of construction materials [9]. By investigating the thermal properties of DWTS-incorporated cement mortars, valuable insights can be gained regarding their suitability for applications where thermal insulation or thermal storage is desired [10]. This knowledge can contribute to the development of innovative and eco-friendly construction materials that not only meet structural requirements but also offer enhanced energy efficiency [11].

In light of this research gap, the present study aims to explore the thermophysical properties of cement mortars containing DWTS. A series of mortar mixtures were prepared, with varying levels of sand substitution, ranging from 0% to 20%. Comprehensive tests were conducted to assess the workability, porosity, density, and most importantly, the thermal properties of the mixtures, including thermal conductivity and thermal diffusivity.

The findings of this study are expected to have far-reaching implications for the construction industry and environmental conservation. Architects, engineers, and policymakers can benefit from these insights as they develop and implement sustainable building practices. Moreover, the incorporation of DWTS into cement mortars may not only provide a viable waste management solution but also lead to the creation of environmentally friendly and energy-efficient construction materials that help combat climate change and reduce the overall ecological footprint of the construction sector.

By pushing the boundaries of our understanding of construction materials' thermophysical properties, this research can pave the way for the widespread adoption of DWTS-incorporated cement mortars and promote a circular economy where waste is turned into a valuable resource. As the world seeks to address environmental challenges, the exploration of alternative methods for waste utilization in construction represents a critical step towards a more sustainable and resilient future.

2 Materials and Methods

2.1 Materials

The drinking water treatment sludge (DWTS) used in this study was sourced from the Bouregreg water treatment plant located in the vibrant city of Rabat, Morocco. The sludge was carefully collected from the settling tank of the plant and subjected to a series of preparatory steps. Firstly, it underwent a meticulous drying process to reduce its moisture content and facilitate subsequent handling. Subsequently, the dried sludge was subjected to a crushing process, breaking it down into smaller particles to ensure uniformity and consistency in the mortar mixtures.

To further refine the DWTS particles, a sieving process was employed, targeting a specific particle size of 2 mm. This sieving step played a crucial role in ensuring that only particles within the desired range were included in the mortar mixtures, thereby optimizing the performance and homogeneity of the final cement mortars.

In crafting the mortar mixtures, CPJ 45 Portland cement was chosen as the binder material. This type of cement boasts a minimum clinker content of 65%, ensuring a high-quality and robust binding agent for the mortar. Its selection was based on its proven track record of performance in construction applications and its compatibility with incorporating supplementary materials like DWTS.

As for the sand component of the mortar mixtures, it was sourced from the Rabat-Sale-Kenitra region, specifically obtained from sea deposits. This natural sand selection was made meticulously, considering factors such as its mineral composition, particle size distribution, and geotechnical properties, all of which contribute to the overall characteristics of the cement mortars.

By using this specific sand variety, the study aimed to explore how the combination of DWTS and natural sea sand influences the thermophysical properties of the resulting cement mortars. This holistic approach not only seeks to enhance our understanding of sustainable construction materials but also provides valuable insights into the potential environmental benefits of utilizing locally available resources.

2.2 Mortar Preparation

In order to comprehensively investigate the impact of Drinking Water Treatment Sludge (DWTS) on various properties of cement mortars, an extensive study was undertaken. The objective was to gain valuable insights into how the incorporation of DWTS affects the workability, density, porosity, and thermal properties of the resulting cement mortars. To achieve this, a series of carefully designed mortar mixtures were meticulously prepared, each involving the substitution of sand with different proportions of DWTS particles.

Table 1 outlines the mix proportions of the various mortar compositions, providing a clear overview of the different combinations used in the experimental study. The mix designations (Mix 0, Mix 5, Mix 10, Mix 15, and Mix 20) represent the varying rates of DWTS substitution, ranging from 0% to 20%. For each mixture, a constant water-to-cement ratio (W/C) of 0.5 was maintained to ensure consistency in the experimental conditions. This precise control of the W/C ratio helped in analyzing the specific effects of DWTS on the mortars without introducing additional confounding factors.

To fabricate the mortar samples for analysis, cubic specimens with distinct dimensions were carefully produced. The selection of different sample dimensions aimed to capture a comprehensive understanding of how DWTS incorporation influenced the properties of cement mortars at varying scales.

The investigation aimed to shed light on several crucial aspects of the cement mortars, including their workability, which reflects how easy the mixture is to handle and apply during construction processes. Additionally, the study delved into the density and porosity of the mortars, exploring their structural characteristics and potential permeability. Understanding these properties provides valuable information about the suitability of the DWTS-incorporated mortars for various construction applications.

Moreover, the research placed significant emphasis on examining the thermal properties of the cement mortars, including their thermal conductivity and thermal diffusivity. These thermophysical properties play a vital role in determining the energy performance and thermal behavior of construction materials. As such, the investigation sought to unravel how the presence of DWTS particles influenced the heat transfer characteristics of the cement mortars, which can have implications for thermal insulation and energy efficiency in building applications.

Table 1. Mix proportions of mortars

Mix designation	W/C	Cement (g)	Sand (g)	DWTS (g)
Mix 0	0.5	450	1350	0
Mix 5	0.5	450	1282.5	67.5
Mix 10	0.5	450	1215	135
Mix 15	0.5	450	1147.5	202.5
Mix 20	0.5	450	1080	270

2.3 Experimental Testing

The workability of the cement mortars was meticulously evaluated through the widely recognized flow table test. This test provides valuable data on how easily the mortar mixture flows and spreads, which is of paramount importance for construction applications. By subjecting the DWTS-incorporated mortars to this test, the researchers gained a clear understanding of how the presence of DWTS affected the ease of handling and application of the cement mortars.

Porosity and density, key structural characteristics of the cement mortars, were determined using the ASTM C642-06 standard, a widely accepted and rigorous testing method. By adhering to this standard, the research team was able to obtain accurate and reliable data on the porosity and density of the mortars with varying proportions of DWTS. This information is crucial for assessing the potential structural performance and permeability of the DWTS-incorporated mortars, offering valuable insights for engineering applications.

Thermal properties, including thermal conductivity and thermal diffusivity, were meticulously measured using the transient planar source (TPS) method. The TPS method is well-regarded for its accuracy and precision in determining these critical thermophysical parameters. Employing a hot disc instrument along with a Kapton sensor, the researchers were able to quantify how the DWTS affected the heat transfer behavior within the cement mortars. This data is of significant importance for applications that require thermal insulation or thermal storage capabilities in construction materials.

3 Results and Discussion

3.1 Workability

Figure 1 examined the workability of cement mortars with different replacement ratios of drinking water treatment sludge (DWTS) particles. Workability was measured using arbitrary units, where higher values indicate better workability. The results showed a consistent trend of decreasing workability as the DWTS replacement ratio increased. The mortar without any DWTS replacement (Mix 0) had the highest workability value of 120, while the workability decreased to 80, 62, 57, and 50 for Mix 5, Mix 10, Mix 15, and Mix 20, respectively. The decrease in workability can be attributed to the presence of fine particles or impurities introduced by DWTS, which hindered the movement and flow of the mortar mixture.

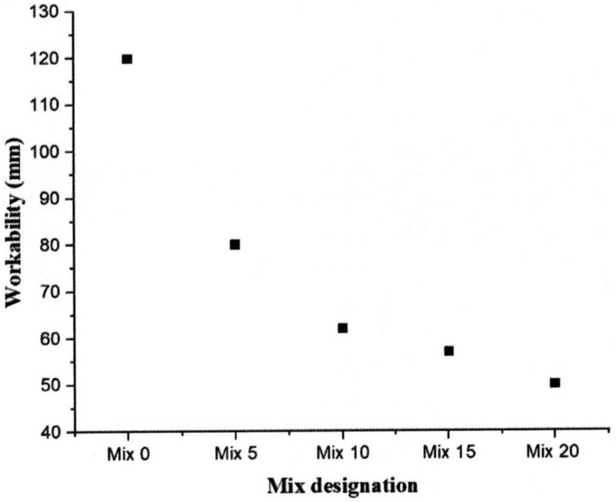

Fig. 1. Workability of samples

3.2 Porosity and Density

Figure 2 illustrates the density and porosity results obtained from the investigation on cement mortars with varying replacement ratios of DWTS particles. As the DWTS replacement ratio increased, the density of the mortars decreased. The decrease in density can be attributed to the lighter composition of DWTS particles compared to natural sand.

In terms of porosity, an increase was observed as the DWTS replacement ratio increased. This increase in porosity is likely due to the presence of additional voids or empty spaces caused by the irregular shape or presence of pores in DWTS particles.

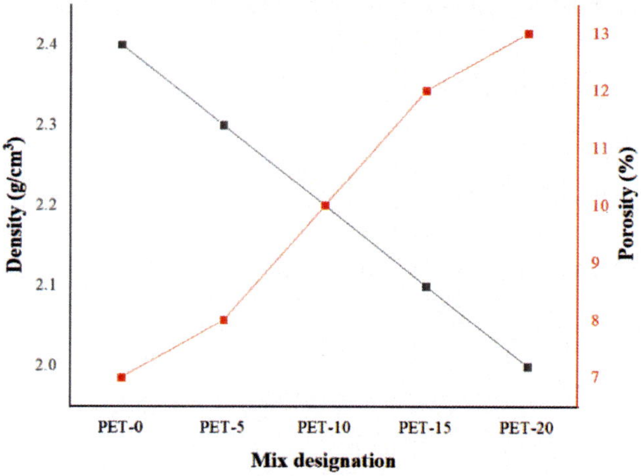

Fig. 2. Porosity and density of samples

3.3 Thermal Properties

Figure 3 shows the thermal properties of cement mortars with varying replacement ratios of drinking water treatment sludge (DWTS) particles. As the DWTS replacement ratio increased, the thermal conductivity decreased, indicating lower heat conduction. The highest thermal conductivity was observed in Mix 0 (1.303 W/(m·K)), while Mix 20 had the lowest value (0.693 W/(m·K)). The thermal diffusivity showed some variation but did not follow a consistent trend. The findings suggest that DWTS inclusion reduces thermal conductivity and highlights the complex nature of heat transfer within the mortar.

4 Conclusions

In this comprehensive study, the focus was on meticulously examining the effects of incorporating Drinking Water Treatment Sludge (DWTS) on various essential properties of cement mortars. The results unveiled insightful trends and correlations, shedding light on how different DWTS replacement ratios impacted the mortar's characteristics.

One of the key findings of this investigation was that as the replacement ratio of DWTS increased in the mortar mixtures, the workability of the cement mortars decreased. This can be attributed to the presence of fine particles and impurities introduced by the DWTS. These particles may have hindered the smooth flow and interlocking of the mortar

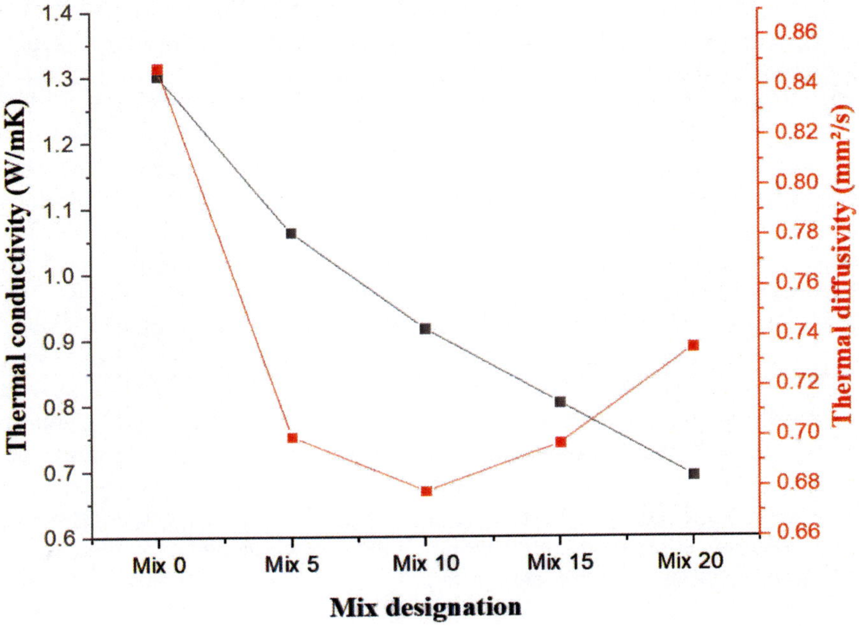

Fig. 3. Thermal conductivity and thermal diffusivity of samples

components, affecting its ease of handling and application during construction processes. This reduction in workability underscores the importance of carefully optimizing the DWTS content to strike a balance between sustainability and practicality in real-world construction applications.

Moreover, the study revealed that the density of the cement mortars experienced a decline with higher DWTS replacement ratios. This decline can be attributed to the specific gravity of DWTS particles being lower than that of natural sand typically used in mortar formulations. As a result, the incorporation of DWTS contributed to a higher number of voids or empty spaces within the mortar matrix, leading to an increase in porosity. This heightened porosity can have implications for the mortar's structural integrity, permeability, and susceptibility to environmental factors, warranting careful consideration in engineering applications.

The investigation further delved into the thermal properties of the DWTS-incorporated cement mortars. The thermal conductivity of the mortars exhibited a consistent decrease with higher DWTS replacement ratios. This trend can be attributed to the lower thermal conductivity of DWTS particles in comparison to conventional sand. As a consequence, the overall thermal conductivity of the mortars was influenced by the proportion of DWTS present in the mixtures. This finding holds great significance for applications that require thermal insulation properties in construction materials, presenting potential opportunities for sustainable and energy-efficient building practices.

Intriguingly, the thermal diffusivity, another vital thermophysical property, did not exhibit a clear trend with increasing DWTS replacement ratios. This observation adds complexity to the understanding of how DWTS affects the heat transfer behavior within cement mortars. Further research and analyses may be warranted to decipher the underlying mechanisms behind this phenomenon and its practical implications for thermal performance in construction applications.

The insights obtained from this study align with previous research on alternative materials in cement mortars, confirming the importance of exploring eco-friendly alternatives and their potential effects on various mortar properties. As these changes in properties can have implications for the mechanical strength, thermal behavior, and long-term durability of the mortar, it is paramount for architects, engineers, and material scientists to weigh these factors carefully when considering the use of DWTS-incorporated cement mortars in real-world construction projects.

In conclusion, this study has contributed valuable knowledge to the field of sustainable construction materials and offers a stepping stone for future research on enhancing the performance and applicability of DWTS-incorporated cement mortars. The findings underscore the importance of a holistic approach to material selection and engineering, considering not only the environmental impact but also the functional properties required for reliable and resilient construction practices. As the construction industry continues to embrace sustainable innovations, the study's findings can serve as a guiding compass for adopting environmentally conscious building practices while maintaining the structural and thermal performance of cement mortars.

References

1. Bourzik, O., et al.: Study of the effects of drinking water treatment sludge on the properties of class F fly ash-based geopolymer. Environ. Sci. Pollut. Res. (2022). https://doi.org/10.1007/s11356-022-21873-9
2. Benlalla, A., Elmoussaouiti, M., Cherkaoui, M.: Characterization and valorization of drinking water sludges applied to agricultural spreading (2015)
3. Foroughi, M., Chavoshi, S., Bagheri, M., Yetilmezsoy, K., Samadi, M.T.: Alum-based sludge (AbS) recycling for turbidity removal in drinking water treatment: an insight into statistical, technical, and health-related standpoints. J. Mater. Cycles Waste Manage. **20**, 1999–2017 (2018). https://doi.org/10.1007/s10163-018-0746-1
4. Bohórquez González, K., Pacheco, E., Guzmán, A., Avila Pereira, Y., Cano Cuadro, H., Valencia, J.A.F.: Use of sludge ash from drinking water treatment plant in hydraulic mortars. Mater. Today Commun. **23**, 100930 (2020). https://doi.org/10.1016/j.mtcomm.2020.100930
5. Dahhou, M., El Moussaouiti, M., Arshad, M.A., Moustahsine, S., Assafi, M.: Synthesis and characterization of drinking water treatment plant sludge-incorporated portland cement. J. Mater. Cycles Waste Manag. **20**, 891–901 (2018). https://doi.org/10.1007/s10163-017-0650-0
6. Li, D., et al.: Reuse of drinking water treatment sludge in mortar as substitutions of both fly ash and sand based on two treatment methods. Constr. Build. Mater. **277**, 122330 (2021). https://doi.org/10.1016/j.conbuildmat.2021.122330
7. Frías, M., Vigil de la Villa, R., de Soto, I., García, R., Baloa, T.A.: Influence of activated drinking-water treatment waste on binary cement-based composite behavior: characterization and properties. Compos. Part B: Eng. **60**, 14–20 (2014). https://doi.org/10.1016/j.compositesb.2013.12.020
8. Messina, F., et al.: Synergistic recycling of calcined clayey sediments and water potabilization sludge as geopolymer precursors: upscaling from binders to precast paving cement-free bricks. Constr. Build. Mater. **133**, 14–26 (2017). https://doi.org/10.1016/j.conbuildmat.2016.12.039

9. Oumaima, B., Nacer, A., Khadija, B., Sana, S., Abderrahman, N.: Experimental study of the thermal and mechanical properties of eco-friendly cement mortar incorporating recycled pet and PP. Rehabend 2022 1600–1608 (2022)
10. Bourzik, O., Baba, K., Akkouri, N., Nounah, A.: Effect of waste marble powder on the properties of concrete. Mater. Today Proc. (2022). https://doi.org/10.1016/j.matpr.2022.07.184
11. Akkouri, N., Oumaima, B., Baba, K., Tayeh, B.: Mechanical and thermal performances of eco-friendly mortar containing recycled PET as partial sand replacement (2021)

Check for updates

Reinforcement of a Rock Mass Slope by Prestressed Anchors and Analysis of It's Behavior Under Tensioning: A Case Study

Rhita Bennouna[1](✉), Latifa Ouadif[1], Ahmed Akhssas[1], Youssef Zerradi[2], Ghizlane Boulaid[1], and Ahmed Skali Senhaji[3]

[1] Laboratory of Applied Geophysics, Geotechnics, Engineering Geology and Environment (L3GIE), Mohammadia School of Engineers, Med V University, Rabat, Morocco
bennounarhita@gmail.com

[2] Laboratory of Analysis and Modelling of Water and Natural Resources (LAMERN), Mohammadia School of Engineers, Med V University, Rabat, Morocco

[3] Setec Maroc, Rabat, Morocco

Abstract. The reinforcement by active anchoring rods technique has been widely used for various cases of studies with approved performances and many researches analyzed this technique of stabilization generally in case of granular soils. The purpose of this study is to analyze the response of a rock formation towards the tensioning of the anchor rods by analyzing the results of the pull-out tests for the case of a schist embankment presenting an alteration on the surface and on which a highly trafficked road had been constructed, and where cracks have appeared, which is explained by the formation of a deep landslide. This work consists, on the first hand, in studying the tensioning mechanism of the anchoring tie rods: its characteristics, the mode of injection, the composition of the grout used, the pressures to be applied for each bearing, etc. On the second hand, we carried out an analysis of the evolution of the tensioning with the displacements on the test tie rods before comparing these displacements with the estimated theoretical displacements, and finally we analyzed the behavior of the tie rods at each level in order to find the critical tensile creep stresses. The tearing tests carried out on the T1 and T2 tie rods revealed a displacement of the head of the tie rods corresponding to the last tension level equal to 94.41 mm and 87.71 mm respectively. The plots of displacements versus logarithm of time are in the form of a linear function at each loading step. After examining the slopes of the regression lines calculated from the pairs of values measured between 5 min and 60 min, it was possible to find the values of the critical creep tensions for each anchor analyzed Tc1 = 886 KN and Tc2 = 996 KN. And to compare them with the theoritical value calculated from the standards. The pull-out tests carried out make it possible to fully understand the behavior of an anchor tie rod and its installation in the ground, and to examine the critical creep tensile force and the ultimate resistance in relation to the ground conditions encountered and the materials used before proceeding with the installation of the final tie rods, in order to avoid any rock mass failure that could cause physical and material damage.

Keywords: Tie rods · Creep · Tensioning · Pull-out tests · Displacements

K. Baba et al. (Eds.): GeoME 2023, SPEES, pp. 172–182, 2023.
https://doi.org/10.1007/978-3-031-49345-4_18

1 Introduction

The stability analysis of slopes have been widely studied using different types of prospections [1, 2, 3, 4] in order to choice the convenient reinforcing method. The reinforcement by prestressed anchors has been adopted for a long time in many research projects, such as dams [5] and slope stabilization [6, 7] or others, for various types of formation and especially for rock formations that have good geotechnical characteristics. This technique is a special one, since it present an ability to transmit a tensile load applied to a rock mass or resistant soil layer in order to reinforce and stabilize slopes, excavations, sheet piles, diaphragm or Berlin walls and retaining walls and is also used in dam engineering [8, 9]. The tie rods have two distinct parts; free and sealed, each of them has a particular function to ensure that the tie rods are held securely and operate correctly when it's under tensioning. The tie rods is made up of a bar or cables with a set of strands, that quantity depends in particular on the expected tensile force to be applied and the elastic limit of the tie rods. Depending on the nature of the structure to be stabilised, either an active (prestressed) or passive anchor could be used, and either a permanent or temporary anchors. As for the choice of the inclination of the tie rods, there is no restriction in this sense, they can be vertical, horizontal or inclined. Many researchers have already studied the application of this technique in rock mechanics for various structures.

This technique take into account several factors to ensure its proper functioning: The steel tendon tensile failure, the grout- tendon interface, the grout—rock interface and finally the nature of the rock formation in place. It is important to meticulously study the whole system to avoid any risk of failure. To do this, a set of tests on the tie rods are carried out in different phases of the project: conformity, inspection and acceptance tests (tensioning). We will focus below on the study of tie rods tensioning during the conformity testing phase and the analysis of the results obtained. The structure, for which this study is carried out, is a road slope resting on a schist formation and presenting a risk of sliding according to two shear surfaces.

2 Materials and Methods

1) Justification of Tie Rods at SLS Serviceability Limit State

To have a stress level of a sealed anchor that remain below its critical creep load during the service life, the calculated value of the tension applied to the anchor in service condition should be lower than the calculated value of the critical creep resistance of the seal (1).

$$F_k \leq R_{cr;d} \tag{1}$$

In order to find the value of $R_{cr;d}$, equations has been developped (guide TA20) (2), (3), (4), (5) & (6). Charts and coefficients should be used which depend on the nature of the geological formations in place and also on the type of sealing grout injection (global unit injection IGU or repetitive and selective injection IRS). We will focus on the case of rock formation where the injection is done in IGU. The pre-sizing chart for anchor rods sealed in weathered and fragmented rocks (Figure H.4 of the guide TA2020 [10]) makes it possible to find the value of the limit lateral friction qs, which will be used in the calculation of $R_{cr;d}$ (5) & (6).

For the serviceability limit state:

$$R_{cr;d} = R_d/\gamma_{serv} \tag{2}$$

With [10, 11] $\gamma_{serv} = 1.35$ and

$$R_d = R_k/\gamma_{a;ELU};\ \gamma_{a;ELU} = 1.1 \tag{3}$$

$$R_k = R_s/\gamma_{Rd};\ \gamma_{Rd} = 1.4 \tag{4}$$

$$R_s = \pi.\emptyset.\alpha_S.q_s.L_S \tag{5}$$

With:

- α_S: a coefficient that depends on the type of geological formation in place, [10, 12].
- qs: limit skin friction,
- Ø: borehole diameter.

For the case of multilayers:

$$R_s = \pi.\emptyset.\int_0^{Ls} \alpha_S.q_s(l)dl \tag{6}$$

The product $\gamma_{serv} * \gamma_{a;ELU} * \gamma_{Rd}$ gives a value close to 2. This value matches the value of the safety factor Ft of the guide Tirant d'ancrage TA20.

2) **Tension losses during the test**

The results of the pull-out test must be as precise as possible, taking into account the various losses likely to occur in order to avoid any over-evaluation which could lead to a permanent excess of tension in the reinforcement or any under-evaluation which could lead to a very low residual tensile value compared to that expected in the test. Both cases can be detrimental to the safety of the structures. There are three categories of losses: during tensioning, during blocking of the tie rod and deferred losses.

Losses during tensioning that occurs in this conformity test are the only losses that will be considered. These losses are due to friction in the cylinder and the anchoring devices.

The percentage of friction in the cylinder is found on the cylinder calibration table presented by its supplier during calibration, and is calculated by the average of the friction values below 50% since the cylinder will work at approximately 50%. As for the friction percentage of the anchor head, it is conventionally taken equal to 1%.

The real pressure applied to each bearing is then calculated following the Eq. (7).

$$P = \frac{T}{S * (1 - f)} \tag{7}$$

With T: the tensile force applied at the head in KN.
S: the section of the piston in cm^2.
f: the percentage of cylinders friction and those of the anchor head in %

3) **Displacements measurement during the test**

It is important to note that before proceeding with the test, the estimated displacement is calculated on the basis of the tie rods characteristics. The loading phase of the test is normally stopped when the tensile force 0.9 × Tp has been reached or when the displacement of the tie rod head has exceeded the estimated breaking displacement Δles. The estimated displacement Eq. (8) is stated in the standard NF P 94 153.

$$\Delta Les\left(the'o\right) = \frac{0.9 * Tp * (Ll + Lsc + Le)}{A * E} + \Delta\,\lg \tag{8}$$

With.

Tp the conventional elastic limit tensile force of the rods steel
Le the external length
Δlg is the displacement corresponding either to the sliding of the rods steel in the sealing grout, or to the sliding of the sealing grout in the ground.
E young's modulus of the materials of the rods steel
A air of the cross-section of the rods steel

On the other hand, in order to obtain displacements conforming to the different bearings and that do not present any problem, their values must be within the elongation zone determined by the two straight lines corresponding to the minimum elongation (LL + Le) and the maximum elongation (LL + Le + Ls/2) relative to the length.

3 Results and Discussions

1) **Presentation of case study**

In this study, we will focus on a road embankment presenting risks of instability. The slope under the road is at limit equilibrium and the movements are favored by probable water flows at the interfaces between more or less permeable formations.

The pressuremeter and cored boreholes carried out revealed the existence of a very altered and friable schist layer which has very weak geomechanical characteristics over a depth of 23 m, resting on the bedrock (Fig. 1).
The inclinometers put in place have shown an evolution of the cracks on the ground of nearly 2 cm per month, with no sign of slowing down, hence the importance of an optimal treatment to slow down the displacement of the embankment as much as possible, while knowing that road traffic is modeled with a load of 10 kPa.
The stability in this case of study is a combination of treatments, stabilization of the slides by piles and prestressed anchors (Fig. 2). A verification of the sliding stability is essential in order to justify the resistance of the structure.

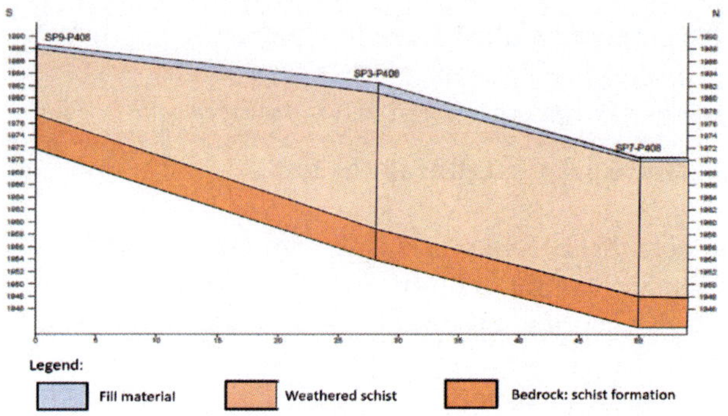

Fig. 1 Lithological model of the study area

The reinforcement by pre-stressed anchors can be carried out in the stable rock mass to a depth of 10 to 40 m, crossing the slip plane. Therefore, we take advantage of the high tensile strength of the rods steel wires to increase the normal force and the shear strength of the rock mass (particularly in the weak structure plane) in order to improve the mechanical behavior of the rock formation to perform the role of reinforcement. Actually, There are two parts in the tie rods: the free and the sealing length. It is through the sealing length, that is anchored in the substratum, that the reinforcement transmits to the slope (needing to be stabilized) the forces which is applied to it. Over its free length, the rods, not sealed to the ground, are protected by a sheath or by a tube inside which it can slide freely, without transmitting any force to the surrounding ground.

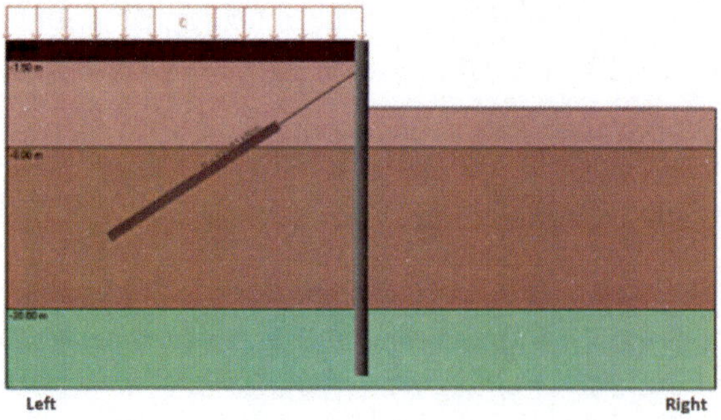

Fig. 2 The adopted geotechnical model

Since it is a reinforcement that is not self-stabilizing, the anchoring method considered was that of prestressed tie rods of 500 kN/anchor. They are made up of strands and their adopted characteristics are listed in Table 1.

The tie rods are linked by a reinforced concrete girder which is supported at the tie rods heads at -2m from the piles head. It allows a longitudinal groupment of the tie rods. The strands of the tie rods are partially sealed with a cement grout that is injected by the IGU unit global injection method [10, 12].

Table 1 Tie rods characteristics

Spécification	Valeur
External diameter	15.7 mm
Number of strands for each tie rod	5
Young's modulus	195000 N/mm^2
Elastic limit	246 KN
Free length	7 m
Sealed length	18 m
Tilt	27°

The execution of the tie rods begins with the preparatory operations; it includes the hole drilling, the installation of the rods, the injection of the grout in the hole in order to seal the tie rods on the ground.
The composition of the grout is $W/C = 0.5$ having a compressive strength greater than 30 MPa at 28 days in order to ensure a suitable tie-ground contact and to support the residual tensile stress in the tie rods.
After sufficient hardening of the seal, we proceed to the conformity tests for three tie rods, to the control tests, and to the tensioning.
In this document, we are interested in conformity tests in order to analyze the movement of the tie rods and the good performance of the structure during its lifetime.
These tests must be continued until the seal breaks. This is why the tie rods used in the test cannot be a part of the projet reinforcement.
The conformity test procedure consists of measuring the displacements of the anchor head during loading. The general process of the tests consists in obtaining, for the first anchor, the displacement curves for increasing loads stages and to determine from these curves the critical creep tension or the limit tension of sealing, then to determine, for the second tie, the value of the loading stages, taking into account the limit tension resulting from the first test.
The third tie rod test will be identical to the first one.

2) **Pull out test results**

The application of tensioning on the test tie rods allow to follow step by step the evolution of the tensioning with the respective displacements illustrated in Fig. 3(a) & (b). We will focus on the results of the most representative test tie rods: T1 and T2.

It should be noted that the tension losses have been calculated and present a percentage of the overall friction of 3.74%, i.e. the sum of 2.74% (percentage of friction in the cylinder) and 1% (percentage of friction of the anchor head).
The measurement of the displacement along the axis of symmetry of the tie rod is made at its free end, in relation to a fixed reference placed outside the effect of the stresses to which the reaction block is subjected.

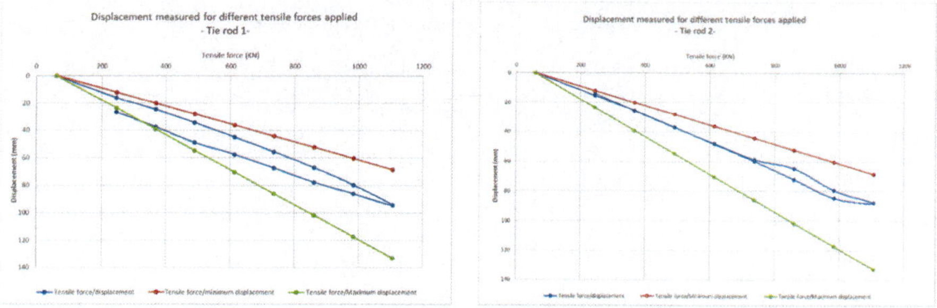

(a) (b)

Fig. 3 The displacements measured for different tensile forces applied for Tie rods T1 (a) and T2 (b)

The pull out tests carried out on the two tie rods T1 and T2 revealed a displacement of the tie rods head during the tensioning which reached maximum values corresponding to the last tension level (0.9*Tp) of 94.41 mm and 87.71 mm respectively.
Even when reaching the tractive effort 0.9*Tp, the recorded displacements remained lower than the estimated displacements that is equal to $\Delta les = 218.10$ mm. That said, the conventional limit tensile force by breaking of the seal is noted greater than or equal to the value of the last loading level. On the other hand, we note that the values of the displacements measured at each intermediate point for the tie rod T1 and T2 are in the zone of the displacements of the minimum and maximum elongations plotted.
By examining the displacements caused by the tensioning for each loading level, we find the critical creep tensile force Tc. To do this, we first obtain the curves of the displacements as a function of the logarithm of time for the two tests T1 and T2 (Fig. 4(a) & (b)). It is noticed that displacements are presented by linear functions for each stage of loading.

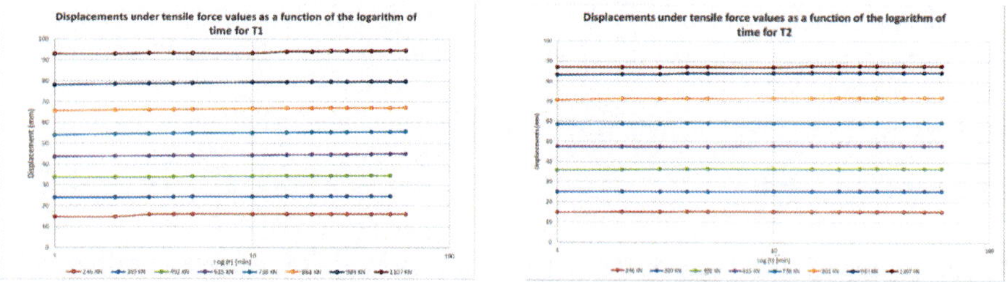

(a) (b)

Fig. 4. The displacements of the tie rods under tensile force values as a function of the logarithm of time for T1 (a) and T2 (b)

Note that the transitions between bearings for the tie rods T1 and T2 are characterized by an average displacement of 9.90 mm and 10.68 mm respectively.

In order to examine the creep phase in more detail, we plot the curve of the slopes αn of the regression lines as a function of the forces applied point by point (Fig. 5(a) &

(b)). αn is the slope of the regression line calculated from the pairs of values measured between 5 min and 60 min expressed in (mm/lgt cycle).

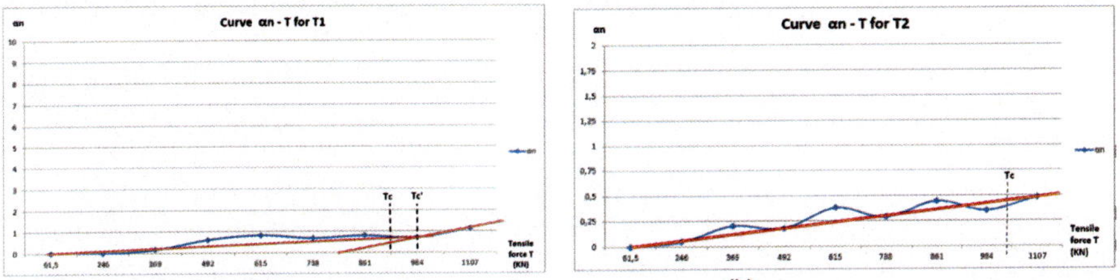

(a) (b)

Fig. 5 The α_n curves as a function of applied tensile force for Tie rods T1 (a) and T2 (b)

For the first stages, a slight increase in the average slope α is observed with the tensile force. Beyond this stages, we can notice that:

- For the test on T1, the slope α increases beyond a characteristic point of inflection, and the relationship α–T then presents a first quasi-linear part extended by a curved part.
- For the test on T2, the slope α continues to increase monotonously until the end of the test (until 0.9 Tp). The α–T relationship in presented in figure W. This type of test illustrates the case where the limit tension TL is located beyond the maximum tension applied. This has already been confirmed in the test results.

The critical creep tensile forces $R_{ELS;m1}$ et $R_{ELS;m2}$ were found from the αn-T graphs (Table 2).

Table 2 The critical creep tensile forces mesured in pull-out tests on T1 & T2.

The critical creep tensile forces	Value (KN)
$R_{ELS}; m1$	886
$R_{ELS}; m2$	996

For a series of tests, it is necessary to define the measured value that represent all the values obtained; It's the smallest measured values of the sealing critical creep tensile forces.

$$R_{cr:k} = (R_{ELS;m})_{min} = 886\ \text{KN}$$

3) **The calculated value of the critical tensile force**

Based on the equations developed (2), (3), (4), (5) & (6), and considering the existing layers (the sound schist and the weathered schist) with the qs values corresponding to each layer, we obtain the calculated value of the critical creep resistance sealing (9) & (10).

$$R_s = \pi * \emptyset * \int_{0}^{Ls1} \alpha_{s1} * q_{s1;k} * dl + \pi * \emptyset * \int_{Ls1}^{Ls2} \alpha_{s2} * q_{s2;k} * dl \tag{9}$$

And

$$R_{cr;d} = R_s/2 \tag{10}$$

With Ls1 correspondind to the sealed length of the anchor in the moderately weathered shale layer.

Ls2 is the total sealed length of the tie rod.

The first integral relates to the tie rods sealing in the moderately altered schist layer and the second corresponds to its sealing in sound schist layer.

With.

$\emptyset = 115$ mm

$\alpha_{s1} = \alpha_{s2} = 1.1$ (for an altered or fragmented rock formation with an IGU type injection)

$q_{s1;k} =$ 100 KPa for moderately weathered schist with a limit pressure of 0.8 MPa

$q_{s1;k} =$ 220 KPa for sound schist with a limiting pressure of 2 MPa.

$L_{s1} = 8.3$ m et $L_{s2} = 18$ m

We can find the calculated value of the critical creep resistance of the seal by resolving the Eqs. (9) and (10).

$$R_{cr;d} = 588.5 \text{ KN}$$

By comparing the measured and calculated values, it can be seen that the critical creep resistance value measured was reached during the pull-out test by presenting an overrun of 50% of the calculated value of this resistance by. The TA2020 guide presents a ratio between these two values in the order of $\gamma_{a:ELS} = 1.2$, i.e. 20% of the calculation value of this resistance for permanent tie rods.

This difference in values may be due to one of the following factors or a combination of factors:

- The borehole theoretical diameter where the tie is placed may be different from its real value, this may be due to the alteration of the rock formation and the impact of the drilling on this formation which could give rise to a diameter greater along the borehole.
- The αn factor is defined according to two parameters: The mode of injection (previously fixed-IGU type-) and the nature of the formation in place. The second element could give rise to a fluctuation in the value of αn that, in our case study, is a soft rock formation presented in two layers: very altered schist (in the upper layer) and sound schist (in the bedrock), both having a sealed part of the ground anchor. It should be noted that the table of αn values (TA2020 guide) does not present a value for a sound rock formation, but only for an altered formation (the value adopted in the study). In

addition, the table B.2.1 of the NF P 94 262 for soil classification standard, presents weathered rocks and fragmented rocks with limit pressures in the range 2.5–4 MPa and greater than 4 MPa respectively. These values exceed those of the soft schist formation present in the case of study, whether for its altered part (Pl = 0.8 MPa) or its sound part (Pl = 2 MPa).

- The limit lateral friction qs of the weathered and the sound schist layers could concretely present real values higher than those adopted and which were obtained by calculation. This calculation values are been found on the basis of the results of pressuremeter tests realized in site and using the abacus relating to altered rocks [10, 13].

4 Conclusion

These pull-out tests analysis makes it possible to fully understand the behavior of an anchor at its serviceability limit and ultimate states. This article has focused on the serviceability limit state to examine the calculated and measured values of the sealing critical creep tensile. The comparison of the values obtained revealed a significant difference that could be due to several reasons given that the calculated value of the critical tensile force involves different parameters.

The value of αn attributed to our formation is relative to an altered or fragmented rock formation with pressuremeter characteristics better than that of our case study. In our case, it is a soft rock formation (part of the Hard Soils—Soft Rocks (HSSR)), with particularities in the geotechnical characteristics that should be taken into consideration in order to assign it a suitable coefficient value. On the other hand, the limiting lateral friction found on the basis of the pressuremeter tests could present real values higher than those obtained by calculation. Similarly, it is interesting to analyze the behavior of this rock formation during the pull-out test at the ultimate limit state until reaching the tensile limit of the seal.

However, some essential elements must be taken into account, such as the overall friction combining that of the cylinder and the anchoring devices as discussed above in order to correct the measurement values found.

This technique is currently very widespread in treatments requiring anchoring in a good soil and has been tested on inconsolidated and rock formations with satisfactory results. In certain cases of particular geological formation, some measures must be taken into account in order to ensure good tie rod—ground contact. In cases where landslides are observed on site, it is interesting to take inclinometric measurements before carrying out the treatment to assess the speed of ground movement, and then, post-treatment measurements to confirm the stability and find the ground movement reduction rate.

References

1. Lahmili, A., Ouadif, L., Akhssas, A., Bahi, L.: Rock stability analysis—a case study. MATEC Web of Conf. (2018). https://doi.org/10.1051/matecconf/201814902072
2. Bouajaj, A., Bahi, L., Ouadif, L., Baba, K.: A methodology based on GIS for 3D slope stability analysis. Int. J. Eng. Tech. **8**(5), 2259–2264 (2016). https://doi.org/10.21817/ijet/2016/v8i5/160805061

3. Hemed, A., Ouadif, L.: Slope stability in a mining environment M'haoudatt-Zouerate site, Mauritania. Am. J. Energy Eng. **9**(3), 85 (2021). https://doi.org/10.11648/j.ajee.20210903.13
4. Zhang, J.: Intelligent identification of rock mass structural plane and stability analysis of rock slope block. Sci. Rep. **12**(1), 16745 (2022). https://doi.org/10.1038/s41598-022-18171-2
5. Brown, E.R.: Rock engineering design of post-tensioned anchors for dams—a review. J. Rock Mech. Geotech. Eng. **7**(1), 1–13 (2015). https://doi.org/10.1016/j.jrmge.2014.08.001
6. Wang, J., Ye, S.: Influences analysis of seismic intensity on dynamic response of slope supported by frame structure with pre-stressed anchors. J. Phys. Conf. Ser. **1670**(1), 012007 (2020). https://doi.org/10.1088/1742-6596/1670/1/012007
7. Tiwari, S., Singh, S.: Analysis of 100T prestressed cable anchors at chenab arch bridge in below S-50. Int. J. Res. Anal. Rev. (IJRAR) **6**(2), 377–385 (2019)
8. Zhu, S., Chen, C., Zhang, G., Du, C.: Theoretical and experimental investigations of anchoring force loss behavior for prestressed ground anchors. Can. Geotech. J. **59**(9), 1587–1601 (2022). https://doi.org/10.1139/cgj-2021-0220
9. Feng, C., Cheng, Z., Zhu, L.: Performance of prestressed anchor cables supporting deep foundation pit of a subway station during spring thaw. Geofluids **2022**, 1–14 (2022). https://doi.org/10.1155/2022/3567816
10. TIRANTS D'ANCRAGE TA 2020—RÈGLES PROFESSIONNELLES relatives à la conception, au calcul,à l'exécution, au contrôle et à la surveillance. CFMS (2020)
11. NF EN 1997-1. Eurocode 7: calcul géotechnique—Partie 1: règles générales. Afnor (2005)
12. NF P 94 262. Justification des ouvrages géotechniques, Normes d'application nationale de l'Eurocode 7 - Fondations profondes. Afnor, Saga Webpour: Fondasol (2012)
13. Clouterre, P.N.: Recommandations CLOUTERRE 1991: pour la conception, le calcul, l'exécution et le contrrôle des soutènements réalisés par clouage des sols. Presses Ponts et Chaussées (1991)

Mitigating Soil Swelling: Exploring the Efficacy of Polypropylene Fiber Reinforcement in Controlling Expansion of Expansive Soils

Ahlam El Majid(✉), Khadija Baba, and Yassine Razzouk

Civil and Environmental Engineering Laboratory (LGCE), Mohammadia Engineering School, Mohammed V University, Rabat, Morocco
ahlam.elmajid@research.emi.ac.ma

Abstract. The reactivity of water in clay and marl soils is well recognized. Their particles become wetter, resulting in swelling. Several variables contribute to this complex phenomenon, including the saturation of the ground's pores, which neutralizes capillary forces, and the absorption of water molecules on the surface of clayey and marly particles. This permits water to pass through the sheets that make up these particles. When the water in the sample evaporates, the volume lowers, causing shrinkage cracks to occur. One of the most common global sources of structural instability is the swelling soil phenomena. These are soils whose volume changes with moisture addition, causing them to expand or contract in response to variations in moisture content. These swelling and shrinkage phenomena provide several issues for civil engineering projects located on the surface of expanding soils and underground structures. The soil cannot be used for construction in this situation, necessitating the implementation of improvement methods. Many stabilizations and reinforcing techniques have been used in civil engineering to address the difficulties and achieve a variety of goals that ensure stability while also improving mechanical strength and decreasing water sensitivity, such as decreasing the number of voids between solid particles (increasing compactness), minimizing gaps, building new connectors, and enhancing existing particle links (mechanical resistance). to increase the bearing capacity and shear strength, as well as to lessen or eliminate the soil's sensitivity to water, soil improvement treatments involve either physically altering the characteristics of the soil (via vibration) or introducing or combining a more resistant substance, such as the PP plastic filament fiber used in this study with a clay and marl soil sample at various contents (0.5%, 1%, 1.5%, and 2%), in order to increase the soil's resistance to liquefaction during an earthquake.

Keywords: Clay · Marl · Swelling · PP · Resistance · Water-sensitivity

1 Introduction

Civil infrastructure and environmental sustainability are severely hampered by expansive soils, which are infamous for changing in volume in response to changes in moisture content. In addition to posing a risk to public safety, soil swelling can seriously harm

K. Baba et al. (Eds.): GeoME 2023, SPEES, pp. 183–192, 2023.
https://doi.org/10.1007/978-3-031-49345-4_19

the foundations of buildings, highways, and other infrastructure. For both geotechnical engineers and researchers, reducing the negative effects of soil swelling is an urgent matter. As a potential method to manage soil expansion and lessen its negative effects, polypropylene fiber reinforcement has gained popularity recently.

Stabilizing expansive soils encompasses both mechanical and chemical treatment methods. The chemical approach primarily involves incorporating chemical admixtures, such as polymers, cement, and lime, into the soil matrix to enhance its engineering properties [1–5]. On the other hand, mechanical treatment entails compacting the soil while introducing reinforcing materials to improve its strength and performance. Among the commonly used reinforcing materials are synthetic fibers like nylon and polypropylene [6, 7], as well as natural fibers such as coconut and coir [8], along with other fibrous substances.

The efficiency of polypropylene fiber reinforcement in limiting the expansion of expansive soils has been thoroughly examined in this scientific work. We examine the effects of several polypropylene fiber parameters, including as length, content, and orientation, on the behavior of the soil swelling using laboratory experiments and numerical studies. We also look at the way polypropylene fibers interact with broad soils and the possible advantages they may have for stabilizing the soil.

The research draws upon a wide range of literature, incorporating seminal works on expansive soil behavior, fiber reinforcement in geotechnical engineering, and the interaction between fibers and soil particles. Notably, the works of Mitchell and Soga [9] provide valuable insights into the nature of expansive soils, while the research by El Majid and BABA [10] offers critical findings on fiber reinforcement techniques. Furthermore, the pioneering investigations of Akbari et al. [11] shed light on the microstructural changes brought about by fiber-soil interaction.

The results of this study contribute significantly to the existing knowledge base on mitigating soil swelling, providing a deep understanding of the potential benefits of polypropylene fiber reinforcement in geotechnical applications. The findings can be instrumental in the design of cost-effective and sustainable solutions for controlling soil expansion, thereby enhancing the stability and longevity of civil infrastructure.

2 Materials and Methods

a. **Soil**

The study focuses on examining the geotechnical characteristics of clay and marl soils indigenous to Morocco's Fez-Meknes region presented in Table 1. The soil samples were taken during the construction of a hospital and a nearby road in TAHLA, and according to GTR 92 criteria, both soils fall within the extremely plastic A3 grade. Experiments revealed that both clay and marl samples have high to very high swelling potential [12–15]. Clay soil has a specific gravity of 2.7, with optimum moisture content at 24.3%, and a CBR of 35% (unsoaked) and 24% (soaked). Its UCS is 110 Kpa, with Atterberg limits of 62% LL, 24% PL, and a PI of 38%. It falls under CH classification in USCS and A-7-6 in AASHTO, indicating very high swell potential. Marl soil has a specific gravity of 2.6, with optimum moisture content at 20.74%, and a CBR of 39% (unsoaked) and 27% (soaked). Its UCS is 240 Kpa, with Atterberg limits of 55% LL, 18% PL, and a PI of 37%. Like clay soil, it also falls under CH in USCS and A-7-6 in AASHTO, indicating very high swell potential.

Table 1. Geotechnical properties of clay and marl soils in fez-meknes region, Morocco

Reference	Lithological section
CS	0.00–0.50:Topsoil 0.50–3.50: Brownish clay
MS	0.00–2.40: Brownish clay 2.40–5.00: Greenish marl

b. **Propylene filament**

Polypropylene (PP) filament is renowned for its exceptional strength, durability, and resistance to abrasion, making it a common material used in the production of sweeping brushes. However, its versatility extends beyond just brushes. In this study, PP filament serves as a reinforcement material for expansive soils, significantly enhancing stability and effectively preventing soil cracking or settling. The advantageous properties of PP fibers, such as high tensile strength, microfine reinforcing capability, and non-corrosiveness, make them an excellent choice for soil reinforcement in various applications.

One of the key advantages of PP fibers is their economic viability, hydrophobic nature, and chemical inertness. Additionally, they possess high melting and ignition points, ensuring stability under elevated temperatures and making them resistant to catching fire at lower temperatures. These properties contribute to their suitability for soil reinforcement, especially in challenging environmental conditions.

The comprehensive physical, chemical, and mechanical properties of the polypropylene (PP) fiber utilized in this research are as follows:

- Specific gravity: 0.89 (relatively lightweight)
- Tensile strength: 0.67 kN/mm^2 (impressive, capable of withstanding considerable loads and stresses without deformation)
- Young's modulus: 4.00 kN/mm^2 (ability to withstand substantial forces without losing shape)
- Melting point range: 160–170 °C (ensuring stability under elevated temperatures)
- Ignition point: 590 °C (resistance to catching fire at lower temperatures)
- Bulk density: 910 kg/m^3 (favorable for easy handling and application)
- Loose density: 250–430 kg/m^3 (suitable for various contexts)
- Available cut lengths: 10, 15, 20, and 25 mm (catering to different requirements)
- Excellent dispersion: Ensuring even distribution within materials like concrete, enhancing their overall strength and durability
- High resistance to acids and salts: Making it a reliable choice for applications where exposure to corrosive substances is a concern

- Chemical-proof nature: Ensuring long-lasting performance in challenging environments.

Due to these exceptional properties, polypropylene fiber finds widespread applications in construction, textiles, and various other industries where strength, resilience, and chemical resistance are paramount. Its use as a reinforcement material for expansive soils is a testament to its versatility and effectiveness in enhancing stability and preventing soil-related issues.

c. **Methodology**

The methodology employed in this study is visually presented in Fig. 1, providing a clear and concise depiction of the research program's specific approach and procedures.

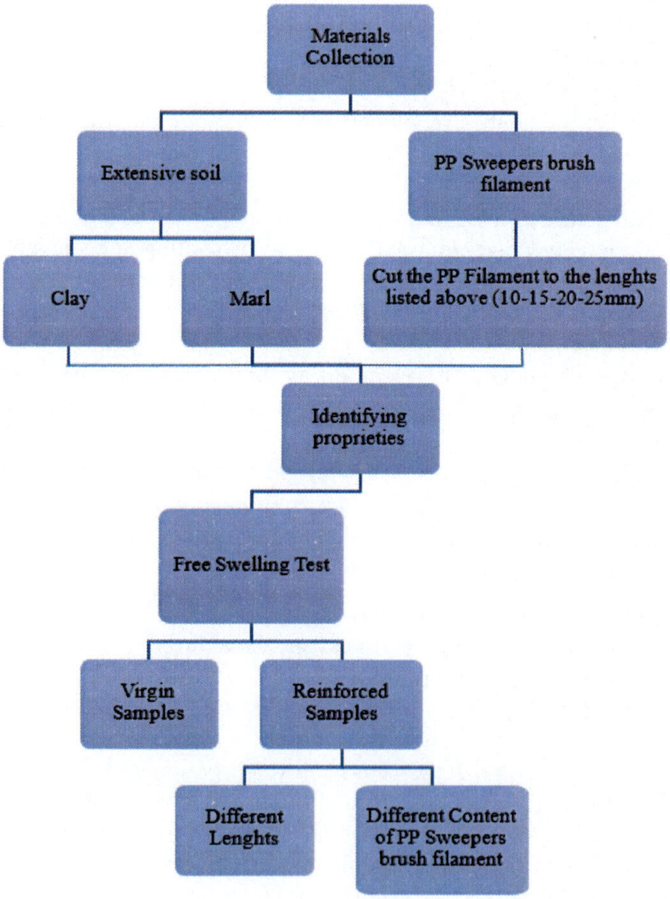

Fig. 1. Flowchart illustrating the utilized methods

3 Results

a. **Swelling Potential**

The research study delves into the impact of fiber content on the swelling potential of two expansive soils treated with varying percentages of polypropylene (PP) fibers. The results provide valuable insights into the behavior of clay and marl soils when reinforced with different fiber levels. Notably, as the fiber content increased from 0.5% to 2%, the swelling potential of both soil types significantly decreased. The clay soil exhibited an impressive 12.25% reduction, reaching an average swelling of 3.14%, while the marl soil showed a reduction of 10.23%, resulting in an average swelling of 3.52%. These findings underscore the effectiveness of PP fibers in mitigating swelling in both soil types. Further investigations explored the influence of immersion duration on swelling behavior, showing a minimal reduction in swelling with extended immersion for both soil types. In the optimal case (2% PP fibers, L = 25 mm), the swelling behavior remained stable over time for both clay and marl samples. The study emphasizes the importance of considering specific soil characteristics when designing fiber reinforcement strategies. As the research progresses, assessing the practical significance of these slight changes in swelling and analyzing their implications for intended applications will be essential. Understanding the long-term behavior of the reinforced soils will offer valuable insights into their suitability for engineering and construction projects. Figure 2 presents a comprehensive analysis of swelling potential, demonstrating promising outcomes and highlighting the potential benefits of incorporating fiber reinforcement for swelling control in geotechnical applications.

b. **Swelling Pressure**

Figure 3 presents compelling evidence of how fiber content influences the swelling pressure of the two expansive soils under study. The results clearly demonstrate the significant effect of fiber inclusions on swelling pressure in both clay and marl soils. Notably, as the fiber content increased from 0.5% to 2%, the swelling pressure experienced a dramatic drop in both soil types. In the case of clay soil, the swelling pressure reduced substantially from 72 kPa in the control soil (no fibers) to 44.5 kPa with 2% PP fiber content. Similarly, the swelling pressure in marl soil decreased from 121 kPa in the control soil to 89 kPa with the same 2% fiber content. The highest swelling pressure was observed in the untreated virgin clay and marl soils, measuring 72 kPa and 121 kPa, respectively. However, the lowest swelling pressure occurred when these soils were reinforced with a 2% fiber mixture, reaching minimal values of 44.5 kPa for clay soil and 89 kPa for marl soil. Notably, the impact of fiber treatment was more pronounced in the clay sample, especially when longer fiber lengths were utilized. The fiber length was found to be extremely significant, as the addition of a 2% fiber mixture with lengths ranging from L1 = 10mm to L3 = 25mm resulted in a reduction of swelling pressure from 69.25 kPa to 44.5 kPa in clay soil and from 117 kPa to 89 kPa in marl soil. These findings underscore the importance of both fiber content and length in effectively reducing swelling pressure in expansive soils. The results reveal the potential of polypropylene fibers to

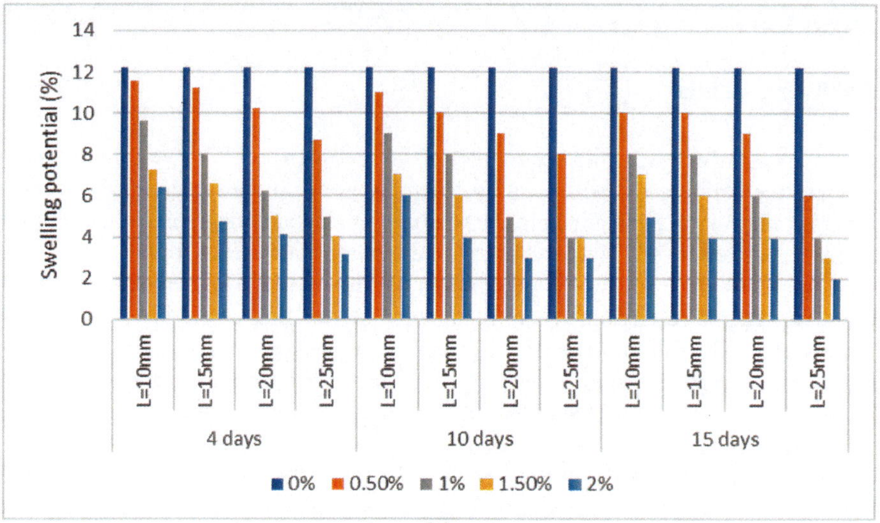

(a)

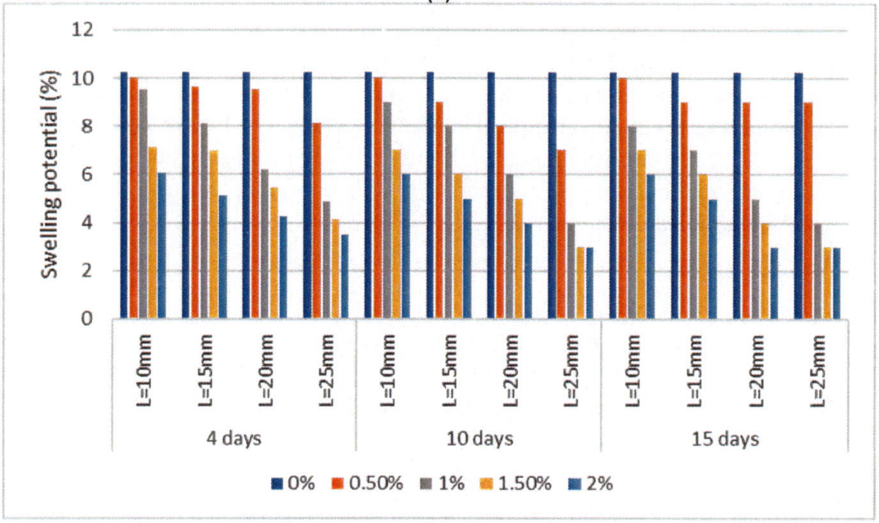

(b)

Fig. 2. Swelling potential reduction in expansive soils with polypropylene fiber reinforcement (a) clay (b) marl

mitigate swelling pressure and suggest that longer fibers may yield more substantial benefits in controlling soil swelling, particularly in clayey soils. Figure 3 provides valuable insights into the significant impact of fiber reinforcement on the swelling pressure of the investigated soils.

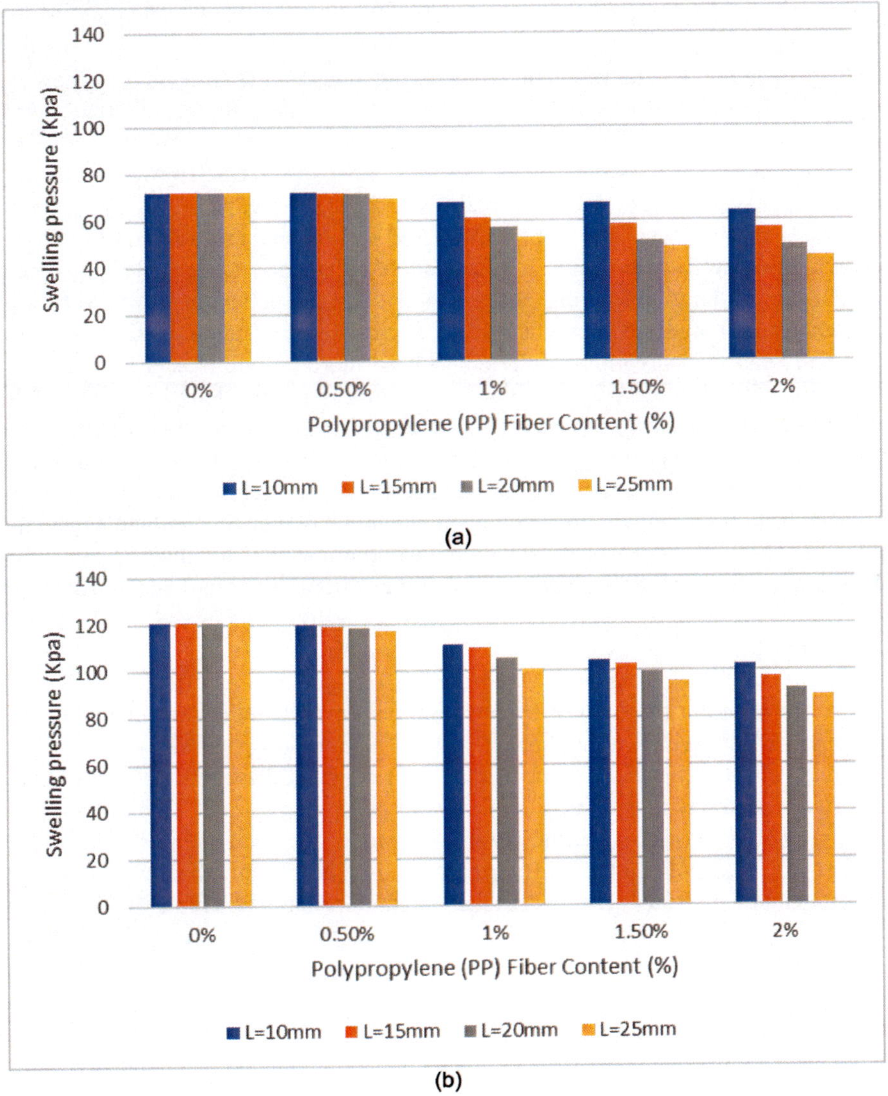

Fig. 3. Variation in swelling pressure with different concentrations and lengths of polypropylene (PP) fibers. (a) clay sample (b) marl sample

4 Discussion

The present study aimed to investigate the influence of fiber content and length on the swelling potential of two expansive soils, namely clay and marl, treated with polypropylene (PP) fibers. The results demonstrate the significant impact of fiber inclusion on reducing swelling pressure in both soil types. As fiber content increased from 0.5%

to 2%, a remarkable decrease in swelling pressure was observed, indicating the effectiveness of PP fibers in mitigating the swelling potential of expansive soils. The lowest swelling pressures were recorded in soils reinforced with a 2% fiber mixture, emphasizing the importance of optimal fiber content for achieving desirable results. The observed reduction in swelling within the fiber-reinforced soil sample can be attributed to the substitution of expansive soil particles with non-expansive fibers. This replacement process leads to a significant decrease in the soil's swelling capacity. By incorporating the non-expansive fibers, the propensity of the soil to undergo volumetric expansion is effectively curtailed, resulting in enhanced stability and reduced swelling potential [17]. Moreover, the influence of fiber length on swelling behavior proved to be highly significant, with longer fiber lengths leading to more pronounced reductions in swelling pressure. Notably, the treatment by fibers showed more pronounced effects on the clay soil compared to the marl soil, particularly with larger fiber lengths. These findings hold vital implications for geotechnical engineering applications, where controlling soil swelling is crucial for ensuring the stability and longevity of structures. As the research progresses, understanding the long-term behavior of fiber-reinforced soils and analyzing their practical significance for specific applications will provide valuable insights for engineering practices in various environmental conditions. Further investigations should consider other relevant factors such as soil composition, fiber characteristics, and reinforcement distribution to develop comprehensive design guidelines for effective fiber reinforcement strategies in expansive soil stabilization.

Fiber reinforcement has been extensively investigated as an effective technique to mitigate the swelling potential of expansive soils. Several studies have demonstrated a significant reduction in swelling behavior upon incorporating non-expansive fibers into the soil matrix. For instance, Hamza et al. [18] conducted a comprehensive study on clayey soils treated with varying percentages of polypropylene (PP) fibers and reported a remarkable decrease in swelling potential as fiber content increased. Similarly, EL Majid et al. [16] examined the impact of fiber length on swelling behavior in marl soils and found that longer fiber lengths resulted in more pronounced reductions in swelling pressure. Additionally, Syed and Guharay [19] investigated the mechanism behind the decrease in swelling, attributing it to the substitution of expansive soil particles by non-expansive fibers. Furthermore, Nguyen and indraratna [20] provided valuable insights into the practical significance of fiber-reinforced soils for geotechnical applications, emphasizing the importance of considering factors such as soil type, fiber characteristics, and reinforcement distribution. These collective findings highlight the potential of fiber reinforcement as a promising solution for controlling swelling in expansive soils, paving the way for safer and more sustainable geotechnical engineering practices.

5 Conclusion

In conclusion, this scientific investigation provides valuable insights into the effectiveness of polypropylene (PP) fiber reinforcement in mitigating soil swelling and controlling the expansion of expansive soils. The study demonstrates a significant reduction in swelling potential by incorporating non-expansive fibers into the soil matrix. The observed decrease in swelling pressure with increasing fiber content emphasizes the importance of optimizing the fiber concentration for effective swelling control. Moreover, the influence of fiber length on swelling behavior underscores the significance of selecting appropriate fiber dimensions to achieve desired outcomes. The mechanism behind the reduction in swelling pressure lies in the substitution of expansive soil particles with non-expansive fibers, which effectively curtails the soil's ability to undergo volumetric expansion. These findings have crucial implications for geotechnical engineering, where controlling soil swelling is essential to ensure the stability and durability of infrastructure projects. Embracing PP fiber reinforcement as a reliable and practical technique, engineers and researchers can develop robust strategies to address challenges posed by expansive soils, leading to more sustainable and resilient construction practices. As the field of fiber-reinforced soils evolves, further research is encouraged to explore the long-term behavior of such stabilization techniques and consider the application of other fiber types and soil compositions. This study contributes to the advancement of geotechnical engineering knowledge and fosters innovative solutions for managing soil swelling in diverse engineering applications. By understanding and harnessing the potential of PP fiber reinforcement, the engineering community can pave the way for more effective and environmentally friendly solutions in soil stabilization.

References

1. Effect of polymers on swelling potential of expansive soils | Proceedings of the Institution of Civil Engineers—Ground Improvement. https://www.icevirtuallibrary.com/doi/https://doi.org/10.1680/grim.2009.162.3.111
2. Al-Rawas, A.A., Hago, A.W., Al-Sarmi, H.: Effect of lime, cement and Sarooj (artificial pozzolan) on the swelling potential of an expansive soil from Oman. Build. Environ. **40**, 681–687 (2005). https://doi.org/10.1016/j.buildenv.2004.08.028
3. Yazdandoust, F., Yasrobi, S.S.: Effect of cyclic wetting and drying on swelling behavior of polymer-stabilized expansive clays. Appl. Clay Sci. **50**, 461–468 (2010). https://doi.org/10.1016/j.clay.2010.09.006
4. Estabragh, A.R., Rafatjo, H., Javadi, A.A.: Treatment of an expansive soil by mechanical and chemical techniques. Geosynthetics Int. **21**, 233–243 (2014). https://doi.org/10.1680/gein.14.00011
5. Tatsuoka, F., Correia, A.G.: Importance of controlling the degree of saturation in soil compaction. Proc. Eng. **143**, 556–565 (2016). https://doi.org/10.1016/j.proeng.2016.06.070
6. Khosrowshahi: Improvement of expansive soils using—Google Scholar, https://scholar.google.com/scholar_lookup?title=Improvement+of+expansive+soils+using+fiber+materials&conference=Proceedings+of+the+11th+International+Congress+on+Advances+in+Civil+Engineering+(ACE+2014)&author=Senol,+A.&author=Khosrowshahi,+S.K.&author=Yildirim,+H.&publication_year=2014
7. Phanikumar, B.R., Singla, R.: Swell-consolidation characteristics of fibre-reinforced expansive soils. Soils Found. **56**, 138–143 (2016)
8. Sivakumar Babu, G.L., Vasudevan, A.K., Haldar, S.: Numerical simulation of fiber-reinforced sand behavior. Geotext. Geomembr. **26**, 181–188 (2008). https://doi.org/10.1016/j.geotexmem.2007.06.004
9. Mitchell, J.K., Soga, K.: Fundamentals of soil behavior. John Wiley & Sons New York (2005)
10. Majid, A., Baba, K.: Assessing the impact of plant fibers on swelling parameters of two varieties of expansive soil. Case Studies Chem. Environ. Eng. **8**, 100408 (2023). https://doi.org/10.1016/j.cscee.2023.100408

11. Effect of polypropylene fiber inclusion in kaolin clay stabilized with lime and nano-zeolite considering temperatures of 20 and 40 °C | SpringerLink. https://link.springer.com/article/https://doi.org/10.1007/s10064-020-02028-x
12. Vijayvergiya, V.N., Ghazzaly, O.I.: Prediction of swelling potential for natural clays. In: Proceedings of the 3rd International Conference on Expansive Soils, pp. 227–236 (1973)
13. Bigot, G., Zerhouni, M.I.: Retrait, gonflement et tassement des sols fins **10** (2000)
14. Van, der M.D.H.: The prediction of heave from the plasticity index and percentage clay fractions of soils—discussion. Civil Eng. = Siviele Ingenieurswese., 226–228 (1964). https://doi.org/10.10520/AJA10212019_17005
15. Dakshanamurthy, V., Raman, V.: A simple method of identifying an expansive soil. 土質工学会論文報告集. **13**, 97–104 (1973). https://doi.org/10.3208/sandf1972.13.97
16. Majid, A., Cherradi, C., Baba, K., Razzouk, Y.: Laboratory investigations on the behavior of CBR in two expanding soils reinforced with plant fibers of varying lengths and content. Mater. Today Proc. (2023). https://doi.org/10.1016/j.matpr.2023.06.395
17. Viswanadham, B.V.S., Phanikumar, B.R., Mukherjee, R.V.: Swelling behaviour of a geofiber-reinforced expansive soil. Geotext. Geomembr. **27**, 73–76 (2009). https://doi.org/10.1016/j.geotexmem.2008.06.002
18. Hamza, M., Ijaz, N., Fang, C., Ijaz, Z.: Stabilization of problematic expansive clays using polypropylene fiber reinforcement. Jordan J. Civil Eng. **16**, 2022 (2022)
19. Syed, M., GuhaRay, A.: Effect of natural fiber reinforcement on strength response of alkali activated binder treated expansive soil: Experimental investigation and reliability analysis. Constr. Build. Mater. **273**, 121743 (2021). https://doi.org/10.1016/j.conbuildmat.2020.121743
20. Nguyen, T.T., Indraratna, B.: Natural fibre for geotechnical applications: concepts achievements and challenges. Sustainability **15**, 8603 (2023)

Assessing Organic Matter of Port Dredged Sediment for Valorization in Civil Engineering

Meryem Bortali[1](✉), Mohamed Rabouli[2], Madiha Yessari[1], and Abdelowahed Hajjaji[1]

[1] Chouaib Doukkali University, National School of Applied Science of El Jadida, 24002 El Jadida, Morocco
Bortali.m@ucd.ac.ma

[2] Ibn Tofail University, National School of Applied Science of Kenitra, 14000 Kenitra, Morocco

Abstract. To ensure proper navigation and safety levels, over one billion cubic meters of sediment are dredged yearly from ports around the world. Beneficial use of port dredged sediment, considered as waste, is a potential alternative for the large quantities of natural resources required in civil engineering. It can solve a joint problem; environmental impacts of dredging and the scarcity of non-renewable materials used in the construction industry. Assessment of the organic matter is a decisive step within the process of valorization, in order to determine the port dredged sediment characteristics and confirm the feasibility of its reuse in the construction sector. The sediment is made up of water, inorganic and organic materials and compounds of anthropic origin. As a substitute of construction material, the presence of organic matter can affect the properties and the behavior of dredged sediment. However, existing research efforts on reusing dredged sediment in civil engineering rarely analyzed organic matter content. Moreover, they relied on the results obtained on the basis of a given method without confirming their findings by other means. This paper, having as main objective the investigation of the possibility of reusing dredged sediment as a substitute of conventional construction aggregate, fills this gap. Organic matter analysis were carried out by chemical method and calcination method. The two methods were applied on numerous samples collected from the port of Safi in Morocco during 2022 dredging campaign. The results of the study reveal that the organic matter levels of dredged sediment are below the reference values. Thus, valorization of dredged sediment in civil engineering is feasible. This study provides valuable information for achieving the sustainable development from waste to resource.

Keywords: Dredged sediment · Valorization · Organic matter · Construction material · Sustainability

1 Introduction

Dredging of ports is necessary to ensure that they remain safe for navigation and access by ships on a permanent basis [1–4]. As sediment would continuously accumulate in the bottom of harbor channels, dredged operations are performed regularly [5, 6]. Sediment is a complex and heterogeneous matrix consisting of water (up to 80%), inorganic

K. Baba et al. (Eds.): GeoME 2023, SPEES, pp. 193–202, 2023.
https://doi.org/10.1007/978-3-031-49345-4_20

materials (rocks, shells and mineral compounds), organic materials (plant debris, humic colloids and microorganisms) and compounds of anthropogenic origin [7]. The organic matter contained in dredged material is composed, in addition to the natural organic constituents, of industrial organic elements; thus representing one of the main contaminants of the excavated sediment [8, 9]. The industrial activity of ports and their location in urban areas seems to increase the levels of organic pollutants in dredged sediment [10]. Besides organic contaminants (polycyclic aromatic hydrocarbons, pesticides, chlorinated solvents and polychlorinated biphenyls), dredged sediment includes inorganic contaminants (nutrients such as phosphorus and ammonia, and heavy metals like copper, lead, cadmium, chromium, mercury, nickel, selenium and arsenic) [11–13]. Once a tolerance threshold has been exceeded, certain components of marine sediment are considered toxic [14]. Hence, the current management method of dumping dredged material at sea has harmful effects on the aquatic environment in general and on human health in particular [15, 16].

On another level, usage of dredged sediment in civil engineering is a potential alternative for natural resources [17]. It can solve a joint problem; environmental impacts of dredging and the scarcity of non-renewable materials used in the construction industry [18]. Indeed, dredging operations generate significant quantities of sediment exceeding 1 billion m^3 of sediment per year worldwide [19]; more than 300 million m^3 of sediments are dredged in the United States, 200 million m^3 in Western Europe and 100 million m^3 in China [20]. A large part of dredged sediment can be valorized in the field of civil engineering, which is experiencing a growing scarcity of natural aggregates [21]. To remedy this situation, several research works have focused on sustainable solutions for waste valorization, including beneficial reuse of dredged sediment [18, 19]. Researches confirmed the feasibility of valorizing dredged sediment in the construction sector [22–25]. Several applications have been studied, such concrete and mortar works [22, 26–29], cement production [20, 30–32], bricks [3, 25] and road techniques [33–36].

The presence of considerable organic matter is one of the challenges to reusing dredged sediment in civil engineering [37]. It is an undesirable component in construction works as it affects the physical and chemical properties of dredged sediment, as well as its mechanical behavior [22, 38]. Therefore, it is imperative to assess the content of organic matter in harbor sediment for a path to valorize dredged material [29, 39]. There are numerous and diverse methods for determining the level of contamination of dredged material by organic matter [38]. However, some of them are time-consuming and require sophisticated equipment; which makes them unsuitable to characterize dredged sediment intended for reuse in civil engineering [38], especially in developing countries. The objective of this work is to determine the organic matter content in the dredged sediment through two rather simple methods adapted to the case study, in order to characterize it for reuse as construction material.

Sediment was extracted from the port of Safi located in west center of Morocco. Six samples were collected in order to compare the results from different areas of harbor basins. Experimental tests to assess the organic matter content were conducted on all samples using two different methods: chemical method and calcination method. The results obtained were then evaluated with regard to the reference levels set by the international standards for the characteristics of aggregates used in civil engineering.

2 Materials and Methods

2.1 Study Area

Sediment was collected from the port of Safi, located in west center of Morocco on the Atlantic coastline (32.312642, − 9.252548). It is one of the oldest Moroccan ports. Safi port is situated at the base of a bay characterized by expansive cliffs, which provide protection to ships from strong winds and storms [40, 41]. It enjoyed a period of thriving operations in the 19th and 20th centuries [42]. The main traffics in transit through Safi harbor are phosphates, fertilizers, ores, cereals, phosphoric acid, sulphur and fishing products. The Safi port is currently facing significant siltation caused by erosion of the cliffs (Cape Ghir-Arhesdis) and the influx of sediments originating from the vast ocean and dry riverbeds. Approximately 250,000 cubic meters of marine sediment are dredged on an annual basis to mitigate this issue [43]. Figure 1 presents the location of the study area.

Fig. 1. Location of the port of Safi, Morocco.

2.2 Data Collection

Six samples were extracted from the port basins, by a diver using a boat, based on a random sampling plan, in order to cover different areas and provide a real informative value. Samples were gathered from various locations: P1 from the commercial basin,

P2 in close proximity to both the commercial and ore basins, P3 from the ore basin, P4 in the pass of the port, P5 in the access channel, and P6 near the breakwater of the main jetty. These samples were carefully placed in isothermal bags and stored in a cold and opaque room at a temperature of 4 °C.Two different bags were collected from each sampling point in order to confirm the characteristics at each point. The storage method was selected based on the previous studies, as a way to conserve the chemical characteristics of dredged sediment. Figure 2 shows the sampling points location.

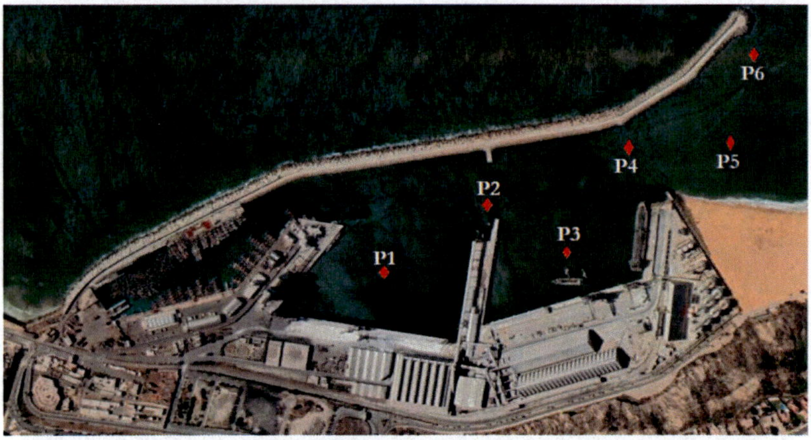

Fig. 2. Location of sampling points from Safi port.

Table 1 summarizes the coordinates, in latitude and longitude of the six sampling points.

Table 1. Coordinates of sampling points.

Sampling point	Latitude	Longitude
P1	9°24′76.9″W	32°30′75.8″N
P2	9°25′03.7″W	32°30′94.4″N
P3	9°24′97.1″W	32°31′12.0″N
P4	9°25′28.9″W	32°31′20.7″N
P5	9°25′36.4″W	32°31′32.0″N
P6	9°25′54.2″W	32°31′37.1″N

2.3 Methods

Two different methods were utilized to measure the organic matter content of each of the six samples:

- Chemical method: according to the standard NF P94-055, each sample is mixed with a solution of potassium dichromate with sulfuric acid. The organic matter in the test sample is oxidized by the solution. The potassium dichromate remaining after the oxidation phase is determined by a solution of double ammonium iron sulfate. The organic content is evaluated in a conventional way by applying a factor to the carbon content thus determined.
- Calcination method: in accordance with the requirements of the standard XP 94-047, the mass content of organic matter is determined by calcination in the granulometric fraction less than or equal to 2 mm of each sample. After subjecting a sediment sample to a series of steps, including crushing, drying at low temperature, and calcination at high temperature (450 °C) for a duration of 3 h, the organic matter within the sample is broken down and released as carbon dioxide and water vapor. Following the combustion process, only the mineral fraction of the soil is left behind in the container. The organic matter content is calculated by mathematical equation.

The literature presents several reference thresholds regarding the admissible content of organic matter for use in civil engineering. Generally, a material can be considered as containing a high level of organic matter when it is over 5% [44]. Each test was repeated three times to confirm the result obtained. The reported value is equal to the average of the results of the three tests.

3 Results and Discussion

3.1 Results

The findings provided in this study represent the mean values derived from six individual tests conducted at each sample point. The initial step involved calculating the average for each of the three tests conducted on every sample, resulting in two values for each sampling point. Afterward, the final result was determined by averaging the two obtained values. This process ensured a consolidated and representative measurement for each sampling point. Figure 3 summarizes results of the tests carried out to measure the organic matter content, by the two chosen methods.

The results obtained vary according to the method used. Indeed, the results obtained through the process of calcination exceed those acquired through the chemical method, for the six samples studied. Based on the chemical method, the organic matter content ranges from 0% for samples P4, P5 and P6 to 3.36% for sample P1, exhibiting 1.59% as average value. As for the results obtained by the calcination method, the average value obtained is 2.37% ranging from 0.3% for sample P6 to 4.57% for sample P1. This variation of results can be explained by the influence of the high temperature used in the calcination method (450 °C) as well as the calcination time [45].

An improvement of the results is also noticed when moving from point P1 to point P6; the average of the organic matter contents for points P1, P2 and P3 is 3.18% by chemical method and 4.27% by calcination method, while for points P4, P5 and P6, the averages of the results obtained by chemical method and calcination method are 0% and 0.47% respectively. Points P1, P2 and P3, situated within the inner basins of the harbor,

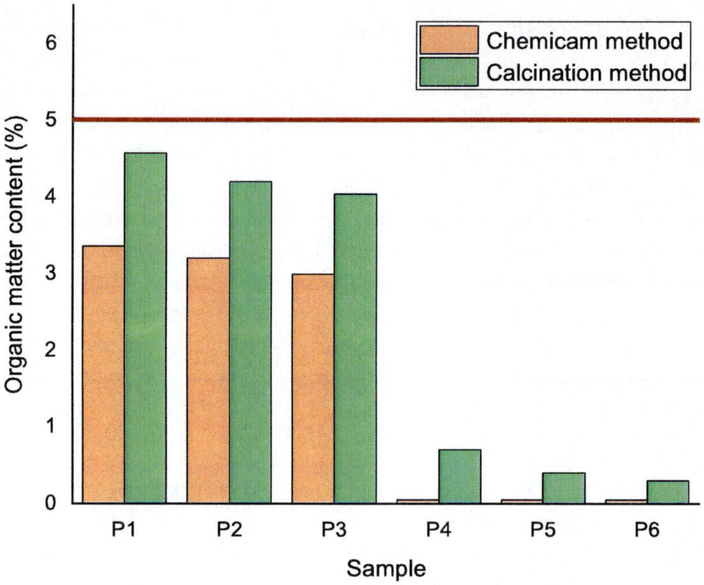

Fig. 3. Organic matter content of dredged sediment.

are more contaminated with organic matter than points P4, P5 and P6 located in the access channel.

This variation by sample can be explained by industrial and urban discharges in the inner basins, linked to port activity.

On average, the organic matter content did not exceed the 5% threshold for both methods. The reference value was exceeded in samples P1 and P2 for the results obtained by the calcination method. In sum, the organic matter content of the dredged sediment of the port of Safi does not constitute an obstacle for its use in civil engineering, especially for the dredged material of the access channel to the port.

3.2 Comparison with Literature Values

When comparing the organic matter content of sediment from Safi port to earlier research that has concentrated on characterizing marine sediment, specifically its potential for reuse within the construction industry, a significant variation in values is observed across different studies. Table 2 provides a summary of the comparison between the results of organic matter characterization, produced during this research, with literature values derived from studies conducted on sediments closely resembling those dredged from the port of Safi.

We observe an average difference of about 67% between the results of the present study and previous research works. This difference is explained by the characteristics of each stady area.

Table 2. Comparison of organic matter content with literature values.

Parameter	Method	Value from this study	Value found in the literature	Difference (%)
Organic matter content (%)	Chemical method	1.59	8.01 [38]	>100
	Calcination method	2.37	9.4 [38]	>100

3.3 Limitations and Directions for Further Research

This study focused only on the analysis of the organic matter content of the dredged material. For future research works, other physico-chemical properties can be studied to complete the feasibility study of the valorization of dredged material in civil engineering.

4 Conclusions

This work was carried out to characterize the organic matter in the sediment dredged from the port of Safi in Morocco. Two measurement methods were used: the chemical method according to the NF P94-055 standard and the calcination method according to the XP 94-047 standard. The results interpretation allowed to draw the following conclusions.

- The results obtained by the calcination method exceed those obtained through the chemical method, which can be explained by the influence of the high temperature used in the calcination method;
- A noticeable improvement in the results is observed when transitioning from the inner basins of the harbor to the access channel. This indicates a reduction in the organic matter content as the sediment moves towards the access channel. This variation can be explained by industrial and urban discharges in the inner basins, linked to port activity;
- On average, the organic matter content did not exceed the 5% threshold for both methods. Hence, the organic matter content of the dredged sediment of the port of Safi does not constitute an obstacle for its use in civil engineering, especially for the dredged material of the access channel to the port.

References

1. Beddaa, H., et al.: Reuse potential of dredged river sediments in concrete: effect of sediment variability. J. Clean. Prod. **265**, 121665 (2020). https://doi.org/10.1016/j.jclepro.2020.121665
2. Lim, Y.C., Shih, Y., Tsai, K., Yang, W., Chen, C., Dong, C.: Recycling dredged harbor sediment to construction materials by sintering with steel slag and waste glass : characteristics, alkali-silica reactivity and metals stability. J. Environ. Manage. **270**(6), 110869 (2020). https://doi.org/10.1016/j.jenvman.2020.110869

3. Hussain, M., et al.: Reuse of harbour and river dredged sediments in adobe bricks. Clean. Mater. **3**(December), 2022 (2021). https://doi.org/10.1016/j.clema.2022.100046
4. Sheehan, C., Harrington, J.: Management of dredged material in the republic of Ireland—a review. Waste Manag. **32**(5), 1031–1044 (2012). https://doi.org/10.1016/j.wasman.2011.11.014
5. Campmans, G.H.P., Roos, P.C., Van der Sleen, N.R., Hulscher, S.J.M.H.: Modeling tidal sand wave recovery after dredging: effect of different types of dredging strategies. Coast. Eng. **165**(1), 103862 (2021). https://doi.org/10.1016/j.coastaleng.2021.103862
6. Zikra, M., Salsabila, S., Sambodho, K.: Toward a better understanding of sediment dynamics as a basis for maintenance dredging in nagan raya port, Indonesia. Fluids **6**(11) (2021). https://doi.org/10.3390/fluids6110397
7. Power, E.A., Chapman, P.M.: Assessing sediment quality. In: Sediment Toxicity Assessment. 1992nd ed. CRP Press, p. 18 (1992)
8. Schneider, G.: Sediment removal from watercourses. Courr. l'Environ. l'INRA. **43**(29/05/2001), 146–147 (2001) [Online]. Available: https://explorable.com/fr/types-de-validite
9. Tessier, E.: Diagnostic de la contamination sédimentaire par les Métaux/Métalloïdes dans la Rade de Toulon et Mécanismes Contrôlant leur Mobilité. Sud Toulon Var (2012)
10. Agostini, F.: Inerting and reuse of marine dredging sediments. Central School Lille (2006)
11. Bocchetti et al., R.: Contaminant accumulation and biomarker responses in caged mussels, Mytilus galloprovincialis, to evaluate bioavailability and toxicological effects of remobilized chemicals during dredging and disposal operations in harbour areas, vol. 89, pp. 257–266 (2008). https://doi.org/10.1016/j.aquatox.2008.07.011
12. Couvidat, J., Chatain, V., Bouzahzah, H., Benzaazoua, M.: Characterization of how contaminants arise in a dredged marine sediment and analysis of the effect of natural weathering. Sci. Total. Environ. **624**, 323–332 (2018). https://doi.org/10.1016/j.scitotenv.2017.12.130
13. Montero, N., Belzunce-segarra, M.J., Gonzalez, J., Menchaca, I., Garmendia, J.M., Etxebarria, N.: Application of toxicity identification evaluation (TIE) procedures for the characterization and management of dredged harbor sediments. Mar. Pollut. Bull. **71**(1–2), 259–268 (2013). https://doi.org/10.1016/j.marpolbul.2013.01.038
14. Bel Hadj Ali, I., Lafhaj, Z., Bouassida, M., Said, I.: Characterization of Tunisian marine sediments in rades and gabes harbors. Int. J. Sediment Res. **29**(3), 391–401 (2015). https://doi.org/10.1016/S1001-6279(14)60053-6
15. Barjoveanu, G., De Gisi, S., Casale, R., Todaro, F., Notarnicola, M., Teodosiu, C.: A life cycle assessment study on the stabilization/solidification treatment processes for contaminated marine sediments. J. Clean. Prod. **201**, 391–402 (2018). https://doi.org/10.1016/j.jclepro.2018.08.053
16. Martins, M., Costa, P.M., Raimundo, J., Vale, C., Ferreira, A.M., Costa, M.H.: Ecotoxicology and environmental safety impact of remobilized contaminants in Mytilus edulis during dredging operations in a harbour area: bioaccumulation and biomarker responses. Ecotoxicol. Environ. Saf. **85**, 96–103 (2012). https://doi.org/10.1016/j.ecoenv.2012.08.008
17. Manni, M., de Albuquerque Landi, F.F., Giannoni, T., Petrozzi, A., Nicolini, A., Cotana, F.: A comparative study on opto-thermal properties of natural clay bricks incorporating dredged sediments. Energies **14**(15):1–17 (2021). https://doi.org/10.3390/en14154575
18. Athira, S., Subaida, E.: A review on sustainable reuse of dredged sediments in earthwork infrastructures. Proc. Int. Web Conf. Civ. Eng. Sustain. Planet 102–108 (2021). https://doi.org/10.21467/proceedings.112.13
19. Bose, B.P., Dhar, M.: Dredged sediments are one of the valuable resources: a review. Int. J. Earth Sci. Knowl. Appl. **4**(2), 324–331 (2022)

20. Amar, M., Benzerzour, M., Kleib, J., Abriak, N.E.: From dredged sediment to supplementary cementitious material: characterization, treatment, and reuse. Int. J. Sediment Res. **36**(1), 92–109 (2021). https://doi.org/10.1016/j.ijsrc.2020.06.002
21. Guerra, B.C., Leite, F.: Circular economy in the construction industry: an overview of United States stakeholders' awareness, major challenges, and enablers. Resour. Conserv. Recycl. **170**(10), 105617 (2021). https://doi.org/10.1016/j.resconrec.2021.105617
22. Aoual-Benslafa, F.K., Kerdal, D., Ameur, M., Mekerta, B., Semcha, A.: Durability of mortars made with dredged sediments. Procedia Eng. **118**, 240–250 (2015). https://doi.org/10.1016/j.proeng.2015.08.423
23. Bagarani, M., De Vincenzo, A., Ievoli, C., Molino, B.: The reuse of sediments dredged from artificial reservoirs for beach nourishment: technical and economic feasibility. Sustain. **12**(17), 1–20 (2020). https://doi.org/10.3390/SU12176820
24. Lim, Y.C., Shih, Y.J., Tsai, K.C., Yang, W.D., Chen, C.W., Di Dong, C.: Recycling dredged harbor sediment to construction materials by sintering with steel slag and waste glass: characteristics, alkali-silica reactivity and metals stability. J. Environ. Manage. **270**(5), 110869 (2020). https://doi.org/10.1016/j.jenvman.2020.110869
25. Slimanou, H., Eliche-quesada, D., Kherbache, S., Bouzidi, N., Tahakourt, A.K.: Harbor dredged sediment as raw material in fired clay brick production : characterization and properties. J. Build. Eng. **28**(10), 101085 (2020). https://doi.org/10.1016/j.jobe.2019.101085
26. Achour, R., Zentar, R., Abriak, N., Rivard, P., Gregoire, P.: Case studies in construction materials. Case Stud. Constr. Mater. **11**, e00244 (2019). https://doi.org/10.1016/j.cscm.2019.e00244
27. Beddaa, H., Ben Fraj, A., Lavergne, F., Torrenti, J.M.: Reuse of untreated Fine sediments as filler: is it more beneficial than incorporating them as sand? Buildings 12(2):1–13 (2022). https://doi.org/10.3390/buildings12020211
28. Chu, S.H., Yao, J.J.: A strength model for concrete made with marine dredged sediment. J. Clean. Prod. **274**, 122673 (2020). https://doi.org/10.1016/j.jclepro.2020.122673
29. Maherzi, W., Benzerzour, M., Mamindy-Pajany, Y., van Veen, E., Boutouil, M., Abriak, N.E.: Beneficial reuse of Brest-Harbor (France)-dredged sediment as alternative material in road building: laboratory investigations. Environ. Technol. **39**(5), 566–580 (2018). https://doi.org/10.1080/09593330.2017.1308440
30. Junakova, N., Junak, J.: Sustainable use of reservoir sediment through partial application in building material. Sustain. **9**(5), 1–13 (2017). https://doi.org/10.3390/su9050852
31. Smith, B.T., Howard, I.L., Vahedifard, F.: Lightly cemented dredged sediments for sustainable reuse. Environ. Geotech. **5**(6), 324–335 (2018). https://doi.org/10.1680/jenge.16.00019
32. Zhao, Z., Benzerzour, M., Abriak, N.E., Damidot, D., Courard, L., Wang, D.: Use of uncontaminated marine sediments in mortar and concrete by partial substitution of cement. Cem. Concr. Compos. **93**(June), 155–162 (2018). https://doi.org/10.1016/j.cemconcomp.2018.07.010
33. Balkaya, M.: Beneficial Use of Dredged Materials in Geotechnical Engineering BT—Recycling and Reuse Approaches for Better Sustainability. In: Balkaya, N., Guneysu, S. (eds.) Recycling and reuse approaches for better sustainability, pp. 21–38. Springer International Publishing, Cham (2019)
34. Ben Abdelghani, F., Maherezi, W., Boutouil, M.: Geotechnical characterization of marine dredging sediments for use in road techniques. Environ. Eng. Dev. N 66-mar (2014). https://doi.org/10.4267/dechets-sciences-techniques.269
35. Loudini, A., Ibnoussina, M., Witam, O., Limam, A., Turchanina, O.: Valorisation of dredged marine sediments for use as road material. Case Stud. Constr. Mater. **13**, e00455 (2020). https://doi.org/10.1016/j.cscm.2020.e00455

36. Miraoui, M., Zentar, R., Abriak, N.: Road material basis in dredged sediment and basic oxygen furnace steel slag. Constr. Build. Mater. **30**, 309–319 (2012). https://doi.org/10.1016/j.conbuildmat.2011.11.032
37. Loudini, A., Ibnoussina, M., Limam, A., Kchikach, A., González, F.D.: Data on characterization of dredging sediment of Safi harbor—Morocco. Data Br. **28** (2020). https://doi.org/10.1016/j.dib.2019.104853
38. Hamouche, F., Zentar, R.: Characterization of organic matter in dredging sediments by thermal and chemical methods for a valorization in the field of road works. AJCE Spec. Issue **35**(1), 218–221 (2018)
39. Zentar, R., Abriak, N.-E., Dubois, V., Miraoui, M.: Beneficial use of dredged sediments in public works. Environ. Technol. **30**(8), 841–847 (2009). https://doi.org/10.1080/09593330902990139
40. Cherfaoui, N.: Ports of Morocco and silting, pp. 1–16 (2000). [Online]. Available: http://www.abhatoo.net.ma/maalama-textuelle/developpement-economique-et-social/developpement-economique/transport/transport-maritime/www.maroc-inge.com/uploads/biblio/ports.pdf
41. Minoubi, A., Elkhalidi, K., Chaibi, M.: Impact of port works on the evolution of the coastline of the bay of Safi (Atlantic coast-Morocco). Moroccan Rev. Geomorphol. 18–35 (2018) [Online]. Available: http://revues.imist.ma/?journal=remageom
42. Bouquerel, J.: Safi, second port of Morocco. Cah. O. M., pp. 217–257 (1965). doi.org/https://doi.org/10.3406/caoum.1965.2393
43. Charrouf, L.: Problems of silting of Moroccan ports on the Atlantic coast: their sedimentological impact on the coast. La Houille Blanche **1**, 49–71 (1991). https://doi.org/10.1051/lhb/1991004
44. Bertrand, M., Dhervilly, P., Levacher, D.: Valorisation des fines de dragages en Manche. In: Paralia—XIIIèmes Journées Nationales Génie Côtier—Génie Civil, pp. 969–976 (2014). https://doi.org/10.5150/jngcgc.2014.107
45. Hamouche, F., Zentar, R.: Effects of organic matter on mechanical properties of dredged sediments for beneficial use in road construction. Environ. Technol. **41**(3), 296–308 (2020). https://doi.org/10.1080/09593330.2018.1497711

Seismic Retrofitting: Analyzing the Effectiveness of RC Shear Walls and CFRP Reinforcement for RC Structures

Adil Ziraoui[1](✉), Benaissa Kissi[1], Hassan Aaya[2], Youssef El mezriahi[1], and Assia Haimoud[1]

[1] University of Hassan 2, Ensam, LISPS2I, Casablanca, Morocco
adil_ziraoui@hotmail.com

[2] International University of Casablanca, LMSI, Casablanca, Morocco

Abstract. In seismic design, engineers analyze the potential seismic hazards in a given area, such as ground shaking, ground rupture, and soil liquefaction. They then incorporate design principles and techniques to ensure that structures can withstand these forces and maintain their structural integrity. Several key considerations are involved in seismic design. These include selecting appropriate building materials, such as reinforced concrete or steel, that can absorb and dissipate seismic energy. Structural elements, such as beams, columns, and walls, are designed to resist lateral forces and prevent collapse during an earthquake. Design reinforced concrete structures are vulnerable to various damages, such as unintentional damage. In the context of seismic retrofitting of existing buildings, the technique of reinforcement by fiber-reinforced polymers (FRP) offers a relevant answer. The reinforcement by FRP for reinforced concrete structures is reproduced within the model by an addition of material and modifications of the parameters of the behavior law of the concrete, justified by experimentation and literature, or by the need to upgrade the structures to comply with new design codes and regulations. This study compares the effectiveness of shear wall reinforcement and carbon fiber reinforced polymer (CFRP) reinforcement on reinforced concrete (RC) elements as two seismic retrofitting techniques. The study examines a reinforced concrete structure with beams and columns using a non-linear static analysis, also known as a pushover analysis.

Keywords: Pushover · Max. displacement · Seismic performance · Shear wall · CFRP · RPS2000

1 Introduction

Large-scale earthquakes over the past 20 years have brought attention to the need to upgrade structures that are subject to seismic loading. According to researchers, the use of conventional techniques is caused by inadequate reinforcement, an uneven distribution of stiffness or mass in the building's plan and elevation, an inadequate foundation system, subpar materials, the deterioration of steel and anchors by corrosion, as well as a potential change in the use of the structure and its equipment [1].

K. Baba et al. (Eds.): GeoME 2023, SPEES, pp. 203–214, 2023.
https://doi.org/10.1007/978-3-031-49345-4_21

An earthquake is the most common phenomenon that causes forces and deformations in the elements of a structure. If these elements are rigid, they move in the same way as the ground, and the forces are then directly associated with the acceleration of the ground according to Newton's second fundamental relation of dynamics. If the elements of the structure are soft or flexible, significant deformations can be created inside the structure, caused by differential movements between the elements. It is, therefore, necessary to design these elements sufficiently resistant or sufficiently deformable, without their internal forces increasing, to resist seismic loads. It is the role of the design regulations of the structures to avoid such damage. Among these codes, are the RPS2000 v2011 (the regulation of seismic construction), the Eurocode 8 (the European regulation), and the ACI318 (the American regulation) [2, 3].

Taking into account the numerous deficiencies that can sometimes be found within the same building, along with the economic costs and societal consequences associated with potentially replacing it, it is crucial to develop a seismic methodology for strengthening or upgrading existing structures. It is important to highlight that reinforcing a building is a technically challenging and intricate process that can be significantly more expensive than constructing new seismic buildings. The formulation of a reinforcement or repair strategy for different stress levels, as well as accurately assessing the structural capacity, are essential and necessary steps in establishing a seismic retrofit strategy [4, 5].

Pushover analysis, also known as non-linear static analysis, involves putting the structure under a series of progressively heavier lateral stresses until it fails. This technique gives a historical comprehension of the structural behavior by concentrating on tracing the production of plastic hinges throughout the process.

Alternatively, shear wall reinforcement involves adding additional reinforced concrete walls. The study will evaluate the effectiveness of this technique in improving the overall performance and seismic resilience of the reinforced concrete structure. Seismic retrofitting using shear walls is a common technique employed to enhance the seismic resistance of structures. Shear walls are vertical structural elements designed to resist lateral forces generated during earthquakes [6, 7].

By employing CFRP reinforcement, the structural elements of the existing reinforced concrete building can be strengthened to enhance their load-carrying capacity, stiffness, and resistance to seismic forces. The CFRP materials, which consist of carbon fibers embedded in a polymer matrix, offer excellent tensile strength and corrosion resistance. These properties make CFRP an effective choice for reinforcing structural elements, as it can help prevent the formation and propagation of cracks during seismic events [8].

In the case of bonding CFRP to RC elements, the carbon fiber fabric is applied to the surfaces of beams and columns using a high-strength adhesive. This technique increases the flexural and shear strength of the structural members, providing improved resistance against the lateral forces generated by earthquakes. The behavior of the reinforced concrete structure with CFRP reinforcement will be analyzed by considering the load-displacement relationship and failure modes under seismic loading.

Through non-linear static analysis, the comparative investigation will assess the performance of the reinforced concrete structure retrofitted with CFRP reinforcement using both bonding and shear wall techniques. The results of the analysis will provide insights into the behavior and effectiveness of each retrofitting technique, helping engineers and researchers make informed decisions regarding the most suitable approach for seismic retrofitting projects [9, 10].

Overall, this study aims to contribute to the understanding of the behavior of reinforced concrete structures when retrofitted with CFRP materials, with a specific focus on their response to seismic forces. By examining the advantages and limitations of bonding and shear wall reinforcement techniques, the research aims to provide valuable insights into the optimal application of CFRP in seismic retrofitting projects [11].

2 Non-Linear Static Method

Pushover analysis, a non-linear static analysis method, is widely used to evaluate the seismic performance of structures. It involves gradually increasing the applied loads until a collapse mechanism is reached, allowing engineers to assess the structure's behavior and identify potential weak points. Pushover analysis provides valuable insights into the capacity, ductility, and energy dissipation capabilities of a structure under seismic forces, aiding in the development of effective retrofitting and design strategies [12] (Fig. 1).

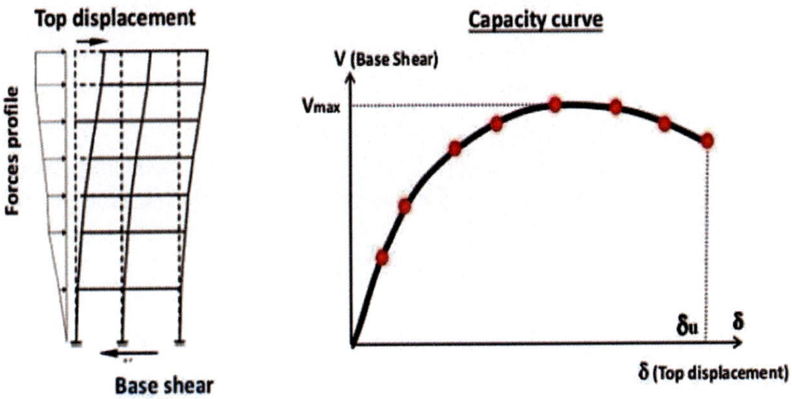

Fig. 1. Capacity curve [12]

3 Description of Seismic Retrofit Approaches

3.1 Seismic Reinforcement with a Shear Wall

Seismic retrofitting with shear walls is a technique used to enhance the seismic resistance of existing structures. Shear walls are rigid vertical elements that are added or reinforced within a structure to withstand lateral seismic forces.

The primary objective of shear wall seismic retrofitting is to increase the structural capacity to withstand seismic forces, thereby reducing the risk of structural damage and failure during an earthquake. Shear walls provide increased strength and stiffness, dissipating seismic forces and limiting excessive deformations.

However, seismic reinforcement using sails can present challenges, particularly in terms of design and installation, depending on the characteristics of the existing structure. Detailed structural analyses and seismic calculations are required to determine strengthening needs and design effective shear walls [12] (Fig. 2).

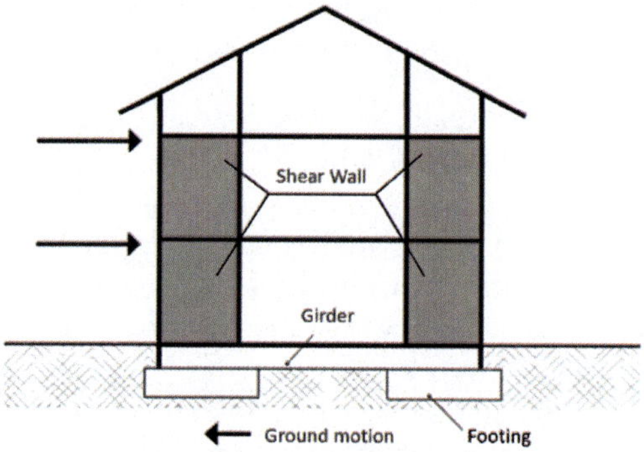

Fig. 2. Application of shear wall techniques [12]

3.2 Seismic Reinforcement with CFRP

Seismic retrofit approaches using Carbon Fiber Reinforced Polymer (CFRP) involve the application of CFRP materials to strengthen existing structures and enhance their resistance to seismic forces. CFRP is a composite material consisting of carbon fibers embedded in a polymer matrix, typically epoxy resin. It offers high strength, excellent durability, and lightweight properties, making it suitable for seismic retrofit applications [13].

Benefits of Carbon Fiber Fabric Reinforcement:

High Strength-to-Weight Ratio: Carbon fiber fabric exhibits a high strength-to-weight ratio, making it an ideal choice for reinforcement applications. It provides excellent tensile strength, allowing for increased load-carrying capacity while minimizing additional weight on the structure [14].
Flexibility and Conformability: Carbon fiber fabric is highly flexible and can easily conform to complex shapes and contours. This attribute enables it to be used for einforcing various structural components, including beams, columns, slabs, and walls, without the need for complex manufacturing processes [15].
Corrosion Resistance: Carbon fiber fabric is inherently resistant to corrosion, unlike traditional reinforcement materials such as steel. This resistance makes it particularly advantageous in environments where corrosion can compromise the structural integrity and durability of reinforced elements [16].

Ductility and Energy Absorption: The use of carbon fiber fabric in reinforcement applications enhances the ductility and energy absorption capacity of structural elements. It helps mitigate the development and propagation of cracks, improving the overall performance and resilience of the structure under dynamic loads, including seismic events [17] (Table 1 and Fig. 3).

Table 1. Carbon fiber properties [17]

Weight per unit volume (t/m^3)	Modulus of elasticity E (MPa)	Poisson's ratio U	Carbon fiber compressive strength f'c (MPA)
18.2	$3,9.10^5$	0.2	2500

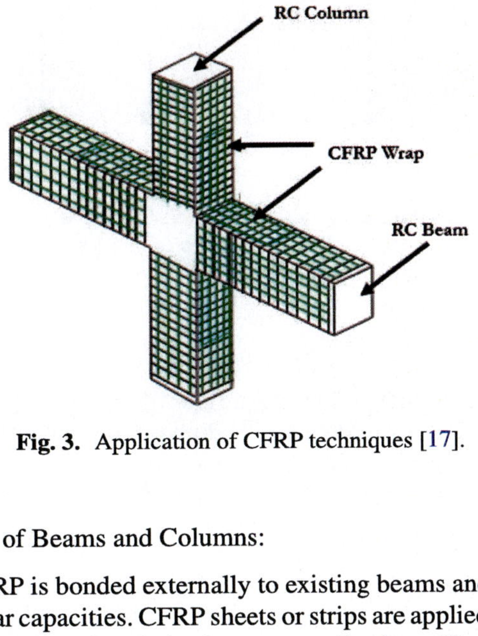

Fig. 3. Application of CFRP techniques [17].

CFRP Strengthening of Beams and Columns:

In this approach, CFRP is bonded externally to existing beams and columns to enhance their flexural and shear capacities. CFRP sheets or strips are applied to the surfaces of the structural elements, and then bonded using epoxy adhesives. The CFRP reinforcement increases the stiffness and strength of the elements, enabling them to better resist seismic loads [18].

4 Study Case

4.1 Geometry

The focus of our study is a six-levels reinforced concrete frame, measuring 18 m in height with each floor spanning 3 m. The structure consists of three bays, with each bay measuring 4 m in width. The slabs have a thickness of 25 cm (20+5), while the beams have dimensions of 25 cm × 25 cm. The columns have dimensions of 30 cm × 30 cm (Fig. 4).

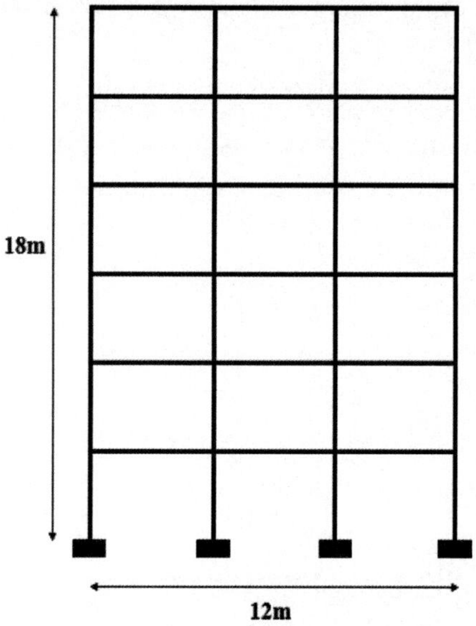

Fig. 4. General view of the 2D model

4.2 Seismic Data

Inserting seismic data into SAP2000 software allows for the comprehensive analysis and design of structures under seismic loading conditions. The process involves incorporating seismic parameters and response spectrum data into the software to accurately simulate the behavior of structures during earthquakes (Table 2).

Table 2. Seismic data

Total weight W(t)	261
Priority factor I	1
Soil factor S	1.2
Horizontal ground acceleration A	0.16
Behavior factor K	2
Amplification factor D	2.31

5 Results

The results of the lateral force applied to each floor are summarized in Table 3.

Table 3. Result of lateral loads for each floor.

Level	V(KN)	H (m)	Fi (KN)
1	578.8	3	27.56
2		6	55.12
3		9	82.68
4		12	110.24
5		15	137.8
6		18	165.37

5.1 Plastics Hinges

Plastic hinges are essential elements in the design of reinforced concrete structures subjected to seismic loads. A plastic hinge is formed when specific sections of the structure reach their ultimate bending and shear capacities. Unlike metal hinges, plastic hinges form in the material itself, mainly in reinforced concrete.

When a structure is subjected to seismic loads, plastic hinges form in critical zones such as columns, beams, and walls. These zones are designed to allow the controlled formation of plastic hinges in order to dissipate seismic energy and protect the rest of the structure.

The formation of plastic hinges in reinforced concrete structural elements is desirable because it enables stress redistribution and efficient energy dissipation. When a plastic hinge is formed, it allows the structure to undergo significant deformation while maintaining its structural integrity.

Plastic hinges are designed to be reversible, meaning that they can deform during a seismic event and return to their original shape once the seismic load has decreased. This reversibility is essential to ensuring the long-term repairability and durability of structures [19] (Figs. 5 and 6).

The subsequent tables present the hinge types assigned to each element within the three structures (Tables 4, 5 and 6).

5.2 Pushover Curves

The pushover curve is typically plotted by incrementally applying lateral loads to the structure until it reaches its ultimate capacity or failure point. The curve starts at the initial elastic range, where the structure deforms elastically without any significant damage. As the load increases, the structure enters the inelastic range, and plastic hinges start to form in the critical elements.

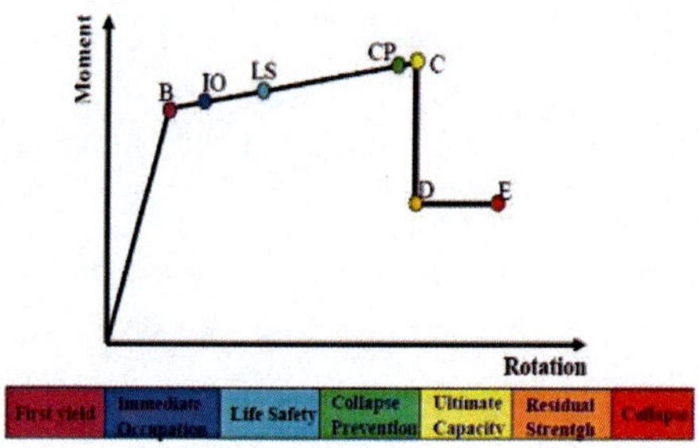

Fig. 5. The plastic hinge behavior [19].

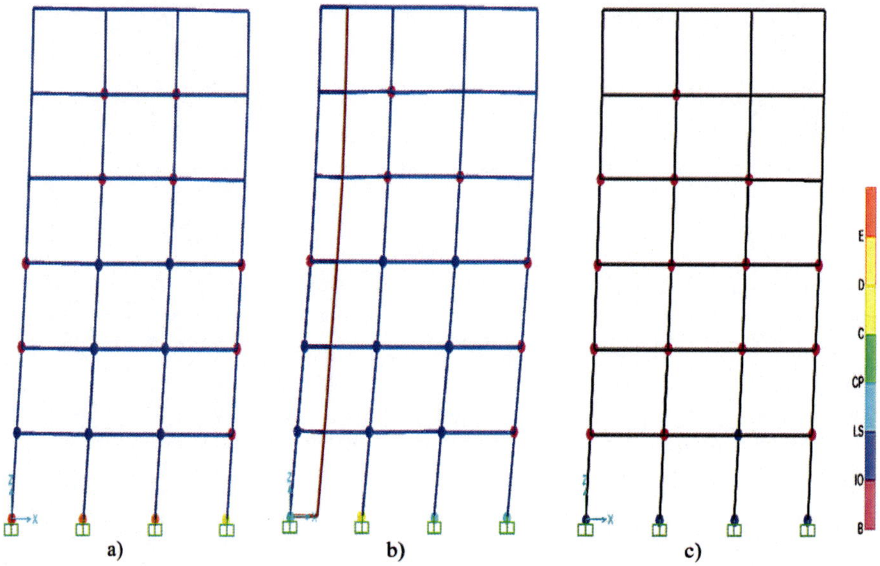

Fig. 6. Plastic hinges formation: a) in the unreinforced structure b) in the reinforced structure with shear wall c) in the structure reinforced with CFRP

The pushover curve depicts the progressive development of plastic hinges and the redistribution of forces within the structure. It shows how the lateral resistance capacity of the structure changes as the load increases. The curve eventually reaches a peak point, indicating the maximum load-carrying capacity of the structure. Beyond this point, the structure may experience significant deformation and ultimately fail.

The capacity curve presented in Fig. 7 illustrates the performance of the three structures. It is evident from the curve that the shear wall reinforced structure exhibits a remarkable reduction of 23% in total displacement. In comparison, the carbon fiber reinforced structure demonstrates a more significant reduction of 54% in total displacement. This indicates that the second reinforcement technique using carbon fiber is considerably more rigid and offers superior resistance to seismic forces when compared to the first technique of shear wall reinforcement.

Table 4. Formation hinges for unreinforced structure

Step	A to B	B to IO	IO to LS	LS to CP	CP to C	C to D	D to E	Beyond E	Total
1	42	0	0	0	0	0	0	0	42
2	41	1	0	0	0	0	0	0	42
3	29	13	0	0	0	0	0	0	42
4	25	14	3	0	0	0	0	0	42
5	21	16	4	0	0	0	0	0	42
6	15	15	12	0	0	0	0	0	42
7	15	13	12	2	0	0	0	0	42
8	15	11	13	2	0	0	0	1	42
9	11	13	14	2	0	1	0	1	42
10	11	13	14	1	0	2	0	1	42
11	9	15	14	1	0	1	1	1	42
12	9	15	14	0	0	1	2	1	42

Table 5. Formation hinges for reinforced structure with shear wall

Step	A to B	B to IO	IO to LS	LS to CP	CP to C	C to D	D to E	Beyond E	Total
1	41	1	0	0	0	0	0	0	42
2	34	8	0	0	0	0	0	0	42
3	30	9	3	0	0	0	0	0	42
4	14	8	4	0	0	0	0	0	42
5	26	11	5	3	0	0	0	0	42
6	26	10	6	4	0	0	0	0	42
7	25	8	7	2	0	0	0	0	42
8	24	6	8	4	0	0	0	0	42
9	23	7	8	3	0	1	0	0	42

Table 6. Formation hinges for reinforced structure with CFRP

Step	A to B	B to IO	IO to LS	LS to CP	CP to C	C to D	D to E	Beyond E	Total
1	41	1	0	0	0	0	0	0	42
2	29	13	0	0	0	0	0	0	42
3	24	14	4	0	0	0	0	0	42
4	23	15	4	0	0	0	0	0	42
5	22	15	5	0	0	0	0	0	42
6	22	12	8	0	0	0	0	0	42

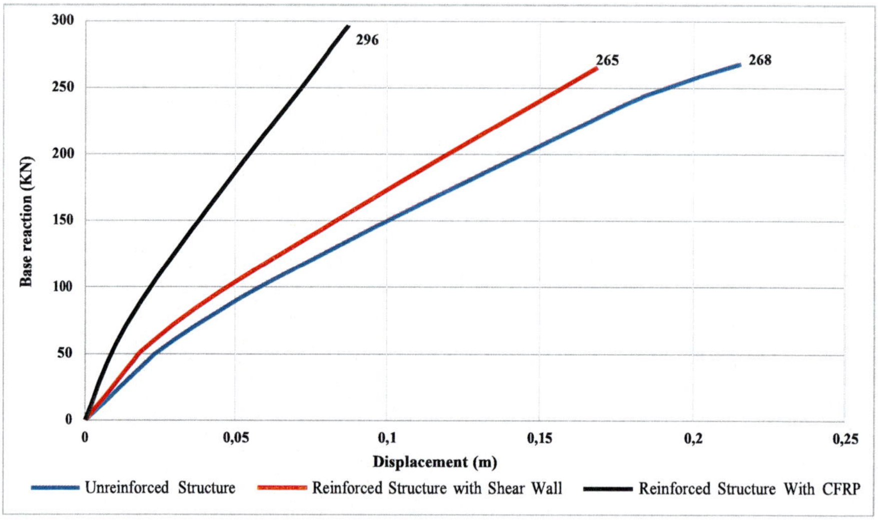

Fig. 7. Pushover curves of the three structures

6 Conclusion

The main objective of this research is to conduct a thorough comparative analysis of two seismic retrofitting methods: the application of shear wall reinforcement and the bonding of carbon fiber reinforced polymer (CFRP) reinforcement to reinforced concrete (RC) elements. This objective is achieved through the implementation of a pushover analysis, which is a non-linear static analysis conducted on a reinforced concrete structure comprising beams and columns.

By employing pushover analysis, the study aims to evaluate the performance and effectiveness of the two retrofitting methods in enhancing the seismic resilience of the structure. This analysis allows for a detailed examination of the load-displacement relationship, as well as the identification of potential failure mechanisms and the behavior of the structural elements under seismic forces.

The comparative investigation aims to provide valuable insights into the advantages, limitations, and overall effectiveness of the bonding of CFRP reinforcement and shear wall reinforcement techniques. The findings of this study can serve as a guide for engineers and researchers in making informed decisions regarding the selection and implementation of appropriate seismic retrofitting methods for reinforced concrete structures.

In summary, the most effective technique for retrofitting is the utilization of Carbon Fiber Reinforced Polymer (CFRP). This method not only minimizes the overall displacement of the building but also offers outstanding properties for reinforcing structural elements. These properties include remarkable strength, a high elastic modulus, exceptional durability, and a lightweight nature. Conversely, reinforcing with shear walls substantially increases the weight of the structure, necessitates a lengthier implementation period, and incurs higher costs in comparison to CFRP.

References

1. Règlement de construction parasismique RPS 2000, Ministère de l'ATUHE, Secrétariat d'État de l'Habitat,Royaume du Maroc (2011)
2. El Ghoulbzouri, A., Kissi, B., Khamlichi, A.: Reliability analysis of reinforced concrete buildings
3. Applied Technology Council, ATC-40. Seismic evaluation and retrofit of concrete Buildings, vol. 1 and 2. California (1996)
4. Federal Emergency Management Agency (FEMA), 1997.NEHRP provisions for the seismic rehabilitation of buildings. Rep FEMA 273 and 274. Washington, DC: FEMA (1997)
5. CEN European Committee for Standardization. Eurocode 8. Design of structures for earthquake resistance. Doc CE N/TC250/SC8/N335. DRAFT No 6. January. Brussels (2003)
6. Marsono, A.K., Subedi, N.K.: Analysis of reinforced concrete shear wall structures with staggered openings part II non-linear finite element analysis (NLFEA). In: Proceeding of the 4th Asian-Pacific Structural Engineering and Construction Conference (APSEC2000). September 13th-15th, 2000. Kuala Lumpur, p. 341–355
7. Fintel, M.: Performance of Buildings With Shear Walls in Earthquakes of the Last Thirty Years. PCI J. **40**(3), 62–80 (1995)
8. Tbatou, T., El Youbi, M.: Dynamic and structural study of a RC building braced by FRP composite materials. Int. Rev. Civil Eng. (IRECE) (2020)
9. Kissi, B., Riyad, Y., Mrani, I., EL Ghoulbzouri, B., Parron, M.A., El khatib, Y.: Reinforcement of RC structure by carbon fibers. MATEC Web Conf. **83**:09009 (2016). https://doi.org/10.1051/matecconf/20168309009
10. Krupa Engineer, P., Shah, H.P.P.D.: Strengthening of reinforced concrete beams with fiber reinforced polymer sheets with different configurations in shear and flexure: a critical review. Int. J. Res. Anal. Rev. **6**(2), 102–106 (2019)
11. Punurai, W., Hsu, C.-T.T., Zhang, Z.: Flexural and shear strengthenings of RC beams using carbon fiber reinforced polymer laminates. ACI Sympos. Publ. **211**. https://doi.org/10.14359/12586
12. Gupta, B., Kunnath, S.K.: Adaptive spectra-based pushover procedure for seismic evaluation of structures. Earthq. Spectra **16**(2), 367–392 (2000)
13. Francesco Bencardino, G.S., Swamy, R.N.: strength and ductility of reinforced concrete beams externally reinforced with carbon fiber fabric. ACI Struct. J. **99**(2). https://doi.org/10.14359/11539

14. Zhang, A.-H., Jin, W.-L., Li, G.-B.: Behavior of preloaded RC beams strengthened with CFRP laminates. J. Zhejiang Univ.-Sci. A **7**(3), 436–444 (2006). https://doi.org/10.1631/jzus.2006.A0436
15. Valivonis, J., Skuturna, T.: Cracking and strength of reinforced concrete structures in flexure strengthened with carbon fibre laminates. J. Civ. Eng. Manag. **13**(4), 317–323 (2007). https://doi.org/10.1080/13923730.2007.9636452
16. Sobuz, H.R., Ahmed, E., Hasan, N.M.S., Uddin, A.: Use of carbon fiber laminates for strengthening reinforced concrete beams in bending. Int. J. Civil Str. Eng. **2**(1), 67–84 (2011)
17. Li, G., Guo, Y., Sun, X.: Investigation on flexural performance of RC beams flexurally strengthened by side-bonded CFRP laminates. Open Civ. Eng. J. **6**(1), 26–32 (2012). https://doi.org/10.2174/1874149501206010026
18. Alferjani, M., Samad, A.A., Elrawaff, B.S., Mohamad, N., Hilton, M., Saiah, A.A.S.: Use of carbon fiber reinforced polymer laminate for strengthening reinforced concrete beams in shear: a review. Int. Refereed J. Eng. Sci. **2**(2), 45–53 (2013)
19. Sharma, A., Reddy, G.R., Vaze, K.K., Eligehausen, R.: Pushover experiment and analysis of a full scale non-seismically detailed RC structure. Eng. Struct.Struct. **46**, 218–233 (2013). https://doi.org/10.1016/j.engstruct.2012.08.006

Effect of Curing Conditions on the Mechanical Properties of Geopolymer Binder Based Natural Moroccan Pozzolan

Khaoula Doughmi(✉) and Khadija Baba

Civil and Environment Engineering Laboratory (LGCE), Mohammadia Engineering School, Mohammed V University, Rabat, Morocco
Khaoula.doughmi@gmail.com

Abstract. Cement is rapidly becoming among the extensively utilized materials. However, its production contributes to approximately 6–8% of the world's CO_2 emissions. Finding ways to reduce the carbon footprint of materials without increasing their cost is quite challenging. One technique to enhance sustainability is to produce alternative materials with low energy consumption, for instance geopolymer. The type of alkali activator, thermal curing environment, and the concentration of alkali hydroxide are only a few of the variables that may have an impact on the physical and mechanical characteristics of geopolymer. This study uses natural Moroccan pozzolan as an aluminosolicate source and sodium silicate as an alkaline solution to investigate the impacts of the curing conditions on the mechanical properties of the geopolymer binder. Natural Moroccan pozzolan is an easily available abandoned natural resource. Alkaline solution to binder ratio and Na_2SiO_3 to NaOH ratio were constant. Same pastes were produced following dissimilar curing conditions (curing at room temperature, in water at 20 °C, oven curing 70 h at 60 °C and continuous curing at 200 °C) and tested for compressive and flexural strengths assessed after 7 and 28 days. The test findings demonstrated a substantial relationship between compressive strength and curing conditions. By subjecting geopolymer to elevated curing temperatures, it is possible to achieve binders with exceptional strength properties due to the rapid geopolimerization.

Keywords: Natural Moroccan pozzolan · Geopolymer binder · Compressive strength · Alkaline solution · Sustainable environment

1 Introduction

The search for alternatives to Portland cement in the construction industry, considering both similar properties and environmental impact, poses a significant challenge. One potential alternative that researchers have explored is geopolymers. Geopolymers are cementitious substances created through the chemical interaction between aluminosilicate precursors and an alkaline activator solution [1]. These materials offer an alternative to traditional cement-based binders, such as Portland cement, and have gained attention in recent years due to their potential for improved mechanical properties, durability, and environmental sustainability [2].

K. Baba et al. (Eds.): GeoME 2023, SPEES, pp. 215–223, 2023.
https://doi.org/10.1007/978-3-031-49345-4_22

Geopolymers are typically synthesized by mixing aluminosilicate-rich materials, such as fly ash, metakaolin, or slag, with an alkaline solution, commonly comprised of sodium or potassium hydroxide and silicate or aluminate compounds. The alkaline activator solution acts as a catalyst, initiating a chemical reaction known as geopolymerization [1]. Geopolymers have exceptional qualities and have the potential to replace various construction materials. They offer numerous advantages such as fast curing, low thermal conductivity and high strength [2]. They also provide economic and environmental advantages due to their affordability, minimal greenhouse gas emissions, and reduced energy consumption during the synthesis process [3].

The utilization of natural pozzolan a powder binder in the production of geopolymer has experienced a noteworthy upsurge in recent times [4–16]. The global distribution of natural pozzolan deposits spans approximately 0.84% of the Earth's soil, which amounts to approximately 124 million hectares and can be found in various regions worldwide [17]. However, it has been observed that natural pozzolan exhibits lower reactivity compared to artificial pozzolan, resulting in inferior mechanical properties and durability of NP-based geopolymer binders. To address this issue, various methods have been explored to enhance the reactivity of Natural pozzolan based geopolymer binders. These methods include calcination of the aluminosilicate source and mechanical activation achieved through extended grinding time [18].

Several studies have investigated the alkali activation of natural pozzolan to develop geopolymers. Researchers have explored the influence of various alkali activators, such as sodium hydroxide (NaOH) [19, 20] and sodium silicate (Na_2SiO_3) [21–23] on the geopolymerization process. It has been observed that alkali activation leads to the formation of a three-dimensional polymeric structure, resulting in improved mechanical properties. However, the effect of curing conditions on these geopolymers has not been thoroughly investigated.

The aim of this study is to evaluate the feasibility and potential of utilizing volcanic natural Moroccan pozzolan in alkali activation processes. The study seeks to explore the mechanical properties and performance of geopolymers produced under different curing conditions. The objective is to assess the suitability of this approach for sustainable construction materials.

2 Materials and Methods

2.1 Materials

Natural pozzolan

The natural pozzolan (PZ) utilized in the study was sourced from the Middle Moroccan Atlas region. It was steamed for 24 h at 105 °C to remove moisture and milled to a thickness of 80 μm. The physical and chemical composition of the pozzolan was assessed through X-ray fluorescence analysis, and the results are presented in Table 1. Additionally, X-ray diffraction analysis was conducted for further characterization (Fig. 1) identified the main phases in the composition of the natural Moroccan pozzolan as Augite, Aluminian, and Forsterite.

Table 1. Chemical and physical properties of natural Moroccan pozzolan

SiO2	Al2O3	Fe_2O_3	CaO	MgO	TiO_2	Na_2O	P_2O_5	K_2O	MnO	LOI
42.54	16.60	10.57	10.29	8.060	2.396	2.006	1.416	0.92	0.167	4.240
Specific density (g/cm^3)						Bulk density (g/cm^3)				
2.22						1.03				

Fig. 1. Mineralogical identification of natural Moroccan pozzolan

Alkaline solution

In this study, the alkaline solution used consists of a mixture of water, sodium hydroxide (NaOH), and sodium silicate. Initially, 8 mol L^{-1} solutions of sodium hydroxide were prepared by dissolving NaOH flakes with a purity of 99% in distilled water. Subsequently, solutions were prepared by combining the NaOH solution with sodium silicate, maintaining a fixed Na_2SiO_3/NaOH ratio of 1.2. These solutions were prepared 24 h in advance of the experimental procedures.

2.2 Methods

Geopolymer synthetis method

The pozzolan was mixed with the solution, following the ratios provided in Table 2. After blending the mixture for 10 min, the resultant slurry was carefully poured into molds of 4 × 4 × 16 cm. After 24 h of ambient temperature polymerization, the geopolymers were demolded, and each one was subjected to a specific curing condition.

Characterization methods

40 × 40 × 160 mm specimens were subjected to flexion testing after 7 and 28 days. The testing was conducted using a three-point bending machine capable of applying loads until failure at a loading speed of 50 N/s.

The half-prisms obtained from the bending test were then subjected to compression testing under a 40 mm × 40 mm section at a loading speed of 2400 N/s until rupture.

Table 2. Proportions of all the mixtures

Code	Pozzolan (%)	NaOH molarity	$Na_2SiO_3/NaOH$	Solution/solid	Curing condition
NP-GPB-A	100%	8M	1.2	0.5	Air
NP-GPB-W					Water
NP-GPB-P					Precured 70 h at 60 °C
NP-GPB-C					Cured at 200 °C

The density and water absorption of the hardened mortars was measured at 28 days according to ASTM C20 and ASTM C642-13, respectively. The measurement was performed using the equations: Bulk density (g/cm^3) = M/V and water absorption (%) = A−M/M × 100, where M represents the oven-dried samples mass, V is the exterior volume and A refers to the weight of a sample when it is saturated in air.

In order to assess the ultrasonic pulse velocity (UPV) following the guidelines outlined in ASTM C597-16, a portable digital indicating tester utilizing ultrasonic non-destructive technology was employed. The tester employs direct transmission and calculates the pulse velocity (Vp) using the formula: $Vp = L / \Delta t$, where L is the distance and Δt is the transit time.

3 Results and Discussion

3.1 Physical Properties & UPV Values

Figure 2 illustrates the impact of various curing conditions on the density, water absorption, and ultrasonic pulse velocity (UPV) of geopolymer binders. It was observed that the values of these properties increased as the curing conditions transitioned from air curing to water curing, precuring, and finally to curing.

With the addition of heat during the curing process, both the density and UPV values exhibited an increase, accompanied by a decrease in water absorption. The application of heat contributed to a more compact microstructure and enhanced hydration process of the cementitious materials due to the availability of sufficient moisture.

Ambient curing, which involves exposing the mortar to normal environmental conditions without additional moisture control, may not provide optimal moisture content for the hydration process. On the other hand, water curing helps maintain a high moisture content in the mortar, facilitating complete and continuous hydration of cement particles. This, in turn, leads to the formation of a denser and more compact microstructure.

By subjecting the mortar to controlled temperature conditions, excess moisture present in the material can be more rapidly removed. This removal of excess moisture can potentially reduce the water content and increase the density of the mortar.

Furthermore, the elevated temperature in the oven accelerates the drying process of the mortar, resulting in the evaporation of water. This evaporation can lead to a reduction in volume and potentially increase the density of the mortar.

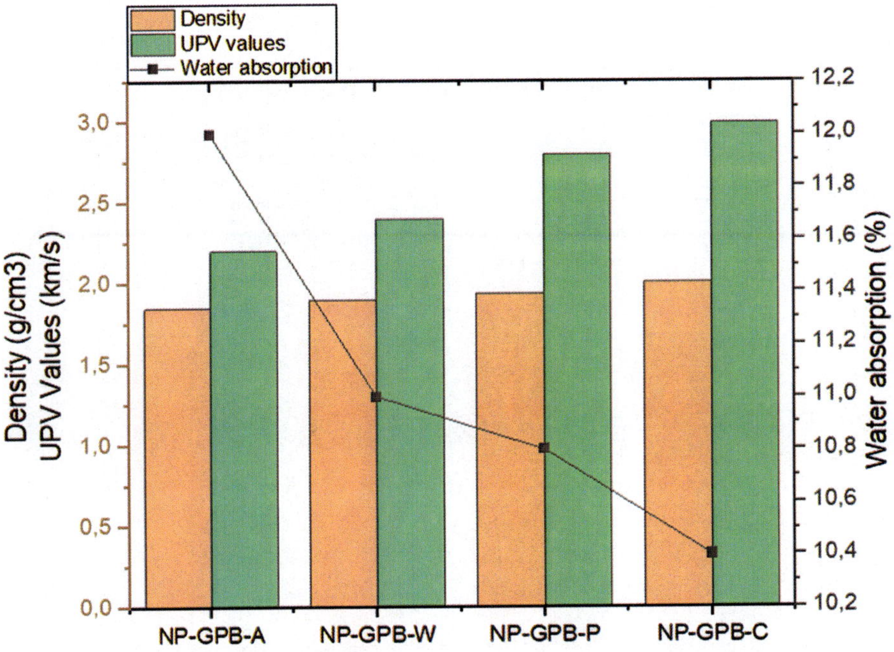

Fig. 2. Density, UPV values and water absorption of 28 days geopolymer binders

3.2 Mechanical Properties

Figures 3 and 4 provide a visual representation of the influence of different curing conditions on the flexural and compressive strengths of the specimens at 7 and 28 days. The results indicate that the both strengths values increased progressively as the curing conditions alternated from air curing to water curing, precuring, and finally to curing. These observations suggest that the choice of appropriate curing conditions significantly affects the development of flexural and compressive strength in the specimens. The progression from air curing to more controlled conditions, such as water curing, precuring, and curing, promotes the enhancement of these mechanical properties over time.

It is evident that the strengths of Geopolymers are influenced by the temperature used during heat curing. This effect is attributed to the acceleration of the geopolymerization process, which involves the reaction between alkali activator and natural pozzolan, necessitating the application of heat promoting the formation of a more cohesive and denser geopolymeric structure [24, 25]. The geopolymerization process is known to be influenced by the curing conditions. Specifically, higher curing temperatures have been shown to enhance the mechanical properties of the material by accelerating the rate of dissolution [17, 26, 27]. However, it should be noted that elevated temperature curing can also result in rapid evaporation of water absorption [6, 28, 29]. This can be observed in the corresponding trends of higher density, lower water absorption, and increased ultrasonic pulse velocity (UPV) values in the specimens subjected to higher curing temperatures.

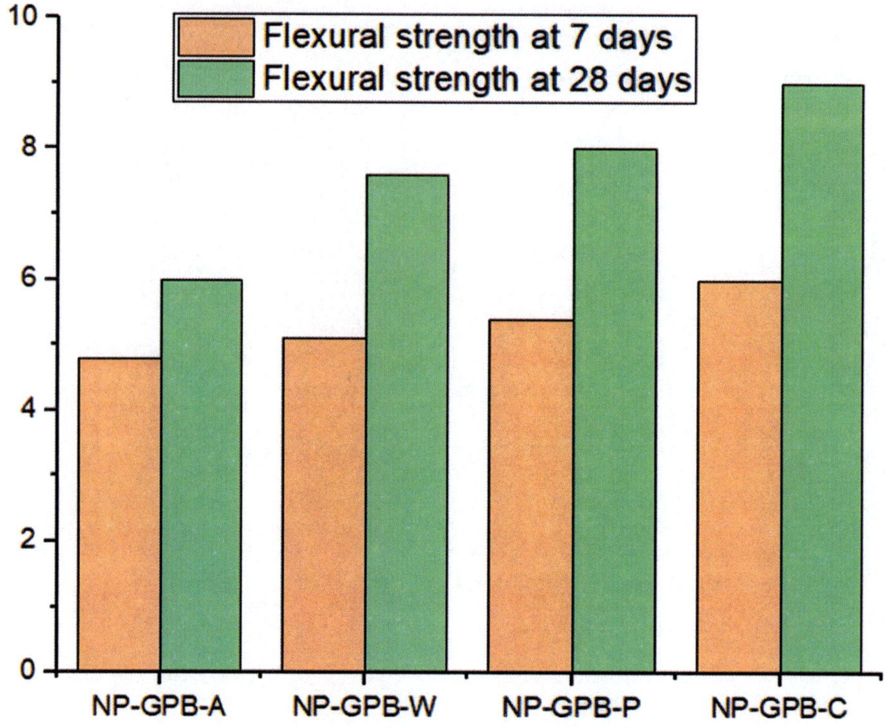

Fig. 3. Flexural strength of geopolymer binders at 7 and 28 days

4 Conclusions

Geopolymer offers a sustainable alternative to conventional cement with lower carbon emissions. In this study, the focus was on examining the impact of different curing conditions on the mechanical properties of geopolymer binders using natural Moroccan pozzolan as an aluminosilicate source. Different curing conditions were tested, including room temperature, water curing, and oven curing at varying temperatures. The findings indicated a significant correlation between curing conditions and both flexural and compressive strength, with higher temperatures promoting rapid geopolymerization and achieving high-strength binders. These findings contribute to the development of sustainable construction materials by optimizing curing conditions and reducing environmental impact.

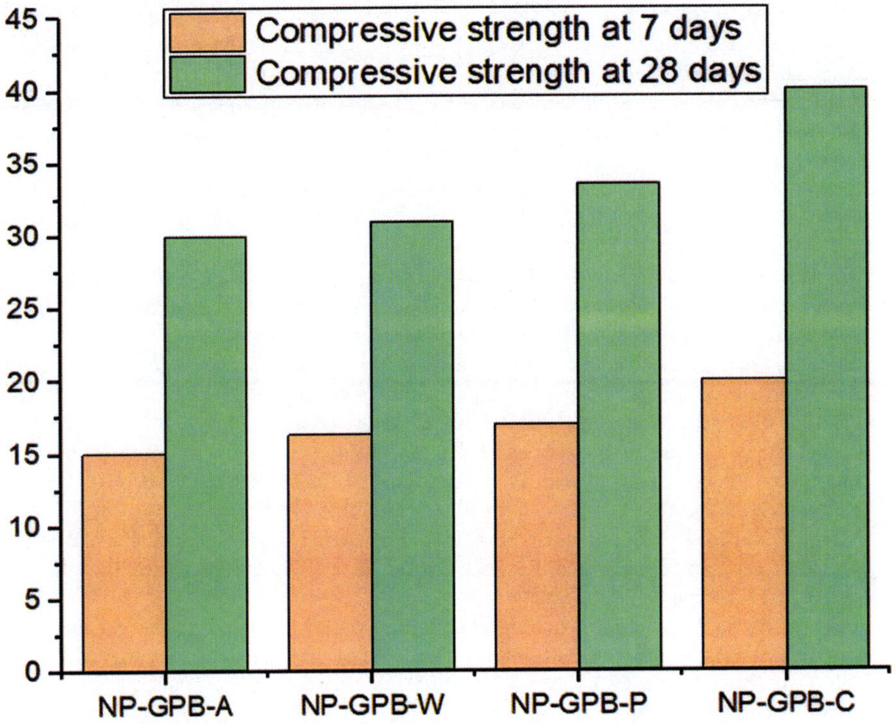

Fig. 4. Compressive strength of geopolymer binders at 7 and 28 days

References

1. Atiş, C.D., Görür, E.B., Karahan, O., Bilim, C., İlkentapar, S., Luga, E.: Very high strength (120MPa) class F fly ash geopolymer mortar activated at different NaOH amount, heat curing temperature and heat curing duration. Constr. Build. Mater. **96**, 673–678 (2015). https://doi.org/10.1016/j.conbuildmat.2015.08.089
2. Duxson, P., Fernández-Jiménez, A., Provis, J.L., Lukey, G.C., Palomo, A., Van Deventer, J.S.J.: Geopolymer technology: the current state of the art. J. Mater. Sci. **42**(9), 2917–2933 (2007). https://doi.org/10.1007/s10853-006-0637-z
3. Xu, M., He, Y., Liu, Z., Tong, Z., Cui, X.: Preparation of geopolymer inorganic membrane and purification of pulp-papermaking green liquor. Appl. Clay Sci. **168**, 269–275 (2019). https://doi.org/10.1016/j.clay.2018.11.024
4. Bondar, D., Lynsdale, C.J., Milestone, N.B., Hassani, N., Ramezanianpour, A.A.: Effect of type, form, and dosage of activators on strength of alkali-activated natural pozzolans. Cement Concr. Compos. **33**(2), 251–260 (2011). https://doi.org/10.1016/j.cemconcomp.2010.10.021
5. Yadollahi, M.M., Benli, A., Demirboğa, R.: The effects of silica modulus and aging on compressive strength of pumice-based geopolymer composites. Constr. Build. Mater. **94**, 767–774 (2015). https://doi.org/10.1016/j.conbuildmat.2015.07.052
6. Haddad, R.H., Alshbuol, O.: Production of geopolymer concrete using natural pozzolan: a parametric study. Constr. Build. Mater. **114**, 699–707 (2016). https://doi.org/10.1016/j.conbuildmat.2016.04.011

7. Vafaei, M., Allahverdi, A.: Influence of calcium aluminate cement on geopolymerization of natural pozzolan. Constr. Build. Mater. **114**, 290–296 (2016). https://doi.org/10.1016/j.conbuildmat.2016.03.204
8. Djobo, J.N.Y., Elimbi, A., Tchakouté, H.K., Kumar, S.: Volcanic ash-based geopolymer cements/concretes: the current state of the art and perspectives. Environ. Sci. Pollut. Res. **24**(5), 4433–4446 (2017). https://doi.org/10.1007/s11356-016-8230-8
9. Kantarcı, F., Türkmen, İ., Ekinci, E.: Optimization of production parameters of geopolymer mortar and concrete: a comprehensive experimental study. Constr. Build. Mater. **228**, 116770 (2019). https://doi.org/10.1016/j.conbuildmat.2019.116770
10. Gultekin, A., Ramyar, K.: Investigation of high-temperature resistance of natural pozzolan-based geopolymers produced with oven and microwave curing. Constr. Build. Mater. **365**, 130059 (2023). https://doi.org/10.1016/j.conbuildmat.2022.130059
11. Aziz et al. A.: Effect of acidic volcanic perlite rock on physio-mechanical properties and microstructure of natural pozzolan based geopolymers. Case Studies Constr. Mater. **15**, e00712 (2021). https://doi.org/10.1016/j.cscm.2021.e00712
12. Aziz et al. A.: Effect of blast-furnace slag on physicochemical properties of pozzolan-based geopolymers. Mater. Chem. Phys. **258**, 123880 (2021). https://doi.org/10.1016/j.matchemphys.2020.123880
13. Safari, Z., Kurda, R., Al-Hadad, B., Mahmood, F., Tapan, M.: Mechanical characteristics of pumice-based geopolymer paste. Resour., Conser. Recycl. **162**, 105055 (2020). https://doi.org/10.1016/j.resconrec.2020.105055
14. Simou, S., Baba, K., Akkouri, N., Lamrani, M., Tajayout, M., Nounah, A.: Mechanical characterization and reinforcement of the adobe material of the chellah archaeological site. E3S Web Conf **150**, 03022 (2020). https://doi.org/10.1051/e3sconf/202015003022
15. Simou, S., Baba, K., Akkouri, N., Lamrani, M., Tajayout, M., Nounah, A.: Mechanical characterization of the adobe material of the archaeological site of chellah. In: Ameen, H., Jamiolkowski, M., Manassero, M., Shehata, H. (eds.) Recent Thoughts in Geoenvironmental Engineering, in Sustainable Civil Infrastructures, pp. 118-130. Cham: Springer International Publishing (2020). https://doi.org/10.1007/978-3-030-34199-2_8
16. Bourzik, O., Akkouri, N., Baba, K., Haddaji, Y., Nounah, A., Assafi, M., Bazzar, K.: Study of the effects of drinking water treatment sludge on the properties of class F fly ash-based geopolymer. Environ. Sci. Pollut. Res. Int. **29**(58), 87668–87679 (2022). https://doi.org/10.1007/s11356-022-21873-9
17. Robayo-Salazar, R., Mejía-Arcila, J., Mejía De Gutiérrez, R., Martínez, E.: Life cycle assessment (LCA) of an alkali-activated binary concrete based on natural volcanic pozzolan: a comparative analysis to OPC concrete. Constr. Build. Mater. **176**, 103–111 (2018). https://doi.org/10.1016/j.conbuildmat.2018.05.017
18. Yankwa Djobo, J.N., Elimbi, A., Tchakouté, H.K., Kumar, S.: Mechanical activation of volcanic ash for geopolymer synthesis: effect on reaction kinetics, gel characteristics, physical and mechanical properties. RSC Adv. **6**(45), 39106–39117 (2016). https://doi.org/10.1039/C6RA03667H
19. Allahverdi, A., Mehrpour, K., Kani, E.N.: Investigating the possibility of utilizing pumice-type natural pozzonal in production of geopolymer cement (2008)
20. Lemougna, P.N., MacKenzie, K.J.D., Melo, U.F.C: Synthesis and thermal properties of inorganic polymers (geopolymers) for structural and refractory applications from volcanic ash. Ceram. Int. **37**(8), 3011–3018 (2011). https://doi.org/10.1016/j.ceramint.2011.05.002
21. Tchakoute, H.K., Elimbi, A., Yanne, E., Djangang, C.N.: Utilization of volcanic ashes for the production of geopolymers cured at ambient temperature. Cem. Concr. Compos. **38**, 75–81 (2013). https://doi.org/10.1016/j.cemconcomp.2013.03.010

22. Djobo, J.N.Y., Elimbi, A., Tchakouté, H.K., Kumar, S.: Reactivity of volcanic ash in alkaline medium, microstructural and strength characteristics of resulting geopolymers under different synthesis conditions. J. Mater. Sci. **51**(22), 10301–10317 (2016). https://doi.org/10.1007/s10853-016-0257-1
23. Robayo-Salazar, R.A., De Gutiérrez, M., Puertas, F.: Study of synergy between a natural volcanic pozzolan and a granulated blast furnace slag in the production of geopolymeric pastes and mortars. Constr. Build. Mater. **157**, 151–160 (2017). https://doi.org/10.1016/j.conbuildmat.2017.09.092
24. Hassan, A., Arif, M., Shariq, M.: Effect of curing condition on the mechanical properties of fly ash-based geopolymer concrete. SN Appl. Sci. **1**(12), 1694 (2019). https://doi.org/10.I007/s42452-019-1774-8
25. Firdous, R., Stephan, D., Djobo, J.N.Y.: Natural pozzolan based geopolymers: a review on mechanical, microstructural and durability characteristics. Constr. Build. Mater. **190**, 1251–1263 (2018). https://doi.org/10.1016/j.conbuildmat.2018.W.191
26. Villa, C., Pecina, E.T., Torres, R., Gómez, L.: Geopolymer synthesis using alkaline activation of natural zeolite. Constr. Build. Mater. **24**(11), 2084–2090 (2010). https://doi.org/10.101d/j.conbui1dmat.2D10.04.052
27. Najafi Kani, E., Allahverdi, A.: Effects of curing time and temperature on strength development of inorganic polymeric binder based on natural pozzolan. J. Mater. Sci. **44**, 3088–3097 (2009). https://doi.org/10.1007/s10853-3811-1
28. Takeda, H., Hashimoto, S., Kanie, H., Honda, S., Iwamoto, Y.: Fabrication and characterization of hardened bodies from Japanese volcanic ash using geopolymerization. Ceramics Int. **40**(3), 4071–4076 (2014). https://doi.org/10.1016/j.ceramint.2D13.08.061
29. Tchakoute, H.K., Elimbi, A., Kenne, B.D., Mbey, J.A., Njopwouo, D.: Synthesis of geopolymers from volcanic ash via the alkaline fusion method: Effect of Al_2O_3/Na_2O molar ratio of soda–volcanic ash. Ceramics Int. **39**(1), 269–276 (2013). https://doi.org/10.101Wj.ceramint.2012.06.021

Sedimentological and Lithostratigraphic Study of Paleocene-Eocene of Foum El Kouss and Imiter, the Southern Flank of the Central High Atlas. Morocco

Said Moujane(✉), Ahmed Algouti, Abdellah Algouti, and Abdelfattah Aboulfaraj

Faculty of Sciences, Department of Geology, Geosciences, Geotourism, Natural Hazards and Remote Sensing Laboratory, Cadi Ayyad University, Bd. BO 2390, 40000 Marrakesh, Morocco
Said.moujane02@gmail.com

Abstract. This work concerns the outcrops of the Paleocene-Eocene series on the meridional side of the Central High Atlas, precisely in the localities of Boumalne-Imiter to the west of the Tinghir city. The main objective of this study is to propose a paleoenvironment and its evolution in time and space. To achieve this objective, we are interested in the lithological and microfacies description along the Imiter and Foum El-Kouss sections. From these two sections, three fundamental formations have been distinguished. From base to top, the Assaghmou and Jbel Guerssif formation (Upper Danian-Lower Thanetian) is dominated by mudstone to packstone limestone with few fragments of bivalves, and it is limited at the top by a maximum flooding surface with nautiloids, indicating an internal carbonate platform (Subdital). Above, the Jbel Talouite formation (Upper Thanetian–Lower Ypresian) is composed of an alternating succession of siltstones and red sandstones with bryozoans, as well as sedimentary figures explaining continental influences. This formation characterizes a continental environment that has undergone periodic marine invasions testified by carbonate passages oolitic grainstone. The whole series is topped by the Ait Ourithane formation (Ypresian) which is formed by marls and packstone to grainstone limestone with bryozoans, echinoderms and sometimes benthic foraminifera (Miliolide), indicating a medium deep and open marine environment (Intertidal to subtidal). According to these results, the lithological and microfacies variations allow us to propose that the paleoenvironmental context had a regime of progradation and retrogradation on a low sloping platform with a less developed barrier.

Keywords: Paleocene-eocene · Paleoenvironment · Central high atlas · Morocco

K. Baba et al. (Eds.): GeoME 2023, SPEES, pp. 224–234, 2023.
https://doi.org/10.1007/978-3-031-49345-4_23

1 Introduction

The Upper Cretaceous (Senonian) corresponds to a stage of tectonic activity that is responsible for the reactivation of the detrital sedimentation by the phenomenon of the subsidence of the basin [1, 2]. At the Maastrichtian-Lower Paleogene, an epicontinental sea flowing from the Atlantic invaded Morocco from west [3, 4]. The Paleocene-Eocene series corresponds to the sedimentological legacy of these great marine extensions and has been called the Subatlasic Group [4, 5]. According to [6] the spatial distribution of the Paleocene-Eocene outcrops allows the individualization of five main domains, namely, Meskala, the Atlasic domain, Middle Atlas, the Souss basin and the Ouarzazate basin. The eastern part of the latter is one of the areas influenced by this Paleogene marine invasion, of which this series was deposited directly on the Senonian red series in unconformity [7], forming a belt along the southern piedmont of the Central High Atlas (CHA). It is separated from its northern equivalent by the High Atlas mountains [8]. This work is focused on the Paleocene-Lower Eocene outcrops in the Boumalne-Imiter areas, the southern edge of the CHA. The principal objective is to characterize the formations from the lithostratigraphic and sedimentological point of view through the macro and microscopic description of the facies in the Imiter and Foum El-Kouss sections. Thus, an interpretation and a reconstruction of a paleoenvironment characterized this targeted area during the Paleocene-Lower Eocene.

2 Geographic and Geological Setting

The study area is located about 15 km east of Boumalne. It is accessible by the national road N° 10 that connects Ouarzazate and Errachidia (Fig. 1). In the geological context, the study area belongs to an intermediate zone between the CHA domain in the north. It is constituted by a Meso-Cenozoic cover affected by the alpine orogeny. in the south, the Anti-Atlasic field is formed by stable terrains attributed to the Precambrian composed volcanic to volcano-sedimentary terrains [9], topped by the Paleozoic cover constituted by shales to sandstone shales [10]. The intermediate part between these two areas corresponds to an intermountain plain occupied by Neogene-Quaternary sediments [11].

3 Methodology

The methodology adopted in this study is based primarily on field investigations to survey sections along the Paleocene-Lower Eocene outcrops in the Boumalne-Imiter regions. Thus a differentiation of several facies macroscopically through the lithological description on field. The sampling was carried out along the targeted sections after the removal of the altered surface coverings. Of which the preparation, processing and analysis of these samples have carried out at the laboratory level 2GRNT and the workshop department of geology, Faculty of Science Semlalia, Cadi Ayyad University.

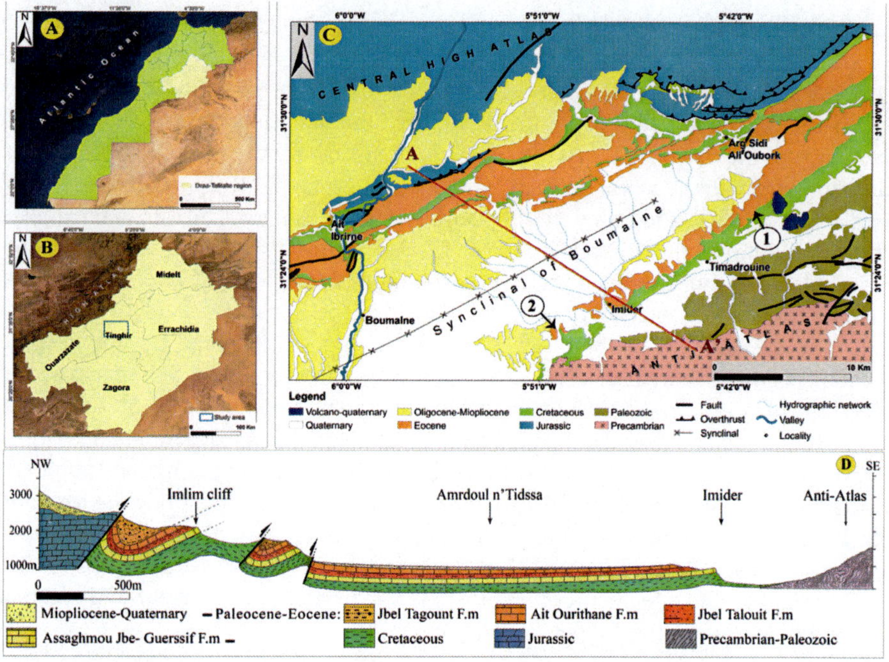

Fig. 1. The situation of the study area. **A**: Location of the Drâa Tafilalat region. **B**: The provinces of Drâa Tafilalate and the study area location. **C**: Simplified geological map of the Boumalne syncline [12, 13] (**1**: Foum El kouss section, **2**: Imiter section). **D**: The geological section (A-A').

4 Results

This study was carried out along the Foum El-Kouss and Imiter sections. The first section is located east of the Timadrouine village (GPS: N31°26′35″ and W5° 40′42″). The second section is situated on the left bank of Assif n'Anou Nizme, west of the Imiter village (GPS: N31°22′28″ and W5°48′14″). Both sections are bounded at the base by siltstones and red sandstones of the Iflite formation attributed to the Senonian and by Neogene-Quaternary alluvium and colluvium at the summit. They are generally constituted by three formations (Fm) which are, from the base to the top, Assaghemou and Jbel Guerssif Fm, Jbel Talouite Fm and Ait Ourithane Fm (Fig. 2).

4.1 Lithostratigraphic Analysis

4.1.1 Asseghmou and Jbel Guerssif Formation

Th Fm of Asseghmou and Jbel Guerssif is started by grayish marls with a thickness which varies between 6 and 8 m, surmounted by a whitish to grayish limestone bar of mudstone /packstone texture with bivalves (Fig. 6/L1) and sometimes dolomitic limestones with geodes occupied by the calcite (Fig. 4/A). It is often intercalated by centimetric marly levels and also passages of crystalline azoic limestone (Fig. 7/L2). Towards the top, the limestone becomes grainstone with bioclasts composed of algae, echinoderms,

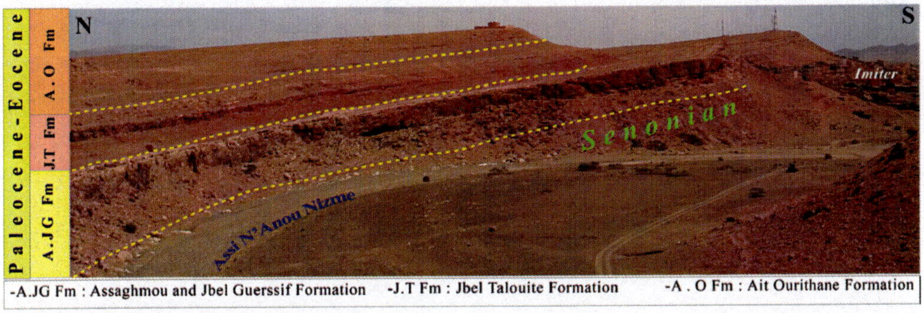

Fig. 2. Panoramic view of the paleocene—eocene succession (the Imiter section).

benthic foraminifera as well as large rounded quartz grains and lithoclasts (Fig. 6/L2). Macroscopically, the fossiliferous content in the summit part shows a condensation of fossiliferous composed generally of lamellibranches fragments, oysters, gastropods (Fig. 4/B) and ammonites of the Nautiloid type (Fig. 5/A) (Fig. 3).

4.1.2 Jbel Ta'louite Formation

The Jbel T'alouite Fm started with friable red silts (50 cm). Towards the top, these silts become more consolidated with microgeodes occupied by quartz as well as concretions of karkoub. They are surmounted by a pink sandstone bed of fine to medium grading limited at the base by an undulating base (erosive base) and at the top by an oxidation surface, that shows a period of exposure to the open air. This succession of siltstones and sandstones are sometimes interspersed with grayish marly limestones (20 cm) oolitic texture oosparite (Fig. 6/L4) showing bioturbations (Fig. 5/B) and sometimes traces of lamellibranches. Then, a stratified fossiliferous red sandstone composed of bivalves and bryozoans (Fig. 6/L3) with a fine to medium grain size with a relatively low hardness. The thickness of the sandstone banks gradually increases towards the top of the Fm. They show sedimentary figures namely oblique laminations with tangential base indicating a WNW direction paleocurrent, to the addition of progradation beds (Fig. 5/C) have the same previous direction and ripples of the current oriented NNE. The passage from this Fm to the Ait Ourithane Fm is marked locally by a whitish sandstone of medium granulometry.

4.1.3 Ait Ourithane Formation

The Fm of Ait Ourithane is constituted at its base by marls whitish alternating with marly azoic limestones of centimetric thickness. Then, they were surmounted by packstone limestones with bioclasts composed of bivalves and gastropods (Fig. 6/L5). The thickness of these limestone bars develop progressively towards the top and they also show a fossiliferous evolution (Fig. 7/L6) composed essentially of bivalves, bryozoans, echinoderms and some benthic foraminifera with a grainstone texture.

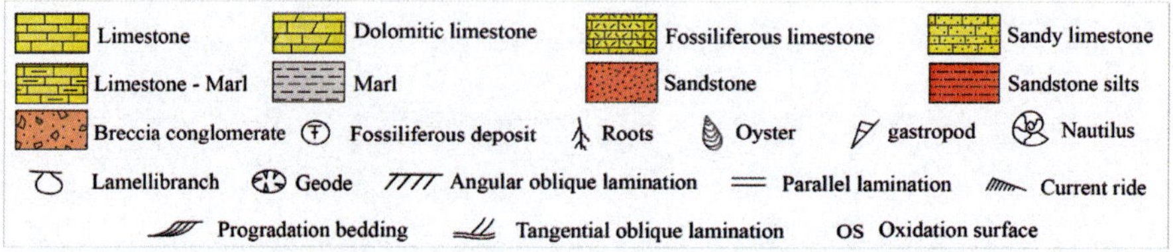

Fig. 3. The legend adopted for both sections below.

5 Discussion

At the base, the Assaghmou and Jbel Guerssif Fm shows the dominance of a mudstone carbonate lithology designating a calm environment [14]. While towards the top, the limestone bars become progressively fossiliferous and of grainstone texture coupled with the appearance of benthic foraminifera (Miliolide) and ammonites of the nautiloids type characterize a deep, open, agitable and highly oxygenated marine environment [15, 16].The intermediate part belongs to the Jbel Talouite Fm. It is dominated by siltstones and sandstones showing sedimentary structures namely, the parallel and sometimes oblique laminations, current ripples and also the lenticular forms of the banks. The assembly of these characteristics related to an environment influenced by the tidal oscillation [17, 18], The existence of fossiliferous sandstone banks and oolitic grainstone carbonate intercalations with bioclastic reworking, especially at the level of the Imiter cut, confirms that the environment has undergone periodic marine upwelling relatively agitated with a medium to shallow depth. While the presence of concretions of Karkoube expressing the emersion phases under a semi-arid climate [19]. All of the above characters argue that the area has been under a subtidal-intertidal regime. The summit part of the series corresponds to the Ait Ourithane Fm, the lithology is composed mainly by packstone-grainstone limestone with a diversified bioclastic composition dominated by bivalves and echinoids and sometimes foraminifera testifying to an oxygenated and moderately agitated environment in communication with the open sea [20]. The correlation of this study with previous work [6, 21] show that the Paleogene sedimentation formed in a transgressive-regressive context beginning at the end of the Senonian.

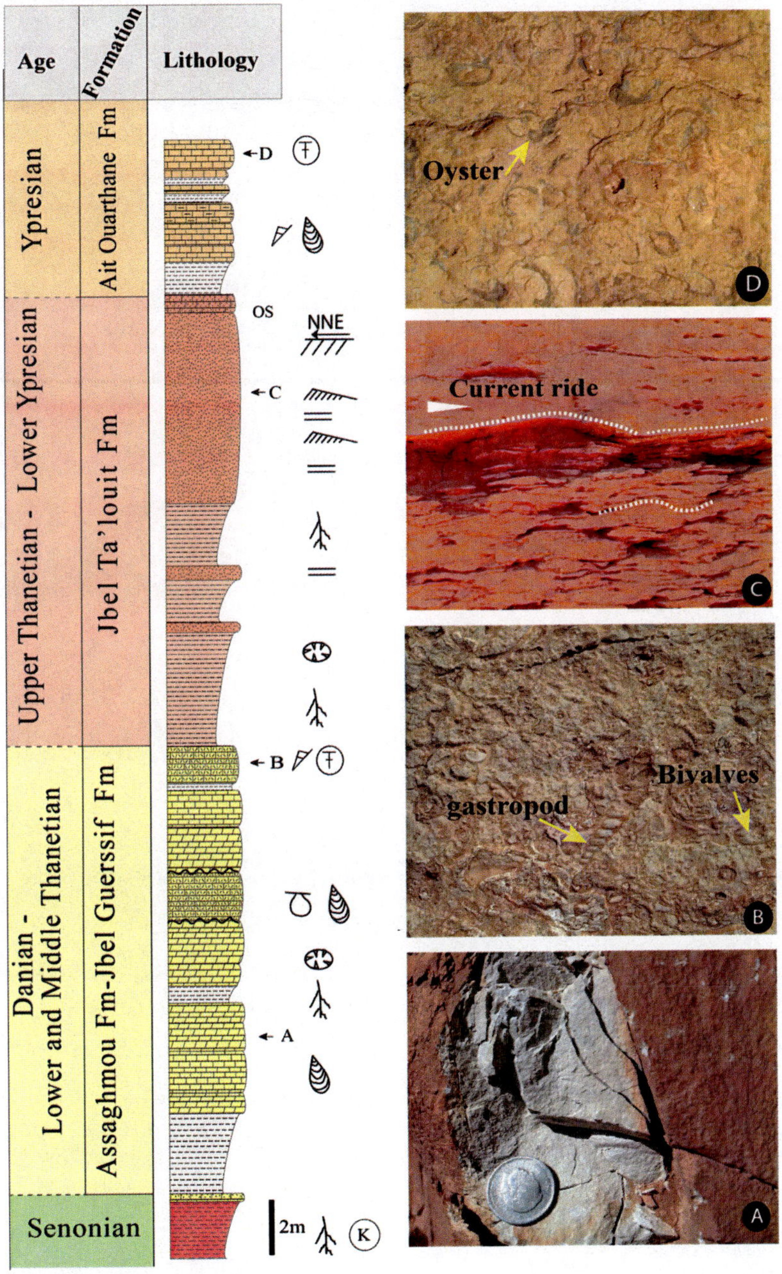

Fig. 4. The foum El kouss section. **A**: dolomitic limestone, **B**: fossiliferous condensation, **C**: current ripple, **D**: lumellic limestone.

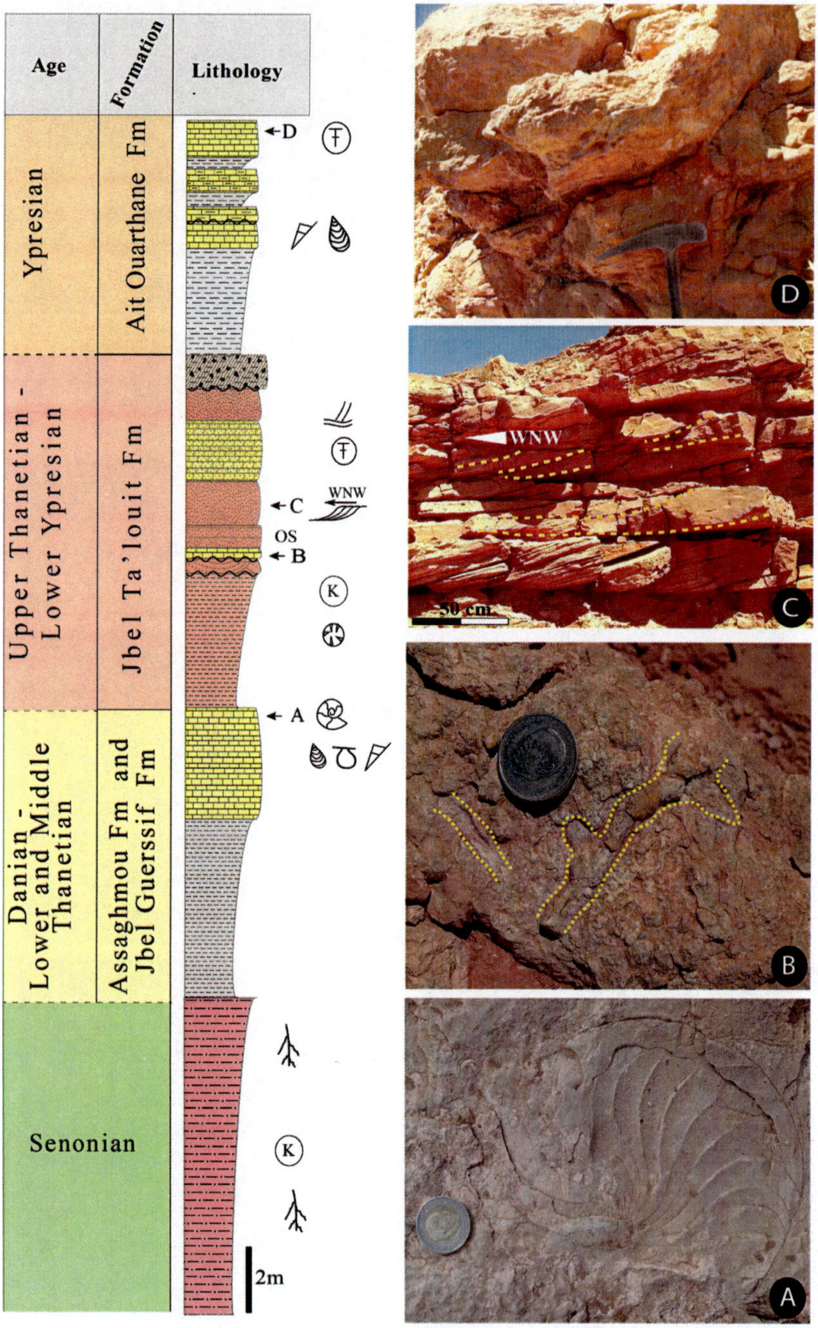

Fig. 5. The Imiter section. **A**: ammonite (nautiloids), **B**: marly limestone with bioturbation, **C**: progradation bedding, **D**: fossiliferous limestone.

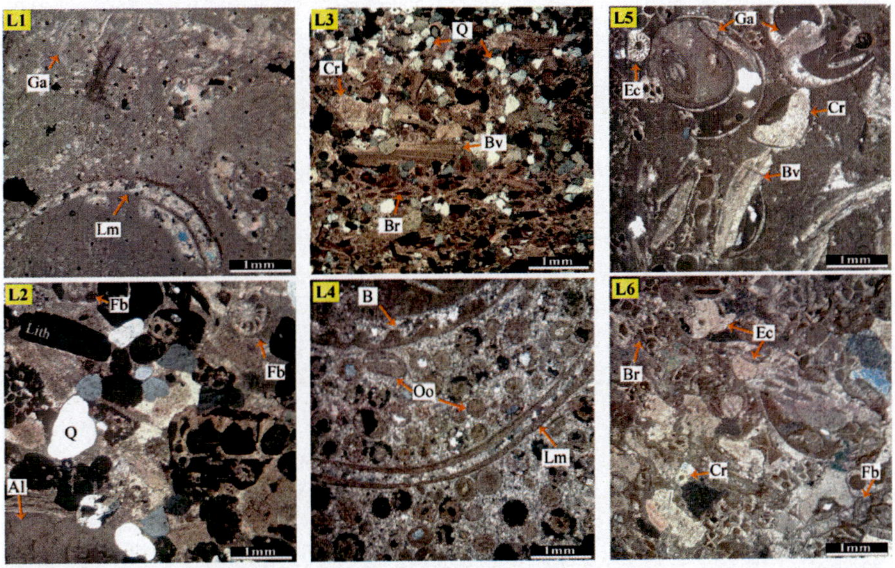

Fig. 6. Some thin sections of the Imiter cup. **Q**: quartz; **Oo**: oolite; **Bv**: bivalves; **Al**: algae; **Ec**: echinoderm; **Lith**: lithoclast; **Fb**: benthic foraminifera; **B**: brachiopod; **Br**: bryozoan; **Cr**: crinoid; **Lm**: lamellibranche; **Ga**: gastropod.

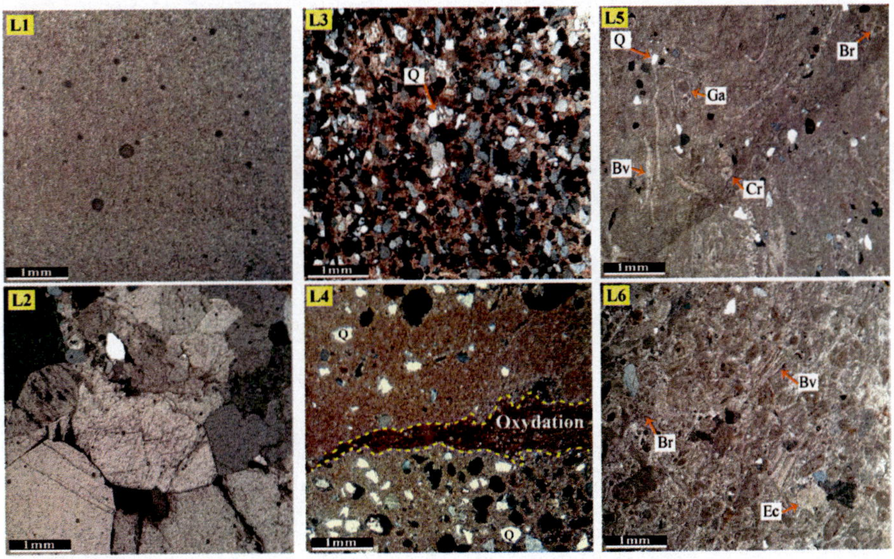

Fig. 7. Some thin sections of the foum El kouss cup. **Q**: quartz; **Bv**: bivalves; **Ec**: echinoderm; **Br**: bryozoan; **Cr**: crinoid; **Ga**: gastropod.

6 Conclusion

The Paleocene-Lower Eocene series in the Boumalne-Imiter regions, the southern side of the Central High Atlas, was deposited under the conditions of marine extensions. The lithostratigraphic variations of which allow the reconstruction of a paleoenvironment and its evolution in 4 stages (Fig. 8). During the Senonian (Fig. 8/A), the environment was under the conditions of a lagoon-continental system generated by synsedimentary faults conducted to the basin subsidence [3, 22]. The Danian-Lower/Middle Thanetian is a period that corresponds to a transgressive phase It is reflected in the sedimentation of limestone Assaghmou and jbel Guerssif Fm, coupled with a faunal evolution indicates an increase in depth (Subtidal) (Fig. 8/B). During the Upper Thanetian-Lower Ypresian, the area underwent a relative regression. It renders the environment under littoral conditions with medium hydrodynamic energy. It is interrupted by emersions periods (intertidal-Subratidal) (Fig. 8/C). In the Ypresian, a second marine advancement is translated by the deposition of fossiliferous limestones of the Ait Ourithane Fm. It makes the area under the effects of an intertidal-subtidal environment (Fig. 8/D).

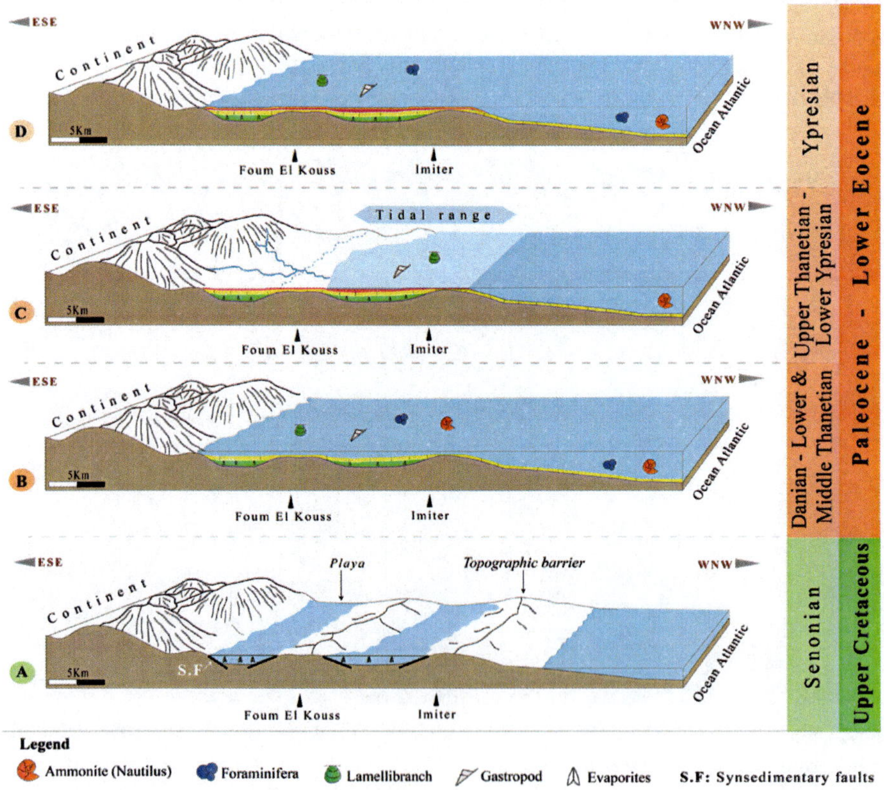

Fig. 8. The stages of paleoenvironmental evolution. **A**: The lagoon-continental system (according to [23]). **B**: the first transgressive stage; **C**: the regressive phase (littoral environment); **D**: the second transgressive stage.

Acknowledgement. Our thanks for the Department of Geology of the Faculty of Sciences Semlalia Marrakech, Cadi Ayyad University, for ensuring all the preparations and treatment of the samples. The authors are very thankful to the anonymous reviewers for their remarks and comments for a careful review of the article that helped to present the work in a lucid manner.

References

1. Laville, E.: Tectonique et microtectonique d'une partie du versant sud du Haut Atlas marocain (boutonnière de Skoura, nappe de Toundout) (1980)
2. Benvenuti, M., Moratti, G., Algouti, A.: Stratigraphic and structural revision of the Upper Mesozoic succession of the Dadès valley, eastern Ouarzazate Basin (Morocco). J. African Earth Sci. **135**, 54–71 (2017). https://doi.org/10.1016/j.jafrearsci.2017.01.018
3. Algouti, A., Algouti, A., Hadach, F., Farah, A., Aydda, A.: Upper cretaceous deposits on the Northern side of the High Atlas range of Marrakesh (Morocco): tectonics, sequence stratigraphy and paleogeographic evolution. Bol. la Soc. Geol. Mex. **74**(1), 1–18 (2022). https://doi.org/10.18268/BSGM2022v74n1a101121
4. Herbig, H.-G.: Das Paläogen am Südrand des zentralen Hohen Atlas und im Mittleren Atlas Marokkos: Stratigraphie, Fazies, Paläogeographie und Paläotektonik, vol. 135. Selbstverlag Fachbereich Geowissenschaften (1991)
5. Choubert, G.S.H.: Essai paléogéographique du sénonien au Maroc. Notes Mémoires du Serv. Géologique du Maroc **3**(3), 13–50 (1950)
6. Herbig, H., Trappe, J.: Stratigraphy of the subatlas group (Maestrichtian—Middle Eocene, Morocco) (1994)
7. Marzoqi, M., Pascal, A.: Séquences de dépôts et tectono-eustatisme à la limite Crétacé/Tertiaire sur la marge sud-téthysienne (Atlas de Marrakech et bassin de Ouarzazate, Maroc). Newsletters Stratigr, pp. 57–80 (2000)
8. Choubert, G., Faure-Muret, A.: Le séisme d'Agadir, ses effets et son interprétation géologique. Service de la carte géologique (1962)
9. Saquaque, A.: Un exemple de suture-arc: le Précambrien de l'Anti-Atlas centre-oriental. These Dr. Es-Sciences. UCA, Marrakech, p. 348 (1992)
10. Tuduri, J.: Processus de formation et relations spatio-temporelles des minéralisations à or et argent en contexte volcanique Précambrien (Jbel Saghro, Anti-Atlas, Maroc). Implications sur les relations déformation-magmatismevolcanisme- hydrothermalisme (2005)
11. Jossen, J.A., Filali Moutei, J.: Bassin d'Ouarzazate, synthèse stratigraphique et structurale. Contrib. à l'étude des aquifères Profond. PNUD–DRPE (Direction la Rech. la Planif. l'Eau) MOR/86/004-Exploration des eaux Profond. Rapp. Inédit, Rabat (1988)
12. Massironi et al., M.: Carte géologique du Maroc au 1/50 000, feuille Boumalne. Notes Mémoires du Serv. Géologique du Maroc **521** (2007)
13. Schiavo et al., A.: Notice explicative. Carte Géologique du Maroc au 1/50.000, feuille Imtir., no. October 2018 (2007)
14. Hadach, F., Algouti, A., Algouti, A., Mourabit, Z.: Aptian-Albian deposits of the ait Ourir basin (high atlas, Morocco): new additional data on their paleoenvironment, sedimentology, and palaeogeography. Open Geosci. **12**(1), 1557–1572 (2020). https://doi.org/10.1515/geo-2020-0214
15. El Bamiki, R., Séranne, M., Chellaï, E.H., Merzeraud, G., Marzoqi, M., Melinte-Dobrinescu, M.C.: The Moroccan high atlas phosphate-rich sediments: unraveling the accumulation and differentiation processes. Sediment. Geol. **403**, 105655 (2020). https://doi.org/10.1016/j.sedgeo.2020.105655

16. Aouissi, R., Salmi-Laouar, S., El Qot, G.M., Ferré, B.: Cenomanian cephalopods of Bellezma-aures mountains, NE Algeria: taxonomy and biostratigraphy. Annales de paléontologie **106**(3), 102409 (2020)
17. de Weger, W., et al.: Contourite channels—facies model and channel evolution. Sedimentology (2022). https://doi.org/10.1111/sed.13042
18. Allen, J.R.L.: Mud drapes in sand-wave deposits : a physical model with application to the folkestone author (s): Allen Source J.R.L. : philosophical transactions of the royal society of London. Series A, published by : royal society. Philos. Trans. R. Soc. London. Ser. A, Math. Phys. Sci. **306**(1493), 291–345 (1982)
19. Hadach, F., Algouti, A., Algouti, A., Mourabit, Z..: Example of paleosebkha littoral deposits of senonian in the basins zone of ait ourir (Marrakech High Atlas, Morocco). Eur. Sci. J. **11**(18) (2015)
20. Chikhi-Aouimeur, F.: Sauvagesiinae du Cénomaniensupérieur de la région de Berrouaguia (Sud d'Alger, Algérie). Geobios **31**, 101–109 (1998)
21. Trappe, J.: Microfacies zonation and spatial evolution of a carbonate ramp: marginal Moroccan phosphate sea during the Paleogene. Geol. Rundschau **81**(1), 105–126 (1992). https://doi.org/10.1007/BF01764543
22. Cappetta, H., Jaeger, J., Sabatier, M., Sige, B., Sudre, J.: Decouverte Dans Le Paleocene du Maroc R6sum6, pp. 257–263 (1978)
23. Farah, A., Algouti, A., Algouti, A., Hadach, F., Mourabit, Z.: Sedimentological and lithostratigraphic study of the Foum El Kous Senonian, Central Hight Atlas of Morocco. E3S Web Conf. 240(4) (2021). https://doi.org/10.1051/e3sconf/202124004001

Conceptual Modelling of the Effects of Moisture on Particle Breakage

Younes Salami[1,2](✉) and Jean-Marie Konrad[3]

[1] National School of Architecture of Fès, Parc Fès-Shore, Sidi Hrazem Road, Fès, Morocco
younes.salami@enafes.ma

[2] Civil Engineering and Environment Laboratory (LGCE), Mohammadia School of Engineering, Mohammed V University, Rabat, Morocco

[3] Département de génie civil et de génie des eaux, Université Laval, Québec, QC, Canada

Abstract. The conceptual model of particle breakage of Konrad and Salami is extended to include the effects of moisture. It is well known that the presence of water increases the intensity of particle breakage in granular materials. This has been observed in laboratory conditions, and in practical projects where the main construction material is of granular nature. This affects the physical structure and integrity of the material, and the mechanical state during loading, which leads to increased strains and collapse of the material. The model under study is developed based on a multiscale approach, where the characteristics of the grain affect the behavior of the granular packing. The model is based on few parameters which can be directly calibrated from crushing tests on dry and wet specimens. When the grain weakening due to the presence of water is introduced in the model, the typical behavior of an unsaturated granular medium is observed: an early onset of particle crushing, and an increased breakage rate. The model is shown to be valid for both saturated and unsaturated conditions. The three-segmented nature of the model proves adequate for the nature of the problem. It is observed that for each saturation level, an associated model line relates the work input to the breakage parameter. The model is then tested against experimental data, and its performance is demonstrated through simulations on different materials.

Keywords: Soil mechanics · Geotechnical engineering · Rockfill · Fracture · Particle breakage

1 Introduction

The behavior of unsaturated granular soils and materials has been a significant focus in geotechnical and civil engineering. Understanding and accurately modeling the mechanical response of these materials is essential for ensuring the stability and performance of various engineering structures. However, the constitutive modeling of unsaturated granular soils has posed considerable challenges due to the complex interactions between the soil particles and water.

K. Baba et al. (Eds.): GeoME 2023, SPEES, pp. 235–243, 2023.
https://doi.org/10.1007/978-3-031-49345-4_24

Water was shown to influence the mechanical behavior of unsaturated granular soils through both physical and physicochemical processes [1]. On one hand, the presence of water impacts various aspect of the mechanical behavior, due to the menisci formed at the contact points between soil particles. On the other hand, affects the integrity of the granular material through processes such as particle breakage and rearrangement, leading to distinct short- and long-term behaviors. These phenomena have significant implications for engineering applications, particularly in the case of rockfill dams, where the stability of the inner structure relies on the rockfill shells.

In practice, rockfill dams commonly experience settlements and collapse during reservoir filling or after heavy rainfall events. Extensive experimental studies have been carried out to comprehend the underlying causes of this collapse behavior [2], with particle breakage recognized as a major contributing factor. It is well known that stressed particles in a granular material can break under specific conditions, resulting in the rearrangement of grains and force chains, as well as the development of plastic deformations. The presence of water has been shown to enhance the occurrence of particle breakage, through various physicochemical processes [3].

To address the complexity of the behavior of granular soils influenced by water, several constitutive models have been proposed. Among these, the influential model by Oldecop and Alonso [4] which describes the material behavior in three stages: an initial stage of plasticity due to grain rearrangement, a clastic yielding stage influenced by suction, and a subsequent clastic hardening stage. This model, limited to oedometric paths, was later extended to triaxial paths and other complex stress-suction paths by Chávez and Alonso [5].

Furthermore, breakage mechanics models, as introduced by Einav [6], consider the coupling of breakage parameters with the mechanics of the material within a thermomechanical framework. The inclusion of hydraulic components by Buscarnera and Einav [7] allows for the calculation of suction based on the soil water retention curve, enabling a more comprehensive understanding of the material response in unsaturated conditions.

In this paper, the engineering framework of Konrad and Salami [8], which was extended to cover the unsaturated/saturated range of conditions [9], is presented, and applied for the modelling of crushable granular materials in unsaturated conditions. The results of one-dimensional compression tests on low quality rockfill reported by Kohgo et al. [10] are simulated using the model.

2 The Engineering Framework of Crushable Granular Materials

Konrad and Salami [8] introduced an engineering framework for modelling the behavior of crushable granular materials in dry conditions. In this framework, breakage is quantified using the parameter ν, which represents the slope of the grain size distribution (GSD) in the log-log plane and quantifies grain breakage. It is noted that the parameter ν is easily related to other breakage parameter used in the field [2, 11, 12]. The evolution of ν during loading is related to the mechanical work per unit volume (W/V) applied by the mechanical loading. The model approximates the relation between W/V and ν using a three-segmented curve. Three characteristic values of this curve with regards to the input work, namely W_{ref}, W_{crit}, and W_u, are defined to delimit the domains of each stage. The first stage ($W_{ref} < W < W_{crit}$) represents low work magnitudes where no grain breakage occurs, while the second stage ($W_{crit} < W < W_u$) corresponds to the initiation of breakage with significant changes in the GSD. The third stage ($W_u < W$) occurs when the work exceeds a stabilization value, resulting the stabilization of the GSD in its fractal distribution [13].

The three-segmented behavior can be characterized by three essential parameters: ν_{ref}, α, and ν_u. ν_{ref} represents a reference value for ν, corresponding to the reference grain size distribution (GSD) associated with the reference work, W_{ref}. The parameter α defines the slope of the model segment for the second stage. Finally, ν_u represents the final value of ν when the work surpasses the ultimate value, W_u, and the GSD stabilizes at a fractal distribution. The model behavior is visually depicted in Fig. 1.

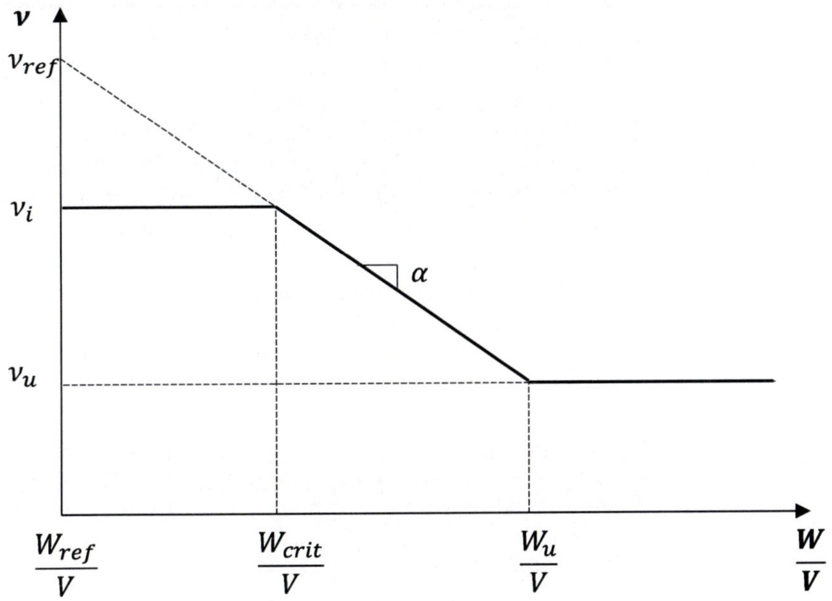

Fig. 1. A schematic representation of the three-segmented behavior introduced by the model.

These model parameters are correlated to the tensile strength σ_{50} of the size fraction d_{50}, which serves as an indicator of material strength and mineralogy. Empirical correlations establish the relationships between the model parameters and the tensile strength:

$$\alpha = -0.061\, ln\left(\sigma_{50}/\sigma_{ref}\right) + 0.65 \tag{1}$$

$$\nu_{ref} = 2.061\, ln\left(\sigma_{50}/\sigma_{ref}\right) + 2.17 \tag{2}$$

$$\nu_u = 0.190\, ln\left(\sigma_{50}/\sigma_{ref}\right) - 0.14 \tag{3}$$

The proposed model provides insights into the behavior of crushable soils, with knowledge of the representative particle strength σ_{50} and initial grading ν_i. The multiscale nature of the model means that it can be calibrated using one parameter: σ_{50}. When water is introduced, the strength of the individual particles is weakened, and the phenomenon of subcritical crack development is activated. This leads to an early onset of particle breakage, and a higher rate of breakage development. From a constitutive point of view, a weakening of the individual particles strengths can be accurately modelled by a linear decreasing function [9, 14]. When the strength is allowed to decrease, Eqs. (1) and (2) show that ν_{ref} decreases, while α increases. This shows that the early development and acceleration of breakage in the presence of water are correctly represented.

3 Modelling of Experimental Results and Discussion

In their study, Kohgo et al. [10] performed one-dimensional oedometric tests on low-quality rockfill. Two types of rockfill, namely material O (tuff) and material S (andesite), were examined. Both initial gradings start from a maximum rock size of 53mm. The experimental investigation encompassed different saturation levels, and the oedometric results were documented for each saturation level. Furthermore, the study presented the strength characteristics of individual particles at various water content levels and reported the grain size distribution (GSD) curves before and after loading.

The details of the tests considered for the simulations are summarized in Table 1. First, the initial and final GSD are approached by linear distributions in the log-log space to obtain the values of the initial ν_i and final ν for all the tests considered. Then, the compressibility curves (void ratio vs. vertical stress) are used to obtain the mechanical work applied by unit volume at the end of the loading. In order to calibrate the model, the results of the single particle crushing test reported by Kohgo et al. [10] are used to calculate the tensile stress for the characteristic size d_{50}. These stress values are then used to obtain the parameters of the model ν_{ref} and α using Eqs. (1), (2) and (3).

Table 1. Calibration of the model's parameters.

Material	Saturation (%)	Initial ν	Final ν	W/V developed (MPa)	σ_{50} (MPa)	ν_{ref}	α
O	Initial	0.4	–	–	0.64	1.24	0.68
	17.8	–	0.32	0.191	0.53	0.88	0.69
	41.7	–	0.28	0.097	0.45	0.50	0.70
S	Initial	0.64	–	–	1.27	2.67	0.64
	25.0	–	0.35	0.288	1.02	2.21	0.65
	48.4	–	0.37	0.310	0.67	1.34	0.67
	57.2	–	0.49	0.449	0.62	1.19	0.68

From Table 1, it is clear that an increase in saturation weakens the individual particles, and results in an increase in α, which means that breakage occurs at an increased rate. ν_{ref} is shown to increase for weaker particles. As a result, the onset of breakage occurs at a lower value of W_{crit}, indicating an earlier initiation of particle breakage. The influence of saturation on the ultimate stage is not well understood and will not be addressed in this discussion.

In the study by Kohgo et al. [10], the original dataset presents the maximum vertical loads plotted against particle size for different water contents. By linearly interpolating this curve, the maximum loads for characteristic sizes $d_{50} = 9.5$mm (for material O) and $d_{50} = 19.0$mm (for material S) are obtained for both dry and wet conditions. These values are then utilized to determine the strengths and construct the strength variations with saturation, assuming a linear relationship. Figure 2 illustrates the linear evolution of strength with saturation, demonstrating the trend.

Furthermore, the values of σ_{50} resulting from fitting the experimental results with a model line are shown to align with the linear relation representing the strength evolution. This observation confirms the assumption of a linear progression of strength with saturation. The plot provides empirical evidence supporting the linear variation of strength with increasing saturation levels.

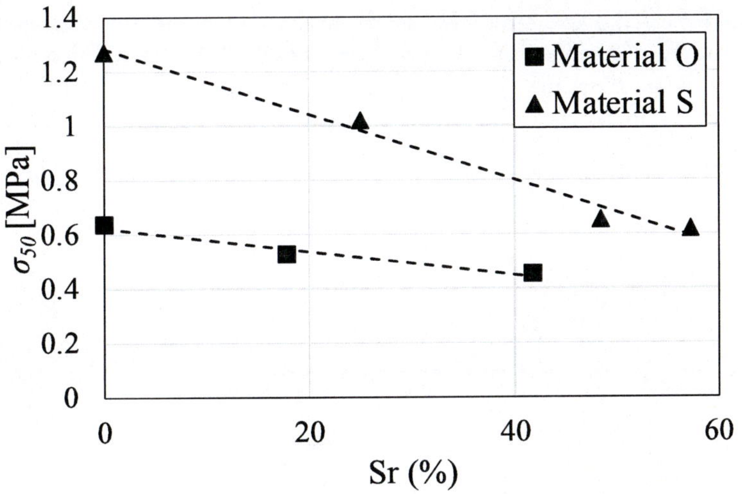

Fig. 2. The evolution of the strength of a representative particle σ_{50} with the saturation for both materials.

The points corresponding the oedometric tests investigated are plotted in a (W/V, v) plot (Figs. 3 and 4). It can be seen that the model lines deriving from the strength values at the corresponding saturation values fall on the data points. This proves the predictive capabilities of the model. In the case of material O (Fig. 3), the energy input during testing at 41.7% saturation is approximately 0.1MPa, whereas the energy input during testing at 17.8% saturation is around 0.3MPa. Surprisingly, a higher degree of breakage is observed in the test conducted at 41.7% saturation compared to the test at 17.8% saturation. This observation can be attributed to the delayed onset of breakage at higher saturation levels, despite the fact that the rate of breakage is ultimately higher for the test with the larger saturation. In the case of material S (Fig. 4), the energy input for saturations of 57.2% and 48.4% is found to be comparable. However, despite this similarity in energy input, a higher degree of breakage is observed for the 57.2% saturation. This disparity can be attributed to the early onset of breakage that occurs at the 57.2% saturation level. On the other hand, for a saturation of 25%, the energy input is higher compared to the previous two tests. Still, the development of breakage is lower in this case. This can be explained by the fact that the critical energy required for breakage development is significantly higher compared to the first two tests. This higher energy requirement causes a delayed onset of breakage and results in relatively smaller breakage development.

When examining the model lines of various saturations for both materials, it becomes evident that higher saturation levels lead to a more pronounced slope in the model line. This means that breakage occurs at an accelerated rate. Additionally, the lines shift towards lower energy values as saturation increases. This means that breakage initiates at smaller values of the critical energy.

By utilizing this information, along with the calibrated parameters of the model, it becomes feasible to simulate the response of the material under different stress paths and saturation conditions. The following stress-saturation paths were tested by Salami and Konrad [9] initial saturation, mid-loading saturation before the critical work, mid-loading saturation after the critical work. The model's ability to describe the crushable behavior of the granular material, taking into account the effects of water on initiation and evolution of breakage, allows for accurate predictions of the final GSD, independent of the stress-saturation path.

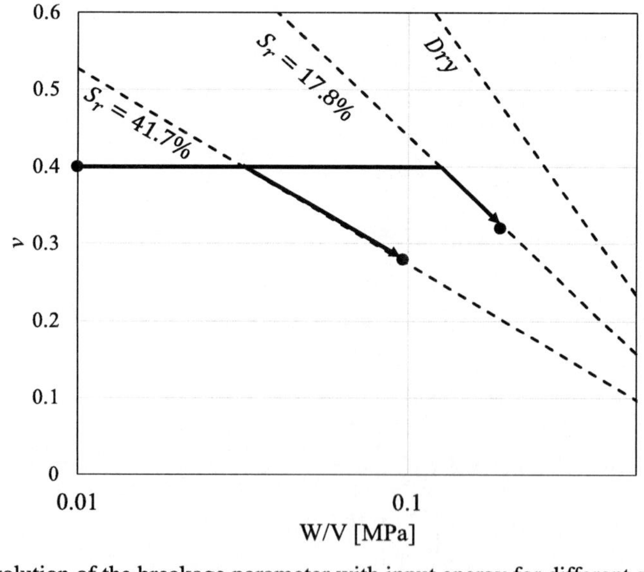

Fig. 3. The evolution of the breakage parameter with input energy for different saturation levels for material O

The model described in the study possesses the capability to predict the behavior of the material under intricate stress-saturation paths. Analogous to a yield function in

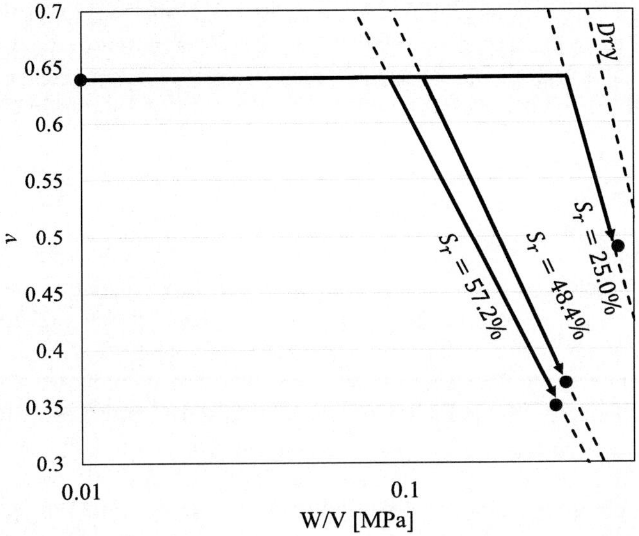

Fig. 4. The evolution of the breakage parameter with input energy for different saturation levels for material S

plasticity development, the model line represents the material's behavior at a specific saturation level. During unloading and subsequent reloading, breakage occurs at the point where the loading line intersects with the model line associated with the given saturation. In cases where the saturation level is altered, breakage develops at the intersection of the loading line with the new model line corresponding to the updated saturation. This characteristic of the model enables it to effectively capture and simulate the evolving behavior of the material under changing saturation conditions. More complex stress paths like drying or mid-loading partial saturation changes require additional experimental developments before they can be investigated.

In order to investigate more complex stress paths such as drying or mid-loading partial saturation changes, additional experimental developments would be necessary. These types of stress paths involve variations in saturation levels that deviate from the straightforward loading-unloading cycles. Therefore, specific experimental procedures and measurements would need to be devised to capture and analyze the behavior of the material under such conditions.

For instance, studying drying conditions would require controlled evaporation or drainage of moisture from the material while monitoring its response. This would involve carefully designed experiments to mimic the drying process and accurately measure parameters such as moisture content, strength, and particle breakage. Similarly, mid-loading partial saturation changes would necessitate experiments that introduce moisture during the loading phase and monitor the resulting changes in the material's behavior and granulometry.

To investigate these more complex stress paths, additional empirical data and insights would need to be obtained. These experimental developments would provide a foundation for further analysis and modeling, allowing for a more comprehensive understanding of the material's response to drying and mid-loading partial saturation changes.

4 Conclusions

This paper demonstrates the applicability of the Konrad and Salami model in describing the behavior of crushable granular materials under the influence of moisture. The model incorporates changes in parameters α and ν_{ref} to account for the weakening of particle strength due to water content, thus considering the effects of water on particle breakage initiation and evolution. The experimental results from previous studies allowed for distinguishing between the behavior of dry and partially saturated samples, enabling the construction of model lines for each saturation state. The study reveals that higher relative humidity results in earlier breakage onset and faster development. By adjusting a single parameter to account for strength variation with humidity, the proposed model successfully captures the crushable behavior of granular materials across a wide range of saturations.

This model represents a valuable and user-friendly tool for predicting the crushable behavior of granular materials under different moisture conditions. It offers simplicity and ease of implementation compared to existing models that require multiple parameters. By directly measuring particle strength through crushing tests on dry and wet specimens, the necessary parameters can be obtained, enabling accurate predictions of the grain size distribution evolution. Its adaptability to various saturation scenarios encountered in engineering applications makes it a versatile tool for studying the behavior of granular materials.

References

1. Alonso, E.E., Oldecop, L.A.: Suction effects on rockfill compressibility. Géotechnique **53**(2), 289–292 (2003)
2. Marsal, R.J.: Large scale testing of rockfill materials. J. Soil Mech. Found. Division ASCE **93**(SM2), 27–43 (1967)
3. Atkinson, B.K.: Subcritical crack growth in geological materials. J. Geophys. Res. Solid Earth **89**(B6), 4077–4114 (1984)
4. Oldecop, L.A., Alonso, E.E.: Theoretical investigation of the time-dependent behaviour of rockfill. Géotechnique **57**(3), 289–301 (2007)
5. Chávez, C., Alonso, E.E.: A constitutive model for crushed granular aggregates which includes suction effects. Soils Found. **43**(4), 215–227 (2003)
6. Einav, I.: Breakage mechanics-Part II: modelling granular materials. J. Mech. Phys. Solids **55**(6), 1298–1320 (2007)
7. Buscarnera, G., Einav, I.: The yielding of brittle unsaturated granular soils. Géotechnique **62**(2), 147–160 (2012)
8. Konrad, J.M., Salami, Y.: Particle breakage in granular materials—a conceptual framework. Can. Geotech. J. **55**(5), 710–719 (2018)

9. Salami, Y., Konrad, J.M.: Modeling effects of moisture on particle breakage. Int. J. Geomech. **21**(9) (2021)
10. Kohgo, Y., Asano, I., Hayashida, Y.: Mechanical properties of unsaturated low quality rockfills. Soils Found. **47**(5), 947–959 (2007)
11. Hardin, B.O.: Crushing of soil particles. J. Geotech. Eng. **111**(10), 1177–1192 (1985)
12. Einav, I.: Breakage mechanics-Part I: theory. J. Mech. Phys. Solids **55**(6), 1274–1297 (2007)
13. Turcotte, D.L.: Fractals and fragmentation. J. Geophys. Res. **91**(B2), 1921–1926 (1986)
14. Salami, Y.: Analyse multi-échelle de l'approche énergétique de la rupture des grains au sein de matériaux granulaires (in French). Ph.D. thesis. Ecole Centrale de Nantes—Polytechnic University of Cataluna (2016)

Enhancing the Carrying Capacity of Friction Anchor Bolts Through Cementitious Concrete Injection for Reinforced Support in Underground Mine

Amine Soufi(✉), Latifa Ouadif, Mohammed Souissi, Youssef Zerradi, and Anas Bahi

Mohammed V University, Rabat, Morocco
soufi@emi.ac.ma

Abstract. Friction anchor bolts are commonly used for providing support in underground structures by relying on frictional forces between the bolt and the surrounding rock. This study proposes a method to enhance the efficiency of these bolts by injecting a cement-based mixture comprising cement, sand, and additives. The injection of this mixture into the bolt results in internal expansion, which reinforces the friction and bearing capacity of the bolt. The increased volume exerts a radial force, leading to improved adherence, load transfer, and void filling. To examine the efficacy of the anchor bolts, pullout tests were performed on various rock masses. The results show that improved rock mass quality and longer cemented bolts result in increased pullout resistance. Furthermore, the addition of a silicate-based additive shortened the cure time of the cement, increasing the strength of the bolts. The study also emphasizes the major impact of groundwater on bolt bearing capacity. These findings support the use of cemented concrete injection in the reinforcement of friction anchor bolts and their anchorage in underground constructions.

Keywords: The carrying capacity · Friction anchor bolt · Cemented concrete injection · Additives · Pullout test

1 Introduction

Friction anchor bolts use frictional forces between the rock bolt and the rock mass to improve support and stability to underground constructions. However, these bolts' performance must be improved in order to improve their bearing capacity (BC) and overall efficacy. This work proposes a novel method for improving the efficiency of friction anchor bolts by injecting a cement-based slurry [1].

The proposed method involves injecting a cement-based mixture comprising cement, sand, and additives into the hollow profile of the friction anchor bolt. This injection leads to internal expansion, resulting in increased volume and reinforcing the frictional interaction between the bolt and the rock. The radial force exerted by the expanded

K. Baba et al. (Eds.): GeoME 2023, SPEES, pp. 244–260, 2023.
https://doi.org/10.1007/978-3-031-49345-4_25

concrete improves the bolt's adherence, load transfer, and void filling capacity, thereby strengthening the anchor bolt and improving the stability of the support system [2].

Pullout tests were conducted on various rock masses to evaluate the performance of the cemented friction anchor bolts. These tests considered different rock mass qualities and bolt lengths to examine their impact on pullout resistance. The results revealed a positive correlation between rock mass quality and pullout resistance, indicating that higher-quality rock masses exhibit improved bearing capacity.

Furthermore, the curing time of the injected cement is critical to the efficiency of cemented friction anchor bolts. To remedy this, a silicate-based addition was added to the cement mixture, which sped up the curing process. This addition decreased the curing time, increasing bolt strength and allowing for rapid support.

The study also looked into the impact of groundwater parameter on the capacity of injected rock bolts. Groundwater was discovered to have a considerable impact on the performance of the bolts, stressing the need of taking groundwater conditions into account in the design and use of cemented friction anchor bolts.

In conclusion, the proposed cemented concrete injection technique offers a promising approach to enhance the strength and anchorage of friction anchor bolts in underground structures. The findings of this study demonstrate the effectiveness of the cemented bolts in improving pullout resistance, bearing capacity, adherence, load transfer, and overall stability.

The incorporation of a silicate-based additive accelerates the curing time, further enhancing the strength of the bolts. Consideration of groundwater conditions is crucial for ensuring optimal performance. These results contribute to the understanding and advancement of friction anchor bolt technologies, offering potential benefits for the design and construction of underground structures.

2 Methods

2.1 Tested Friction Anchorage Bolt

Friction bolts are hollow and thin metal profiles brought into intimate contact with the rock along their entire length, allowing for friction-based anchoring, their effectiveness is immediate. The Split Set (Fig. 1) is a friction anchorage bolt for rocks that is inserted within seconds by simple percussion into a hole slightly smaller in diameter than the tube.

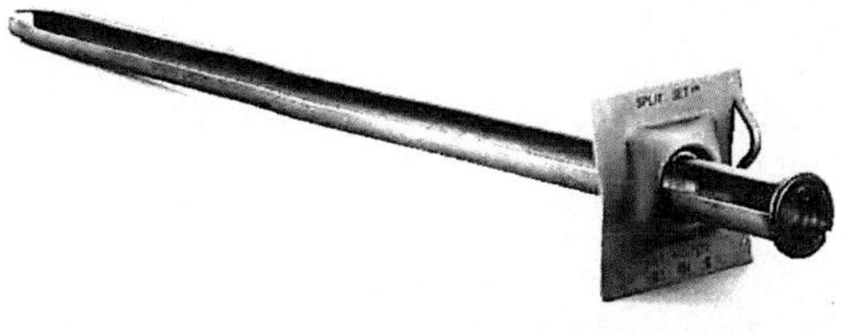

Fig. 1. A split set rock bolt

The straightness of the hole is not important. The bolt conforms to the irregularities of the terrain, provides significant friction through radial pressure along its entire length (Fig. 2), and ensures immediate anchoring [3].

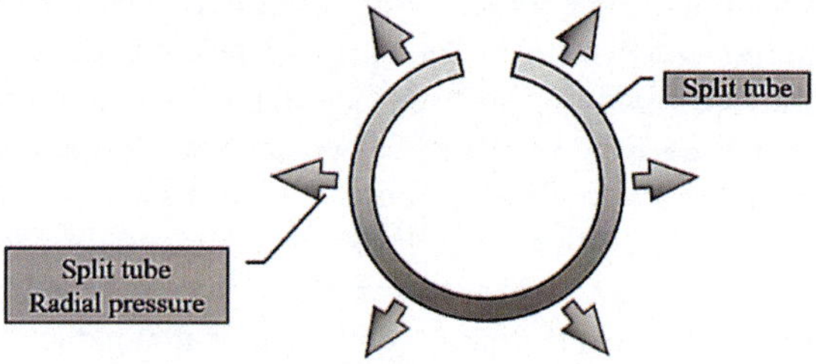

Fig. 2. Radial pressure of the friction bolt

2.2 Principle of the Test

The anchoring force of the anchorage bolt is commonly assessed using an extraction test, and the anchorage behavior of the anchorage bolt has been examined using the extraction test's load-displacement curves. An extraction load is applied to the anchoring bolt's outer end, and the displacement of the anchorage bolt is measured [4].

The traction jack (Fig. 3) is made up of two major components: a mechanical element (grip, bell, threaded bar, spacer, and nut) and a hydraulic part (jack, pump, and pressure gauge). The process adheres to the AFTES GT6R4F1 (French Association of Tunnels and Underground Space) standards, which present the general principles of bolt traction [5].

An anchorage bolt hole is usually drilled perpendicular to the tunnel excavation surface. A 1.8 m long drilling hole for the split set was drilled, and a friction bolt was fitted. The grout used was a 1:1 mixture of standard Portland cement and sand. Sand with a particle size of no more than 2 mm was used.

3 Theory

3.1 Friction Bolt Adherence

By combining the pressure generated through the diameter difference between the holes and the bolt, our proposal aims to enhance the bolt's adherence to the rock mass by introducing a cementitious mixture consisting of cement, sand, and two additives. This mixture will be injected into the friction bolt to ensure internal expansion, with the objective of improving friction and consequently the load-bearing capacity of the bolt. When the pressures generated by the expansion of the cemented mixture and the diameter gap between the bolt and the hole are combined (Fig. 4).

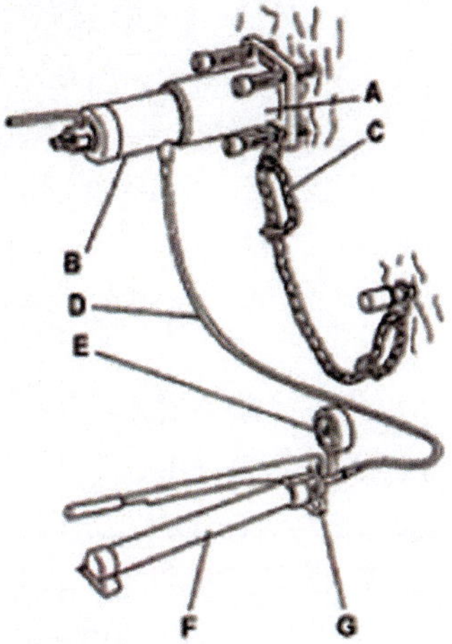

Fig. 3. Pull-out test setup (A: Charging case - B: Jack - C: Safety chain - D: Hydraulic hose - E: Manometer - F: Hand pump - G: Valve)

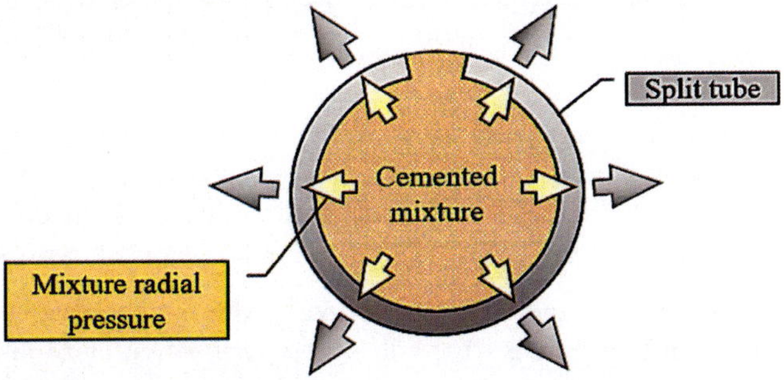

Fig. 4. Radial pressure of cementation mixture

Our tensile strength is naturally enhanced by confinement. When a friction anchor bolt is inserted into a hole in a rock mass, the rock exerts a radial confinement force on the bolt when it is tensioned. This force results from the elastic deformation of the surrounding rock (Fig. 5). The radial confinement increases the pressure exerted on the bolt, thus improving the roughness between the rock bolt and the surrounding rock mass. This increased friction enhances the bolt's resistance to pull-out. Consequently, the anchorage is more stable and can better withstand external loads.

3.2 Cementitious Mixture Injection

In the present study, we propose the injection of friction bolts with concrete and two types of additives for two objectives. Firstly, to accelerate the consolidation of concrete in a very short time, and secondly, and most importantly, to enhance the increase in volume of concrete as a function of consolidation over time. These additives are designed to induce controlled expansion of the concrete, leading to a specific volume increase.

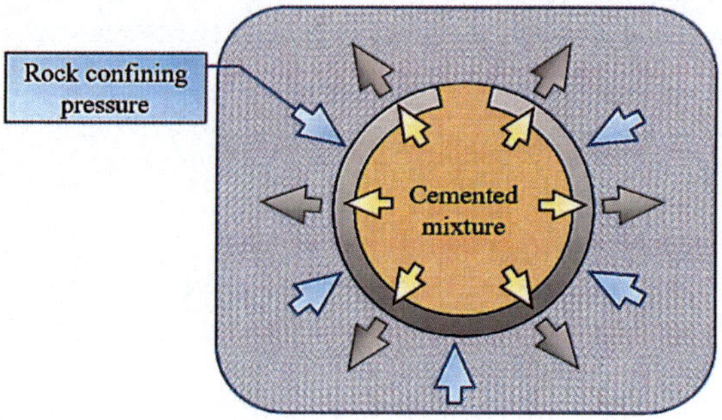

Fig. 5. Radial confining pressure of rock mass

- **Silicate-based additives**: Alkali silicates can also be used to accelerate concrete consolidation. These additives react with cement components to form reaction products that promote faster setting.
- **Gypsum-based expansive additives**: These additives contain gypsum (calcium sulfate dihydrate), which reacts with water present in the concrete to form expansive crystals. These crystals create internal pressure that increases the volume of the concrete. Gypsum-based additives are often used to compensate for subsequent plastic shrinkage of the concrete.

When our concrete is injected inside a bolt anchored in rock, its volume increases over time due to chemical hydration reaction and the gypsum-based additive. This increase in concrete volume exerts a radial force inside the bolt, directed towards the rock mass. This radial force has several beneficial effects on the bolt's pull-out resistance:

- **Increased adhesion**: The radial force exerted by the increased volume of concrete promotes better adhesion between the bolt and the rock mass. This improved adhesion strengthens the bolt's ability to resist pull-out forces.
- **Load transfer**: The radial force transmitted by the concrete expansion helps transfer applied loads on the bolt to the rock mass. This reduces local stresses on the bolt and contributes to a better distribution of forces, enhancing overall pull-out resistance.
- **Void filling**: During concrete injection, voids may exist inside the bolt and between the bolt and the rock. Concrete expansion helps fill these voids and ensure closer contact between the bolt and the rock, thus improving anchorage efficiency.
- **Increased stability**: The increase in concrete volume can also contribute to greater overall stability of the support system. By reinforcing the bolt anchorage, the concrete reduces undesired movements or deformations of the bolt and rock mass, thereby improving the strength and durability of the support structure.

In summary, the volume increase of concrete injected into an anchored bolt promotes adhesion to the rock mass, transfers loads, fills voids, and enhances the overall stability of the support system. These combined effects contribute to strengthening the bolt's pull-out resistance and ensuring effective anchorage in the rock.

4 Results and Discussion

4.1 Behavior of Bolts in a Dry Rock Mass

Several field tests were undertaken at the Imiter mining location in Morocco to study the anchoring and traction behavior of an anchorage bolt in dry rock (Fig. 6). This section contains 9 pull-out tests on three different rock mass qualities: RMR = 40, RMR = 60, and RMR = 80. The principal rock component is sandy siltstone, with dense joint spacing and soft infill materials.

Fig. 6. Pullout test

The load-displacement relationship in the extraction test was used to investigate the behavior of anchorage bolts. The extraction resistance and displacement at the yield limit were compared to determine the effect of each dry rock mass on the behavior of the anchorage cemented bolt. The relationship between extraction load and displacement for each rock condition is depicted in Fig. 7. The extraction resistance was evaluated in accordance with ASTM D443513e1.

The test findings show that when rock mass quality improves, the slope of the load-displacement curve increases, resulting in increased pull-out resistance. When the tested dry rock mass got more competent (stiffer), this trend became more evident.

The detection of the pullout failure point of cemented friction bolts was achieved by observing the change in slope of the load-displacement curve (Fig. 8). The upper (abc) and lower (a′b′c′) boundaries were plotted, correlating the two inflection points (bb′), in order to obtain the failure line:

$$BC\,(KN) = 1.32\,RMR + 60 \tag{1}$$

4.2 Behavior of Bolts in a Dry Rock Mass

In 2017, Liu examined the influence of cemented anchor rock bolt length on tensile behavior in soft rocks, employing lengths ranging from 1.5 to 3 m. The results showed that the bolts' maximum tensile load and pull-out strength increased with length [6].

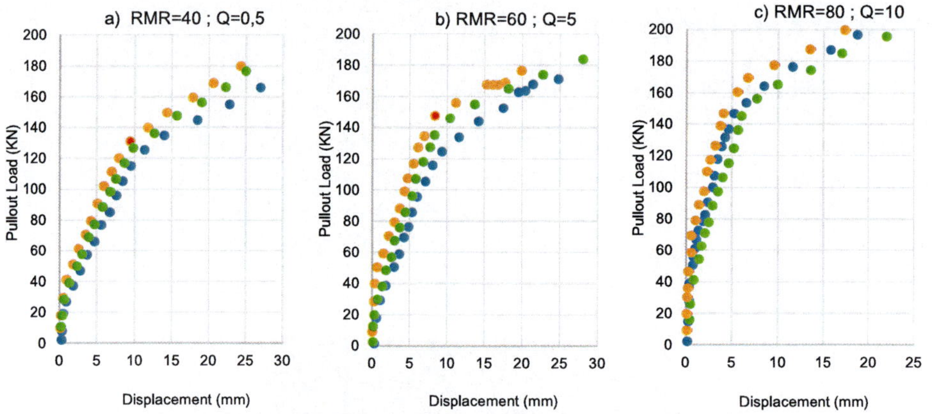

Fig. 7. a) Pullout tests chart for RMR = 40; b) Pullout tests chart for RMR = 60; c) Pullout tests chart for RMR = 80

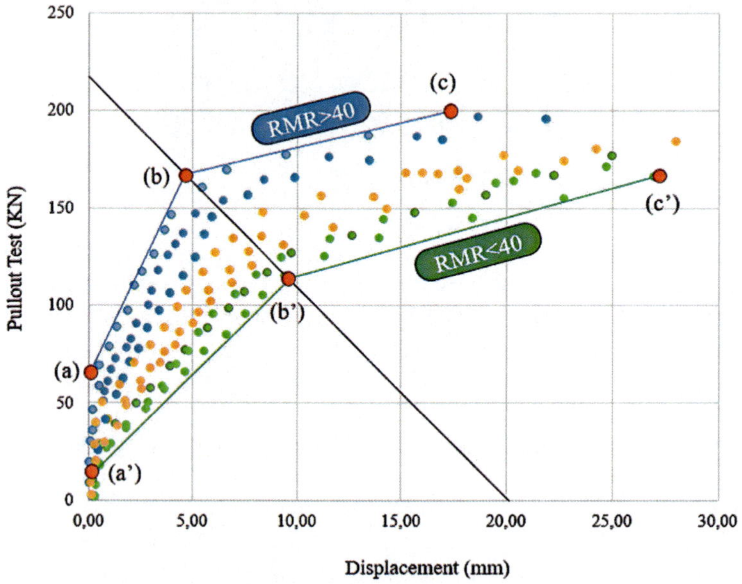

Fig. 8. Pullout load-displacement curves for RMR = 60, RMR = 60 and RMR = 80

In 2019, Chen inspected the effect of cemented anchor bolt length on tensile behavior in soft rocks, employing lengths ranging from 1.5 to 3 m. The results showed that the bolts' maximum tensile load and pull-out strength increased with length [7].

The influence of cemented anchorage bolt length on traction behavior was investigated in moderate and soft rocks with lengths ranging from 2 to 4 m. Figure 9 depicts the relationship between pull-out load and displacement for various anchorage bolt lengths.

The slope of the load-displacement curve increased with the length of the anchorage bolt in rock masses of quality RMR = 40 and Barton Q-system index (Q = 0.5) and

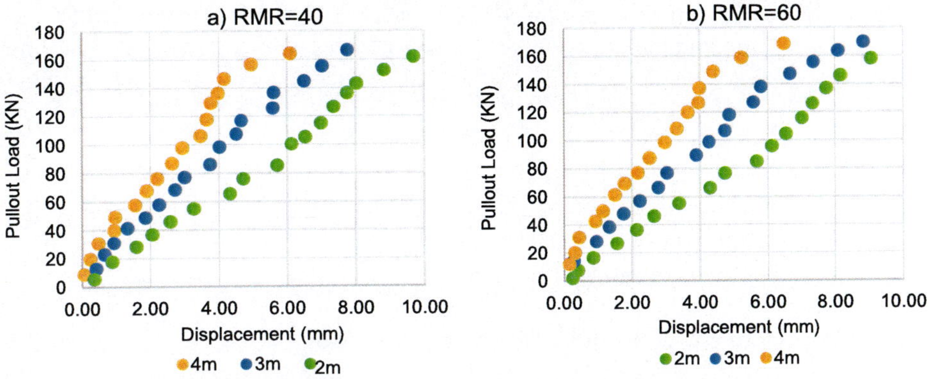

Fig. 9. Pullout load-displacement with lengths ranging for a) RMR = 40 b) RMR = 60

RMR = 60 (Q = 5). In other words, increasing the length of the anchorage bolt increased the yield limit load while decreasing the extraction resistance (Fig. 10).

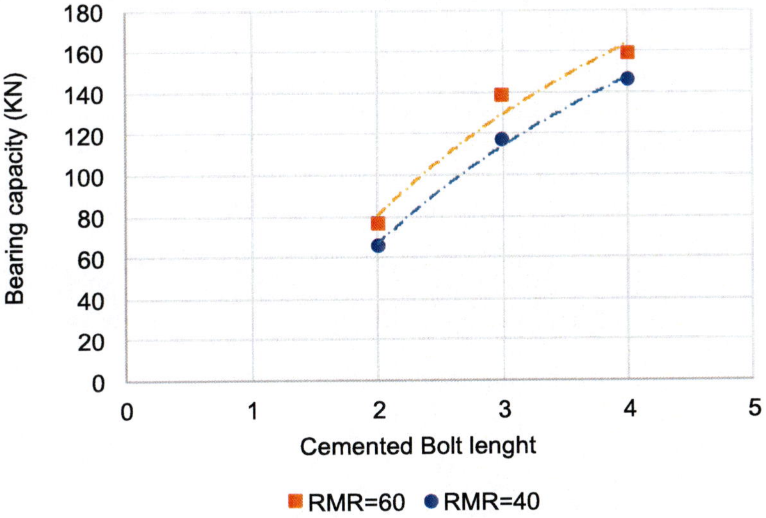

Fig. 10. Pull-out bearing capacity with lengths ranging

For RMR = 40:

$$BC_{40}\,(KN) = 120.6\ln(L) - 3.2;\quad R^2 = 0.96 \tag{2}$$

For RMR = 60:

$$BC_{60}\,(KN) = 117.2\ln(L) - 14.7;\quad R^2 = 0.99 \tag{3}$$

To estimate the carrying capacity of cemented friction bolts, taking into account the cemented bolt length and the quality of the rock mass, the following expression is proposed:

$$BC\,(KN) = 5.6BL + 2.36\,RMR - 18 \quad (4)$$

where

BC: bearing capacity (KN)
BL: cemented bolt length (m).

4.3 Effect of Curing Time on Cemented Anchor Bolts

In the use of cemented anchor bolts, the setting time of injected cement is critical. It has an impact on the bolts' ability to stabilize. Because cement takes time to solidify and set, cemented bolts cannot be utilized for instant support.

Kilic et al. tested traction on eight sets of identical bolts and mortar with a water-to-cement ratio of 0.4. They were tested to see how curing time affected bolt bond strength. Each bolt group had a varied cure time [8].

To solve the issue of setting time, it was decided to use a silicate-based additive to speed up the consolidation process. Following that, a series of pull-out tests were performed on six cemented bolts distributed in two distinct types of rock masses with quality indices RMR = 40 and RMR = 60. The installed bolts' displacements were then measured at three different time intervals: after 24, 48, and 72 h (Fig. 11).

When the length of the anchorage bolt was increased from 2 to 4 m, the anchoring force and pull-out resistance rose significantly. Each additional meter of length increased tensile strength by at least 15%. This pattern was detected in both RMR = 40 and RMR = 60 rock types.

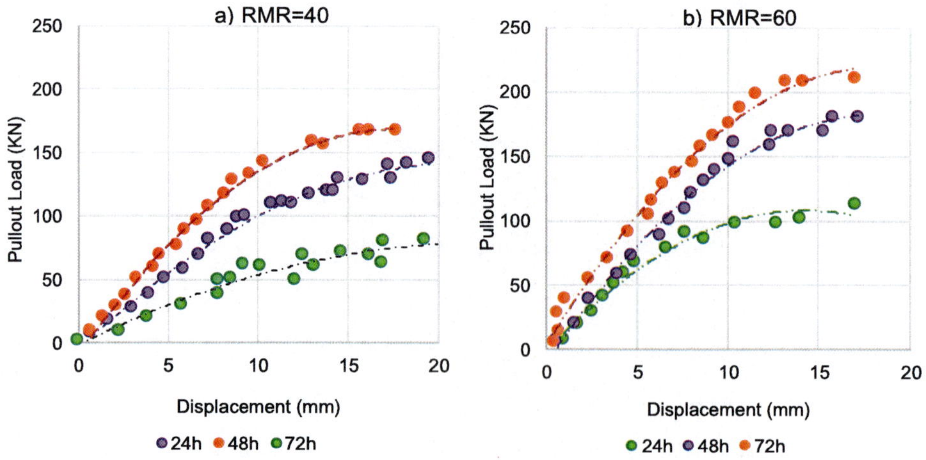

Fig. 11. Pull-out bearing capacity with lengths ranging a) RMR = 40 b) RMR = 60

For an RMR = 40 rock mass, it is evident that the carrying capacity experienced a significant increase, rising from 80 to 160 kN after a period of 72 h following the installation of the cemented bolt.

The results obtained clearly demonstrated a significant increase in bearing capacity after the installation of the cemented bolt over a 72-h period. Initially, the carrying capacity was 100 kN, but it increased significantly to reach 200 kN. In other words, we observed a 200% improvement in bearing capacity between the second and third day.

The diagram depicted Fig. 12 illustrates the correlation existing between the carrying capacity and curing time for two distinct classes of Rock Mass Rating (RMR), to estimate the carrying capacity of cemented friction bolts, taking into account the curing time and the quality of the rock mass, the following expression is proposed:

$$BC\,(KN) = 1.7Ct + 1.3\,RMR - 25 \tag{5}$$

where

BC: bearing capacity (KN)
Ct: curing time (hours).

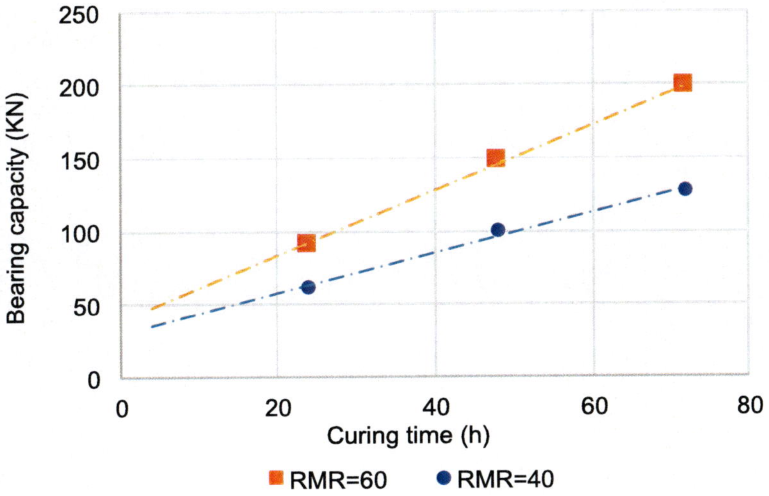

Fig. 12. Pull-out bearing capacity with curing time

These results indicate that the setting time plays a crucial role in enhancing the strength of the cemented bolt. The analysis of the obtained data significantly confirms the positive effect of setting time on the strength of the cemented bolt.

4.4 Impact of Groundwater on Bearing Capacity

In 2017, Zhang examined the impact of groundwater parameter on the load-bearing ability of anchor bolts in a coal mine, employing the RMR classification's rating of water condition (Rw) index to estimate water flow in the mine's underground structures. The findings revealed that the presence of groundwater has a substantial impact on the anchor bolts' load-bearing capacity [14].

Li et al. inspected the effect of groundwater on anchor bolts' load carrying capacity, used numerical models to determine water flow within the mine's underground structures. The findings showed that the presence of groundwater had a substantial effect on anchor bolts' load carrying capacity [10].

The presence of water in rock masses can significantly affect the carrying capacity of friction anchor bolts used in underground structures. To quantify the influence on bolt bearing capacity, we assessed the water flow rate in the mine's subterranean structures using the Rw index of the RMR classification, as given in Table 1.

Table 1. Parameter of water presence in the RMR index

Inflow per 10 m length (l/m)	0	< 10	10–25	25–125	> 125
Condition	Dry	Damp	Wet	Dripping	Flowing
Rw rating	15	10	7	4	0

We recorded the pull-out displacements of four bolts installed in four mining sectors with varied amounts of groundwater presence to explore the impact of groundwater presence in imiter rock mass on cemented bolts' load carrying capacity. The four mining zones under investigation have the same overall rating of RMR = 60; however, the rating of the Rw parameter, which represents the existence of groundwater in the mass, differs across the zones (Fig. 13).

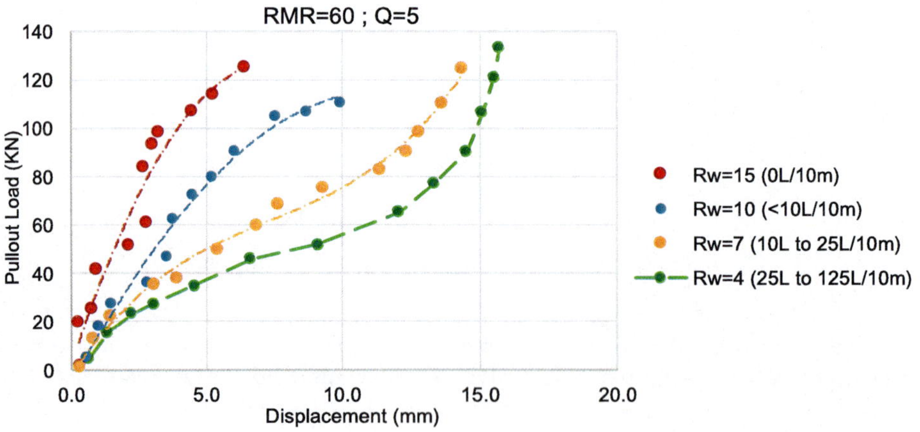

Fig. 13. Pull-out bearing capacity with curing time

To establish a relationship between the capacity to bear load of cemented bolts and the presence of water in the rock mass, we graphically represent the load-bearing capacity values as a function of the Rw parameter rating, which characterizes the presence of groundwater according to the RMR index (Fig. 14).

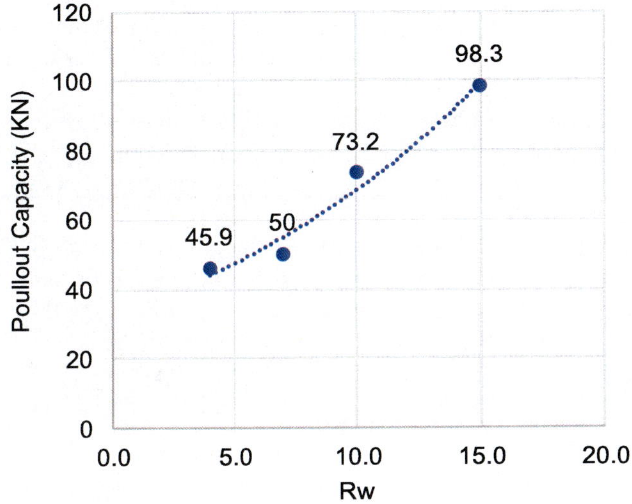

Fig. 14. Pull-out bearing capacity with groundwater factor for RMR = 60

To quantify the correlation between the tensile strength and the Rw index, the following expression is proposed:

$$BC\,(KN) = 32.83\,e^{0.074\,.\,R_w};\quad R^2 = 0.97 \tag{6}$$

The analysis of the results leads to the conclusion that the maximum pull-out capability of cemented friction bolts decreases proportionally with the presence of water in the rock mass. Here are some proposed explanations for this impact:

- **Friction reduction**: When water infiltrates rock fissures and joints, it can reduce friction between the anchored bolt and the rock block. This can reduce the bolt's holding capacity, resulting in a reduction in load-bearing capacity.
- **Rock erosion**: Water can cause erosion of the surrounding rock, weakening the overall structure of the rock wall. This can decrease the rock's ability to support the load exerted by the anchor bolt, thus reducing its load-bearing capacity.
- **Corrosion of the anchor bolt**: Water can also lead to corrosion of the anchor bolt, especially if it is made of steel. Corrosion weakens the bolt, diminishing its strength and load-bearing capacity.
- **Rock swelling**: In certain situations, when water penetrates into the rock, it can cause the rock to swell. This swelling can exert pressure on the anchor bolt, thus reducing its load-bearing capacity.

Several precautions can be taken to limit the impacts of water on the load-bearing capability of friction anchor bolts, such as employing corrosion-resistant materials and implementing drainage systems to remove water.

4.5 Effect of Drilling Speed on Pull-Out Resistance

Tomory studied the effect of drilling speed on the pull-out resistance of anchor bolts in hard rocks, providing a graphical summary of their findings. They observed a positive correlation between drilling time and ultimate capacity [11].

Nicholson and Hadjigeorgiou examined the effect of drilling speed on the pull-out resistance of anchor bolts in hard rocks, while differentiating between bolt suppliers. They found stronger relationships (higher regression coefficients) between drilling time and pull-out resistance [12].

To determine the effect of drilling speed on the pull-out test of 34, 36, and 38 mm diameter bolts, we measured the drilling advancement time to correlate it with the ultimate load capacity. A test campaign was conducted on bolts of the three diameters in a rock mass of moderate quality with RMR = 60 and Q = 5 (Fig. 15).

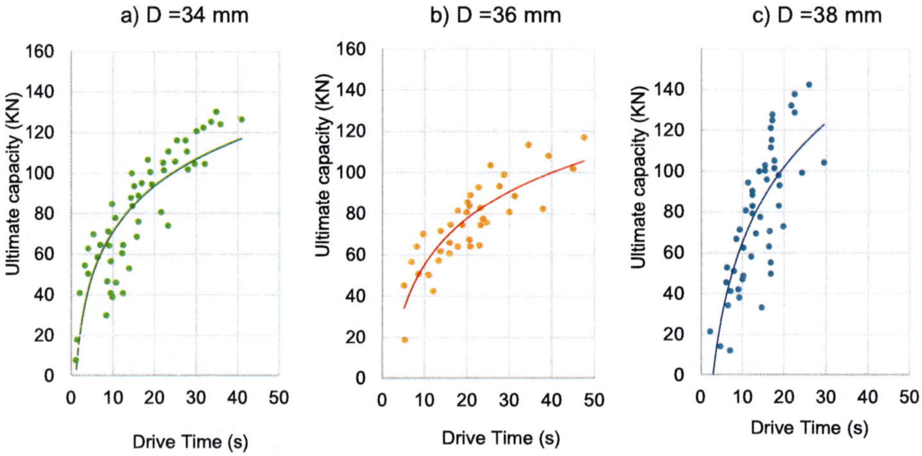

Fig. 15. Pullout load-displacement with Drive time for a) D = 34 mm, b) D = 36 mm and c) D = 38 mm

The time required to drive a given bolt into the borehole indicates the difficulty encountered by the equipment during the bolt installation process. This can then be related to the force required to remove the bolt from the borehole.

The regression curves of the drive time for each drilling diameters 34mm, 36mm, and 38mm are given by the following expressions:

$$UC_{34\,mm} = 53\ln(Dt) - 56.3 \tag{7}$$

$$UC_{36\,mm} = 32.56\ln(Dt) - 4 \tag{8}$$

$$UC_{38\,mm} = 32.1\ln(Dt) - 18.7 \tag{9}$$

Figure 16 depict the relationship between driving time and ultimate capacity based on the diameters of SS34, SS36, and SS38 bolts. When outliers are excluded, there appears to be a relationship between ultimate capacity and driving time. Overlaying the regression curves of the different diameters on a single graph provides a clear understanding of the pull-out resistance domains.

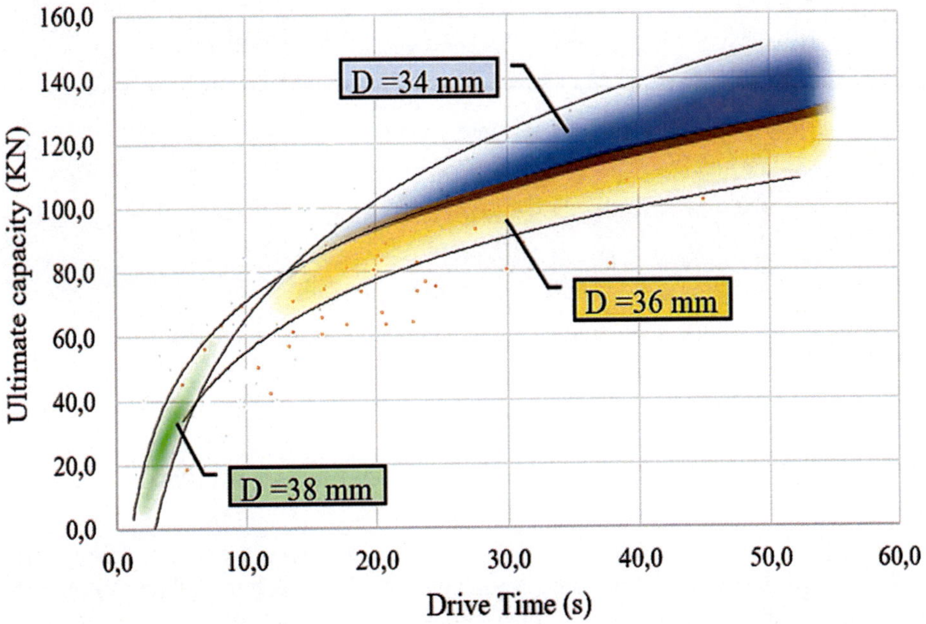

Fig. 16. Pullout load-displacement with drive time

4.6 Impact of Water-to-Cement Ratio on Rock Bolt Resistance

In 2005, Moosavi examined the variation of the water-cement proportion on anchor bolt strength in moderate rocks, focusing on the bolt-mortar interface. They discovered that when the water-to-cement ratio grew, so did the shear strength of the bolt-mortar interface [13].

In 2017, Zhang explored the variation of the water-cement proportion on anchor bolt strength in moderate rocks, using numerical simulations to determine bolt strength. They discovered that increasing the water-to-cement ratio boosted bolt strength up to a point where it began to decline [14].

The most significant factor in concrete mix design is the water-to-cement ratio (w/c), which influences the mechanical qualities and durability of hardened concrete. The water-to-cement ratio is the mass ratio of water to cement in freshly formed concrete. It is calculated by dividing the amount of water in the mix by the weight of cement, both of which are fixed in the same concrete mix.

$$\mathrm{WCR} = \frac{\mathbf{w}}{\mathbf{c}} \tag{10}$$

The pullout test was conducted on four bolts installed with four different water-to-cement ratios, after 28 days from the date of their installation. In order to avoid the influence of water presence and the quality of the rock mass, it was decided to install the bolts in dry blocks with an RMR rating of 60 (Fig. 17).

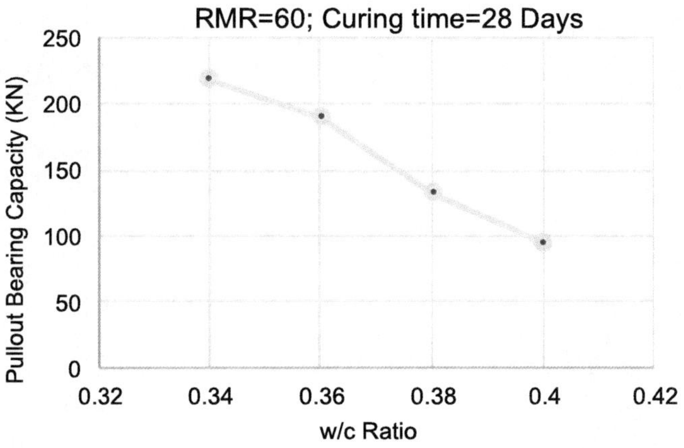

Fig. 17. Pull-out bearing capacity with water-to-cement ratio

Experimentally, a ratio of 0.34 provides the best bolt-rock resistance; however, this ratio has significant drawbacks. The grout's pumpability declines, and many complications develop during application. Pumpability increases as the ratio increases, making it easier to fill the boreholes. This, however, has an effect on the connection strength.

5 Conclusion

In conclusion, the study demonstrates the effectiveness of cementitious concrete injection in enhancing carrying capacity of friction anchor bolts and reinforcing their anchorage in underground structures. The injection of a cement-based mixture comprising cement, sand, and additives into the bolt leads to internal expansion, resulting in increased volume and improved frictional interaction between the bolt and the rock mass. The main conclusions and advantages of this method can be summarized as follows:

- **Increased Adherence**: The radial force exerted by the expanded concrete enhances the adhesion between the bolt and the rock mass, improving the bolt's ability to resist pull-out forces.
- **Improved Load Transfer**: The expanded concrete's radial force aids in transmitting applied loads from the bolt to the rock mass, resulting in reduced local stresses on the bolt and better force distribution, hence improving overall pull-out resistance.

- **Void Filling**: The expansion of the concrete helps fill voids inside the bolt and between the bolt and the rock, ensuring closer contact between the bolt and the rock. This improves the efficiency of anchorage.
- **Overall Stability**: The increase in concrete volume contributes to the overall stability of the support system by reinforcing the bolt anchorage. It reduces undesired movements or deformations of the bolt and rock mass, enhancing the strength and durability of the support structure.
- **Silicate-Based Additive**: The incorporation of a silicate-based additive accelerates the curing time of the cement, enhancing the strength of the bolts. This allows for immediate support after the injection process.
- **Influence of Groundwater**: Groundwater Influence: The study emphasizes the major influence of groundwater on the carrying capacity of cemented bolts. It underlines the significance of taking groundwater conditions into account while designing and installing cemented friction anchor bolts.

The pullout tests conducted on various rock masses demonstrate that higher rock mass quality and longer cemented bolts result in increased pullout resistance. The length of the anchorage bolt plays a significant role in the carrying capacity, with longer bolts exhibiting higher tensile strength and pull-out resistance. The proposed empirical equations provide estimates for the carrying capacity based on the rock mass quality and cemented bolt length.

Overall, the cemented concrete injection technique offers a promising approach to enhance the strength and anchorage of friction anchor bolts in underground structures. It provides specific advantages such as increased adhesion, load transfer, and overall stability of the support system. The use of a silicate-based additive accelerates the curing time, further enhancing the strength of the bolts. Considering groundwater conditions is crucial for optimal performance. These findings contribute to the advancement of friction anchor bolt technologies, offering potential benefits for the design and construction of underground structures.

References

1. Li, S.C., Wang, H.T., Wang, Q., Jiang, B., Wang, F.Q., Guo, N.B., Ren, Y.X., et al.: Failure mechanism of bolting support and high-strength bolt-grouting technology for deep and soft surrounding rock with high stress. J. Cent. South Univ. **23**(2), 440–448 (2016)
2. Liu, J., Li, H., Li, Y., Yang, Y., Sun, T., Song, R., Sun, R.: Study on the effect of isotropic initial stress on the anchoring performance of self-expanding bolts. Adv. Civ. Eng., 1–19 (2021)
3. Li, C.C.: Principles of rockbolting design. J. Rock Mech. Geotech. Eng. **9**(3), 396–414 (2017)
4. Malhotra, V.M., Carino, N.J.: Handbook on Nondestructive Testing of Concrete. CRC Press (2014)
5. Serratrice, J.F.: Une méthode d'interprétation des essais effectués sur site sur des boulons à ancrage réparti sollicités en traction. Rev. Fr. Géotech. **171**, 4 (2003)
6. Liu, G., Xiao, M., Chen, J., Zhou, H.: Study on mechanical characteristics of fully grouted rock bolts for underground caverns under seismic loads. Math. Probl. Eng. (2017)
7. Chen, J., Yang, S., Zhao, H., Zhang, J., He, F., Yin, S.: The analytical approach to evaluate the load-displacement relationship of rock bolts. Adv. Civ. Eng. **2019**, 1–15 (2019)

8. Kılıc, A., Yasar, E., Celik, A.G.: Effect of grout properties on the maximum pull-out capability of fully grouted rock bolt. Tunn. Undergr. Space Technol. **17**(4), 355–362 (2002)
9. Zhang, Y., Li, X., Liu, Y.: Effect of groundwater on the carrying capacity of friction anchor bolts used in underground structures. Adv. Civ. Eng. (2017)
10. Li, X., Zhang, Y., Liu, Y.: Numerical simulation of the effect of groundwater on the carrying capacity of friction anchor bolts used in underground structures. Adv. Civ. Eng. (2018)
11. Tomory, M.J., Stacey, T.R.: The effect of drilling speed on pull-out resistance. Int. J. Rock Mech. Min. Sci. **35**(3), 315–320 (1998)
12. Nicholson, D., Hadjigeorgiou, J.: The effect of drilling speed on the pull-out resistance of fully grouted rock bolts. Tunn. Undergr. Space Technol. **70**, 1–7 (2017)
13. Moosavi, M., Sharifzadeh, M., Ahmadi, M.: Effect of confining stress and water/cement ratio on the shear strength of bolt-grout interface. Tunn. Undergr. Space Technol. **20**(4), 362–369 (2005)
14. Zhang, Y., Li, X., Liu, Y.: Numerical simulation of the effect of water-cement ratio on the pull-out resistance of fully grouted rock bolts. Adv. Civ. Eng. (2017)

Foundations on Rocky Sites: Behavioral Differences and Optimization of Shallow Foundation Systems with and Without Connecting Sleeper Beams on Sandstone Formation

Othmane Hniad[1] (✉) and Jamila El Brahmi[2]

[1] Laboratoire de Géophysique, Géotechnique, Géologie de l'Ingénieur et de l'Environnement (L3GIE), Mohammed V University in Rabat, Rabat, Morocco
hniad@emi.ac.ma

[2] Laboratoire d'Ingénierie Civile, Hydraulique et Environnement (LINCHE), Mohammed V University in Rabat, Rabat, Morocco

Abstract. When it comes to construction sites, having several operating at a time can put a toll on both human and material resources of a company, especially when one of them presents more time-sensitive restrictions than others. Companies tend to seek solutions to this problem by reducing both time and work loads, making their schedules extremely optimized and leaving very little room for further improvements unless they have to do with the work amount itself, which would lead to a building redesign. This study, which centers around a group of residential buildings founded on sandstone, focuses on one of these possible optimization points: foundations. In rocky formations such as sandstone, soils have both an advantage when it comes to the footing's dimensioning by allowing multiple reductions (Souloumiac and Despeyroux in Rev Fr Géotech 37:13–16, 1986), but also mandate additional costs and time related to additional earthworks that come naturally with excavating in said formations. The lack of a single standard regrouping all necessary data for this research made it necessary to find the best way to take into account and use in combination three paraseismic calculation standards as well as two different foundation calculus standards. Two goals were set: outline the necessary checks to account for all standards while studying the feasibility of a foundation system without connecting beams on one hand, and sizing both approaches in order to obtain final values of quantity and budget, allowing a valuation of both systems as well as an estimation of how much optimization was achieved. The soil mechanics' study where the residential area is projected allowed us to identify and distinguish two separate zones within, each with its own stratification. One of these two zones was mainly fractured porous and sand-like sandstones, both of which have a limit pressure value close to 725 psi. In our study, after sizing both variants, length of connecting beams in foundation systems was reduced by 78%. In terms of costs, budgets were also cut by the same percentage. Such an alteration to the buildings' design would not impact its foundation system's efficiency, but would dramatically improve both resources and time allotted to that construction phase. This would, in turn, allow the company to divert its resources without taxing the study's worksite and consequently suffer no delays.

K. Baba et al. (Eds.): GeoME 2023, SPEES, pp. 261–273, 2023.
https://doi.org/10.1007/978-3-031-49345-4_26

Keywords: Foundation · Connecting beams · Standards · Sandstone · Optimization

1 Introduction

Rocky terrain, known for its excellent bearing capacity and to allow the optimization of foundation systems, also usually brings the difficulty of carrying out earthworks to accommodate said systems' different constituents. For civil engineering enterprises, workload optimization is mainly based on calendar saves and workload restriction, as well as the best management possible of available human and material resources. In the case of this study, the nature of the terrain and resource constraints forced us to find optimizations outside the framework of on-site operations, hence the suggestion to reduce the foundation system (in both quantity and workload/allocated times) considered for the residential buildings which were part of the project.

The optimization of the foundation system consists in our case of removing the connecting beams laid in between footings in the area which is dominated by rocky soil. Said removal also presents a major technical issue given that part of the foundation soil is sandstone that is topped with lower quality embankment clayey sands. However, such a reduction in the foundation system is not completely prescribed by the current-use paraseismic standards, which had us include both Eurocode standards and PS92 standards into consideration in order to exhaustively check the validity and security of a potential connecting beam removal, be it partial or total.

Our objectives will therefore be, after exposing our research's context as well as different standardized checks required for foundation footings and connecting beams, to proceed towards said checks, determine where it is possible to remove connecting beams and compare both solutions in two distinct terms: the first being technical based on the currently applied regulations, and the second being financial, i.e. comparing both fully connected foundations and a system where connecting beams have been removed accordingly.

2 Context

This project consists of several buildings that are part of a residential area, of which two distinct geotechnical zones are distinguished. The first, which our study focuses on, is comprised of 8 5-story buildings one common underground parking lot. According to the laboratory's geotechnical reports, the recommended foundation system for this project was isolated footings and connecting beams.

Stability vis-à-vis vertical and horizontal static loads is ensured by a rigid bracing system (central core at the elevator and stairs shaft). The basement's upper floor is a post/beam frame system, with 20-cm-thick reinforced concrete slab. The higher upper floors, for their part, will be a posts/concrete wall system supporting post-tensioning prestressed slabs that are 20 cm thick.

3 Standard Legislation on Foundation Connecting Beams

3.1 RPS2011

On the seismic construction standard RPS2000/V2011 in use in Morocco, the connecting beams are mentioned in chapter IV, discussing basic design regulations. On paragraph IV-4.4 we find the following directives:

- The foundation system must ensure the structure is embedded in the ground and transmit all forces from the superstructure to the ground, as well as limit differential compactions and/or relative horizontal displacements which could reduce the structural system's rigidity and/or resistance.
- The support points of each of the blocks making up the structure must be joined together by a two-dimensional network of connecting beams (Fig. 1) or any other equivalent tending to oppose their relative movement in the horizontal plane. Such fastening of footings is not required if these are suitably anchored in unfractured and undisintegrated rocky ground.

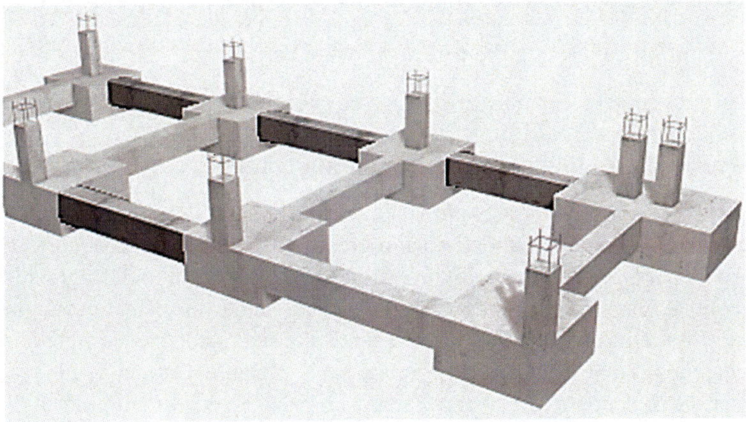

Fig. 1. A foundation system where footings are linked using horizontal connecting beams

3.2 PS92

According to section 4.3.3 of the NF P 06-013 standard, the support points of the same building block must generally be joined together by a two-dimensional network of connecting beams (or any other equivalent system) tending to oppose their relative displacement in the horizontal plane. However, no particular precaution is required in the case of footings suitably engraved in a rocky ground or of rocky consistency, not fractured and not disintegrated (Souloumiac and Despeyroux 1986). This directive is similar to the one prescribed in the Moroccan paraseismic regulations.

4 Necessary Checks for Foundation Footings

In each European country, the Eurocode 7 uses a national appendix which gives the partial coefficients' values to be applied for each structural calculus. All of said partial coefficients provided in appendix A of the EC7 are accepted as long as there's no contradiction with one of six nationally accepted and used standards supplementing Eurocode 7, including the NF P 94-261 standard designed for shallow foundations. This standard proposes a new format for verification of the soil's bearing capacity based on applied force and on an updated calculus method of said bearing capacity which relies on pressuremeter and penetrometer data. Other aspects of the dimensioning of shallow foundations are also addressed such as footing slide, overturning, along with limitation of eccentricity and linking checks with verification under seismic load.

The DTU 13.12, more commonly known as Booklet 62 title V, predecessor of the NF P 94-261 standard, adopts a similar verification principle. This verification is based on Ménard's fundamental formula of the effective breaking stress under a vertically loaded footing.

5 Study Area's Soil Mechanics

For the recognition campaign, the following were carried out:

- Seven twenty-meter-deep core drillings with sampling, pressuremeter tests conducted every 1.5 m in the core drillings;
- Six manual well diggings using a mechanical excavator.

In light of the carried-out recognizance tests, in-situ soils are mainly characterized by the presence of fractured sandstone and gresified sand. Pettijohn et al. (2012) define sandstone as the aggregation of a mixture of mineral grains and rock fragments resulting from the natural disintegration of pre-existing rocks under the effect of natural erosion. These rock fragments are mostly grains of sand-like sizes (not exceeding 2 mm in diameter). Sandstones can be either very coherent or very friable. Okada (1971) recommends a classification of sandstones on three main criteria (given the number and complexity of previous classifications): grain size, clay matrix and composition of framework constituents.

In the study area, two types of lithological sections were distinguished, dividing the project into two zones:

- Zone 1: sandstones surmounted by reddish clayey sands 3–5 m deep.
- Zone 2: outcrop of the sandstone horizon on the surface which is surmounted by a thin layer of clayey sand fill of variable thicknesses.

From the results of the geotechnical tests, it can be observed that our project's surface can be divided into two main areas:

- An area with healthy, unfractured and undisintegrated rock;
- A second area dominated in its upper layer by clayey sands.

Table 1. Test results for manual excavations and pressuremeter drillings

Sample	Drilling spot	SM1	SM2	SM3	SP2	SP3	SP4	SM4
		0.3–1.3 m	0.3–1.3 m	0.2–1.0 m	3.5–4.0 m	6.5–6.8 m	0.3–1.3 m	0.2–1.0 m
	Nature	Embankment – red clayey sands			Clayey sands	Sandstone		
Water content	W (%)	7.7	3.9	5.2	13.2	2	1.6	1.1
Particle size analysis	D_{max}	25	63	40	1	–	–	–
	> 20 mm (%)	4	21	24	0	–	–	–
	> 2 mm (%)	24	47	42	0	–	–	–
	0.08–2 mm (%)	47	41	29	13	–	–	–
	< 0.08 mm (%)	29	12	29	87	–	–	–
Atterberg limits	W_L (%)	29	–	33	32	–	–	–
	W_P (%)	17	–	18	18	–	–	–
	IP	12	–	15	14	–	–	–
Humid density	γ_h (kg/m^3)	–	–	–	–	1740	1680	1810
Dry density	γ_d (kg/m^3)	–	–	–	–	1710	1650	1790

Some of the recognizance results tests carried out made available to us are shown in Table 1.

Sandstone is known to have an axial compressive strength of between 20 and 170 MPa, which places them in a medium to very high strength category. However, in our case, the SP3, SP4 and SP7 soundings returned values between 3.5 and 6.6 MPa at test depths between 3.5 and 6.8 m.

6 Study Area's Seismic Context

6.1 RPS2011

According to the Moroccan RPS2011, the site of our research is located in seismic zone 2, a zone with low seismicity where the maximum acceleration is 0.1 g.

6.2 PS92 and EC8

Since Morocco is not included in the maps intended for the use of these regulations, we will rely on the value of ground acceleration obtained by the RPS2011 in order to locate the study area. This allows us to place our research site in zone 3 for EC8 (zone with moderate seismic hazard) and in zone Ia for the PS92 rules (zone with very low seismicity). Both regulations classify the site in areas with a maximum acceleration of 1.1 m/s^2.

The various geotechnical test points of the project align with the position of its various blocks. This makes it possible to categorize the soils under each of our eight buildings, thus making it possible to confirm or not the possibility of removing the connecting beams for the footings. Two buildings located in the northeast quarter of the project's site are located on the area dominated by clayey gravelly sands on its surface, while the remaining six are located on the rocky sandstone soils.

Thus, and relying on the values of the limit pressures of the pressuremeter tests carried out, it was possible to draw up the following table to distinguish between the two categories of foundation soils (Table 2).

Table 2. Soil nature, category, and influence on use of connecting beams

Buildings position	Limit pressure (MPa)	Soil category (PS92)	Use of connecting beams necessary?
On gresified soils (altered rock)	Around 5 MPa	Rock	Not if some specific conditions are properly met
On clayey/gravelly sands	0.5–2 MPa	B category	Yes

In our case, the connecting beams required for the two buildings standing out as being founded on loose soil, as well as the ones that will connect to the other buildings in vis-à-vis, must be built. For the rest, and in order to introduce our optimization suggestion to the foundation system, the connecting beams will be removed.

However, we were aware that such an optimization wasn't possible without introducing the necessary condition of limiting our footing's horizontal displacements, recommended by both the RPS2011 and (in greater detail) the PS92. Said limitation can be ensured by anchoring the footings to a depth of 50 cm below excavation level, thus preventing lateral displacements. It was recommended on our part to pour the concrete directly in full excavation and the installation of a bed of ballast under the lowest concrete slab and over the footing in order to offer additional resistance to lateral displacements.

7 Checking the Foundation Footing

For the following checks, we chose to place our footings over the SP7 drilling spot. The measured total and dry densities are respectively equal to 18.1 and 17.9 kN/m^3 at depths of between 3.5 and 3.8 m. These same densities increase to 17.9 and 17.7 kN/m^3 at depths between 5 and 5.5 m.

The checks will be carried out on a two-by-two square-meter isolated footing laid at an earthwork level provided at 3.5 m from the natural terrain, and a 50 cm embed from earthworks top level. It was recommended by the engineering firm in charge of design to load the footings at 280 t for the ultimate limit state and 170 t for the serviceability limit state.

In order to meet the requirements of seismic regulations regarding the use of foundations without horizontal connections, the anchoring will act as physical restriction along with the absence of transverse load from the posts (centered load without bending moment). Thus, both slip and load eccentricity checks were not amended.

7.1 EC7

The depth H_r under the foundation base is equal to 1.5B in the case of this footing, so 3 m. The equivalent net limit pressure p^*_{le} calculus was made on the basis of the calculation of the net limit pressures between 4 and 7 m deep and gives a value of 4.94 MPa. The equivalent embedding depth has a value of 49.9 cm.

The calculation of the pressuremeter lift factor depends on the footing's geometry and the soil itself. In case of a square footing on weathered rocky ground, the calculation of k_p gives a value of 0.945. Thus, the bearing capacity q_u becomes 4.66 MPa.

The stressing load will thus be the difference between the ultimate load taken into account, equal to 280 t, and the mass of soil displaced by the volume of the foundation, equal to 2 m * 2 m * 0.5 m * 18 kN/m^3, the value soil density taken as an average around 4 m of depth.

The design value of the net soil resistance $R_{v,d}$ which depends on the effective seating surface area (equal to that of the foundation in the centered load case) and the overall combined safety factor (equal to 1.44 at the seismic ultimate limit state), gave 6.75 MN. The bearing capacity condition $Q_{v,d} - R_0 \leq R_{v,d}$ thus becomes verified.

The compaction under the footing can be estimated under serviceability limit state combinations using the pressuremeter method. The computing of the E_c and E_d moduli from the harmonic averages (Fig. 2) gives the following values: $E_c = 467.9$ MPa and $E_d = 435.7$ MPa.

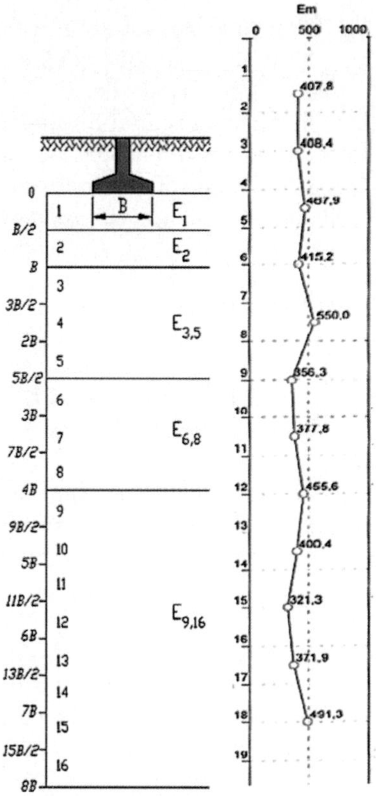

Fig. 2. Determination method of harmonic average moduli for E_c and E_d calculus

Corrective coefficients λ_c and λ_d values as well as that of the rheological coefficient α are taken respectively equal to 1.1, 1.12 and 0.66. The value of q_0, mass of soil before modification at the footing's top level, and with an average total density taken equal to 18 kN/m^3, is equal to 72 kPa.

With all the data present, the value of the compaction under the footing can be calculated. The value obtained is 0.38 mm.

7.2 DTU 13.12

Using the same interval as the previous method, we determine $p^*_{l\,min}$, equal to 4936.5 MPa. Thus, an approximation of the evolution of p^*_l by the least squares method gives the following function: $p^*_l(z) = 4937.4 + 4.89z$. We therefore find the value p^*_{le} at the coordinate D + 0.66B, equal to 4.94 MPa.

The embedding depth being identical to that obtained previously, one then proceeds to the determination of k_p, of which the formula is directly obtained according to the soil's nature (itself determined by the classification of the soil from the value of p^*_{le}, this being a fragmented rock in our case). The value of k_p obtained is 1.06.

Verification can thus be carried out by calculating the admissible load and that provided by the structure:

$$q_{adm} = \frac{k_p P^*_{le}}{3} + q'_0 = \frac{5.27}{3} + 0.5 \times \gamma = 1.76\,\text{MPa} \quad (1)$$

$$q_{structure} = \frac{2.8\,\text{MN}}{2\,\text{m} * 2\,\text{m}} = 0.7\,\text{MPa} \quad (2)$$

The bearing capacity condition is thus verified. Checking compaction under the footing is done in the same way as for EC7, and will not have to be reproduced.

8 Connecting Beams Dimensioning

In this part, we will only focus on the stringers that can be removed (those linked to the footings whose lateral displacements are limited by good anchoring in the sandstone-supporting soil) in order to evaluate the difference in terms of costs and delays between the variant with outriggers and that without outriggers.

In our case, the tensile force for which a sill must be sized is obtained by the following formula:

$$F = \frac{a_N}{g} \,.\, \tau \,.\, \alpha \,.\, W \quad (3)$$

where

- $a_N/g = 0.15$ in the gresified area of this project;
- $\tau = 1$ due to the original flat topography;
- $\alpha = 0.4$ for this study's sandstone;
- A load at the column's foot valued at 28 tons.

The dimensioning force of the connecting beams is equal to 168 kN. In order to satisfy the minimum conditions on the dimensions of the beams (a 20-cm minimum for each side), we will consider 25 * 35 beams for our study. The tensile concrete being neglected, the entire tensile force will be handled by the reinforcements, the section of which must respect the following minima:

- At ultimate limit state:

$$A_u \geq \frac{N_u}{f_e/\gamma_s} \quad (4)$$

- At serviceability limit state:

$$A_s \geq \frac{N_s}{\overline{\sigma_s}} \quad (5)$$

where, in the case of very detrimental cracking:

$$\overline{\sigma_s} = \min\left(\frac{1}{2}f_e;\, 90\sqrt{f_{tj}}\,\text{MPa}\right) \quad (6)$$

- Non-fragility condition:

$$A_f \geq B\frac{f_{tj}}{f_e} \tag{7}$$

where

- N_u is the ultimate traction load, equal to 1.35 times the design traction load;
- N_s is the serviceability traction load, equal to the design traction load;
- f_e is the rebar steel's elasticity limit;
- γ_s is the rebar steel's partial security coefficient;
- f_{tj} is the mechanical resistance of concrete under traction load at j days.

The transversal rebars must comply with spacing less than the smallest dimension of the connecting beam, for they play no role in resistance and are only put in place for the assembly of the longitudinal rebars.

For a 30 MPa limit concrete and a 500 MPa reinforcing steel, the following minimal sections are calculated: $A_{u\,min} = 5.2\ cm^2$; $A_{s\,min} = 9.53\ cm^2$; $A_{f\,min} = 4.2\ cm^2$.

For transverse reinforcement, a minimum diameter of 6mm is prescribed by the rules PS92. Also, and with a view to confining the concrete, the same rules recommend spacing that must respect the following condition: $s_t < \min(24\Phi_t, 8\Phi_l, 0.25d)$. This leads to the following reinforcement scheme:

- 9 rebars of 12 mm in diameter arranged around the perimeter and a central rebar;
- A spacing between concrete surface and rebar surfaces set at 5 cm;
- Transversal reinforcements of 8mm spaced by 7.5 cm, arranged in a perimeter frame and two pins connecting the central rebar to the others (Fig. 3).

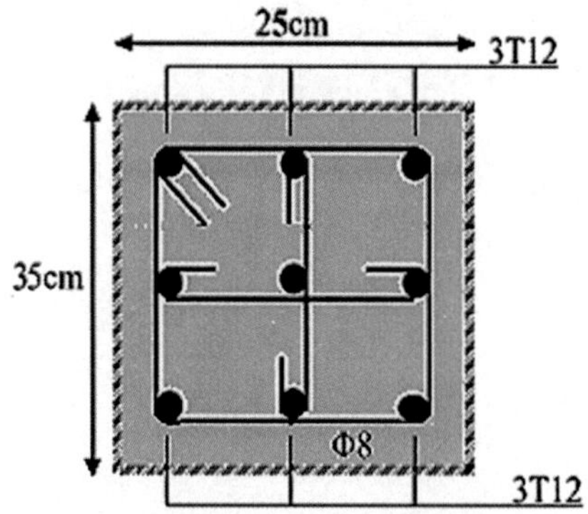

Fig. 3. Steel rebar disposition in the studied connecting beam's transversal section

9 Comparison of Foundations with and Without Connections

In order to compare the two variants, we will carry out a detailed measurement of the stringers with the following elements:

- Total distance covered by connecting beams' length;
- Total concrete volume;
- Steel rebar weight at a ratio of 0.12 t/m^3;
- Formwork quantity required.

9.1 Costs

If connecting beams are expected for the whole foundation system, we estimated the use of 549 cubic meters of concrete, 5960 m^2 of formwork surface, 65 m^3 of mud slab and 65.8 tons of rebar steel. If we consider a unit cost per cubic meter of beam without formwork of 1543.7 dhs (price generated by the CYPE structural calculation software), the total cost of said beams without formwork is 798,000 dhs.

After removing the connecting beams wherever authorized, a new estimate gives us 120 m^3 of concrete, 1307 m^2 of formwork surface, 10.7 m^3 of mud slab and 14.5 tons of rebar steel. At the same unit price per beam concrete shown above, the total cost of connecting beams without formwork is, after optimization, 185,800 dhs.

9.2 Timelapse

Connecting beams' build (including formwork, reinforcement on site, pouring and stripping as soon as possible) takes an average of one and a half hour per cubic meter. The difference in terms of time between the solutions with and without said beams on rocky ground becomes a 77-day time save if we consider working days typically last between 8 and 9 h.

9.3 Total Optimization

By removing the connecting beams wherever authorized, we were able to reduce workloads and costs by a percentage of 75% (Fig. 4).

10 Discussion

The prescriptions of the RPS2011 do not offer a straight directive on skipping the use of connecting beams apart from the limitation of movements. In our research, we have, in addition to having consulted the PS92 rules which offered a better perspective on the question, asked the service provider to incorporate this footing anchoring and a top layer of ballast in the construction.

These recommendations subsequently made it possible to satisfy the seismic requirements and to proceed with connecting beam removal, thus optimizing all aspects of the construction of the foundation system. While taking care to limit said displacements. Our research took place on a mostly rocky support, which offered both a good degree of bearing capacity as well as the possibility of anchoring the footings fifty centimeters into the rock, embedding it in the rock, thus forcing the blocking of potential displacements. Such displacements must be kept under certain thresholds that are influenced by several factors such as structure's size and usage, the construction material's characteristics and, most notably, the soil-structure interaction (Ausilio and Zimmaro 2015).

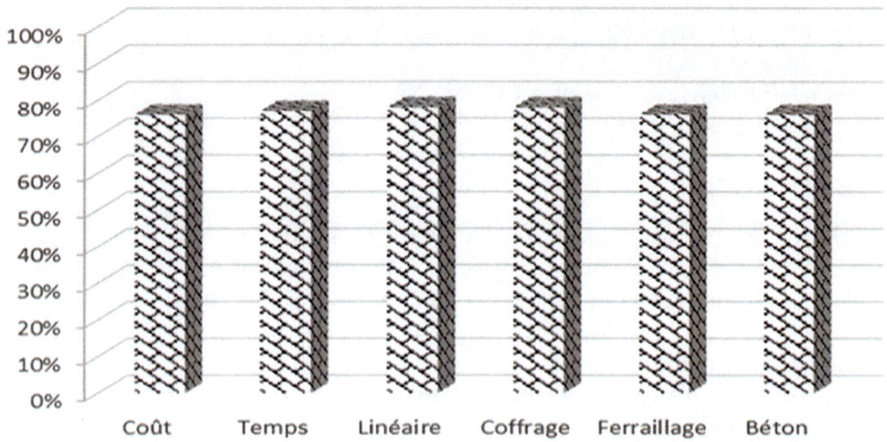

Fig. 4. Reduction percentages resulting in the studied foundation system optimization by connecting beam removal (from left to right, in terms of: cost, time, length, framework, rebar, concrete)

This anchoring, along with sufficient roughness of the footing, should have positive effects on both bearing capacity (Keshavarz and Kumar 2018) and lateral displacements by blocking the latter.

It is also possible, for the fractured sandstone that was identified on site, to be artificially cemented with clay or silty clay (Chang et al. 2008), in order to properly assess the failure mechanism of shallow footings in cases of hard rocky soil and its difference from our case study which had a softer soil.

The use of various seismic standards also prompted the verification of the foundations according to their counterparts, which proved to be satisfactory. It's worth noting that a better estimate of time and costs saved can be made by also evaluating the time per linear meter of trenching in the mass of sandstone, not evaluated in this research. Having to dig trenches through soil of rocky structure for the beam's reservations, as well as inserting a mud slab layer underneath would add considerably to computed costs and timelines.

11 Conclusions

The use of connecting beams for footings in foundation systems is a general requirement in all seismic regulations. As the lateral displacement of the foundations is a potential risk during seismic events, any building must either have a reinforced concrete anchorage designed to limit the relative displacements of the surface footings in the horizontal plane, or incorporate horizontal connecting beams linking footings in both directions. However, under certain conditions, it is possible to do without these beams, if a certain set of requirements, among which proper anchoring of foundations or proper displacement limitation.

That particular case presented itself in our research, as we had the possibility of embedding the footings into the rocky soil, ensuring limited displacements as well as excellent bearing capacity. This, in turn, allowed us to reach our objective or optimizing workloads and timelines. This type of optimization is highly solicited by companies which have high workload densities coupled with multiple construction sites, which puts a toll on resources and could impose several delayed tasks which would cost considerable amounts in the long haul.

In our research, such delays were averted and the company in charge was able to quickly divert its material resources towards other construction sites as well as gain an early lead into the construction phase following foundations.

References

Ausilio, E., Zimmaro, P.: Displacement-based seismic design of a shallow strip footing positioned near the edge of a rock slope. Int. J. Rock Mech. Min. Sci. **76**, 68–77 (2015)

Chang, J.C., Liao, J.J., Pan, Y.W.: Failure mechanism and bearing capacity of shallow foundation on poorly cemented sandstone. J. Mech. **24**(3), 285–296 (2008)

Keshavarz, A., Kumar, J.: Bearing capacity of foundations on rock mass using the method of characteristics. Int. J. Numer. Anal. Meth. Geomech. **42**(3), 542–557 (2018)

Okada, H.: Classification of sandstone: analysis and proposal. J. Geol. **79**(5), 509–525 (1971)

Pettijohn, F.J., Potter, P.E., Siever, R.: Sand and sandstone. Springer Science & Business Media (2012)

Souloumiac, R., Despeyroux, J.: Dispositions technologiques concernant la protection parasismique des ouvrages de fondation. Rev. Fr. Géotech. **37**, 13–16 (1986)

Effect of Overstress on Slope Stability in a Fractured Massif

Ahmed Hemed[1](✉), Latifa Ouadif[1], and Khadija Baba[2]

[1] 3GIE Laboratory, Mohammedia Engineering School, Mohammed V University, Rabat, Morocco
ouldhemed@gmail.com

[2] GCE Laboratory, High School of Technology-Sale, Mohammed V University, Rabat, Morocco

Abstract. Natural fracturing is known in the deposit based on geological observations of the area. This one comprises a stratification of the layers sheared by joints of rough discontinuities and families of diaclases whose orientations are widely dispersed. The orientation of the fracturing on the bench front affects its stability state by changing the dip relative to the direction of the exploitation, as well as by the length and type of filling. The M'HAOUDATT pit has stability problems, influencing the production chain and affecting the strip ratio rate of the mine. As a result of the voluminous blasting, the pit suffers from the absence of safety platforms able to absorb the charge exerted by the massif, which damages the pit edges that play a primordial role, in order to recover a maximum of ore without incurring excessive risks as to the stability of the slopes. This paper aims to analyze the effect of stresses on the pit walls located in a highly fractured rock mass. This approach will be exhaustive for the mine in order to eliminate the source of the large opening fractures exposed on the pit fronts. Due to the absence of an experienced hand, the M'HAOUDATT pit has never been a subject of such a relevant study. Regular monitoring of the fractures was done through topographic stations on a two-month scale, plus analytical simulations of the pit fronts and empirical calculations estimating the components of the applied stresses. This study pointed out that the erroneous geological interpretation of the ground lithology and the incomplete geotechnical model for the exploited pit play a primordial role in the stability of the fronts during the exploitation phase, also the realization of vertical fronts in a ground full of fracturing without any horizontal berms accelerates its sliding in favor of the voids created by the extraction. It is important to note that the under-caving of the loading engines will cause slope failures because of the scaling of the families of discontinuities. Fracturing activity leads to water passage through the joint spacing, which promotes land-sliding. It is always recommended to adopt a well detailed structural model for the pit in order not to exceed the safety screen formed by the families of discontinuities, and to pay attention to the realization of horizontal berms to minimize the load on the pit slopes.

Keywords: M'HAOUDATT pit · Open pit · Slope failure · Slope stability · Hematite · Mining operations · Zouerate

K. Baba et al. (Eds.): GeoME 2023, SPEES, pp. 274–289, 2023.
https://doi.org/10.1007/978-3-031-49345-4_27

1 Introduction

Failures resulting from fracturing always lead to significant stability problems. This justifies the importance of the distinction between fracturing induced by mining operations, and natural fracturing [1]. Discontinuous fracturing and the influence of massive pressure can present numerous risks of instability due to the development of various failure modes [2]. The phenomenon of instabilities affects the strip ratio of the mine by the addition of extra tonnage, which is considered a loss of production especially in the deposit of poor ground holding [3]. Many slope stability problems require a good knowledge of the fracturing state to which the rock mass is subjected for their resolution [4]. The slopes of the M'HAOUDATT pit are unstable due to the effect of the non-control of natural fracturing and the absence of observations described underground [5]. The brutal release of stresses in a fractured massif generates a weakness in a surface zone parallel to the excavated surface, due to the excavation and the vibrations induced by the construction machines and the repeated use of explosives [6]. Depending on the depth, it represents a certain thickness more or less large where the rock is decompressed [7]. A specific mechanical behavior must then be progressively taken into account for the deformation calculations, during excavation and in the long term [8]. Indeed, the permanent unfavorably oriented discontinuity at low stress levels in the M'HAOUDATT pit generates a complete collapse of the rock face. The collapse of the rock face is exposed on the West wall of the pit following planar failure mechanisms. The observed failure mechanisms will be analyzed later in order to identify the main causes of this type of slide. The process in this approach is to present the method generally used for the determination of existing families, trying to predict possible problems of instabilities related to discontinuities, and then we will try to synthesize the statements made [9].

2 Study Area

Located 55 km north of the Zouerate city (see Fig. 1), the M'AHOUDATT deposit is part of a complex chain formed almost entirely of BIF (Banded Iron formation) naturally rich in iron (Hematite), with some quartzite units and schists. Along the M'HAOUDATT Chain, the BIF forms large-scale boudins and constriction zones with a frequency of a few kilometers. This is interpreted as the result of boudinage and circulation of oxidizing fluids concentrated in the strangulation zones which also form the privileged sites of the enriched BIF. The focus of this study is oriented to the condition of the pit wall, which is the object of several instabilities and which constitutes the zone of weakness (Footwall sector 045 DDR_MH3 Pit) (see Fig. 2).

3 Structural Analysis of the M'HAOUDATT Pit Wall

3.1 Fracturing Wall Fronts State

The structural measurements along the pit wall fronts aimed at collecting the maximum number of families, influencing the stability of the pit benches. In areas of high deflection, measurements were made remotely to minimize errors. The collected structural information was compiled and spatially plotted on stereonets to identify orientation groups for the area where schistose lithology is dominant (see Fig. 3).

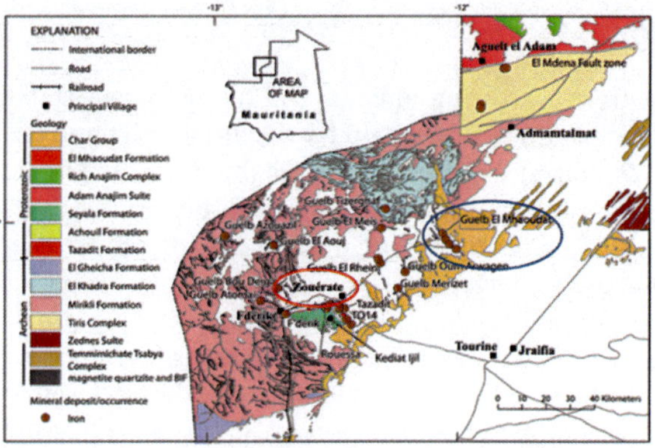

Fig. 1. Regional and geographical situation of the study area

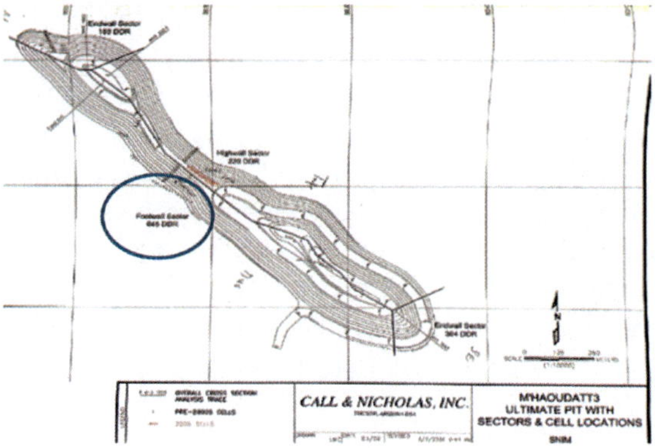

Fig. 2. M'HAOUDATT pit

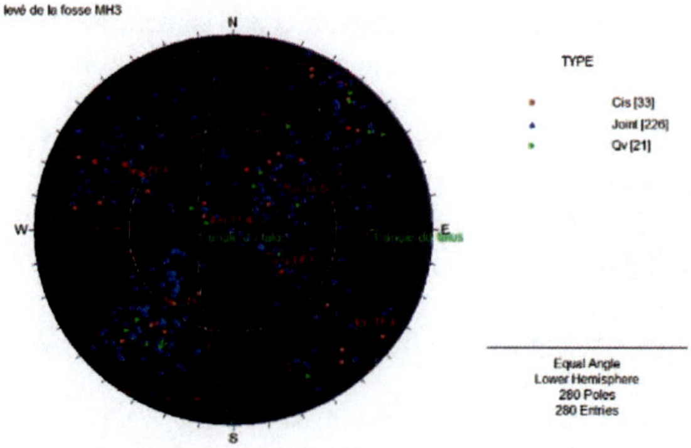

Fig. 3. Stereographic treatment of the schist area

The following structures were defined during mapping:

- Cis: Locally observed shear planes.
- Qv: Quartz veins are found in association with banded rocks and mica-schists, presented in the pit wall. Their dimensions vary in width and length, from centimeters to decimeters thick and from decimeters to several meters long. The quartz veins are mainly fracture related and have obvious contacts with their local host rocks.
- Joint: Fracture plane.

The M'HAOUDATT units are closed to the southwest and northeast by significant fault zones. The approach of the faults to each other leads to the pinching out of the QV families on the M'HAOUDATT wall to the north and south. The faults along the perimeter of the wall are complex, hosting BIF and shale units, as well as zones of secularity, sometimes associated with epidote. Within the QV units, shear zones are present, with kinematic indicators showing a sinister direction of movement. Along these shear zones, quartz veins as well as zones of hematite platelets indicate mobilization of the fronts during shearing [10]. The stratification and schistosity planes (superimposed) are presented on the surface along a mean direction N 123 and a dip of about 53° towards the SW (see Fig. 4). The fracturing treatment shows that the scale of the M'HAOUDATT pit is affected by a series of fractures that have the same geometric characteristics both on the surface and at depth.

Fig. 4. A. Stratification plans in the schist zone; B. Bending of the QV families

The massive fracturing of the M'HAOUDATT pit represents a context favorable to different instability modes. The kinematic analysis aims to identify the families causing these different modes of failure on the pit wall fronts (see Fig. 5).

The safety screen formed by the discontinuity families is summarized in Table 1.

It is recommended for this wall, an exploitation through bench face angles lower than the discontinuities to avoid unstable ruptures.

3.2 Exploited Benches Condition

The M'HAOUDATT pit has experienced several modes of failure during its exploitation. This corresponds to the influence of the blasting on the poor ground and the difficulty of controlling the discontinuity families in the schistose wall of the pit (see Fig. 6).

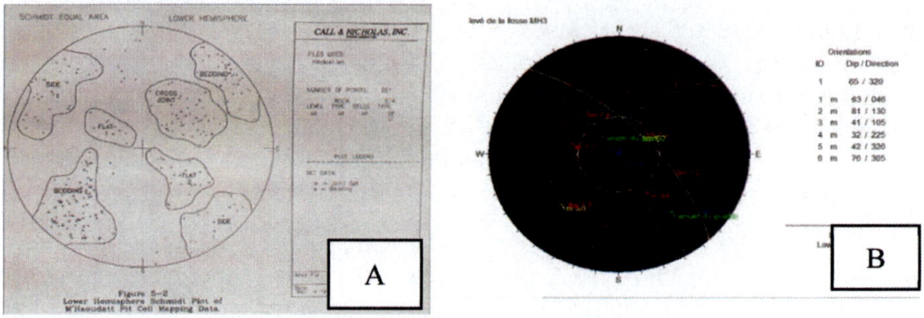

Fig. 5. A. Plot of pit wall discontinuity families; B. Stéreonet of the discontinuity families of the pit wall

Table 1. Classification of discontinuity families

Sector	F1	F2	F3	F4	F5	F6
Foot-wall sector 045 Dip/DDR_MHDT Pit	63/046	81/130	41/105	32/225	42/328	76/305

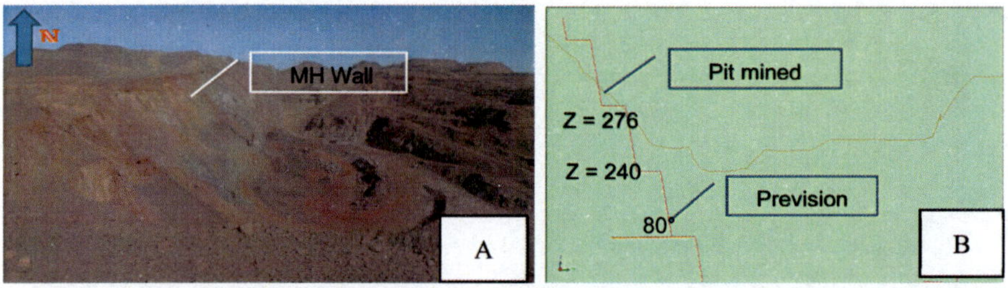

Fig. 6. A. Condition of the MHDTT wall; B. Cross section on the unstable wall

It is clear that the pit benches are mined by angles above the wall structural families.

The design of pit slopes involves the determination of the pit's stability condition. Therefore, it is necessary to study the commonly observed aspects and assumed failure mechanisms.

3.3 Pit Wall State

During its exploitation, instabilities began to appear on the M'HAOUDATT pit wall, such as the roc detachment and fractures of very large openings and depths (see Fig. 7). With the absence of any safety structure on the pit wall, the fracture develops under the overload exerted by the massif. This fracture development was monitored through prisms installed on the pit wall and they were regularly checked for 2 months (see Fig. 8).

Table 2, summarizes the results of the prism monitoring:

Failure may have been generated by the continuous progression of maximum shear strain and the degree of deformation localization around the sliding surface after the

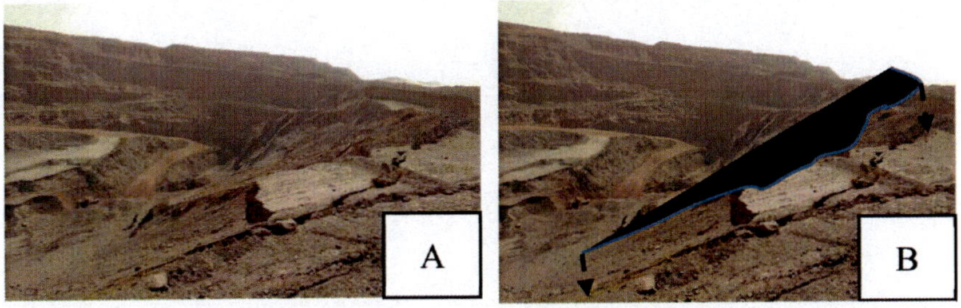

Fig. 7. A. Nature of the fracture on the wall of M'HAOUDATT; B. Form of the fracture development

Fig. 8. Monitoring the fracture's evolution

local failure first manifested itself. Therefore, a considerable link between fracturing and deformation can be used to define the slope failure mechanism [11]. This result demonstrates that the direction of fracture displacement was influenced by the stresses imposed by the massif, discontinuous joints, blast of the shot, and is mainly determined by the excavation and slope inclination (see Fig. 9). At the end of this follow-up, it can be seen that the fracture changes linearly with the excavation of the pit, while the main causes of this type of instability are the combination of the nature of the rock, its alteration, the geological structure of the massif, the orientation of the excavation walls and the damage from the influence of the blast on the final pit walls. This instability resulted in the destruction of the main pit road, necessitating the construction of a new road, resulting in a critical increase in the strip ratio. Significant issues that remain to be resolved in the M'HAOUDATT pit are safety berms, blasting management, and product loading methods. The processing of fracturing records shows that the M'HAOUDATT pit is affected by a series of fractures which have the same geometrical characteristics both at the surface and at depth. The fractured massif of the M'HAOUDATT pit presents a context favorable to different modes of instability. An analytical analysis of the fracture was carried out to reflect the reality of the stability of the pit walls (see Fig. 10).

Table 3, summarizes the results of the modeling.

This analysis has shown that the pit wall in question has a tendency to become unstable, which affects the safety of the human and material resources of the mine. In what follows, we will analyze the pit slopes prior to the occurrence of this fracture. In order to evaluate the slopes of the M'HAOUDATT pit wall, an analytical examination of

Table 2. Survey table

Points	01/12/21	05/12/21	10/12/21	15/12/21	20/12/21	25/12/21	30/12/21	05/01/22	10/01/22	15/01/22	20/01/22	25/01/22	30/01/22
A1_B1	1.147	1.103	2.175	2.202	2.256	2.288	3.174	3.311	3.189	3.214	3.27	3.32	3.359
A2_B2	0.378	0.353	1.808	1.801	1.825	1.89	2.84	2.865	2.89	2.897	2.91	3.11	3.214
A3_B3	0.83	0.756	2.204	2.147	2.181	2.237	3.177	3.184	3.246	3.256	3.312	3.346	3.385
A4_B4			1.515	1.557	1.578	1.589	2.519	3.123	3.215	3.313	3.551	3.621	3.733

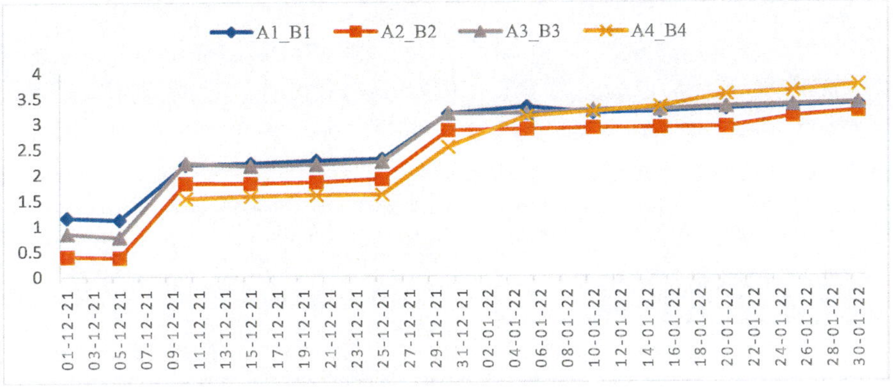

Fig. 9. Fracture evolution

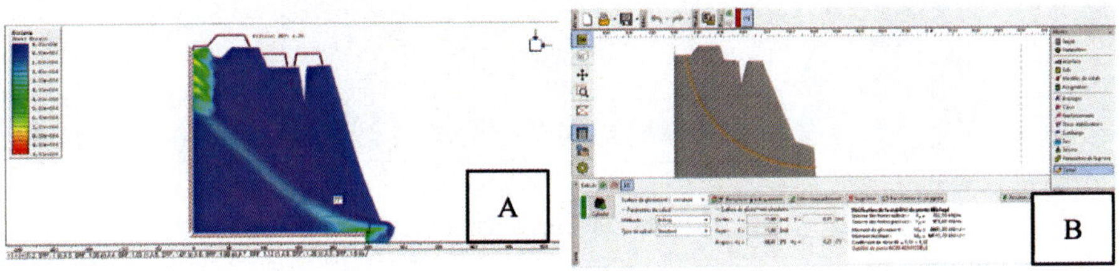

Fig. 10. A. State of the fracture by finite element method; B. State of the fracture by bishop method

Table 3. Modeling results

Sector name	SSR method	Bishop method
FOS recommended	1.2	1.5
MHD Pit footwall FOS	1.06	1.32

the slopes was carried out by comparing the current state of the slopes with the planned project in order to determine the source of this instability [12].

4 M'HAOUDATT Pit Wall Slopes Analysis

For this analysis, a set of sections were created to evaluate the current pit wall slopes and compare them with the planned design (see Fig. 11).

4.1 Absence of Safety Berms

Figure 12 shows the nature of the current pit wall faces on section 1400, compared to the planned design.

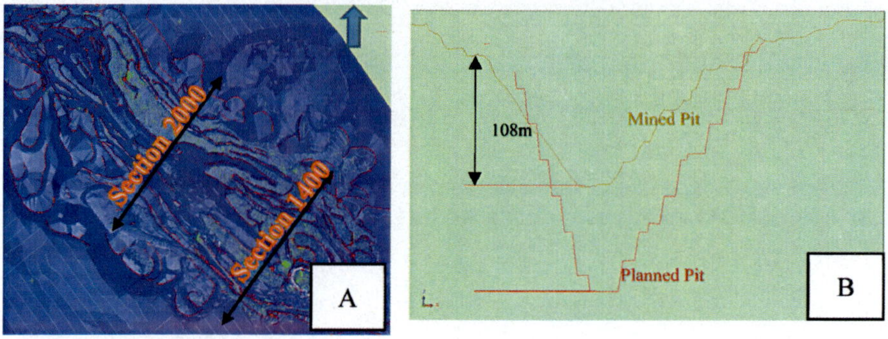

Fig. 11. A. Location of sections on the pit plan; B. M'HAOUDATT pit front state (section 1400)

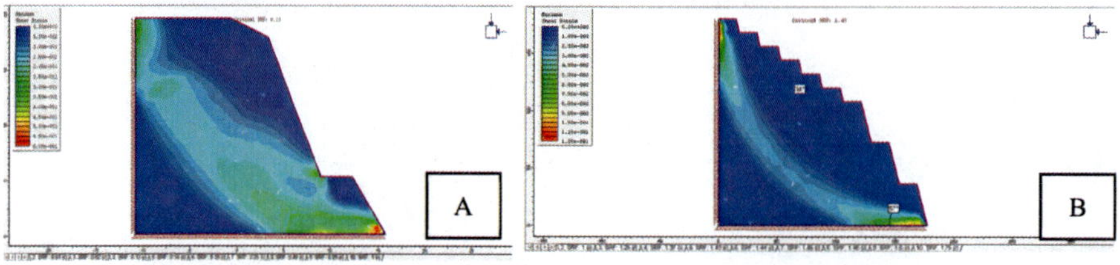

Fig. 12. A. State of current fronts; B. State of planned fronts

Table 4. Simulation results

Sector		2D finite element analysis (FOS recommended)	FOS evaluated
Footwall sector 045 DDR_MH3 Pit	Current fronts (no berms)	1.5	0.13
	Planned fronts (with berms)	1.5	1.45

Table 4, summarizes the simulation results.

The absence of safety berms accelerates the instability of the pit fronts.

4.2 Effect of the Principal Constraint

The wall of the M'HAOUDATT pit suffered a variable stress field corresponding to the change in stress condition after the excavation (see Fig. 13).

Natural stresses affect the pit wall in a variety of ways. They accelerate the deformation of rock slopes and cause their failure when they exceed their strength [13]. During this period of analysis, we tried to estimate the natural stress before the beginning of the excavation which is in this case: the gravitational stress [14]. The vertical component of

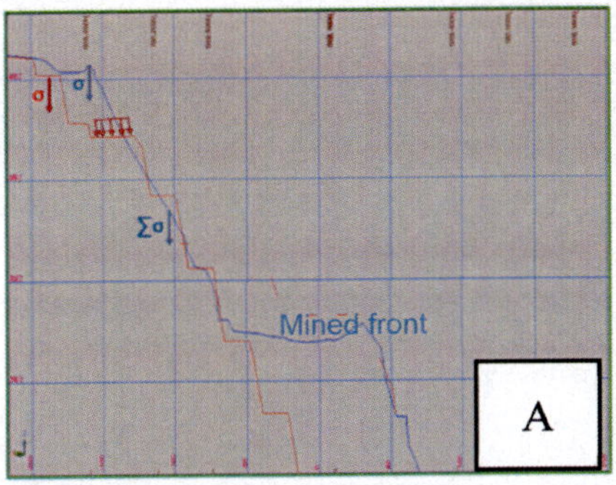

Fig. 13. A. Effect of stresses on pit slopes; B. Height of the fronts without safety berms

this stress exerted on a given point is estimated by (1) as follows [15]:

$$\sigma_v = \int_0^z \rho g dz, \tag{1}$$

where ρ = density, g = gravitational acceleration and z = depth

The measured density of the rock of the M'HAOUDATT wall is 2670 kg/m^3.

The vertical component at a depth of 100 m will be calculated by (2):

$$\sigma_v = \frac{Density\left(\frac{\text{kg}}{\text{m}^3}\right) \times Acceleration\left(\frac{\text{m}}{\text{s}^2}\right) \times S\left(\text{m}^2\right) \times depth(\text{m})}{S\left(\text{m}^2\right)}$$

$$\sigma_v = 26 \times 10^5\left(\frac{\text{N}}{\text{m}^2}\right) = 379.5psi = 26.68\left(\frac{\text{kg}}{\text{cm}^2}\right) \tag{2}$$

The horizontal component, considering that the material is strictly elastic and that no horizontal displacement is possible, is evaluated in the (3):

$$\sigma_h = \frac{\nu}{1-\nu}\sigma_v \tag{3}$$

where $\nu = Poisson\text{'}s\ ratio\ catch\ of$ 0.25

$$\sigma_h = 8.89\left(\frac{\text{kg}}{\text{cm}^2}\right)$$

Note that it is difficult to determine the horizontal component of the gravitational stress due to the effect of the boundary conditions and the effect of the rock properties [16].

Along with gravitational stress, tectonic stress can be mentioned as a component of normal stress.

The tectonic forces have given rise to phenomena influencing the stability of the slopes such as the presence of faults, shear zones, and fractures [17]. The stress tensors were analyzed by the company (operator of the M'HAOUDATT mine), with the elastic constants by rock type and the method used for the determination of stresses (Table 5).

Table 5. Stress tensors analysis

Drill hole	Depth (m)	Rock type	Stress (kg/cm^2)		Average Poisson's ratio
			σ_v	σ_h	
MH017	93	Hematite quartzite	25.74	7.18	0.23
MH23	110	BIF	30.3	8.52	0.21
MH32	129	Schist	35.31	10.08	0.2
MH41	138	Hematite ore	38.2	11.15	0.21
MH53	205	Schist quartzite	42	17.25	0.2
MH62	212	Hematite quartzite	46.25	19.87	0.24
MH78	220	Schist	52.6	21.96	0.19

4.3 Effect of Waste Rock Embankment

The exploitation of the M'HAOUDATT pit's fronts has been exceeded its ultimate limits.

This exceeding of the limits reduces the safety distance between the pit and the waste rock embankments, which puts the pit under the weight of the embankment charge. The 2000 section (see Fig. 11) is analyzed with the current pile location compared to the prediction (see Fig. 14).

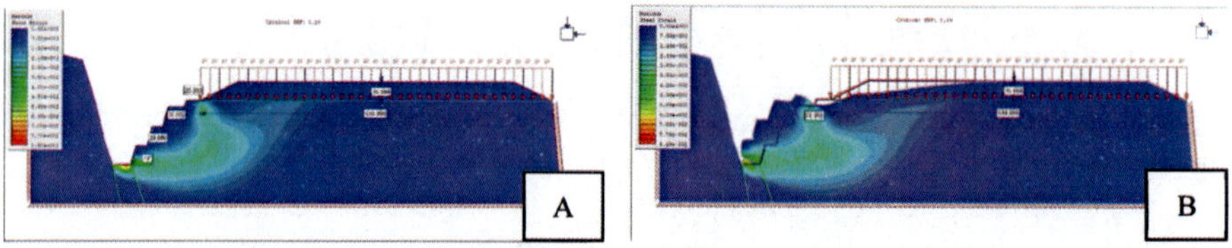

Fig. 14. A. Embankment distance from the pit; B. Potential Embankment failure

The reduction of the distance between the pile and the pit increases the weight exerted by the pile on the pit's fonts. It is always recommended to place the embankment at a safe distance from the pit slopes (see Fig. 15).

Table 6, summarizes the results of safety distance between embankment and the pit analysis.

4.4 Water Condition

Another factor that significantly increases the number of possible failures is the application of hydraulic pressure: For the Back-break analysis from the company, a dry water condition was assumed.

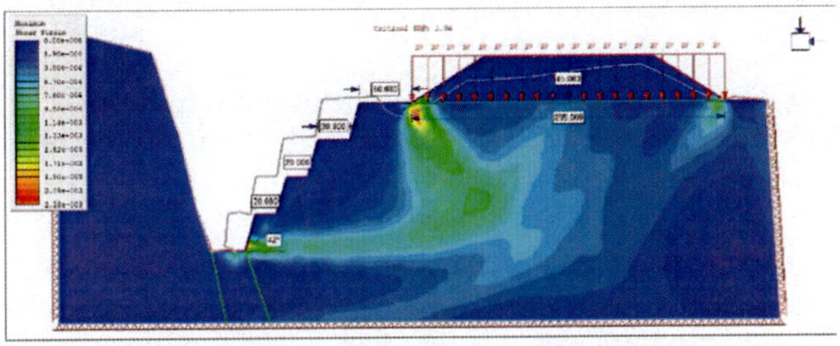

Fig.15. New Embankment distance

Table 6. Impact of embankment safety distance

Distance (m)		FOS recommended	FOS evaluated
Pit wall	25	1.5	1.19
	50	1.5	1.34

The dry water condition assumes the existence of pore pressures acting on the structures in the bench-face. Even in saturated pit slopes, this type of water condition is commonly observed, since rock dilation due to mining and blasting provides many paths for water to flow, and effectively dewaters the bench-face. Seepage will usually exist from the toe of the bench leaving the crest area relatively dry. This means that the mined bench faces will be dry in the area where Back-break is occurring. During the field visits we observed water curves on the pit walls. The sources of this water are still unknown to the mine (see Fig. 16).

Fig. 16. Water curves on the pit walls

5 Discussion

The main influence of this fracturing on the mechanical behavior of the M'HAOUDATT massif results in the spacing of the joints, the infiltration and the presence of water between the fissures [18].

The joints of the stratification show an uncontrolled dispersion on the pit walls which makes it difficult to control the blasting at the pit fronts. As the pit is deepened, the main joints found at the bottom of the pit cause significant stability problems. It should always be noted that it is strictly forbidden that the slope angle must not exceed the dip of the safety screen formed by the natural fracturing.

On the pit wall of M'HAOUDATT, the slopes intersect the joints with significant dips, which generates failures in the massif and then the presence of collapsing fractures on the high slopes of the pit.

With the poor performance of the pit wall and the angle of the benches superior to that of natural fracturing, the mining operations have only increased the spacing between the joints and the disorientation of the slump towards the void created by the exploitation, which leads to the destruction of the majority of the final access roads of the pit [19]. During the exploitation works, phenomena such as fractures in very large openings, destruction of the rolling roads, etc., started to appear on the walls of the pit. We have observed that this damage resides in the part where the dip of the natural fracturing was not taken into account. The most important problem for the stability of the slope of the pit wall is the large-scale sliding of the ground. This phenomenon is favored when there are continuous shots on the wall without any preservation or a reliable shoulder due to the protection of the pit fronts [20]. Analyses have shown that the fracture plane of the pit wall must be traversed (undercut) in the slope. In other words, the inclination of the fracture plane must be less than the inclination of the pit slope. This is not the case for the M'HAOUDATT pit. The presence of some detached rocks or irregularities on the joint surface may prevent sliding. The interpretation of structural measurements made at the mining company of the pit, showed that the joints at the origin of the majority of risks of sliding are the joints foliated and superimposed on fault families [21]. The discontinuity families forming the problem zone are present in the end walls along contours whose dips and friction angles are difficult to control. The planar rupture created by the families subparallel to the wall of the M'HAOUDATT pit causes problems of slope stability due to its alteration and dip towards the pit [22]. First of all, the slopes are greater than the dips of the bedding, which leads to the overloading of the soil to the point of significant planar rupture instabilities. These ruptures are mainly located on the pit wall. These slides are caused by:

- The slopes of the discontinuities are lower than those of the fronts.
- Shooting on the bedding generates a loss of cohesion of the discontinuities which causes the detachment of blocks.
- The absence of safety berms at the planned levels.
- The non-control of product loading creates zones favorable to instability.

The realization of berms increases the critical factor of safety, by decreasing the effect of the maximum shear deformation, and the crossing of the slopes of discontinuities accelerates the planar mode of failure, by creating the effect of the stress of the mass of the voids for the detachment of blocks between the upper levels. Experience has shown that when the horizontal joint is inside the void, a planar-type failure can also occur, although the average case is impossible. From this regression analysis, we can affirm that the principal compressive stresses increase with depth according to a linear relationship. This increase as the pit deepens requires us to prepare horizontal catchment platforms to absorb the stress field in order to avoid its critical influence on the stability of the fronts, which is not the case for the M'HAOUDATT pit wall. As the pit deepens, the effect of the stresses becomes a source of risk for the pit walls.

The absence of safety benches contributes to the transformation of this risk to failure mechanisms that are difficult to control. Large-scale pits that are operated by the push-back system often suffer from this type of risk due to the effect of insufficient temporary pit widths and the additional strip ratio [23].

Field observations have shown that gravity is involved in the majority of reported failures at the M'HAOUDATT pit, and this stress in the rock masses is not systematically measured for slopes and its effects are largely unknown. The weight of the mass exerted on the abrupt orientation of the pit wall bedding generates damage to the stratification of the slopes and consequently to the slope angles reached, which are different from those expected. Horizontal stress may tend to slightly decrease stability and reduce the depth of significant shear compared to a hydrostatically stress condition. The explanation is that the lower horizontal stress causes a slight decrease in normal stress on potential shear surfaces and/or joints within the slope. If this is not achieved, instabilities can occur due to the weight of the pile on the pit fronts.

The embankment load exerted on the pit fronts, whose lithology is composed of low hardness rocks, participates in the development of the shear stress and consequently in the appearance of fractures on a large opening scale [24]. The most significant problem for the stability of the pit wall slopes is the large-scale toppling. This effect will be favored when there is a continuous shot on the wall without any preservation or a reliable shoulder following the alteration, the blasting and the bad cleaning of the berms. The interpretation of the structural measurements carried out on the pit showed that the joints that cause the majority of the sliding risks are the foliation and the stratification joints with the diaclases families. It was previously proven during this study that the discontinuity families forming the problematic zone are present in the final walls following contours that are difficult to control their dips and their friction angle. Another factor that dramatically increases the number of rupture possibilities is the application of hydraulic pressure [25].

6 Conclusions

The natural fracturing state is difficult to interpret, due to the complexity and the change of direction with depth. Our concern in this analysis paper is to point out the risk of neglecting the orientation of the discontinuities in the deposit and the impact of the voids created by the exploitation as well as the multiple orientations encountered. The systematic structural study with the deepening of the pit should have led to the identification of some problems related to the instability of the slopes.

It is always recommended not to neglect small instabilities, and to extend the investigations on the state of fissuring of the ground to raise the probable causes of the instabilities of the slopes in order to avoid or to prevent possible movements of larger scale. The objective of this research work was to evaluate the stability of the slopes in a pit which does not present safety structures by taking the M'HAOUDATT pit as an example. The rock mass of the pit is characterized by soft and relatively heterogeneous rock. The intact rock present at the MHAOUDATT pit has a low compressive strength. Poor mining practices generate instability phenomena that affect the working fronts according to numerical simulations. Another point for further study in this article would be the characterization of the natural stresses present around the wall of the M'HAOUDATT mine. The presence of compression on the eastern wall could develop warping. The deformation mechanism and deformation behavior of a slope varies with geology, structural geological conditions, slope geometry, rock mass properties, groundwater conditions and other transient environmental influences. It is therefore important to characterize the natural stresses present in M'HAOUDATT to determine its magnitude and orientation. This should be included in a more detailed study to fully assess the state of stresses affecting the M'HAOUDATT massif.

References

1. Wang, Z., Lin, M.: Finite element analysis method of slope stability based on fuzzy statistics. Earth Sci. Res. J. **25**(1), 123–130 (2021). https://doi.org/10.15446/esrj.v25n1.93320
2. Zhang, H., et al.: Study on stability of permafrost slopes during thawing. Res. Cold Arid Reg. **14**(5), 293–297 (2022). https://doi.org/10.1016/j.rcar.2022.12.004
3. Takahashi, Y., et al.: Study on prediction of ground vibration in consideration of damping effect by fragment in the rock mass. J. Geosci. Environ. Protect. **6**(6), 1–11 (2018). https://doi.org/10.4236/gep.2018.66001
4. Zhao, J., et al.: A further study of P-wave attenuation across parallel fractures with linear deformational behavior. Int. J. Rock Mech. Min. Sci. **43**(5), 776–788 (2006). https://doi.org/10.1016/j.ijrmms.2005.12.007
5. Yang, J.H., Jiang, Q.H., Zhang, Q.B., Zhao, J.: Dynamic stress adjustment and rock damage during blasting excavation in a deep-buried circular tunnel. Tunnel. Undergr. Space Technol. **71**, 591–604 (2018). https://doi.org/10.1016/j.tust.2017.10.010
6. Hemed, A., Ouadif, L.: Pre-split performance evaluation. JP J. Heat Mass Transf. **31**(1), 19–31 (2023). https://doi.org/10.17654/0973576323002
7. Yingguo, H., Wenbo, L., Xinxia, W., Liu, M., Li, P.: Numerical and experimental investigation of blasting damage control of a high rock slope in a deep valley. Eng. Geol. **237**, 12–20 (2018). https://doi.org/10.1016/j.enggeo.2018.01.003
8. Sainoki, A., Mitri, H.S.: Instantaneous stress release in fault surface asperities during mining-induced fault-slip. J. Rock Mech. Geotech. Eng. **8**(5), 619–628 (2016). https://doi.org/10.1016/j.jrmge.2016.05.003
9. Zhu, W.B., Xu, J.M., Xu, J.L., Chen, D.Y., Shi, J.X.: Pier-column backfill mining technology for controlling surface subsidence. Int. J. Rock Mech. Min. Sci. **96**, 58–65 (2017). https://doi.org/10.1016/j.ijrmms.2017.04.014
10. Driouch, A., Ouadif, L., Benjmel, K., Bhilisse, M., Ilmen, S.: Determining the regional tectonic stress field by remote sensing in the Bou Azzer inlier, Central Anti-Atlas, Morocco. Min. Miner. Depos. **16**(2), 49–54 (2022). https://doi.org/10.33271/mining16.02.049
11. El Janati, M., Soulaimani, A., Admou, H., Youbi, N., Hafid, A., Hefferan, K.P.: Application of ASTER remote sensing data to geological mapping of basement domains in arid regions: a case study from the Central Anti-Atlas, Iguerda inlier, Morocco. Arab. J. Geosci. **7**(6), 2407–2422 (2013). https://doi.org/10.1007/s12517-013-0945-y

12. Chimmani, K.V., Lokhande, R.D.: Study the behaviour of underground oil cavern under static loading condition. Geotech. Geol. Eng. **40**, 995–1007 (2022). https://doi.org/10.1007/s10706-021-01939-0
13. Lin, F., Ren, F., Luan, H., Ma, G., Chen, S.: Effectiveness analysis of water-sealing for underground LPG storage. Tunn. Undergr. Space Technol. **51**, 270–290 (2016). https://doi.org/10.1016/j.tust.2015.10.039
14. Liu, J., Zhao, X.D., Zhang, S.J., Xie, L.K.: Analysis of support requirements for underground water-sealed oil storage cavern in China. Tunn. Undergr. Space Technol. **71**, 36–46 (2018). https://doi.org/10.1016/j.tust.2017.08.013
15. Corbyn, J.A.: Stress distribution in Laminar Rock during sliding failure. Int. J. Rock Mech. **15**(3), 113–119 (1978). https://doi.org/10.1016/0148-9062(78)90005-0
16. Zhipeng, X., Zhao, Z., Sun, J., Ming, L.: Determination of hydraulic conductivity of fractured rock masses: a case study for a rock cavern project in Singapore. J. Rock Mech. Geotech. Eng. **7**(2), 178–184 (2015). https://doi.org/10.1016/j.jrmge.2014.10.006
17. Öge, İ.F.: Assessing rock mass permeability using discontinuity properties. Proc. Eng. **191**, 638–645 (2017). https://doi.org/10.1016/j.proeng.2017.05.373
18. Xiong, F., Jiang, Q., Chaoshui, X., Zhang, X., Zhang, Q.: Influences of connectivity and conductivity on nonlinear flow behaviours through three-dimension discrete fracture networks. Comput. Geotech. **107**, 128–141 (2019). https://doi.org/10.1016/j.compgeo.2018.11.014
19. Alaoui, H.E., Moujahid, E., Ibouh, H., Bachnou, A., Babram, M.A., Harti, A.E.: Mapping and analysis of geological fractures extracted by remote sensing on Landsat TM images, example of the Imilchil-Tounfite area (Central High Atlas, Morocco). Estud. Geol. (Madrid) **72**(2), 12 (2016). https://doi.org/10.3989/egeol.42328.394
20. Adiri, Z., et al.: Comparison of Landsat-8, ASTER and Sentinel 1 satellite remote sensing data in automatic line-aments extraction: a case study of Sidi Flah-Bouskour inlier, Moroccan Anti Atlas. Adv. Space Res. **60**(11), 2355–2367 (2017). https://doi.org/10.1016/j.asr.2017.09.006
21. Driouch, A., Ouadif, L., Lahmili, A., et al.: Evaluation of the compression potential of serpentine rock masses of the Bou Azzer mining district in the central anti-Atlas of Morocco. Min. Metall. Explor. **39**, 189–200 (2022). https://doi.org/10.1007/s42461-021-00517-5
22. Farhadian, H., Nikvar-Hassani, A.: Development of a new empirical method to tunnel squeezing classification (TSC). Q. J. Eng. Geol. Hydrogeol. **53**(4), 655–660 (2020). https://doi.org/10.1144/qjegh2019-108
23. Driouch, A., Ouadif, L., Lahmili, A., Belmi, M.A., Benjmel, K.: Geotechnical modeling of the method for mining cobalt deposits at the Bou Azzer Mine, Morocco. Min. Miner. Depos. **17**(1), 51–58 (2023). https://doi.org/10.33271/mining17.01.051
24. Azadeh, A., Osanloo, M., Ataei, M.: A new approach to mining method selection based on modifying the Nicholas technique. Appl. Soft Comput. **10**(4), 1040–1061 (2010). https://doi.org/10.1016/j.asoc.2009.09.002
25. Soufi, A., Bahi, L., Ouadif, L.: Contribution of geostatistical analysis for the assessment of RMR and geomechanical parameters. ARPN J. Eng. Appl. Sci. **13**(24), 9359–9374 (2018). https://www.semanticscholar.org/paper/CONTRIBUTION-OF-GEOSTATISTICAL-ANALYSIS-FOR-THE-OF-Soufi-Bahi/03ff56e3481619a6d988facd28d91dbc1f52f558

Analyzing the Influence of Bracing Types on the Overall Displacement of Reinforced Concrete Buildings

Yassine Razzouk, Khadija Baba(✉), Mohamed Ahatri, and Ahlam El Majid

Civil Engineering and Environment Laboratory (LGCE), Mohammadia School of Engineering, Mohammed V University, Rabat, Morocco
khadija.baba@est.um5.ac.ma

Abstract. For several decades, bracing systems have been employed to enhance the structural integrity of buildings and protect them against seismic risks. These systems are designed to provide the necessary rigidity and ductility based on the specific characteristics of the structure and the construction location, particularly in regions with high seismic activity, such as the northern part of Morocco. The objective of this study is to conduct a comprehensive analysis and comparison of three types of bracing systems: columns, shear walls, and combined bracing. The analysis considers both a standardized earthquake scenario defined by the Moroccan seismic regulation (RPS 2000 revised in 2011) and a real earthquake recorded at the ZGH station near Nador, Morocco. To examine the performance of these bracing systems, structures with varying numbers of stories (2-story, 4-story, 6-story, 8-story, and 10-story) are analyzed using finite element software. This software enables the simulation of structural behavior and response under seismic loads, allowing for a detailed evaluation of the effectiveness and performance of each bracing type. By studying the behavior of these bracing systems under both standardized and real earthquake events, this research aims to provide valuable insights into their suitability, effectiveness, and limitations. The findings will contribute to the understanding of optimal bracing system selection and design considerations for buildings in seismically active areas.

Keywords: Earthquake · Bracing system · Seismic analysis · Spectral analysis · Response spectrum

1 Introduction

Different load-bearing systems, including columns, shear walls, and combined systems, can be utilized in earthquake-resistant construction, provided that appropriate construction provisions are followed. However, their behavior varies when subjected to seismic events [1, 2]. The implementation of earthquake-resistant construction techniques is crucial in safeguarding structures and human lives from the devastating impact of earthquakes. Seismic forces generated by earthquakes can cause severe damage to buildings, leading to structural failure, injuries, and loss of life. Bracing systems, such as columns,

K. Baba et al. (Eds.): GeoME 2023, SPEES, pp. 290–304, 2023.
https://doi.org/10.1007/978-3-031-49345-4_28

shear walls, and combined systems, are specifically designed to enhance a structure's ability to withstand the horizontal forces during an earthquake. These systems distribute and dissipate these forces, reducing structural deformation and minimizing the risk of catastrophic failures. Neglecting earthquake-resistant construction practices can result in significant property damage, economic losses, and, most importantly, the loss of human lives. Therefore, a thorough study of different bracing systems is essential for engineers and architects to make informed decisions, ensuring the safety and resilience of buildings in seismically active regions. While earthquake-resistant construction provisions improve the seismic resistance of various structures, their efficacy can vary [3, 4]. It is therefore important to make relevant choices regarding the structural system during the architectural design phase. The selection of the appropriate bracing system, whether it is a column, shear wall, or combined system, is influenced by architectural, functional, and technical considerations [5–7].

This study aims to analyze the three types of bracing systems using multiple analysis methods, as mentioned in the abstract [8]. For the evaluation of real earthquake conditions, the Mediterranean earthquake that occurred on January 25th, 2016, recorded at the ZGH station near Nador, in the seismic-prone northern region of Morocco, will be utilized [9]. This area is known for its significant seismic activity [10] and has experienced devastating earthquakes throughout history [11, 12]. Dynamic studies were conducted for each bracing type, considering both seismic and spectral analysis [13]. It was observed that shear wall bracing offers higher security and is more suitable for high-rise buildings, while column bracing is more economically viable for structures of lesser height.

By studying the behavior and effectiveness of different bracing systems, this research contributes to the knowledge base of engineers and architects, aiding them in making well-informed decisions for earthquake-resistant construction. The findings highlight the importance of selecting appropriate bracing systems to ensure the structural integrity and safety of buildings in seismic-prone areas.

2 Bracing Systems

Bracing systems are an integral component of earthquake-resistant construction. They are specifically designed to enhance the structural integrity of buildings and minimize damage caused by seismic activity. Bracing systems provide additional rigidity and ductility to structures, enabling them to withstand the lateral forces exerted during an earthquake. There are various types of bracing systems available, including column bracing systems, shear wall bracing systems, and mixed bracing systems. Each system offers unique advantages and characteristics that make them suitable for different types of structures and seismic risk profiles. By implementing the appropriate bracing system, engineers and architects can significantly improve the resilience and safety of buildings in areas prone to earthquakes.

The bracing systems analyzed in this study fall into the following categories.

2.1 Column Bracing Systems

Column bracing systems utilize reinforced concrete columns, with or without masonry in-fill, as the primary bracing elements. Columns, being vertical structural elements, bear the vertical loads of a building and transmit them to the foundation. In earthquake-resistant construction, columns play a crucial role in withstanding the lateral forces generated during seismic activity. By providing vertical support, columns ensure the stability and integrity of a structure, minimizing excessive movement and deformation [14, 15]. In this case, the bracing is solely provided by the columns, which bear both the vertical and horizontal loads [16].

2.2 Shear Wall Bracing System

Shear wall bracing systems consist of vertical, rigid walls integrated within a building's structural framework. These walls, designed to resist lateral forces, transfer them to the foundation. Shear walls effectively distribute seismic loads throughout the structure, reducing displacement and strain. They enhance the stiffness and strength of the building, improving its resistance to earthquakes. This system can consist of shear walls alone or a combination of shear walls and columns. When shear walls and columns are combined, the shear walls bear at least 20% of the vertical load solicitations [17]. It is assumed that horizontal forces are solely resisted by shear walls.

2.3 Mixed Bracing System

Mixed bracing systems combine elements of both columns and shear walls to enhance seismic resistance. This approach harnesses the strengths of both systems to optimize structural performance. Combined systems may employ columns for vertical support while strategically incorporating shear walls to enhance lateral stability and mitigate seismic forces. By integrating these two systems, engineers can achieve a synergistic effect and design structures better equipped to withstand earthquakes. In this case, shear walls bear a maximum of 20% of the vertical load solicitations [18].

The comparative study of these three bracing systems aims to evaluate their effectiveness in minimizing structural displacement and ensuring overall stability during seismic events. Understanding the behavior and performance of each system enables engineers and architects to make informed decisions when selecting the appropriate bracing system for a specific building design and the corresponding seismic risk profile.

3 Moroccan Seismic Construction Regulations (RPS) 2000 Revised in 2011

The seismic construction regulations in Morocco, known as the RPS 2000 revised in 2011, provide criteria for evaluating the impact of earthquakes on structures. These criteria consider factors such as maximum soil acceleration, maximum soil velocity, site influence, and response spectrum of vertical motion.

One of the key parameters is the maximum soil acceleration, denoted as Amax, which represents the maximum acceleration experienced by the soil during an earthquake [19, 20]. The values of Amax vary depending on the seismic zone, with higher values indicating greater seismic activity, as shown in Table 1. Similarly, the maximum soil velocity, represented as Vmax, is another important factor that measures the maximum velocity of the soil during an earthquake, as shown in Table 2.

Table 1. The maximum soil acceleration for each seismic zone [16]

Seismic zone acceleration (Za)	A max (%g)
0	4
1	7
2	10
3	14
4	18

Table 2. The maximum soil velocity for each seismic zone [16]

Seismic zone velocity (Zv)	V max (m/s)
0	0.05
1	0.07
2	0.10
3	0.13
4	0.17

The site influence is also considered in the regulations, taking into account the type of soil and the local geological and geotechnical conditions. Different site classes, such as rock, very dense soil and soft rock, stiff soil, soft soil, and special conditions, are assigned specific coefficients based on their influence on earthquake intensity, as shown in Table 3.

Additionally, the regulations specify a response spectrum of vertical motion, which accounts for the vertical movement of the ground during an earthquake. This spectrum complements the horizontal response spectrum and provides a comprehensive understanding of the seismic forces acting on a structure.

Table 3. The site coefficient [16]

Seismic zone velocity (Zv)	V max (m/s)	V max (m/s)
S1	Rock	1
S2	Very dense soil and soft rock	1.2
S3	Stiff soil	1.4
S4	Soft soil	1.8
S5	Special conditions	Determined by an expert

While there are other characteristics involved, the maximum acceleration and response spectrum are considered sufficient for the application of the Moroccan seismic regulations. These criteria provide guidelines for assessing and designing structures to withstand seismic forces in accordance with the RPS 2000 revised in 2011.

4 Study Data

4.1 Normalized Earthquake Following the Moroccan Seismic Construction Regulations RPS 2011

The study area, the town of Nador, falls within seismic zoning 4 (Z4) according to the Moroccan seismic regulations. This zone is characterized by a maximum soil acceleration (Amax) of 0.18 g and a maximum soil velocity (Vmax) of 0.17 m/s. The structural ductility is defined as ND1, indicating that the seismic response of the structure should remain within the elastic domain. The behavior factor of the structure depends on the type of bracing, as indicated in Table 4.

Table 4. Behavior factor values for concrete buildings [16]

Type of bracing	ND1	ND2	ND2
	Behavior factor		
Columns	2	3.5	5
Shear walls and columns	2	3	4
Shear walls	1.4	2.1	2.8

* *ND* Level of ductility

4.2 Earthquake Data—ZGH Station

The earthquake that occurred on January 25, 2016, had an impact on the Mediterranean region, specifically the coasts of Nador and El Hoceima. This earthquake, with a moment magnitude of 6.3, had an epicenter located at coordinates (35.586, − 3.690) and a depth of 26 km. The United States Geological Survey (USGS) initially assessed the earthquake as having a low probability of loss and damage (see Fig. 1). The selection of the Mediterranean earthquake of January 25th, 2016, as a case study holds significant importance for studying earthquake-resistant construction. This real-life event allows researchers to analyze the performance of different bracing systems under actual seismic conditions. By examining the effects of this earthquake on structures, valuable insights can be gained regarding the effectiveness of bracing systems in mitigating damage and enhancing structural resilience.

Furthermore, providing an overview of the seismic activity in the region reinforces the necessity of this study. The Mediterranean region is known for its susceptibility to earthquakes due to its location along tectonic plate boundaries. The occurrence of earthquakes in this area highlights the urgency of implementing robust earthquake-resistant construction practices. Understanding the specific characteristics of seismic activity in the region, including frequency, intensity, and potential impact, enables engineers to tailor their design strategies and select appropriate bracing systems to address the unique challenges posed by regional seismicity.

In conclusion, by selecting the Mediterranean earthquake of January 25th, 2016, as a case study and providing an overview of the seismic activity in the region, this research gains practical relevance and significance. It allows for the evaluation of bracing systems in a real-life context while addressing the specific seismic challenges encountered in the Mediterranean region. Ultimately, this study contributes to the advancement of earthquake-resistant construction practices, assisting engineers in designing safer and more resilient structures in seismically active areas.

4.3 Analysis of Response Spectrum

To analyze the response spectrum of the Mediterranean earthquake, a custom MATLAB program was created. The program utilizes the time records obtained from the earthquake event [21] to solve the ordinary differential equation and generate the response spectra. The response spectra, depicted in Fig. 2, provide valuable information about the structural response to different frequencies and amplitudes of ground motion. By studying the response spectra, engineers can gain insights into the dynamic behavior of structures and make informed decisions regarding their design and performance under seismic conditions.

4.4 Structural Design and Modeling

To facilitate a meaningful comparison between different bracing types, the design of the studied structures was carefully executed in accordance with the Moroccan seismic regulation (RPS 2011 [16]). The regulation emphasizes the importance of appropriately placing bracing elements during the design phase. The recommendations of the RPS were followed in the structural designs shown in Fig. 3.

The Moroccan seismic regulation specifies that, at each story, the distance between the center of mass and the center of rigidity, measured perpendicularly in the direction of seismic action, should not exceed 0.20 times the square root of the ratio of torsional

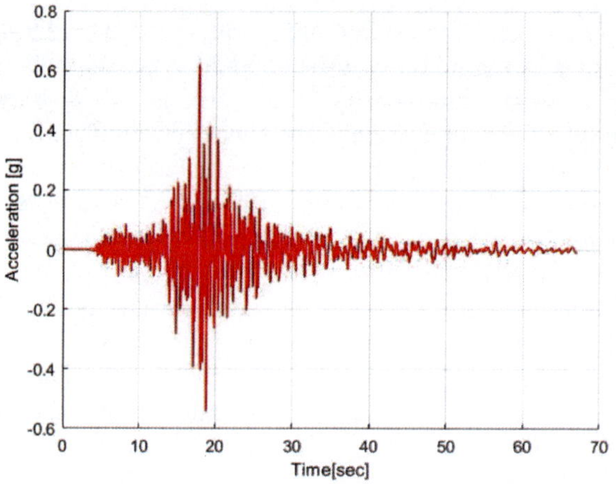

Fig. 1. Acceleration of the 2016 south Alboran earthquake

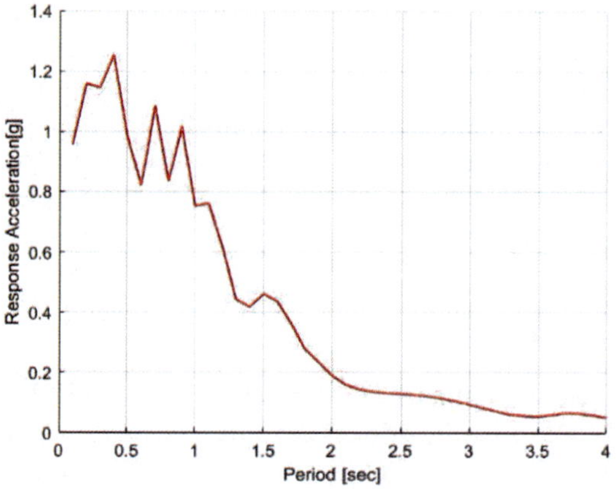

Fig. 2. Spectral acceleration (SA)

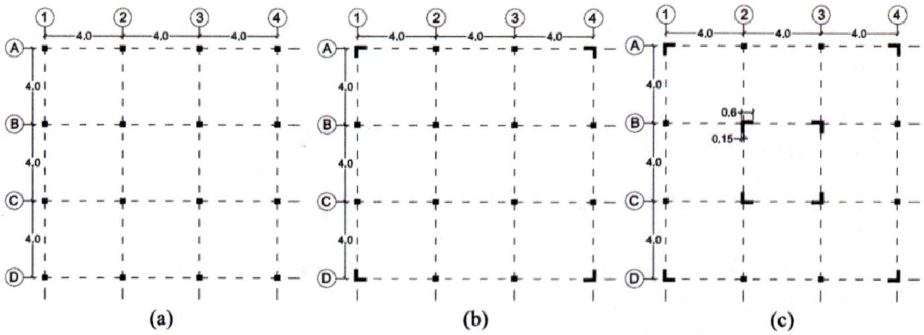

Fig. 3. Structural design of (a) column bracing (b) combined bracing (c) shear wall bracing

stiffness to translational stiffness [16].

$$r^2 = \frac{torsion\ stiffness}{translation\ stiffness}$$

$$r_x^2 = \frac{\sum I_{xi}.x_i^2 + \sum J_{yi}.y_j^2}{\sum I_{xi}} \text{ and } r_y^2 = \frac{\sum I_{xi}.x_i^2 + \sum J_{yi}.y_j^2}{\sum J_{yi}}$$

The models were designed to ensure that the center of mass and the center of rigidity coincide. Additionally, a thorough assessment of the distribution of vertical forces on columns and shear walls was performed to validate the selection of the specific bracing type for each model [4] (Table 5).

Table 5. Force distribution on the columns and shear walls

Design	*Fz (T)	Fz (T) on the columns	Fz (T) on the shear walls	% Fz (T) on the shear walls	Bracing type
Design (a)	106.56	106.56	0.00	0.00	Column
Design (b)	110.09	92.23	17.85	16.21	Combined
Design (c)	113.61	43.75	69.86	61.49	Shear wall

* Fz (T) vertical forces in tons

The designs maintained plane symmetry and elevation continuity, as prescribed by the regulation [16].

In this study, ETABS software was employed as the primary tool for structural analysis and modeling. ETABS, which stands for "Extended Three-Dimensional Analysis of Building Systems," is a highly regarded software package specifically developed for the analysis and design of building structures [22, 23]. It offers a wide range of features and capabilities for simulating and evaluating the behavior of structures under various loading conditions, including seismic forces. Leveraging the advanced functionalities of ETABS, we conducted finite element analysis on structures of varying heights, ranging from 2 to 10 floors, to investigate the performance of different bracing systems. The utilization of ETABS software enabled us to obtain accurate and detailed results, providing valuable insights into the response and behavior of the examined structures during seismic events. Figure 4 showcases the models of different bracing types generated using the ETABS Structure software

4.5 Material and Geometry

Multiple models were designed to address our research objectives, divided into two parts. In the first part, we adhered to the minimum section requirements specified by the Moroccan seismic regulation, utilizing column sections of 25 cm × 25 cm and shear walls with a thickness of 15 cm. In the second part, we adjusted the sections of load-bearing elements to ensure that the models remained within acceptable limits and kept structural damages at an acceptable level.

The concrete structures in the various models exhibited the following characteristics:

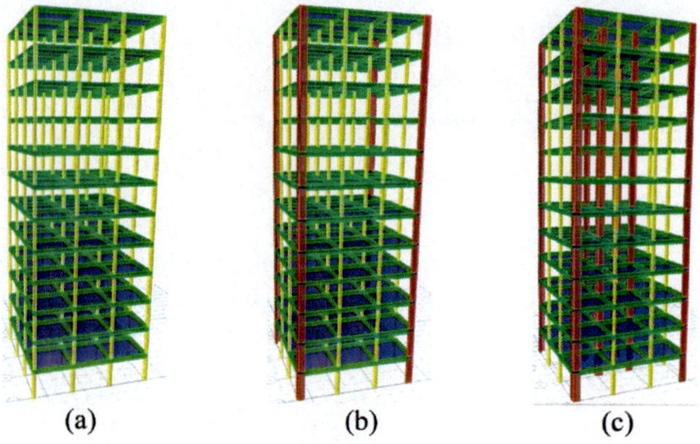

Fig. 4. ETABS 3D models of (a) column bracing (b) combined bracing (c) shear wall bracing

- Column cross-sections ranged from the minimum section of 25 cm × 25 cm to larger sections up to 100 cm × 100 cm.
- Shear wall thickness varied from 15 to 25 cm.

For the remaining elements used in the design, the following characteristics were adopted for all models:

- Beam cross-section: 25 cm × 45 cm.
- Thickness of the solid slab: 20 cm.
- Story height (h): 3 m.
- Concrete rigidity value (E): 32,000 MPa.
- Poisson coefficient: 0.2 (considered in the calculations).
- Specific weight: 2.5 t/m^3.

These design considerations, in terms of materials and geometry, ensure the representation of realistic structural configurations and provide a solid basis for evaluating the performance of different bracing systems in the subsequent analyses.

5 Results and Discussion

5.1 Global Lateral Displacement

The quantification of structural protection against damage in seismic areas is primarily determined by the magnitude of global lateral displacements. These displacements play a crucial role in the design and sizing of structures to ensure their resilience. In our study, we initially considered the minimal sections required by the RPS 2011 regulation [16], while varying the bracing type and the building's story height, ensuring compliance with the regulation's recommendations. Subsequently, we conducted seismic and spectral analyses for each model to assess the impact of the bracing type on the spectral and seismic response.

Figure 5 illustrates a notable distinction between the seismic analysis based on RPS 2011 and the spectral analysis based on the ZGH station accelerogram, highlighting

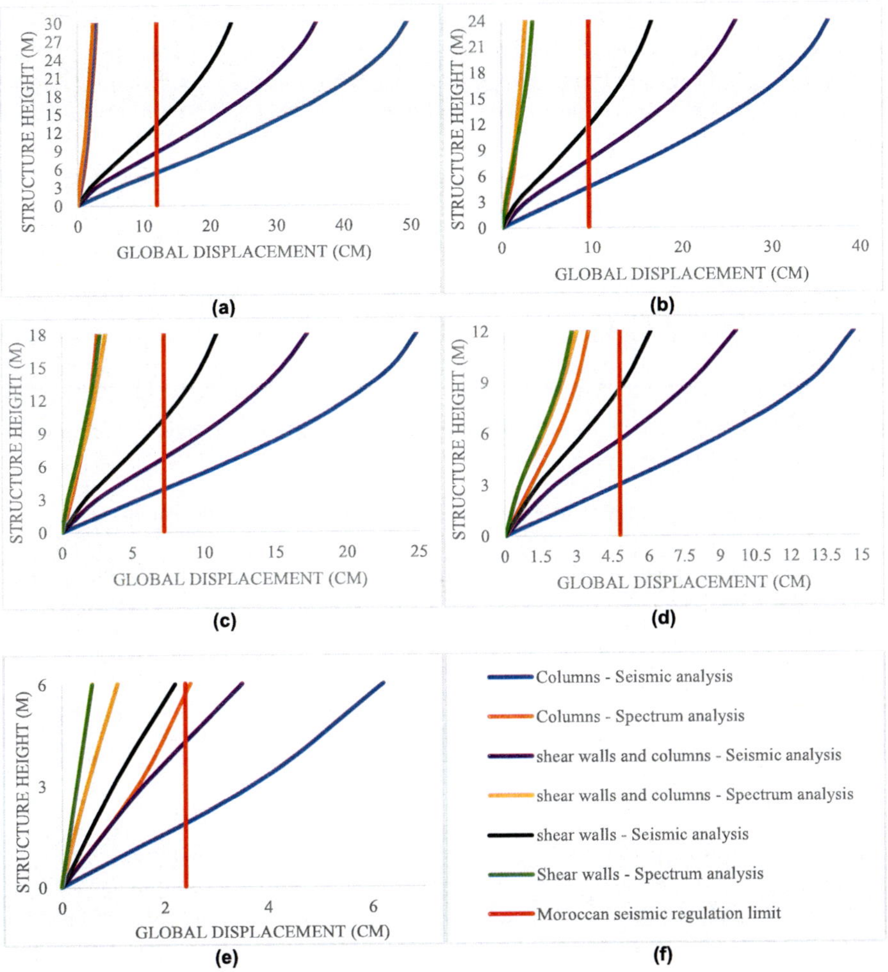

Fig. 5. Global displacements according to both "seismic and spectral" analysis of (a) 10-story buildings (b) 8-story buildings (c) 6-story buildings (d) 4-story buildings (e) 2-story buildings (f) legend

the stringent safety provisions of the RPS 2011 regulation. Across buildings of different heights, it is observed that shear wall bracing offers greater safety than combined bracing, while combined bracing is safer than column bracing.

Furthermore, the graphs demonstrate that the displacements obtained from the spectral analysis fall within the regulation's limits. However, when considering the seismic analysis, taller buildings (above 7 stories) exhibit displacements within the limits only up to the 4th story for shear walls, up to the 3rd story for combined bracing, and up to the 2nd story for columns. In the case of medium and low-height buildings (below 7 stories), the displacements are within the limits up to the 3rd story for shear walls, up to the 2nd story for combined bracing, and at the 1st story level for columns.

These findings highlight the varying performance of different bracing systems in managing structural displacements during seismic events. It emphasizes the importance of selecting an appropriate bracing type based on the building's height and the desired level of structural safety, as defined by the regulatory requirements. By understanding and considering these displacements, engineers and designers can make informed decisions regarding the adoption of suitable bracing systems to ensure the structural integrity and safety of buildings in seismic areas.

5.2 Comparison of Structural Masses

To enhance the precision of our analysis, we made adjustments to the dimensions of the load-bearing elements to ensure that the overall displacements of the structures comply with the displacement limit specified in RPS 2011, which corresponds to 0.4% of the total height. The results of these modifications are presented in Tables 6, 7, 8, 9 and 10.

Table 6. Mass before and after verification of 10-story buildings

Bracing type	Total initial displacement (cm)	Initial mass (kg)	Total displacement after verification (cm)	Verified structure mass (kg)
Column	49.3	6,163,162.02	12	9,470,965.6
Combined	35.8	6,233,700.48	11.8	8,114,225.84
Shear wall	23.3	6,304,238.94	11.7	6,856,540.08

Table 7. Mass before and after verification of 8-story buildings

Bracing type	Total initial displacement (cm)	Initial mass (kg)	Total displacement after verification (cm)	Verified structure mass (kg)
Column	36.4	4,930,529.61	9.4	6,366,512.58
Combined	26	4,986,960.38	9.5	5,611,300.81
Shear wall	16.6	5,043,391.15	9.6	5,377,173.15

Tables 6, 7, 8, 9 and 10 provide the initial mass of the structures in kilograms, considering minimal dimensions and structure masses that meet the regulatory limit of RPS 2011 for global displacements. Figure 6 summarizes these findings, illustrating the relationship between the bracing type and the required mass to comply with the regulations.

Our analysis reveals that for buildings taller than 21 m (7 stories and above), choosing shear wall bracing is both suitable and cost-effective. The global displacements remain within the limit with a moderate increase in mass, ranging from 6 to 8% compared to the initial mass. In contrast, the other types of bracing result in a higher mass increase,

Table 8. Mass before and after verification of 6-story buildings

Bracing type	Total initial displacement (cm)	Initial mass (kg)	Total displacement after verification (cm)	Verified structure mass (kg)
Column	24.8	3,697,897.21	7	4,216,580.02
Combined	17.2	3,740,220.29	7.1	3,986,054.33
Shear wall	10.9	3,782,543.36	6.9	3,941,029.78

Table 9. Mass before and after verification of 4-story buildings

Bracing type	Total initial displacement (cm)	Initial mass (kg)	Total displacement after verification (cm)	Verified structure mass (kg)
Column	14.7	2,465,264.81	4.7	2,645,363.01
Combined	9.7	2,493,480.19	4.8	2,608,743.04
Shear wall	6.1	2,521,695.58	4.8	2,579,327

Table 10. Mass before and after verification of 2-story buildings

Bracing Type	Total initial displacement (cm)	Initial mass (kg)	Total displacement after verification (cm)	Verified structure mass (Kg)
Column	6.2	1,232,632.4	2.4	1,279,457.94
Combined	3.5	1,246,740.1	2.3	1,275,555.81
Shear wall	2.2	1,260,847.79	2.2	1,260,847.79

ranging from 11 to 34%. For buildings with heights between 15 and 21 m, both shear wall bracing and combined bracing demonstrate comparable adequacy. However, combined bracing offers additional advantages and more design flexibility, making it an attractive option. Lastly, for buildings with heights below 15 m (less than 5 stories), all three types of bracing yield similar building masses. Therefore, opting for column bracing is a prudent choice, as it provides architects with greater design freedom while imposing fewer constraints on engineers.

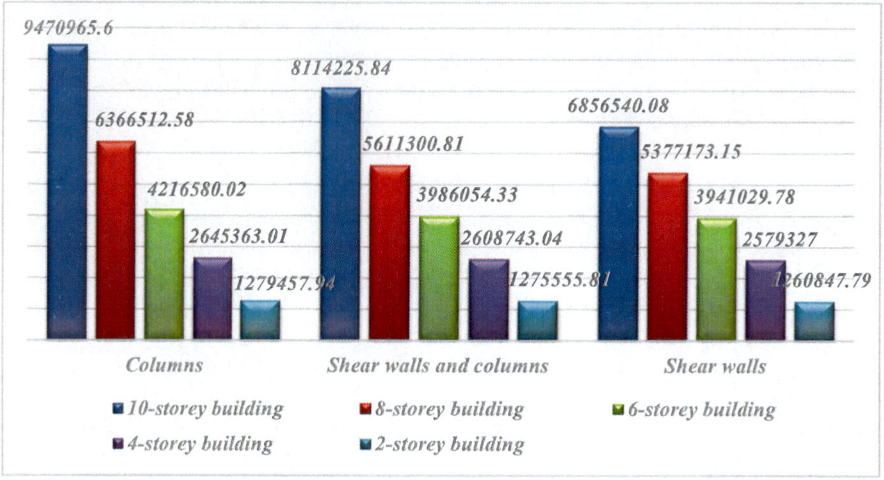

Fig. 6. Synthesis: influence of the type of bracing on weight of structure (kg)

6 Conclusion

The study focused on the selection of bracing systems for buildings of varying heights (2, 4, 6, 8, and 10 stories), considering compliance with the Moroccan seismic code. Two types of analyses were conducted: a seismic analysis based on the Moroccan seismic regulation RPS 2011 and a spectral analysis using the ZGH station accelerogram. A comprehensive analysis was carried out for all models, considering both the minimum sections specified in the regulations and a structural optimization study to compare different bracing systems and determine the most suitable option.

The key findings of the study are as follows:

- The RPS 2011 seismic analysis demonstrated more rigorous criteria, indicating a higher level of safety compared to the spectral analysis.
- In seismic areas, even with the use of shear walls, it was found that buildings cannot exceed two stories when utilizing minimum sections (i.e., 25 * 25 cm sections for columns and a 15 cm thickness for shear walls).
- Columns as a bracing system exhibited higher vulnerability, as the global displacement exceeded the limit of the Moroccan regulation more quickly, specifically up to the second story.
- Shear wall bracing proved to be most suitable for taller buildings with heights exceeding 21 m, while combined bracing was more appropriate for medium-height buildings ranging from 15 to 21 m. Column bracing remained the preferred choice for buildings with heights below 15 m, offering advantages such as aesthetic appeal due to the absence of obstructions and the presence of larger openings.

These findings provide valuable insights for architects and engineers when selecting the optimal bracing system based on building height, ensuring both structural integrity and adherence to regulatory requirements.

References

1. Karsaz, K., Tosee, S.V.R.: A comparative study on the behavior of steel moment-resisting frames with different bracing systems based on a response-based damage index. Civ. Eng. J. **4**(6), 1354 (2018). https://doi.org/10.28991/cej-0309178
2. Afzali, A., Mortezaei, A., Kheyroddin, A.: Seismic performance of high-rise RC shear wall buildings subjected to ground motions with various frequency contents. Civ. Eng. J. **3**(8), 568–584 (2017). https://doi.org/10.28991/cej-2017-00000113
3. Elghazouli, A. (ed.): Introduction: seismic design and Eurocode 8. In: Seismic Design of Buildings to Eurocode 8, pp. 17–21. CRC Press (2016). https://doi.org/10.1201/9781482266177-8
4. Association F.S.: Earthquake resistant construction rules. In: Earthquake Resistant Rules Applicable to Buildings, Called Ps 92. Afnor 1995. Afnor Editions (1995)
5. Razzouk, Y., Ahatri, M., Baba, K., El Majid, A.: Optimal bracing type of reinforced concrete buildings with soil-structure interaction taken into consideration. Civ. Eng. J. **9**(6), 1371–1388 (2023). https://doi.org/10.28991/CEJ-2023-09-06-06
6. Razzouk, Y., Ahatri, M., Baba, K., Majid, A.E.: The impact of bracing type on seismic response of the structure on soft soil. CEA **11**(5), 2706–2718 (2023). https://doi.org/10.13189/cea.2023.110534
7. Razzouk, Y., Ahatri, M., Baba, K.: The impact of the seismic area on the bracing type choice of reinforced concrete buildings. J. Southwest Jiaotong Univ. **57**(6), 899–912 (2022). https://doi.org/10.35741/issn.0258-2724.57.6.77
8. Ahmad, B., Najar, I.A.: Comparative seismic analysis of EL Centro and Japan earthquakes using response spectra method. Int. J. Curr. Eng. Technol. **6**(5), 1859–1864 (2016)
9. Ahatri, M., Baba, K., Touijrate, S., Bahi, L.: Characteristics of spectral responses for a ground motion from mediterranean earthquake—ZEGHANGHANE Station (6.3Mw) in Morocco, and its influence on the structures. In: Diouri, A., Boukhari, A., Ait Brahim, L., et al. (eds.), MATEC Web Conferences, vol. 149, p. 02041 (2018). https://doi.org/10.1051/matecconf/201814902041
10. Cherkaoui, T.E.: Sismicité de La Région Taza-Al Hoceima-Taounate, Cas Du Séisme d'Al Hoceima Du 24 Février 2004 (2004)
11. Alami, S.O.E., Tadili, B.A., Cherkaoui, T.E., et al.: The Al Hoceima earthquake of May 26, 1994 and its aftershocks: a seismotectonic study. Ann. Geophys. **41**(4) (1998). https://doi.org/10.4401/ag-3801
12. Bezzeghoud, M., Buforn, E.: Source parameters of the 1992 Melilla (Spain, MW = 4.8), 1994 Alhoceima (Morocco, MW = 5.8), and 1994 Mascara (Algeria, MW = 5.7) earthquakes and seismotectonic implications. Bull. Seismol. Soc. Am. **89**(2), 359–372 (1999). https://doi.org/10.1785/BSSA0890020359
13. Razzouk, Y., Baba, K., Ahatri, M.: The influence of spectral responses on the structures heights: case of the Rhiss river earthquake in morocco (6.3 mw)-seismogenic source 4 (Rif oriental-Al Hoceima-Alboran). ARPN J. Eng. Appl. Sci. **17**(6), 645–651 (2022)
14. Simou, S., Baba, K., Akkouri, N., Lamrani, M., Tajayout, M., Nounah, A.: Mechanical characterization of the adobe material of the archaeological site of Chellah. In: Ameen, H., Jamiolkowski, M., Manassero, M., Shehata, H. (eds.) Recent thoughts in geoenvironmental engineering. Sustainable civil infrastructures, pp. 118–130. Springer International Publishing (2020). https://doi.org/10.1007/978-3-030-34199-2_8
15. Simou, S., Baba, K., Akkouri, N., Lamrani, M., Tajayout, M., Nounah, A.: Mechanical characterization and reinforcement of the adobe material of the Chellah archaeological site. In: Akhssas, A., Baba, K., Bahi, L., et al. (eds.) E3S Web Conference, vol. 150, p. 03022 (2020). https://doi.org/10.1051/e3sconf/202015003022

16. Housing M of, Policy U. Seismic Building Code RPS 2000—Version 2011 (2011)
17. Housing M of, Urbanism. Algerian Paraseismic Rules RPA 99/Version 2003. CGS (2003)
18. Trombly, B.: The International Building Code (IBC). CMGT 564-Term Paper. Published Online (2006)
19. El Majid, A., Cherradi, C., Baba, K., Razzouk, Y.: Laboratory investigations on the behavior of CBR in two expanding soils reinforced with plant fibers of varying lengths and content. Mater. Today Proc. S2214785323037884. https://doi.org/10.1016/j.matpr.2023.06.395
20. Majid, A.E., Baba, K.: Assessing the impact of plant fibers on swelling parameters of two varieties of expansive soil. Case Stud. Chem. Environ. Eng. **8**, 100408 (2023). https://doi.org/10.1016/j.cscee.2023.100408
21. Ahatri, M., Baba, K., Touijrate, S., Bahi, L.: The seismic motion parameters effects on response spectra: comparison between El Centro 1940 and imperial valley 1979 earthquakes. Int. J. Civ. Eng. Technol. (IJCIET)-Scopus Index. **9**(10), 1610–1618 (2018)
22. Syed, E.U., Manzoor, K.M.: Analysis and design of buildings using Revit and ETABS software. Mater. Today Proc. **65**, 1478–1485 (2022). https://doi.org/10.1016/j.matpr.2022.04.463
23. Abbas, D.H.S., Kharnoob, D.M.M., Atia, D.N.S., Bassam, D.B.F.: Analysis or design of composite column by using Etabs software. Webology. **19**(1), 5289–5301 (2022). https://doi.org/10.14704/WEB/V19I1/WEB19355

Comparing Design Methods and Code Structures: A Case Study Analysis

Yassine Razzouk[1](✉), Ali Azizi[2], Khadija Baba[1], and Mohamed Ahatri[1]

[1] Civil Engineering and Environment Laboratory (LGCE), Mohammadia School of Engineering, Mohammed V University, Rabat, Morocco
yassine.razzouk@gmail.com

[2] Department of Roads and Bridges, Hassania School of Public Works, Casablanca, Morocco

Abstract. The design of structures has evolved over time to reflect advances in engineering knowledge and technology. Early design methods were limited in scope and did not account for the complex behavior of structures under different load situations. The concept of limit states was introduced in the early twentieth century, providing a more rigorous framework for structural design. Load and resistance factor design (LRFD) and the Eurocode followed, providing comprehensive standards for structural design. More recently, performance-based design methods have emerged, taking into account the actual performance of structures under different load situations, and using computer models and simulations to predict their behavior. In this paper, the results of tests to failure of an experimental column were compared with those of calculations using the Working Stress Design (WSD), the Limit State Design (LSD) and the Eurocodes calculation. A security coefficient of 4.6–3.4 and 4.1 was obtained for the three design codes respectively. Since the column is not very slender and the centering and formwork uncertainties are reduced compared to columns made on site, the buckling effect can be neglected. The failure load of the experimental column was also calculated under this assumption. The results obtained are much closer to the experimental failure load. Furthermore, the best approximation was obtained in the case of Limit States Design followed by the code using the Eurocodes.

Keywords: Column · Comparison of code structures · Working stress design · Limit state design · Eurocode

1 Introduction

The approaches to structural design have undergone significant evolution over time, driven by advancements in engineering knowledge and technology. Initially, early design methods relied on empirical formulas derived from field observations of structural performance. These methods, though valuable in their time, had limited scope and failed to account for the intricate behavior of structures under various load situations.

In the early twentieth century, the concept of limit states revolutionized structural design. Limit state design (LSD) introduced a more rigorous framework that emphasized the need for structures to withstand loads up to a specific limit state, beyond which

K. Baba et al. (Eds.): GeoME 2023, SPEES, pp. 305–315, 2023.
https://doi.org/10.1007/978-3-031-49345-4_29

failure would occur. LSD considered both serviceability and ultimate limit states, and incorporated partial safety factors to address uncertainties related to loads, material properties, and other factors. This approach provided engineers with a comprehensive and reliable methodology for designing structures [1].

Building upon the principles of LSD, the United States introduced the concept of load and resistance factor design (LRFD) in the 1980s. LRFD shared the same fundamental principles as LSD but employed a different approach for determining design loads and capacities of structural elements. Instead of partial safety factors, LRFD utilized load and resistance factors. This method gained widespread acceptance not only in the United States but also in various parts of the world [2] (Figs. 1 and 2).

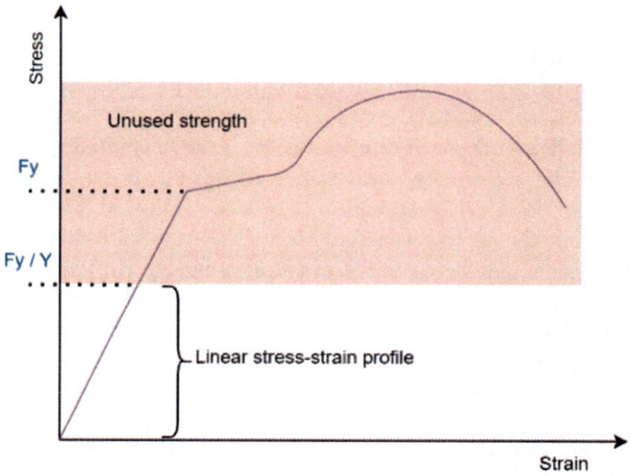

Fig. 1. WSD stress–strain main relationship [3]

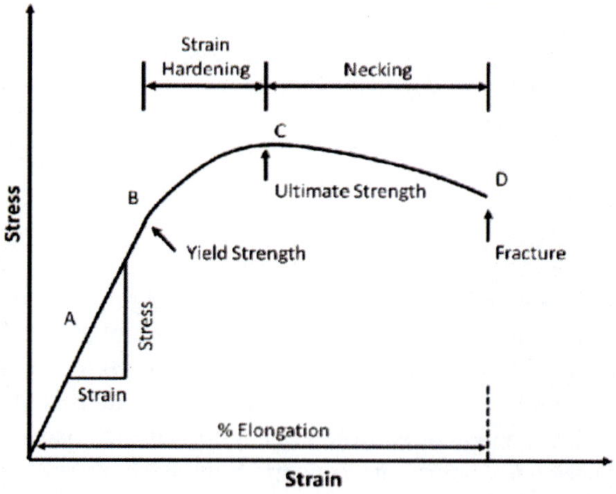

Fig. 2. LSD stress–strain main relationship [4]

In Europe, the 1990s saw the introduction of the Eurocode, a comprehensive set of standards for structural design. Based on the LSD approach, the Eurocode went a step further by considering a broader range of factors. These factors included fire resistance, fatigue, and seismic resistance, among others. The Eurocode also provided detailed guidance on the design of individual structural elements and incorporated provisions for utilizing new materials and construction techniques [5] (Fig. 3).

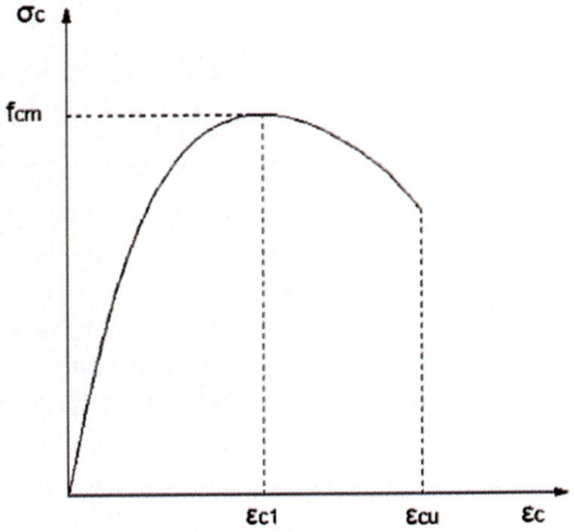

Fig. 3. Eurocode 2 stress–strain main relationship [6]

Advancements in computing power and simulation technology have opened new avenues for structural design. Performance-based design methods have emerged, taking into account the actual behavior of structures under different load situations. By utilizing computer models and simulations, performance-based design predicts how a structure will behave, offering a promising approach for the future of structural design and construction.

In this article, we will explore three distinct approaches to structural design: Working Stress Design (WSD), Limit State Design (LSD), and the Eurocode. WSD, an older method predominantly used in North America until the 1980s, considered only serviceability limit states and did not account for variability in material properties or diverse load situations. In contrast, LSD, a more advanced design method, addressed uncertainties by employing partial safety factors and accounted for both serviceability and ultimate limit states. The Eurocode, a comprehensive set of European standards, offered detailed guidance on structural design, incorporating the latest research and technological advancements. It surpassed LSD by considering additional factors such as fire resistance, fatigue, and seismic resistance.

Our particular focus in this article is the calculation of the resisting force of an experimental column [7] using Working Stress Design (WSD) [8], Limit State Design (LSD) [9], and the Eurocodes [10]. Subsequently, we compare the theoretical results to the experimental findings [7], with and without considering buckling. The obtained results demonstrate a significantly better agreement with the experimental failure load.

Moreover, the most accurate estimation was achieved when using Limit State Design, followed by the Eurocodes.

To summarize, as engineering knowledge and technology have advanced, the design approaches for structures have evolved. While the Working Stress Design is no longer widely used, Limit State Design and the Eurocode currently dominate structural design methodologies worldwide, particularly in Europe. These modern approaches provide enhanced reliability, comprehensive guidelines, and account for a broader range of factors, ensuring the safety and efficiency of structural designs.

2 Resistance Force of the Column with Consideration of Buckling

The resistance force of a column refers to the capacity of the column to withstand and resist external loads or forces acting upon it without undergoing significant deformation or failure. It represents the maximum force that the column can withstand before reaching its limit state and experiencing structural failure. The resistance force of a column is a crucial parameter in structural design as it determines the adequacy and safety of the column in supporting the intended loads and maintaining structural stability [11, 12]. It is typically determined through calculations and analysis based on the material properties, dimensions, and design parameters of the column, considering factors such as compressive strength, buckling resistance, and load distribution [13, 14]. The resistance force of a column is compared to the applied loads to ensure that the column can safely support the anticipated forces without compromising structural integrity [15].

In this article, we will proceed to calculate this resistance using the three codes mentioned earlier: Working Stress Design (WSD), Limit State Design (LSD), and Eurocode.

Calculating the resistance of the column according to Working Stress Design (WSD) involves utilizing an approach that primarily considers serviceability limits. This code, predominantly used in North America until the 1980s, incorporates safety factors for materials and load situations. By determining the maximum stress that the column can withstand under anticipated loads, the resistance force can be evaluated.

For Limit State Design (LSD), the calculation of the resistance force involves a more advanced methodology. LSD takes into account both serviceability and ultimate limit states, considering uncertainties in material properties, loads, and other factors. Partial safety factors are utilized to account for these uncertainties and ensure the structural integrity of the column.

In accordance with the Eurocode, the calculation of the resistance force includes a comprehensive analysis. The Eurocode provides detailed guidelines for structural design, incorporating a wide range of factors such as fire resistance, fatigue, and seismic resistance. By considering these additional factors, the Eurocode ensures a more robust assessment of the resistance force, resulting in a safer and more reliable design.

By applying the respective methodologies outlined in the WSD, LSD, and Eurocode, we will determine the resistance force of the column in this article. The calculated values will be compared to each other and to experimental results to assess the accuracy and effectiveness of each design approach. This analysis will provide valuable insights into the performance and applicability of these codes in predicting the resistance force of the column under different loading conditions.

2.1 Model with Working Stress Design

The column has a rectangular section 0.30 * 0.40 m^2, free length 3.00 m, supposed to be articulated at both ends and reinforced with six HA20 bars, as mentioned in Table 1.

$$N_r = 0.3(B + \eta A)f_{c28} \tag{1}$$

Table 1. The characteristics of the column

Parameter	Value
Cross-section	0.30×0.40 m^2
Free length	3.00 m
Reinforcement	HA20 bars

The resistant force Nr is calculated by:

B: Cross section of the column,
A: Longitudinal reinforcement section,
$f_e = 400\,\text{MPa}$: yield strength of the steel used
$f_{c28} = 25\,\text{MPa}$

$$N_r = 0.3\left(0.3 * 0.4 + 15 * 18.84 * 10^{-4}\right) * 25$$

$$N_r = 1.112\,MN$$

2.2 Model with Limit State Design

The resistant force N_r is calculated by:

$$N_r = \alpha(\lambda)\left(\frac{B_r f_{c28}}{0.9\gamma_b} + A_s\frac{f_e}{\gamma_s}\right) \tag{2}$$

$$\lambda = \frac{2\sqrt{3}l_0}{a} = \frac{2\sqrt{3} * 3}{0.3} = 34.64$$

$$\alpha(\lambda) = \frac{0.85}{1 + 0.2\left(\frac{\lambda}{35}\right)^2} = 0.71 \tag{3}$$

$$N_r = 0.71\left(\frac{0.28 * 0.38 * 25}{0.9 * 1.5} + 18.84 * 10^{-4} * \frac{400}{1.15}\right)$$

$$N_r = 1.864\,MN$$

2.3 Model with Eurocodes

N_r is given by:

$$N_r = \alpha(\lambda)K_b\left(A_c * f_{cd} + A_{sr} * f_{yd}\right) \tag{4}$$

$$f_{cd} = \frac{f_{c28}}{\lambda_b};\quad f_{yd} = \frac{f_e}{\gamma_s}$$

$$\lambda = \frac{2\sqrt{3}l_0}{a} = \frac{2\sqrt{3} * 3}{0.3} = 34.64$$

$$(\lambda) = \frac{0.86}{1 + \left(\frac{\lambda}{62}\right)^2} = 0.655 \tag{5}$$

$$K_b = (0.75 + 0.5a)\left(1 - 0.6\frac{A_{sl}}{A_c}\right) \tag{6}$$

$$a = 0.30\,\text{m}$$

$$K_b = (0.75 + 0.5 * 0.3)\left(1 - 0.6\frac{18.84 * 10^{-4}}{0.3 * 0.4}\right)$$

$$K_b = 0.8915$$

$$N_r = 0.656 * 0.8915\left(0.3 * 0.4 * \frac{25}{1.5} + 18.84 * 10^{-4} * \frac{400}{1.15}\right)$$

$$N_r = 1.551\,MN$$

3 Calculation of Ultimate Load Q

The ultimate load, denoted as Q, refers to the maximum external load that a structure or a structural component can sustain before experiencing failure or collapse. It represents the critical point at which the structure reaches its ultimate limit state, surpassing its capacity to resist the applied forces.

The ultimate load is determined through rigorous analysis and calculations, taking into account factors such as the material properties, dimensions, and design parameters of the structure or component. It considers various failure modes, such as yielding, buckling, or excessive deformation, depending on the type of structure and the nature of the load [16].

In structural design, it is essential to ensure that the ultimate load capacity of a structure or component is significantly higher than the anticipated applied loads. This safety margin, often referred to as the factor of safety, accounts for uncertainties, variations in loads, material properties, and other factors to ensure the structural integrity and safety of the design [17, 18].

Determining the ultimate load is crucial in assessing the structural stability and performance of a system. It provides valuable information for engineers and designers to make informed decisions regarding the structural adequacy and reliability of the design under anticipated loading conditions [19].

3.1 Calculation Using Working Stress Design

The ultimate load is calculated by:

$$Q = \frac{N_r - G}{1.2} \tag{7}$$

G represents the self-weight of the column

$$Q = \frac{1.112 - (0.3 * 0.4) * 25 * 10^{-3}}{1.2}$$

$$Q = 0.919\,MN$$

3.2 Calculation Using Limit State Design

$$Q = \frac{N_r - 1.35G}{1.5} \tag{8}$$

$$Q = \frac{1.864 - 1.350 * 0.3 * 0.4 * 3 * 25 * 10^{-3}}{1.5}$$

$$Q = 1.026\,MN$$

4 Safety Factor

The safety factor, also known as a factor of safety, is a numerical value used in engineering to account for uncertainties and ensure the reliability and safety of a design [12]. It represents the ratio of the ultimate load capacity or strength of a structure or component to the expected or design load. The safety factor is calculated by dividing the ultimate load capacity by the design load [20].

In the context of the Working Stress Design (WSD) method and Limit State Design (LSD) Eurocode, the safety factor is determined as part of the design process.

In the WSD method, the safety factor is applied by comparing the allowable stress in the material to the actual stress induced by the applied load. The allowable stress is typically derived from empirical formulas or industry standards, and the safety factor ensures that the actual stress remains below the allowable limit.

In the LSD Eurocode approach, the safety factor is incorporated through the use of partial safety factors. These factors account for uncertainties in material properties, loads, and other variables. By multiplying the characteristic loads with the appropriate partial safety factors, designers ensure that the design loads remain safely below the ultimate limit states of the structure.

The specific values of the safety factors used in WSD and LSD Eurocode may vary depending on the specific design requirements, regulations, and risk levels associated with the project. The determination of safety factors involves considering factors such as the consequences of failure, the reliability of material properties, and the level of uncertainty in the design process.

By calculating the safety factor using the WSD method and LSD Eurocode, engineers can assess the structural adequacy and ensure that the design provides an appropriate margin of safety to withstand anticipated loads and uncertainties. This helps to ensure the overall structural integrity and reliability of the design.

The safety factor noted C_s is calculated by:

$$C_s = \frac{Q_{expérimental}}{Q_{calculé}} \tag{9}$$

$Q_{expérimental} = 4.2\,MN$ [7].

C_s is equal to 4.6–3.4 and 4.1, calculated using the allowable stress design, limit state design, and Eurocode, respectively.

5 Calculating the Resisting Force, Neglecting Buckling

When calculating the resisting force of a structural element without considering buckling, several factors need to be taken into account. These factors include the material properties, dimensions, and design parameters of the element [21].

The resisting force can be determined by evaluating the maximum load that the element can withstand before failure occurs. This involves assessing the ultimate strength or capacity of the material and considering its behavior under applied loads. The specific calculations will depend on the design method being used, such as Working Stress Design (WSD), Limit State Design (LSD), or the Eurocode.

In the case of the WSD method, the resisting force can be calculated by comparing the applied load to the allowable stress or strength of the material. The allowable stress is determined based on empirical formulas or industry standards, and the resisting force is obtained by dividing the applied load by the calculated stress.

In the LSD approach, the resisting force is evaluated by considering both serviceability and ultimate limit states. Partial safety factors are employed to account for uncertainties, and the design loads are multiplied by these factors to determine the resisting force.

If the calculation is based on the Eurocode, the resisting force is determined by considering a broader range of factors, such as fire resistance, fatigue, and seismic resistance. The Eurocode provides detailed guidelines and equations that consider these additional factors, resulting in a more comprehensive assessment of the resisting force.

It is important to note that when neglecting buckling effects in the calculation, the obtained resisting force may not reflect the actual behavior of the structural element in real-world conditions. Buckling is a critical failure mode that needs to be considered, especially for slender or compression-loaded elements. Ignoring buckling effects can lead to underestimated load capacities and compromised structural safety [22].

5.1 Calculation Using Working Stress Design

$$N_r = (B + 15A)f_{c28} \tag{10}$$

The actual value obtained on the cylinder is used.
For f_{c28}, we take the actual value obtained on the cylinder [7], which is $f_{c28} = 31$ MPa.

$$N_r = \left(0.3 * 0.4 + 15 * 18.84 * 10^{-4}\right) * 31$$
$$N_r = 4.596\,MN$$

5.2 Calculation Using Limit State Design and Eurocodes

$$N_r = Bf_{c28} + Af_e \tag{11}$$

$$N_r = 0.3 * 0.4 * 31 + 18.84 * 10^{-4} * 400$$
$$N_r = 4.474\,MN$$

6 Discussions of the Results

By neglecting the self-weight of the column, equal to 0.009 MN, the actual buckling coefficient α can be calculated by:

$$\alpha = \frac{N_{r\,expérimental}}{N_{r\,calculé}} \tag{12}$$

According the limit states design and the Eurocodes, we obtain:

$$\alpha = \frac{4.2}{4.474} = 0.94$$

As for the values of α taken into account in the calculation of Nr, they are equal to 0,71 for Limit States Design and 0,655 in the case of Eurocodes.

Regarding the Working Stress Design, since $\frac{l_f}{a} = 10 < 14.4$, no conditions for buckling are recommended by the regulations.

The results obtained when neglecting the buckling effects demonstrate good agreement with the experimental findings. This can be attributed to several factors, such as the relatively short length of the column and the minimal centering and formwork defects present in the laboratory setting. Unlike columns constructed on-site, these laboratory conditions contribute to a more controlled and consistent behavior of the column.

However, when considering the buckling phenomenon, it becomes evident that the Limit State Design (LSD) regulation provides the closest approximation to the failure load of the experimental column. The LSD methodology, with its incorporation of partial safety factors and comprehensive consideration of limit states, proves to be effective in capturing the impact of buckling on the structural behavior.

The Eurocodes, which also account for a wide range of factors including buckling, exhibit a deviation of -17% compared to the Limit State Design calculation. While the Eurocodes demonstrate a slightly lower accuracy compared to the LSD approach, they still offer a reasonable estimation of the failure load.

On the other hand, the Working Stress Design (WSD) calculation yields a larger deviation of approximately – 25.6% when compared to the Limit State Design calculation. This discrepancy can be attributed to the limitations of the WSD approach, which primarily considers serviceability limit states and does not fully account for the uncertainties and variability associated with buckling effects.

Overall, these results highlight the significance of considering buckling effects in the design of structural elements, particularly in scenarios where column height and construction conditions may increase the susceptibility to buckling failure. The findings emphasize the superiority of the Limit State Design approach in accurately predicting failure loads, followed by the Eurocodes. Meanwhile, the Working Stress Design method demonstrates relatively lower accuracy in capturing the true behavior of the column under buckling conditions.

References

1. MacGregor, J.G., Wight, J.K., Teng, S., Irawan, P.: Reinforced Concrete: Mechanics and Design, vol. 3. Prentice Hall Upper Saddle River, NJ (1997)
2. Chen, W.-F., Kim, S.-E.: LRFD Steel Design Using Advanced Analysis, vol. 13. CRC Press (1997)
3. Lysay, G.J.: Comparison of Limit States Design with Working Stress Design for Shallow Foundations. University of British Columbia (1999)
4. Manaenkov, I., Korenkov, P., Grezeva, A.S., Zinoveva, E.: Calculation of deformations of concrete with indirect reinforcement according to limit state design. In: IOP Conference Series: Materials Science and Engineering, p. 052033. IOP Publishing (2020)
5. Calgaro, J.-A., Gulvanessian, H.: Management of reliability and risk in the Eurocode system. In: Safety, Risk, and Reliability—Trends in Engineering. International Conference, pp. 155–160. Malta (2001)
6. Jędrzejczak, M., Klempka, K.: Limitation of stresses in concrete according to Eurocode 2. In: MATEC Web of Conferences, EDP Sciences, p. 00067 (2017)
7. Dreux, G. : Calcul Pratique du Béton Armé: Règles BAEL 80. Editions Eyrolles (1981)
8. Calgaro, J.-A., Virlogeux, M.: Projet et construction des ponts. In: Presses de l'École nationale des ponts et chaussées (1987)
9. Perchat, J., Roux, J.: Pratique du Bael 91-Cours Avec Exercices Corriges (1993)
10. Multon, S.: Béton Armé à l'Eurocode 2 (2018)
11. Razzouk, Y., Ahatri, M., Baba, K., El Majid, A.: Optimal bracing type of reinforced concrete buildings with soil-structure interaction taken into consideration. Civ. Eng. J. **9**(6), 1371–1388 (2023). https://doi.org/10.28991/CEJ-2023-09-06-06
12. Razzouk, Y., Ahatri, M., Baba, K., Majid, A.E.: The impact of bracing type on seismic response of the structure on soft soil. CEA **11**(5), 2706–2718 (2023). https://doi.org/10.13189/cea.2023.110534

13. Simou, S., Baba, K., Akkouri, N., Lamrani, M., Tajayout, M., Nounah, A.: Mechanical characterization of the adobe material of the archaeological site of Chellah. In: Ameen, H., Jamiolkowski, M., Manassero, M., Shehata, H. (eds.) Recent Thoughts in Geoenvironmental Engineering. Sustainable Civil Infrastructures, pp. 118–130. Springer International Publishing, Cham (2020). https://doi.org/10.1007/978-3-030-34199-2_8
14. Simou, S., Baba, K., Akkouri, N., Lamrani, M., Tajayout, M., Nounah, A.: Mechanical characterization and reinforcement of the adobe material of the Chellah archaeological site. E3S Web Conf. **150**, 03022 (2020). https://doi.org/10.1051/e3sconf/202015003022
15. Hanson, N.W., Connor, H.W.: Seismic resistance of reinforced concrete beam-column joints. J. Struct. Div. **93**(5), 533–560 (1967)
16. Ahatri, M., Baba, K., Touijrate, S., Bahi, L.: Characteristics of spectral responses for a ground motion from mediterranean earthquake—ZEGHANGHANE Station (6.3Mw) in Morocco, and its Influence on the structures. MATEC Web Conf. **149**, 02041 (2018). https://doi.org/10.1051/matecconf/201814902041
17. El Majid, A., Cherradi, C., Baba, K., Razzouk, Y.: Laboratory investigations on the behavior of CBR in two expanding soils reinforced with plant fibers of varying lengths and content. Mater. Today Proc. S2214785323037884 (2023). https://doi.org/10.1016/j.matpr.2023.06.395
18. Majid, A.E., Baba, K.: Assessing the impact of plant fibers on swelling parameters of two varieties of expansive soil. Case Stud. Chem. Environ. Eng. **8**, 100408 (2023). https://doi.org/10.1016/j.cscee.2023.100408
19. Taylor, R.: A note on a possible basis for a new method of ultimate load design of reinforced concrete slabs. Mag. Concr. Res. **17**(53), 183–186 (1965)
20. Poutanen, T., Länsivaara, T., Pursiainen, S., Mäkinen, J., Asp, O.: Calculation of safety factors of the eurocodes. Appl. Sci. **11**(1), 208 (2020)
21. Razzouk, Y., Baba, K., Ahatri, M.: The influence of spectral responses on the structures heights: case of the RHiss river earthquake in Morocco (6.3 Mw)-seismogenic source 4 (Rif oriental-Al Hoceima-Alboran). ARPN J. Eng. Appl. Sci. **17**(6), 645–651 (2022)
22. Yuan, W., Kim, B., Li, L.: Buckling of axially loaded castellated steel columns. J. Constr. Steel Res. **92**, 40–45 (2014)

Comparative Study of Two Reinforcement Layouts for Improved Efficiency and Load Distribution in Prestressed Concrete Elements

Ali Azizi[1](✉), Yassine Razzouk[2], Khadija Baba[2], and Mohamed Ahatri[2]

[1] Department of Roads and Bridges, Hassania School of Public Works, Casablanca, Morocco
ali.azizi.4research@gmail.com

[2] Civil Engineering and Environment Laboratory (LGCE), Mohammadia School of Engineering, Mohammed V University, Rabat, Morocco

Abstract. The use of prestressing in civil engineering has become widespread due to its numerous benefits, including increased strength, reduced weight, improved durability, and increased resistance to environmental factors such as moisture and corrosion. Prestressing is a technique used in civil engineering to improve the performance and load-carrying capacity of concrete structures. The process involves introducing a compressive stress in the concrete before it is subjected to external loads. The design of a prestressed concrete structure is complex and requires careful consideration of several factors, including the magnitude and location of the prestressing force, the properties of the materials used, and the geometry of the structure. The installation and tensioning of the prestressing tendons must also be precise to ensure that the prestressing force is applied evenly and does not cause damage to the structure. While prestressing offers many benefits, including increased strength, reduced weight, and improved durability, it involves a significant amount of time and expertise to execute. In this perspective, our study is based on a comparison between 2 reinforcement layouts. The first reinforcement layout studied allows to reduce the work time used during the calculations which is often excessive. This layout was compared to a second layout by assuming for both layouts that the connection of the different branches of the parabolas constituting the layout is made on the neutral axis of the prestressed element. an edge span of a continuous beam, it has been shown that in the usual cases, the total prestressing moment from the second plot remains negative and increases rapidly, whereas in the case of the first plot this moment is always positive and decreases very slowly. As a result, the first reinforcement layout, unlike the second, reduces the moments on the supports of the prestressed element in service and offers a better distribution of the stresses along the element. In contrast, the second layout offers the possibility of using a concordant layout and the calculations will be carried out as in isostatic.

Keywords: Prestressing · Reinforcement layout · Prestressed concrete calculation · Civil engineering · Load distribution · Continuous beam

K. Baba et al. (Eds.): GeoME 2023, SPEES, pp. 316–325, 2023.
https://doi.org/10.1007/978-3-031-49345-4_30

1 Introduction

Prestressing in civil engineering involves complex calculations and a significant amount of time because the technique requires careful consideration of several factors.

The design of a prestressed concrete structure involves the determination of the magnitude and location of the prestressing force, which is based on the loads that the structure will be subjected to during its service life. This requires a thorough understanding of the principles of structural mechanics and the behavior of concrete under different loading conditions [1].

In addition, the design of the prestressing system must take into account the properties of the materials used, such as the strength of the concrete and the steel, and the geometry of the structure. The design process also involves ensuring that the prestressing force does not exceed the capacity of the materials, which requires a detailed analysis of the stresses and strains in the structure [2, 3].

The process of installing and tensioning the prestressing tendons also requires careful planning and execution. The installation of the tendons must be precise, and the tensioning process must be controlled to ensure that the prestressing force is applied evenly and does not cause damage to the structure [4].

Overall, while prestressing offers many benefits in civil engineering, it involves a complex design process and precise execution, which requires significant time and expertise.

Our study aims to compare two different reinforcement layouts. The first layout aims to reduce the time required for calculations, which is often excessive. The comparison is made with a second layout assuming that the connections of the different branches of the parabolas that form the layout are made on the neutral axis of the prestressed element. The study was conducted on an edge span of a continuous beam, and it was found that in usual cases, the total prestressing moment from the second layout remains negative and increases rapidly. On the other hand, the first layout always yields a positive moment that decreases slowly. This results in a reduction of moments on the supports of the prestressed element in service and offers a better distribution of the stresses along the element, which is not the case for the second layout that offers the possibility of using a concordant layout and calculations that can be carried out as in isostatic. The layouts of prestressing reinforcement will be composed of branches of parabolas connected to the neutral axis of the element. Specific conditions on the deflections of the different parabolas will be imposed [5, 6]. The total prestressing moment on the supports will be calculated using the direct method [7]. The calculation principle for an intermediate span of a continuous beam will be presented, while the calculations for an end span will be detailed.

Through this study, we aim to deepen understanding of prestressing reinforcement layouts and their effects. The findings will contribute to optimizing design processes, enhancing the structural performance of prestressed concrete elements, and potentially reducing construction costs. Implementing more efficient reinforcement layouts will enable engineers to achieve improved load distribution, enhanced structural stability, and overall durability of prestressed concrete structures. Moreover, our study also aims to investigate the influence of the two different reinforcement layouts on the behavior and performance of prestressed concrete elements. By conducting a comprehensive analysis, we seek to gain valuable insights into their effects on the structural response and load distribution.

2 Reinforcement Drawings Used

Reinforcement drawings are detailed engineering drawings that provide specifications for the placement and arrangement of reinforcing steel bars (rebars) in a concrete structure. These drawings guide the construction team in accurately installing the rebars according to the structural design.

In this article, we specifically used two types of reinforcement drawings: "Simply-supported beam: intermediate span" and "Fixed–fixed beam on one end and simply-supported on the other: Edge span." These drawings depict the reinforcement details and arrangement for these specific types of beams. By comparing the reinforcement layouts of these beam configurations, we aim to gain insights into their respective structural behaviors and performance.

2.1 Simply-Supported Beam: Intermediate Span

A simply-supported beam with an intermediate span refers to a structural element that is supported at its ends and has a clear span between the supports. This type of beam is commonly used in various construction projects, including bridges, buildings, and other structures [8]. As shown in Fig. 1. The reinforcement design for this type of beam aims to ensure sufficient strength and stiffness to support the expected loads and prevent excessive deflection or failure [9]. By analyzing and comparing the reinforcement layouts for simply-supported beams with intermediate spans, valuable insights can be gained into their structural behavior, load distribution, and overall performance. This information is essential for optimizing the design and construction of such beams to meet the required structural integrity and safety standards [10].

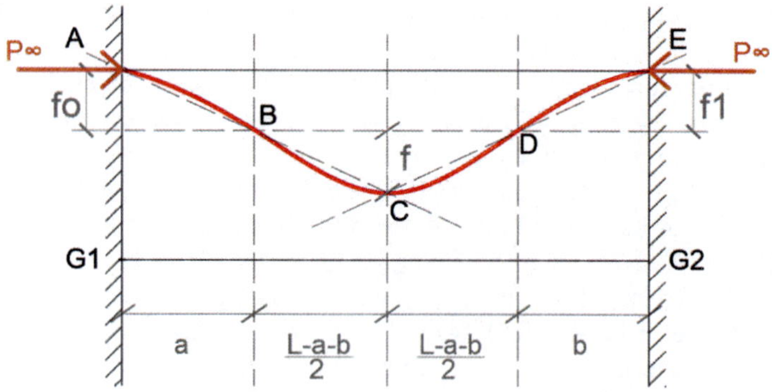

Fig. 1. The first reinforcement layout "1", inspired by Courtot [5]

This reinforcement layout is inspired by Courtot [5]. Points A, B, C, and points C, D, E are assumed to be aligned respectively, and the end parabolic branches have the same radius of curvature. This is expressed by the following relationships:

$$\frac{f_0}{a^2} = \frac{f_1}{b^2}$$
$$\frac{f_0}{a} = \frac{2f}{l-a-b}$$
$$\frac{f_1}{b} = \frac{2f}{l-a-b} \quad (1)$$

The following plot is due to Albiges [6], who assume that (Fig. 2):

$$\frac{f}{a} = \frac{2f_0}{l-a-b}$$
$$\frac{f}{b} = \frac{2f_1}{l-a-b} \quad (2)$$

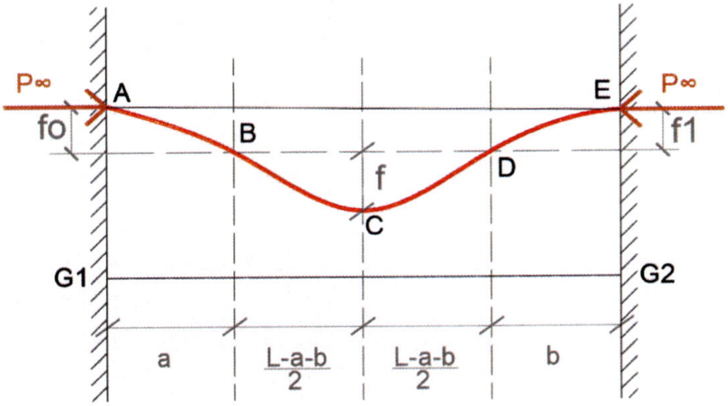

Fig. 2. The second reinforcement layout "2", inspired by Albiges [6]

P_∞ denotes the residual prestress after all losses, the moment of inertia I of the beam is assumed to be constant,

If M_1 and M_2 denote the total prestress moment at G1 and G2, respectively, then by the direct method [5],

$$M_1 = \frac{4EI}{l}\omega_2' + \frac{2EI}{l}\omega_2''$$
$$M_2 = \frac{4EI}{l}\omega_2'' - \frac{2EI}{l}\omega_2' \quad (3)$$

ω_2' and ω_2'' represent the support rotations of the isostatic span under the effect of the prestressing force, given by:

$$\omega_2' = -\frac{q_0 a^2 (2l-a)^2}{24EIl} - \frac{q_1 b^2 (2l^2 - b^2)}{24EIl}$$

$$+\frac{q}{24EIl}(l^4+b^4-a^4+4a^3l-4a^2l^2-2b^2l^2$$

$$\omega_2''=+\frac{q_0a^2(2l^2-a)^2}{24EIl}-\frac{q_1b^2(2l-b)^2}{24EIl}$$

$$+\frac{q}{24EIl}(b^4-a^4-l^4+4b^3l+2a^2l^2+4b^2l^2 \quad (4)$$

q_0, q, and q_1 are the equivalent distributed loads on the AB, BCD, and DE parabolic branches, respectively. They are given by the following equations according to the first design.

$$q_0=q_1=\frac{2f_0P_\infty}{a^2}$$

$$q=\frac{8fP_\infty}{(l-a-b)^2};\quad q_0=\frac{2f_0P_\infty}{a^2} \quad (5)$$

For the second layout, we have,

$$q_0=\frac{2f_0P_\infty}{a^2}$$

$$q_1=\frac{2f_1P_\infty}{b^2}$$

$$q=\frac{8fP_\infty}{(l-a-b)^2};\quad f=\left(\frac{l-a-b}{2a}\right)*f_0 \quad (6)$$

2.2 Fixed–Fixed Beam on One End and Simply-Supported on the Other: Edge Span

A fixed–fixed beam on one end and simply-supported on the other, also known as an edge span configuration, refers to a structural beam that is supported in a fixed manner at one end and supported as simply-supported at the other end. This beam arrangement is commonly encountered in various construction projects, including buildings, bridges, and other structures [11]. In this configuration, one end of the beam is firmly fixed, preventing rotation and allowing for both vertical and horizontal restraint, while the other end is supported as simply-supported, allowing for vertical movement and rotation [12]. This type of beam configuration is often employed to accommodate differential movements, thermal expansion, or to provide stability against lateral forces. The reinforcement design for this type of beam aims to ensure adequate strength, stiffness, and resistance against bending moments and shear forces. The reinforcement drawings play a critical role in guiding the construction team to accurately install and secure the rebars in accordance with the specified reinforcement detailing [13]. Analyzing and comparing the reinforcement layouts for fixed–fixed beams on one end and simply-supported on the other in an edge span configuration allows for valuable insights into their structural behavior, load distribution, and overall performance. This information is vital for optimizing the design and construction of such beams to meet the required structural integrity and safety standards [14, 15].

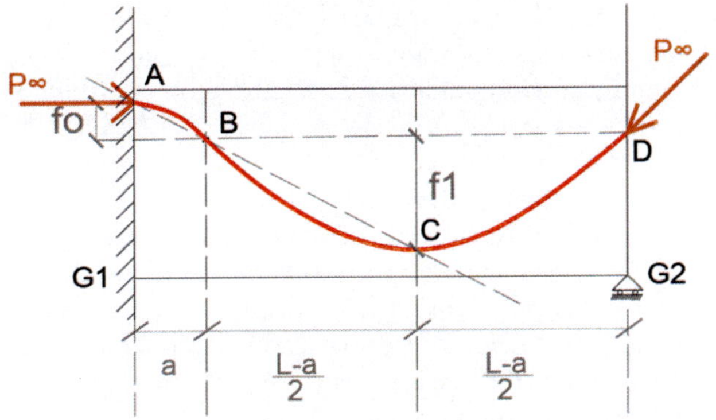

Fig. 3. Fixed–fixed beam on one end and simply-supported on the other: Edge span. Inspired by Courtot [5]

According to Courtot [5], the plot is as follows (Fig. 3).

Points A, B, and C are aligned, which results in:

$$f_0 = 2f_1 * \left(\frac{a}{l-a}\right) \tag{7}$$

The plot according to Albiges [6] is identical, but it must satisfy the condition:

$$f_0 = \left(\frac{l-a}{2a}\right)f_1 \tag{8}$$

The direct method gives the total prestress moment at G1, remaining M_0, namely:

$$M_0 = \left[\left(1 + 2\beta - \beta^2\right)f_1 - \left(1 - \frac{\beta}{2}\right)^2 f_0\right] * P_\infty \tag{9}$$

By using the specific conditions for each type of plot, we obtain, respectively for the first and the second, the relationships:

$$M_0 = \left[\left(1 + 2\beta - \beta^2\right) - \frac{(1-\beta)(2-\beta)^2}{2(1-\beta)}\right]f_1 P_\infty$$

and

$$M_0 = \left[\left(1 + 2\beta - \beta^2\right) - \frac{(1-\beta)(2-\beta)^2}{8\beta}\right]f_1 P_\infty \tag{10}$$

where, $\beta = \frac{a}{l}; 0 < \beta < 1$.

The relationships (10) result from relationship (9) and relationships (7) and (8), respectively.

The results of our analysis reveal interesting findings regarding the behavior of the two reinforcement layouts, particularly in relation to the prestressing moment, denoted as M_0.

In the case of the second layout, which corresponds to a typical configuration, we observed that M_0 remains negative throughout the analysis and exhibits a rapid increase. This indicates a significant compressive prestressing moment in the structure, which can have implications on its overall behavior and load distribution. The negative trend and rapid increase in M_0 for this layout suggest higher compressive forces and potential limitations in terms of structural performance.

On the other hand, the first layout demonstrates a contrasting behavior. For this layout, M_0 remains positive throughout the analysis and exhibits a gradual decrease. The positive prestressing moment indicates a tensile force in the structure, which counteracts the compressive forces and contributes to improved load distribution. The gradual decrease in M_0 implies a reduction in the tensile prestressing moment over time, suggesting a more stable and balanced structural response.

Importantly, we found that the decrease in M_0 for the first layout does not exceed 4.5% for typical values of the coefficient β. This indicates a relatively small reduction in the tensile prestressing moment, which confirms the effectiveness and stability of the layout. This limited decrease in M_0 highlights the robustness and consistency of the first reinforcement layout in maintaining its positive prestressing moment, ensuring better structural performance.

Overall, these results emphasize the importance of selecting an appropriate reinforcement layout, as it directly affects the prestressing moment and, consequently, the behavior and load distribution within the structure. The findings suggest that the first layout offers advantages in terms of maintaining a positive prestressing moment and achieving a more balanced and stable structural response compared to the second layout.

3 Finding a Layout of Wedges that Match in the Case of an End Span

When designing and constructing end spans of structural elements such as beams or bridges, it is important to find a layout of wedges that match. In this context, wedges refer to the tapered or sloped elements that are used to transition between the main structural component and the support or abutment. The purpose of these wedges is to provide a smooth and gradual transfer of forces and loads from the end span to the support [16, 17]. Finding a layout of wedges that match involves ensuring that the geometry, dimensions, and angles of the wedges align properly with the adjacent components. This ensures a seamless connection and efficient load transfer without creating any stress concentrations or structural discontinuities [18]. The process of finding a matching layout of wedges typically involves careful analysis and consideration of the design requirements, including factors such as the type of support, the load distribution, and the expected forces and movements in the end span. It may also involve conducting structural calculations and simulations to assess the performance and stability of the wedge layout. By achieving a matching layout of wedges, the structural integrity and performance of the end span can be enhanced. It helps to minimize any potential vulnerabilities, such as localized stress concentrations or uneven load distribution, which could lead to premature failure or structural issues [19]. Overall, finding a layout of wedges that match is a crucial aspect of designing and constructing end spans to ensure the overall stability, durability, and functionality of the structural element.

The cable concordance condition is given by the following equation [7]:

$$a^2(2l-a)^2 = \frac{q_1}{q_0}(l-a)^2\left[2l^2-(l-a)^2\right] \tag{11}$$

where q_0 and q_1 represent the equivalent distributed loads for the parabolic branches AB and BCD, respectively, using the coefficient $\beta = \frac{a}{l}$.

For the type 1 plot [5]

$$\beta(2-\beta)^2 = 2(1-\beta)\left(1-\beta^2+2\beta\right) \tag{12}$$

And for the type 2 plot [6]

$$(1-\beta)(2-\beta)^2 = 8\beta\left[2-(1-\beta)^2\right] \tag{13}$$

Based on the obtained results, we can draw several significant conclusions regarding the two plots, labeled as "1" and "2", and their respective coefficient values, denoted as β.

For plot "1", we found that the coefficient β has a value of approximately 0.689. This value falls outside the usual range of variation for β, indicating a deviation from the typical values observed. This finding is noteworthy as it suggests that plot "1" exhibits a unique behavior that differs from the common trends observed in prestressed structures. The relatively high value of β indicates that the cable concordance might not be feasible or may require special considerations.

In contrast, for plot "2", we observed that the coefficient β has a value of approximately 0.221. This value is within the range of current values typically observed for β in prestressed structures. This implies that plot "2" aligns with conventional design practices, allowing for the possibility of achieving cable concordance. Cable concordance refers to the arrangement of prestressing tendons in a manner that promotes a balanced and efficient distribution of forces throughout the structure. The fact that β falls within the expected range for plot "2" suggests that this layout can be more easily implemented, and the desired cable concordance can be achieved without significant challenges or deviations from standard procedures.

These results emphasize the importance of considering the coefficient β when designing and implementing prestressed structures [20]. The deviation of β in plot "1" from the usual range suggests the need for careful evaluation and potential modifications in the reinforcement layout to ensure the structural integrity and performance of the system. In contrast, the consistent value of β in plot "2" indicates a more straightforward design process and a higher likelihood of achieving the desired cable concordance.

4 Conclusion

In summary, our analysis has yielded significant findings regarding the behavior of two reinforcement layouts in relation to the prestressing moment (M_0) and the coefficient β. For the second layout "2", we observed a negative M_0 with a rapid increase, indicating a significant compressive prestressing moment and potential limitations in structural performance. In contrast, the first layout "1" exhibited a positive M_0 with a gradual decrease, suggesting a stable and balanced structural response with a reduced tensile prestressing moment. Furthermore, we found that the layout "1" deviated from the typical range of variation for the coefficient β, indicating unique behavior and the possibility of challenging cable concordance. In contrast, the layout "2" aligned with current design practices, making cable concordance more achievable.

These results underscore the importance of selecting an appropriate reinforcement layout to influence the prestressing moment and overall load distribution in a structure. The layout "1" showed advantages in terms of maintaining a positive prestressing moment and achieving a stable response, while the layout "2" aligned with conventional design practices. Considering the coefficient β is crucial in the design and implementation of prestressed structures. Deviations in β, as observed in the layout "1", may necessitate careful evaluation and potential modifications in the reinforcement layout to ensure structural integrity. Conversely, consistent values of β, as seen in the layout "2", facilitate a more straightforward design process and enhance the likelihood of achieving desired cable concordance.

These findings contribute to the understanding of the behavior and performance of different reinforcement layouts in prestressed structures, providing valuable insights for structural engineers in optimizing their designs and ensuring efficient load distribution and structural integrity.

References

1. Vollmer, M., et al.: Novel prestressing applications in civil engineering structures enabled by Fe Mn Al Ni shape memory alloys. Eng. Struct. **241**, 112430 (2021). https://doi.org/10.1016/j.engstruct.2021.112430
2. Razzouk, Y., Ahatri, M., Baba, K., El Majid, A.: Optimal bracing type of reinforced concrete buildings with soil-structure interaction taken into consideration. Civ. Eng. J. **9**(6), 1371–1388 (2023). https://doi.org/10.28991/CEJ-2023-09-06-06
3. Razzouk, Y., Ahatri, M., Baba, K., El Majid, A.: The impact of bracing type on seismic response of the structure on soft soil. CEA **11**(5), 2706–2718 (2023). https://doi.org/10.13189/cea.2023.110534
4. Ghallab, A., Beeby, A.W.: Calculating stress of external prestressing tendons. Proc. Inst. Civ. Eng. Struct. Build. **157**(4), 263–278 (2004). https://doi.org/10.1680/stbu.2004.157.4.263
5. Courtot, P.: Vues nouvelles sur le calcul de la précontrainte dans les ouvrages hyperstatiques (1963). https://doi.org/10.5169/SEALS-66295
6. Albiges, M., Coin, A.: Resistance des Materiaux Appliquee-Tomes 1 et 2–3e Edition (1978)
7. Chaussin, R., Fuentes, A., Lacroix, R., Perchat, J.: La précontrainte (1992)
8. Razzouk, Y., Baba, K., Ahatri, M.: The influence of spectral responses on the structures heights: case of the RHiss river earthquake in Morocco (6.3 Mw)-seismogenic source 4 (Rif oriental-Al Hoceima-Alboran). ARPN J. Eng. Appl. Sci. **17**(6), 645–651 (2022)
9. Kurniawan, C.W., Mahendran, M.: Elastic lateral buckling of simply supported LiteSteel beams subject to transverse loading. Thin-Wall. Struct. **47**(1), 109–119 (2009)
10. Michaltsos, G., Sophianopoulos, D., Kounadis, A.N.: The effect of a moving mass and other parameters on the dynamic response of a simply supported beam. J. Sound Vib. **191**(3), 357–362 (1996)

11. Ahatri, M., Baba, K., Touijrate, S., Bahi, L.: The seismic motion parameters effects on response spectra: Comparison between El Centro 1940 and imperial valley 1979 earthquakes. Int. J. Civ. Eng. Technol. (IJCIET)-Scopus Index. **9**(10), 1610–1618 (2018)
12. Bassily, S.F., Dickinson, S.M.: Buckling and lateral vibration of rectangular plates subject to inplane loads-a Ritz approach. J. Sound Vib. **24**(2), 219–239 (1972)
13. Jacobucci, C.: Automating Structural Stress Analysis: Beam Deflection, Shear, and Moment Diagram Generator for Single and Multi-Span Beams. Massachusetts Institute of Technology (2017)
14. Majid, A.E., Baba, K.: Assessing the impact of plant fibers on swelling parameters of two varieties of expansive soil. Case Stud. Chem. Environ. Eng. **8**, 100408 (2023). https://doi.org/10.1016/j.cscee.2023.100408
15. El Majid, A., Cherradi, C., Baba, K., Razzouk, Y.: Laboratory investigations on the behavior of CBR in two expanding soils reinforced with plant fibers of varying lengths and content. Mater. Today Proc. S2214785323037884 (2023). https://doi.org/10.1016/j.matpr.2023.06.395
16. Simou, S., Baba, K., Akkouri, N., Lamrani, M., Tajayout, M., Nounah, A.: Mechanical characterization of the adobe material of the archaeological site of Chellah. In: Ameen, H., Jamiolkowski, M., Manassero, M., Shehata, H. (eds.) Recent Thoughts in Geoenvironmental Engineering. Sustainable Civil Infrastructures, pp. 118–130. Springer International Publishing, Cham (2020). https://doi.org/10.1007/978-3-030-34199-2_8
17. Simou, S., Baba, K., Akkouri, N., Lamrani, M., Tajayout, M., Nounah, A.: Mechanical characterization and reinforcement of the adobe material of the Chellah archaeological site. E3S Web Conf. **150**, 03022 (2020). https://doi.org/10.1051/e3sconf/202015003022
18. Souza Junior, O.A., de Oliveira, D.R.C.: Influence of the cable's layout on the shearing resistance of prestressed concrete beams. Rev. Ibracon Estrut. Mater. **9**, 765–795 (2016)
19. Roca, P., Marí, A.R.: Numerical treatment of prestressing tendons in the nonlinear analysis of prestressed concrete structures. Comput. Struct. **46**(5), 905–916 (1993). https://doi.org/10.1016/0045-7949(93)90152-4
20. Jiang, R.J., Kwong Au, F.T., Xiao, Y.F.: Prestressed concrete girder bridges with corrugated steel webs: review. J. Struct. Eng. **141**(2), 04014108 (2015). https://doi.org/10.1061/(ASCE)ST.1943-541X.0001040

Seismic Performance Investigation of RC Building Using Nonlinear Static Analysis

Sana Afnzar[1(✉)], Mina Derife[1], Mohamed Mouhine[2], Abderrahman Atmani[1], El Hassan Ait Laasri[1], and Driss Agliz[1]

[1] GMES Laboratory, National School of Applied Sciences, Ibn Zohr University, Agadir, Morocco
Sana.afnzar@gmail.com

[2] ASE Laboratory, National School of Applied Sciences, Ibn Tofail University, Kenitra, Morocco

Abstract. The seismic efficiency of existing buildings refers to their ability to withstand and respond to earthquake forces. Evaluating the eathquake performance of existing structures is essential to identify potential vulnerabilities, evaluate the level of risk, and determine the need for retrofitting or strengthening measures, especially for buildings built before the application of the seismic construction regulations as well as for masonry and adobe structures. Regarding this point, the current paper takes a case study of a reinforced concrete structure to assess its seismic behavior and validate the relevance of the method used, it's about a school building. As a method, we used a non-linear response analysis, termed as Pushover analysis. This method provides a simplified but effective way to understand the behavior of structures under the effect of seismic forces. It allows us to identify potential weak points, to evaluate the effectiveness of design and retrofit measures, and to evaluate the overall seismic behavior of the structure. Finally, the results obtained show that the method used gives us the information about the comportment of the studied structure under the effect of seismic forces, it is represented as a curve, it's a capacity curve.

Keywords: Performance · Seismic · Pushover analysis · RC structure · Capacity curve

1 Introduction

Agadir is situated in southwestern Morocco, near the boundary between the African and Eurasian tectonic plates. This tectonic interaction gives rise to seismic activity in the region. The convergence between these plates creates stress and strain accumulation, which can be released in the form of earthquakes. Therefore, it is important for residents, engineers, and authorities in Agadir to remain vigilant about seismic hazards, implement appropriate measures for seismic safety, and regularly review and update building practices to enhance the resilience of structures against earthquakes.

A number of researchers have been interested in this issue, and some of them have introduced various studies on the seismic performance of structures. Poluraju and

K. Baba et al. (Eds.): GeoME 2023, SPEES, pp. 326–334, 2023.
https://doi.org/10.1007/978-3-031-49345-4_31

Nageswara Rao [1] analyzed G + 3 building to evaluate its performance in the event of future earthquakes using a non-linear pushover static analysis by Sap 2000 software. The authors mentioned that the pushover analysis is a relatively simple way to examine the non-linear comportment of structures, indeed, the results achieved in terms of demand, capacity and plastic hinges gave an overview of the actual behavior of the structures. Sayin et al. [2] treated the methodology for assessing the seismic performance of masonry-filled reinforced concrete buildings before and after the retrofitting practice based on nonlinear analyses, the results obtained shows that the existing building does not achieve the "Controlled Failure" performance level for a ground shaking (DD-2), In fact, the strengthening suggestions have been represented on the finite element method. In conclusion, the authors noted that when comparing the before and after intervention ("modernization") states, an improvement of around 20% in terms of structural performance points was achieved. The choice of material models and analysis methods for assessing the seismic performance of reinforced concrete and dual-system masonry buildings is important and essential for capturing real-world behavior. Milani and Valente [3] investigated seven seismically damaged stone churches and evaluated their resistance to seismic loading. The authors used three methods, namely analytical finite element pushover analysis, kinematic limit analysis and finite element kinematic limit analysis. After comparing the three alternatives, the results show that finite element limit analysis is the most appropriate method, as it is more precise in determining failure processes. Asteris et al. [4] investigated masonry buildings in three countries: Greece, Portugal, and Cyprus. They introduced a methodology to evaluate the seismic vulnerability of these cases, creating fragility curves for both the original and repaired structures. These curves are particularly valuable as they enable the ranking and benchmarking of the structures' seismic vulnerability. They assist civil authorities in optimizing their decision-making process when faced with a multitude of structures, helping them identify the most vulnerable ones requiring immediate reinforcement. Suwondo and Alama [5] conducted an evaluation of the nonlinear response of existing constructions in four distinct seismic areas (Makkah, Gizan, Jeddah and Haql) by employing Pushover function. The findings indicate that the building is generally secure across all regions. However, certain critical points within the structure have exceeded the specified limits in the Haql region. This indicates that the structure has incurred some damage and there is a possibility of localized failure occurring. Güneş et al. [6] executed a comprehensive assessment of the seismic resistance of a masonry structure situated in Istanbul's historical peninsula. The building incorporates several reinforced concrete structural elements. The researchers employed two distinct analytical approaches, one considering the material as linear and the other accounting for nonlinearity, to evaluate the structural integrity of the building.

The purpose of this paper was to study the response of reinforced concrete (RC) buildings when exposed to earthquake forces. The study primarily focused on conducting pushover analysis on RC framed structures, subjecting them to lateral forces until reaching the designated performance point.

2 Pushover Analysis

Pushover analysis involves subjecting a structure to a series of lateral loads applied at various locations and directions. These loads are incrementally increased until the structure reaches its ultimate capacity or a predetermined limit state, as depicted in Fig. 1. Pushover analysis is an effective means of assessing the overall behavior of structures, providing insights into the displacement ductility and the base shear capacity of the structure. It can also be used to determine the performance of the structure under a specific earthquake intensity level [7]. The output of this analysis method is a performance point, which is obtained by crossing the capacity curve over the demand curve. The demand curve represents the corresponding response spectrum for the seismic zone, damping level of the structure and soil type (see Fig. 2).

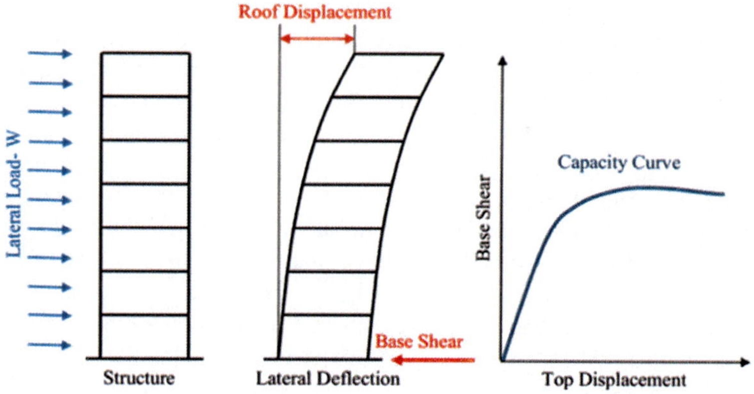

Fig. 1. Pushover analysis method based on FEMA-356 [8]

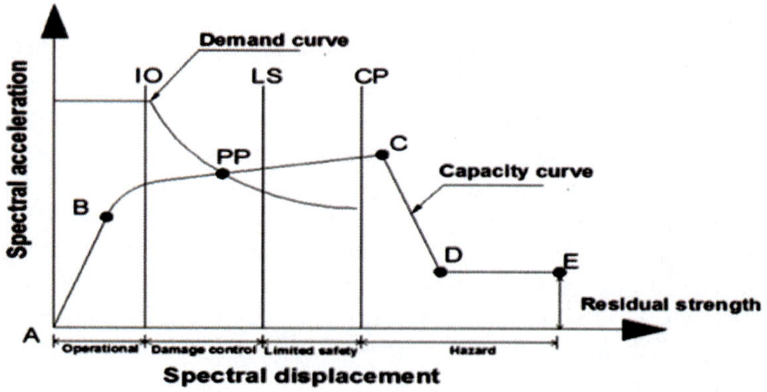

Fig. 2. Pushover analysis involves plotting capacity and demand curves

3 Description of the Structure Studied

The tested structure is located in the National School of Applied Sciences of Agadir (see the location in Fig. 3), The building in question is a low-rise school building comprising three floors. Each floor is dedicated to classrooms and teacher offices, as illustrated in Fig. 4. The floor area measures 38 m by 31 m. It is a reinforced concrete structure composed of structural elements such as beams, columns and RC walls with different cross-sections (see Table 1).

It is worth mentioning that our team has previously conducted an operational modal analysis of this structure. This analysis involved studying the dynamic behavior and response of the building under operational conditions [9].

Fig. 3. The location of structure studied

4 Structure Modeling and Analysis

The structural model has been developed in a modeling software (Fig. 5), and the various structural elements are modelled in reinforced concrete with the following characteristics ($E = 25{,}000{,}000$ kN/m^2; Poisson's ratio $v = 0.2$).

In the static pushover procedure, a force of 1 kN was incrementally applied to every point in the structure's nodes, considering the assumption of plastic material behavior. This analysis aimed to evaluate the structural response and deformation patterns as the applied forces increased gradually.

Fig. 4. Real image of the structure examined

Table 1. Details of structure elements

Element type	Section/diameter (cm)
Rectangular column	30 × 30
	40 × 30
	30 × 60
Circular column	40
	50
Beam	20 × 20
	20 × 40
	20 × 50
	20 × 60
	20 × 65
	20 × 70
	20 × 80
	25 × 100
Thickness of RC walls	25

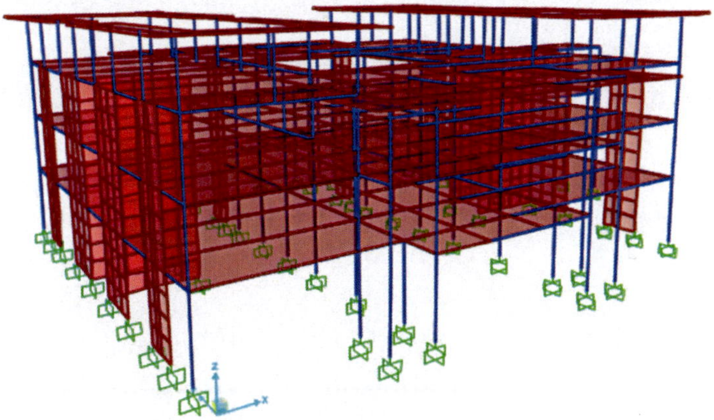

Fig. 5. Numerical model

5 Results and Discussion

The results of the Pushover analysis method are presented as capacity curves. Figure 6 illustrates the capacity plot in the X direction, while Fig. 7 shows the capacity graph in the Y direction. These curves provide insights into the overall performance of the building when subjected to lateral loads.

Initially, the pushover curves display a linear increase, indicating that the building is operating within the elastic range. As the applied forces continue to increase, the slope of the pushover curves progressively decreases. This behavior signifies the deformation of structural elements and a loss of rigidity in the building. The reduction in slope indicates the onset of plastic deformation and nonlinear behavior in the structure.

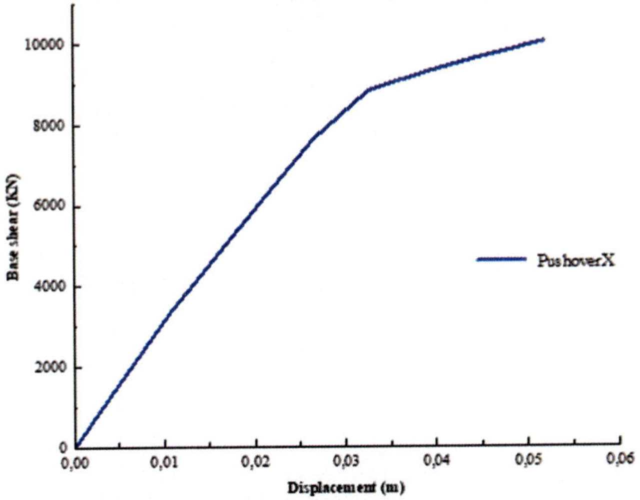

Fig. 6. Capacity plot in X direction

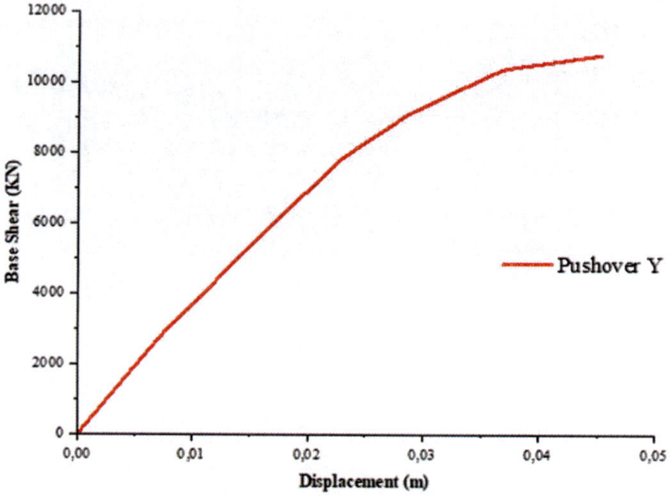

Fig. 7. Capacity graph in Y direction

To evaluate the seismic vulnerability of the structure and determine the performance point, it is necessary to convert the capacity curves from Figs. 6 and 7 into capacity spectra. These spectra are then overlaid on the demand curve, which is obtained using the ATC40 method [10]. The following figures depict the performance points in both the x and y directions:

- Figure 8 shows the performance point in the y direction, corresponding to a demand displacement of 0.94 cm and a base shear force of 3486.58 kN. This performance point is located at the Immediate Occupancy (IO) performance level.
- Figure 9 presents the performance point in the x direction, which has a demand displacement of 1.09 cm and a shear force of 3397.56 kN. This point is also situated at the Immediate Occupancy (IO) performance level.

By determining these performance points, we can evaluate the structural response and its ability to withstand seismic forces, ensuring the safety and functionality of the building under earthquake conditions.

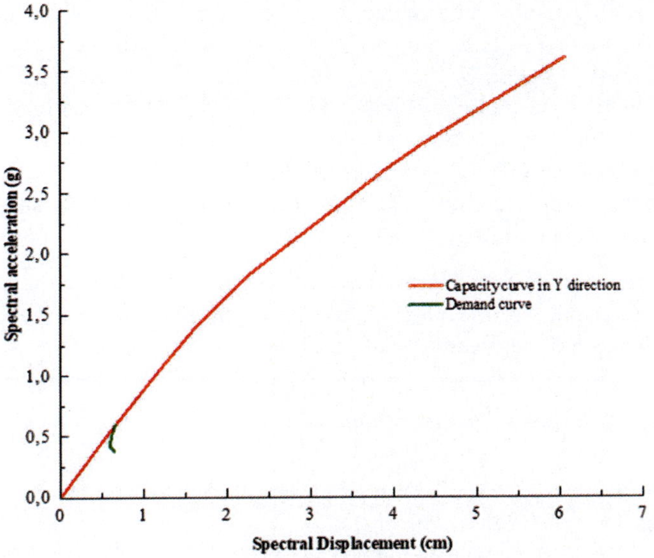

Fig. 8. Capacity-demand curve in Y direction

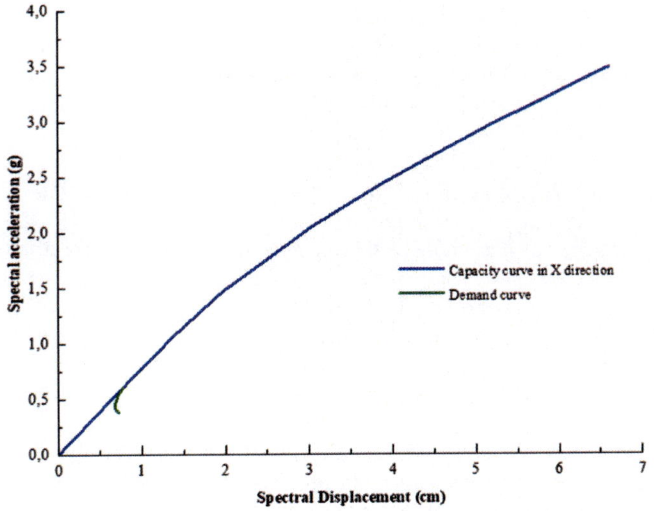

Fig. 9. Capacity-demand curve in X direction

6 Conclusion

Drawing from the discoveries of this investigation, the ensuing conclusions can be drawn concerning the structural configuration:

- The 3D modeling results indicate that the construction remains within the elastic field in both directions. The behavior of the structure under seismic forces is similar in both the X and Y directions.
- Pushover analysis proves to be a straightforward method for investigating the nonlinear behavior of buildings. It is also effective in evaluating the earthquake susceptibility of existing constructions. By studying the structural response to earthquake forces, pushover analysis provides crucial information regarding ductility, elastic limits, and plastic limits of the building.

Overall, these conclusions highlight the importance of understanding the structural behavior under seismic action and the valuable insights that pushover analysis can offer in assessing the seismic vulnerability of buildings.

References

1. Poluraju, P.: Pushover analysis of reinforced concrete frame structure using SAP 2000. Int. J. Res. Eng. Technol. **04**(28), 173–181 (2011)
2. Gunes, B., Mangir, A., Cosgun, T., Sayin, B., Akcay, C.: Seismic performance assessment of a historical masonry-infilled RC building located in the historical peninsula of Istanbul (1940s). Structures **45**(May), 951–968 (2022). https://doi.org/10.1016/j.istruc.2022.09.074
3. Milani, G., Valente, M.: Comparative pushover and limit analyses on seven masonry churches damaged by the 2012 Emilia-Romagna (Italy) seismic events: possibilities of non-linear finite elements compared with pre-assigned failure mechanisms. Eng. Fail. Anal., **47**(PA), 129–161 (2015). https://doi.org/10.1016/j.engfailanal.2014.09.016
4. Asteris, P.G., et al.: Seismic vulnerability assessment of historical masonry structural systems. Eng. Struct. **62–63**, 118–134 (2014). https://doi.org/10.1016/j.engstruct.2014.01.031
5. Suwondo, R., Alama, S.: Seismic assessment of RC building designed by local practice. IOP Conf. Ser. Earth Environ. Sci. **426**(1) (2020). https://doi.org/10.1088/1755-1315/426/1/012047
6. Güneş, B., Mangır, A., Okumuş, V.: Assessment of an existing masonry structure using linear and nonlinear material models. In: Proceeding of the Fourth Conference on Smart Monitoring, Assessment and Rehabilitation of Civil Structures (2017)
7. Mallick, M., Sahu, R.B., Mandal, K.K.: 4. A Case Study of Liquefaction-Induced Damages to a Port Building Supported on Pile Foundation and 6. Analysis of Laterally Loaded Single Pile in Cohesionless Soil Considering Non ... Theme 11 Dynamics of Soils and Foundations Index, Mar 2021 (2020)
8. Marabi, B., Marsono, A.K.: A Numerical and Analytical Study on Optimization and Efficiency of Structural Forms by Two-Outrigger in Efficiency of Structural Forms by Two—Outrigger in Tall Buildings, Sept 2016 (2017)
9. Uzdil, O., Cosgun, T., Sayin, B., Akcay, C.: Seismic performance evaluation and strengthening proposal for a reconstruction project of a historic masonry building demolished in the 1940s. J. Build. Eng. **66**(January), 105914 (2023). https://doi.org/10.1016/j.jobe.2023.105914
10. Atc, A.: 40, Seismic evaluation and retrofit of concrete buildings. Appl. Technol. Counc. **1**, 334 (1996)

A Strategic Approach in Order to Manage and Conserve Historic Buildings, Using GIS and 3D Technologies

Sana Simou(✉), Khadija Baba, and Abderrahman Nounah

Civil Engineering and Environment Laboratory, Mohammadia School of Engineering, Civil Engineering, Water, Environment and Geosciences Centre (CICEEG), Mohammed V University, Rabat, Morocco
sana_simou2@um5.ac.ma

Abstract. Cultural heritage institutions have a crucial role in enabling the community to access, generate, and share information while possessing the necessary tools for preserving historical heritage. This represents a significant opportunity for the community at large. In light of this, it is crucial to prioritize the availability of tools that streamline the efficient and prompt management of risks associated with deterioration. This prioritization enables preventive conservation and the safeguarding of cultural assets. The study investigates the utilization of modern digital technologies as well as Geographic Information Systems (GIS) to enhance the conservation planning of heritage cities. Specifically focusing on historic buildings in Rabat, Morocco, this research explores how the integration of these tools can facilitate the preservation efforts. Contemporary digital technologies provide innovative methodologies and tools that have the potential to revolutionize heritage conservation planning. By examining the application of GIS and digital technologies in this context, the study aims to uncover new approaches for effectively preserving the historical buildings in Rabat, Morocco. The aim is to develop a strategy for effective management and preservation of historic buildings by utilizing these tools. The study involves collecting and managing various forms of information, including building materials, textual records, icons, and photographs, to create a comprehensive database of historical heritage. Thematic maps are generated to provide visual representations of these properties.

Keywords: Heritage buildings · GIS management · Degradation · 3D technology · Thematic maps

1 Introduction

Cultural heritage is a significant aspect of human civilization, representing the historical, artistic, and architectural achievements of societies across time. Preserving and conserving this heritage is essential to maintain a connection with our past and to pass on these valuable assets to future generations [1]. In recent years, the integration of digital technologies and Geographic Information Systems (GIS) has offered new opportunities for effective management and conservation of historic buildings and cultural

K. Baba et al. (Eds.): GeoME 2023, SPEES, pp. 335–345, 2023.
https://doi.org/10.1007/978-3-031-49345-4_32

assets [2–4]. Cultural heritage institutions play a vital role in safeguarding and disseminating information related to historical heritage. They have the responsibility of keeping information available, generating and sharing knowledge, and utilizing the necessary tools for the conservation of cultural assets [5]. As technology advances, these institutions have access to new methodologies and tools that facilitate the management of risks associated with deterioration, enabling preventive conservation and preservation efforts [5, 6].

Digital technologies, including close-range digital photogrammetry, have emerged as powerful tools for documenting and analyzing historic buildings. Close-range digital photogrammetry involves capturing high-resolution photographs of architectural elements and using specialized software to create accurate three-dimensional models [7]. These models provide detailed information about the materials, ornamentation, and structural components of historical structures. Digital documentation plays a crucial role in evaluating the condition of historic buildings, identifying areas of deterioration or damage, and assisting in the formulation of suitable conservation plans [8, 9].

Geographic Information Systems (GIS) provide a spatial framework for organizing and integrating diverse data types related to historic buildings, such as architectural plans, historical records, and geographic features. GIS allows for the creation of detailed digital maps that can visualize and analyze the relationships between buildings, their surroundings, and other relevant information [10]. These spatial representations enable a holistic understanding of heritage sites and aid in decision-making for conservation planning. In the context of Morocco, a country with a rich cultural heritage, there is a need to investigate how the integration of digital technologies and Geographic Information Systems (GIS) can be applied in the field of heritage conservation planning [11, 12].

This study aims to address this gap by investigating the potential of these tools in the management and preservation of historic buildings in Morocco, with a focus on the city of Rabat. The primary goals of this study entail gathering and organizing various types of information and documentation, including building materials, textual records, icons, and photographs. These efforts aim to establish an extensive and inclusive database of historical heritage. This database will be utilized to generate thematic maps and provide views of the degradation. The strategy includes conducting a thorough inventory of monuments and sites in Rabat, evaluating their historical significance and identifying potential conservation needs. Additionally, heritage protection zones and the preservation status of material buildings will be assessed [12–16]. The mapping of degradation in stone and masonry heritage is typically a labour-intensive process that is often performed manually. However, there is a recognized need for more affordable solutions to conduct degradation mapping in digital environments for traditional masonry conservation purposes. To address these challenges, this work aims to recommend a combination of techniques. The primary focus is on proposing a methodology that integrates photogrammetric and GIS tools to generate thematic maps of degradation in an efficient manner.

This approach involves utilizing Geographic Information System (GIS) software, which enables quantitative analysis of degraded assets through spatial and statistical evaluation. Additionally, this methodology facilitates the visual representation of the preservation state. By combining these techniques, the presented methodology offers a more efficient and accessible way to map stone and masonry heritage degradation. It reduces the labour intensity associated with manual mapping processes and provides a means to evaluate and visualize the extent of degradation on assets surfaces. Ultimately, this approach contributes to the conservation efforts by facilitating better monitoring and understanding of the preservation state of stone heritage.

2 Materials and Methods

2.1 Study Area

The study area emphasizes on the historical monuments of Rabat, the capital city of Morocco. Rabat is renowned for its rich cultural heritage, including numerous historical sites and architectural landmarks that reflect its historical significance. It includes notable historical sites such as the Kasbah of the Udayas, Hassan Tower, Chellah Necropolis, and various mosques, palaces, and traditional houses. The analysis of these monuments involves documenting their architectural details, understanding their historical context, and assessing their conservation needs (Fig. 1).

Fig. 1. Satellite photo of the study area (Rabat, Morocco)

By conducting field surveys, archival research, and data collection, the study seeks to create a comprehensive database of historical monuments in Rabat. The database acts as an invaluable asset for researchers, conservationists, and policymakers engaged in the safeguarding and administration of cultural heritage. Through this study, a deeper understanding of the historical significance of Rabat's monuments can be gained, contributing to their preservation, promotion, and sustainable management. The findings and recommendations of the study aim to support informed decision-making regarding the conservation and future development of the historical monument sites in Rabat.

2.2 Data Collection

Within the domain of cultural heritage, inventories are widely recognized by national and international organizations as crucial tools for defence and management [17]. The management strategy implemented by the cultural sector has played a crucial role in the notable development and progress of the inventory of historical sites in Morocco, as depicted in Fig. 2. The present approach relies on the adoption of standardized documentation standards, specifically the cultural heritage inventory sheet. This sheet serves as a valuable tool to facilitate the harmonization of diverse databases, preferably in digital format.

The cultural heritage inventory cards play a crucial role in consolidating all relevant information pertaining to the legally protected heritage and the properties being studied. These cards streamline the registration process for a broad spectrum of data and grant access to additional information of various kinds. This comprehensive approach to inventorying cultural heritage in Morocco has contributed to the organization, preservation, and accessibility of valuable information, enabling effective management and conservation of the country's rich historical legacy.

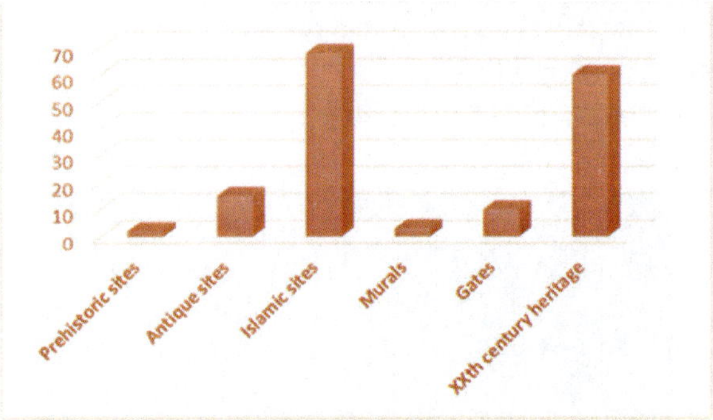

Fig. 2. Quantifying the heritage assets in Rabat City

2.3 Photogrammetric Survey

Digital close-range photogrammetry encompasses a variety of recording techniques, selected depending on the object's complexity and size during the survey. In the case of simple geometries, employing digital image rectification proves to be a cost-efficient approach for producing ortho-images. These ortho-images are then utilized as backdrops for deterioration mapping purposes. Homography-based image rectification is well-suited for approximately planar surfaces, allowing for accurate alignment and correction. On the other hand, polynomial rectification approaches are more beneficial when dealing with complex surfaces that require higher-order transformations for proper rectification. In this study, Agisoft Metashape software was used to perform the rectification process on RGB and NIR images. The software supports the processing of various types of images and can perform geometric rectification to align and correct distortions in images captured by RGB and NIR cameras.

In cases involving complex architectural surfaces or entire historical structures, ortho-rectification may not be sufficient for generating comprehensive photorealistic representations. Instead, simultaneous processing of a large number of images is required to cover the entire object, ensuring consistent geometric results and ortho-images. The effectiveness and capabilities of photogrammetric solutions have significantly improved due to advancements in dense image matching techniques and the development of more advanced camera sensors. These advancements in technology have led to notable progress in the field, enhancing the accuracy and overall performance of photogrammetric processes [18, 19].

The process of architectural recording using dense multiview reconstruction (DMVR) methods involves several steps. It begins with capturing overlapping images that cover all parts of the object under study. By employing structure-from-motion (SfM) techniques, camera positions and 3D point coordinates are estimated through the detection and description of image features, resulting in a sparse point cloud [2, 20]. This sparse point cloud is then refined using dense-matching algorithms to generate a detailed 3D model using triangulation techniques. The model undergoes a process of texturing wherein colors or intensities are interpolated from the image dataset. This texturing step enhances the visual appearance of the model. Furthermore, the textured model can be utilized to generate ortho-image mosaics, providing comprehensive and detailed representations of the surveyed area. The images processed in software like Metashape undergo radiometric correction, ensuring consistent color and intensity across the dataset. High-resolution ortho-images are generated using a mosaicking approach, preserving the radiometric quality of the results instead of averaging. This comprehensive process enables accurate and detailed representations of architectural surfaces, serving various purposes in the field of architectural documentation [2].

2.4 Methodology and Database Creation

The application of Geographical Information Systems (GIS) in the field of heritage conservation and urban revitalization offers a wide range of benefits. With advancements in technology and GIS tools, urban planners, architects, conservationists, managers, and other experts involved in cultural heritage management have greatly benefited. Both local authorities and central governments responsible for cultural heritage management and protection are increasingly relying on information systems and implementing GIS technologies for the management and exchange of spatial data [6, 12].

Accurate spatial information and detailed documentation play a crucial role in the development of restoration programs and the management of historic areas. Detailed and precise documentation is the initial and most important phase in preservation and conservation programs, as it provides a realistic representation of the condition and location of historical objects. Technical documentation and assessment of the current situation enable the development of plans and strategies for protection and future sustainable development [21].

The creation of documentation for historic areas serves three main purposes. Firstly, it serves as a record or backup of historical buildings in the event of destruction or demolition. Secondly, it allows for the preparation of conservation plans and management programs for future development. Lastly, it facilitates monitoring and detection of changes that occur over time or due to external factors [22]. Overall, GIS technology and detailed documentation play a vital role in the effective management, preservation, and sustainable development of historic areas, providing valuable information and tools for decision-making and monitoring.

The methodology employed in this study for creating thematic degradation maps varied depending on the specific types of damage observed on the stone surfaces. Digital image processing tools were utilized to extract features and generate thematic data. The approach for analyzing degradation patterns depended on the complexity of the observed phenomena. Two distinct procedures were employed based on the characteristics of the surfaces. In cases where the surfaces exhibited material loss and discolorations, specific techniques were utilized for feature extraction. This involved manipulating the histogram of the images, applying thresholding to distinguish areas of interest, and performing edge detection to highlight the boundaries of the degraded regions. The output of this process was binary images, where the degraded areas were represented as foreground (white) and the unaffected areas as background (black). By employing these techniques, the aim was to effectively isolate and identify the regions of material loss and discoloration on the surfaces, facilitating further analysis and assessment of the degradation. In cases where surfaces exhibited more complex degradation, such as biological colonization, autonomous image classification using FIJI software was employed to generate labeled image results. The ICOMOS Glossary was used to identify and categorize degradation patterns during monument inspections. It provided standardized definitions and terminology for accurate documentation and communication [23].

Thorough thematic maps depicting the degradation of architectural surfaces in the investigated monuments were developed using ArcGIS 10.8.1. The rectified images and ortho-image mosaics functioned as foundational maps in the GIS environment, allowing annotation of the morphology of building stone elements. In order to conduct a comprehensive analysis of degradation at the structural element level, a thorough examination and analysis are carried out for each individual asset. The degradation patterns were represented as spatial data, and the raster results obtained from digital image processing were automatically converted into a vector format. Separate layers were created for each degradation form. It was important to preserve the topology of the thematic data to establish digital connections between spatial entities and facilitate spatial analyses. Hence, after the conversion of deterioration forms into polygons, the resulting spatial features underwent thorough examination and correction to ensure accuracy and topological consistency.

The resulting digital maps included distinct thematic layers illustrating the outlines of structural elements and various observed degradation types. These visually represented the condition of architectural surfaces, allowing for a comprehensive assessment of damage. Numerical information was extracted from these maps to evaluate the overall state of the historic masonry. Thematic layers were utilized to conduct spatial and logical analyses, enabling the derivation of quantitative data and calculation of deterioration indexes. Specifically, the thematic layers related to vegetation, biological crusts, and material loss were spatially joined to create a consolidated layer capturing overlapping areas.

Using the resulting joined layer, the percentage of coverage for each stone was determined. This information was then used to categorize the stones into five severity levels, indicating the extent of the damage. This categorization allowed for a quantitative assessment of the severity of the degradation observed on each stone block, providing valuable insights into the overall condition of the architectural surfaces.

3 Results and Discussions

The digital techniques mentioned above were implemented for many application case studies (Monuments in Rabat City), focusing on different surface degradation, in archaeological sites located at Chellah archaeological site (Rabat, Morocco).

The study has developed and validated an effective methodology for mapping stone and masonry degradation on historical monuments and remains, utilizing close-range sensing technologies and Geographic Information System (GIS). The methodology prioritizes practicality and flexibility by incorporating affordable and accessible instruments. However, the utilization of low-cost sensor technology requires rigorous data validation and correction to ensure accuracy and reliability. It is essential to thoroughly review and rectify the collected data to maintain its quality. Additionally, verifying the accuracy fundamental cartographic data is crucial for reliable spatial and visualization analysis results. By implementing these measures, the methodology can provide reliable and accurate information for the visualization and analysis of material degradation on historical structures.

3.1 Thematic Maps of Marinid Madrasa Elements

The thematic data related to specific surfaces at the Chellah archaeological site were acquired by digitally processing near-infrared (NIR) reflectance images. These images were chosen due to their consistent radiometric properties, which proved vital in extracting meaningful thematic information [2]. The inclusion of NIR reflectance images played a critical role in avoiding misinterpretation of other degradation forms, as illustrated in the accompanying Fig. 4. This approach ensured more precise identification and analysis of the targeted surfaces at the archaeological site, contributing to enhanced accuracy in the assessment of degradation. Considering the extensive cleaning conducted on the architectural surfaces of the Chellah archaeological site, the visible degradation categories were primarily restricted to material loss, cracks, and localized discoloration. After rectifying and correcting the NIR images, binary images were created to represent these specific categories. To identify areas of discoloration, thresholding techniques were employed following flat-field correction, effectively addressing any remaining concerns related to uneven lighting. Cracks were identified through the application of edge extraction and subsequent binarization techniques projected onto the first principal component of the images. Moreover, material loss was detected by employing histogram equalization and thresholding methods.

3.2 Thematic Map of Roman Assets

For this particular study, visible and near-infrared (NIR) spectrum images were chosen to account for the presence of biological colonization (biogenic crusts, lichens, and

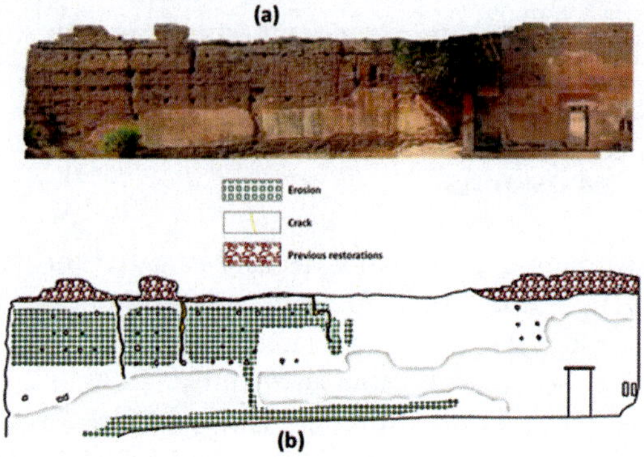

Fig. 3. Orthoimage image of Madrasa Marinid wall (a), and corresponding degradation map (b)

vegetation) at the Roman assets. This decision was based on the unique reflectance characteristics displayed by these colonization forms at different frequency, as illustrated in Fig. 4. To generate orthoimage mosaics of the Roman assets sections, a photogrammetric approach was employed using both RGB and NIR-reflectance images (Fig. 4). These mosaics were blended together in the Agisoft Metashape software, utilizing NIR bands, the Red and the Green to create pseudocolored orthophotos mosaics. The obtained base maps were segmented into distinct zones showing various forms of degradation using an unsupervised clustering method called k-means. The clustering centroids were initialized using an alternative software, such as the DBSCAN algorithm. The segmented images were then digitized and deterioration patterns were labelled in ArcGIS by associating them with corresponding thematic layers, as depicted in Figs. 3, 4, 5 and 6. Additionally, a map highlighting the types of damage specifically related to the remains of the Roman part in the archaeological site of Chellah was produced Fig. 6. Figures 5 and 6 present statistical information regarding the extent of surface coverage of historical remains by different forms of degradation. Specifically, the studied surfaces were either fully covered with biogenic materials, discoloured, or lacked material.

Fig. 4. Orthoimages and near-infrared images of Roman assets at the Chellah archaeological site

Fig. 5. Deterioration index map of Roman remain at the Chellah archaeological site

Fig. 6. Degradations map of Roman remains at the Chellah archaeological site

The results obtained from the study highlighted the vulnerability of the monuments remains within the archaeological site Chellah. These monuments were identified as the most vulnerable elements of the cultural heritage due to the lack and absence of programs focused on preventive conservation in place. This finding emphasizes the need for proactive conservation efforts to mitigate the risks faced by these vulnerable elements.

4 Conclusion

The integration of digital technologies and GIS in heritage conservation planning proved to be beneficial, offering effective tools for the management of deterioration risks and the preservation of cultural assets [24]. The study's results contribute to the field of heritage conservation by demonstrating the potential of these technologies in supporting conservation planning efforts. The findings also highlight the importance of implementing preventive conservation programs to protect vulnerable elements of cultural heritage and ensure their long-term preservation.

By utilizing digital technologies and GIS, the study successfully created a comprehensive database of historical heritage. Thematic maps were generated, providing valuable insights into the distribution and characteristics of cultural assets at the Chellah archaeological site. Furthermore, the utilization of 2D and 3D views enhanced the visualization and understanding of these properties. The integration of digital technologies and GIS in heritage conservation planning proved to be beneficial, offering effective tools for the management of deterioration risks and the preservation of cultural assets. The findings also highlight the importance of implementing preventive conservation programs to protect vulnerable elements of cultural heritage and ensure their long-term preservation.

Acknowledgements. I would like to express my gratitude and appreciation to the Higher School of Technology in Salé and the Rabat-Salé-Kénitra region for their valuable collaboration. Their support and cooperation have been instrumental within the project 3D reconstruction of the Marinid Madrasa located in the Chellah archaeological site (Rabat, Morocco).

References

1. Nilson, T., Thorell, K.: Cultural Heritage Preservation: The Past, the Present and the Future (2018). Consulté le: 7 juin 2023. [En ligne]. Disponible sur: https://www.diva-portal.org/smash/get/diva2:1224014/FULLTEXT01.pdf
2. Adamopoulos, E., Rinaudo, F.: Combining multiband imaging, photogrammetric techniques, and FOSS GIS for affordable degradation mapping of stone monuments. Buildings **11**(7) (2021). Art. no. 7. https://doi.org/10.3390/buildings11070304
3. Bruno, N., Rechichi, F., Achille, C., Zerbi, A., Roncella, R., Fassi, F.: Integration of historical GIS data in a HBIM system. ISPRS Int. Arch. Photogram. Remote Sens. Spat. Inf. Sci. **XLIII-B4-2020**, 427–434 (2020). https://doi.org/10.5194/isprs-archives-XLIII-B4-2020-427-2020
4. Simou, S., Baba, K., Nounah, A.: 3D reconstruction of the Marinids Site Located at the Chellah Archaeological Area (2021). https://doi.org/10.1038/srep01183
5. Brumann, C.: Cultural heritage. In: Wright, J.D. (ed.) International Encyclopedia of the Social & Behavioral Sciences, second edn, pp. 414–419. Elsevier, Oxford (2015). https://doi.org/10.1016/B978-0-08-097086-8.12185-3
6. Sánchez-Aparicio, L., et al.: Web-GIS approach to preventive conservation of heritage buildings. Automat. Constr. **118**, 103304 (2020). https://doi.org/10.1016/j.autcon.2020.103304
7. Salagean-Mohora, I. Anghel, A.A. Frigura-Iliasa, F.M.: Photogrammetry as a digital tool for joining heritage documentation in architectural education and professional practice. Buildings **13**(2) (2023). Art. no. 2. https://doi.org/10.3390/buildings13020319
8. Chen, Y.,Wu, Y., Sun, X., Ali, N., Zhou, Q.: Digital documentation and conservation of architectural heritage information: an application in modern Chinese architecture. Sustainability **15**(9) (2023). Art. no. 9. https://doi.org/10.3390/su15097276
9. Murphy, M., Mcgovern, E., Pavía, S.: Historic building information modelling (HBIM). Struct. Surv. **27**, 311–327 (2009). https://doi.org/10.1108/02630800910985108
10. Santos, B., et al.: GIS in architectural teaching and research: planning and heritage. Educ. Sci. **11**(6) (2021). Art. no. 6. https://doi.org/10.3390/educsci11060307
11. Nocca, F.: The role of cultural heritage in sustainable development: multidimensional indicators as decision-making tool. Sustainability **9**(10) (2017). Art. no. 10. https://doi.org/10.3390/su9101882
12. Simou, S., Baba, K., Nounah, A.: A GIS-based methodology to explore and manage the historical heritage of Rabat City (Morocco). J. Comput. Cult. Herit. **15**(4), 74:1-74:14 (2022). https://doi.org/10.1145/3517142
13. Simou, S., Baba, K., Nounah, A.: The integration of 3D technology for the conservation and restoration of ruined archaeological artifacts. Hist. Sci. Technol. **12**(1) (2022). Art. no. 1. https://doi.org/10.32703/2415-7422-2022-12-1-150-168

14. Razzouk, Y., Baba, K., Ahatri, M.: The inflence of spectral responses on the structures heights: case of the RHiss river earthquake in Morocco (6.3 MW)—seismogenic source 4 (Rif oriental—Al Hoceima—Alboran). ARPN J. Eng. Appl. Sci. **17**(6), 645–651 (2022)
15. Razzouk, Y., Ahatri, M., Baba, K., Majid A.E.: Optimal bracing type of reinforced concrete buildings with soil-structure interaction taken into consideration. Civ. Eng. J. **9**(6), 1371–1388 (2023). https://doi.org/10.28991/CEJ-2023-09-06-06
16. Razzouk, Y., Ahatri, M., Baba, K., El Majid, A.: The impact of bracing type on seismic response of the structure on soft soil. Civ. Eng. Arch. **11**(5), 2706–2718 (2023). https://doi.org/10.13189/cea.2023.110534
17. Dudley, N.: Guidelines for applying protected area management categories. IUCN (2008). https://doi.org/10.2305/IUCN.CH.2008.PAPS.2.en
18. Kortaberria, G., Mutilba, U., Gomez-Acedo, E., Tellaeche, A., Minguez, R.: Accuracy evaluation of dense matching techniques for casting part dimensional verification. Sensors **18**(9) (2018). Art. no. 9. https://doi.org/10.3390/s18093074
19. Panagiotopoulou, A., et al.: Super-resolution techniques in photogrammetric 3D reconstruction from close-range UAV imagery. Heritage **6**(3) (2023). Art. no. 3. https://doi.org/10.3390/heritage6030143
20. Ioannakis, G.A., Koutsoudis, A., Arnaoutoglou, F., Chamzas, C.: Enhancing Multi-Image Photogrammetric 3D Reconstruction Performance on Low-Feature Surfaces (2015)
21. Muhammad, S.: Documentation and monitoring of built heritage in Abu Dhabi, UAE. ISPRS Int. Arch. Photogram. Remote Sens. Spat. Inf. Sci. **XL-5/W2**, 437–442 (2013). https://doi.org/10.5194/isprsarchives-XL-5-W2-437-2013
22. Lourenco, P., Peña, F., Amado, M.: A document management system for the conservation of cultural heritage buildings. Int. J. Arch. Herit. **5**, 101–121 (2011). https://doi.org/10.1080/15583050903318382
23. ICOMOS: ICOMOS, Charte internationale sur la conservation et la restauration des monuments et des sites (2018)
24. Mendoza, M.A.D., De La Hoz Franco, E., Gómez, J.E.G.: Technologies for the preservation of cultural heritage—a systematic review of the literature. Sustainability **15**(2) (2023). Art. no. 2. https://doi.org/10.3390/su15021059

Road Distress Detection and Classification: Harnessing the Synergy of Deep Learning and Transfer Learning Approaches

Oumaima Khlifati(✉) and Khadija Baba

Civil Engineering and Environment Laboratory (LGCE), Mohammadia School of Engineering, Mohammed V University in Rabat, Rabat, Morocco
khlifatioumaima@gmail.com

Abstract. Pavement maintenance is an essential step to maintain good service quality, pavement structure, durability and sufficient safety. The auscultation allows to obtain the necessary information on the pavement condition in order to apply the appropriate corrective actions at the right time. In the past, the assessment was based on the visual survey which took a lot of time and required domain expertise. Therefore, an automatic detection is very necessary to stimulate this progress. Deep learning has proven to be remarkably effective in the classification of pavement degradation. Nevertheless, achieving the highest level of predictive accuracy often requires extensive annotated image datasets. To tackle this challenge, we propose the utilization of a Deep learning model with transfer learning to identify and classify distress in pavement. The dataset utilized in this study comprises a limited number of asphalt pavement images to evaluate the efficacy of the model. In this paper, both VGG-16 and VGG-19 models are evaluated for binary classification to differentiate between cracked and non-cracked pavement images. Additionally, we conducted further testing using multi-label classification to classify pavement images into four distinct categories: alligator crack, transverse crack, longitudinal crack, and pothole. The performance evaluation of both models is conducted using various metrics, including accuracy, F1 score, recall, precision. According to the findings from the experiments, it is evident that VGG-16 is a robust model for binary classification, as it optimizes computational time while achieving a high level of accuracy. On the other hand, VGG-19 proves to be a powerful model for classifying different types of pavement degradation while ensuring optimizing computational time, and achieving a higher accuracy level without the need for a large dataset.

Keywords: Transfer learning · Deep learning · Road distress · Automatic classification

1 Introduction

Road infrastructure plays a vital role as a valuable public asset by promoting economic development and growth, and delivering essential social benefits. It serves as a vital link connecting communities and businesses, granting access to healthcare services, employment and education. Nevertheless, the pavement gradually erodes and deteriorates over

K. Baba et al. (Eds.): GeoME 2023, SPEES, pp. 346–355, 2023.
https://doi.org/10.1007/978-3-031-49345-4_33

a period due to various factors such as climatic conditions, geographical position and traffic flow. Consequently, understanding the extent of its degradation becomes crucial to ensure effective and economical maintenance. Initially, road damage detection relied on inspectors visually assessing the roads, resulting in labor intensity, traffic disruption, inaccuracy, significant level of risk and subjective outcomes. To address these limitations, the concept of automated monitoring of pavement distress was introduced by Cheng et al. [1]. This approach involved the utilization of video cameras to capture road images, which were subsequently subjected to analysis using diverse image processing techniques by Li et al. [2]. Currently, the evaluation of pavement distress has evolved significantly, transitioning from manual visual inspections to the utilization of high-speed digital cameras mounted on data collection vehicles [3]. However, in most cases, the gathered images still require manual processing.

The initial effort towards automated classification was carried out using thresholding methods. In 1999, Cheng et al. introduced a crack identification method that incorporated fuzzy thresholding [4]. However, the effectiveness of this method was found to be limited and it did not yield satisfactory results. Kirschke and Velinsky implemented a methodology where they divided the image into numerous sub-blocks [5]. They utilized the gray histogram for threshold segmentation to detect pavement crack. However, the overall accuracy of this approach was found to be subpar. Oliveira and Correia employed a dynamic thresholding technique in conjunction with morphological filters to enhance the detection and segmentation of cracks [6]. Cheng and Shi proposed an alternative approach involves determining the threshold through the reduction of the sample space and interpolation by utilization of the average and variance of pixel grayscale values [7]. However, the segmentation outcomes of this approach remain inadequate, characterized by a high false detection rate, as well as the presence of isolated noise points along the crack's edges.

Different methods for edge detection, such as the application of Canny, Sobel detector and Gaussian filter have been suggested for the purpose of road crack detection (Ayenu-Prah and Attoh-Okine) [8]. The application of multi-scale resolution wavelet techniques has garnered considerable interest, particularly in relation to pavement crack analysis (Subirats and Dumoulin) [9]. Ying and Salari proposed a method for road crack detection and classification, which relies on the beamlet transform as a fundamental technique [10]. This approach is claimed to exhibit enhanced robustness in identifying linear features, even when noise is present. According to the review conducted by Koch et al., computer vision has emerged as a leading approach for crack detection in image-based systems, leveraging its robust capabilities [11]. Lin and Liu employed a non-linear Support Vector Machine (SVM) for the identification of potholes within a targeted region [12]. In a similar vein, Kapela et al. employed Support Vector Machine (SVM) in conjunction with histogram features to effectively detect cracks [13].

In the aforementioned traditional image recognition methods, manual intervention is required for feature engineering, which involves extracting relevant characteristics and descriptors from the images. The advancement of convolutional neural network (CNN) algorithms has experienced rapid progress, enabling the automated extraction of image features. Zhang et al. introduced a compact deep convolutional neural network (DCNN) architecture tailored specifically for the purpose of pavement crack detection [14]. The model consisted of four blocks, with each block consisting of a convolutional layer followed by a max-pooling layer and two fully connected layers. A concrete crack detection model was presented by Cha et al., which utilizes a DCNN model that eliminates the need for explicit calculation of defect features, offering a more streamlined approach to crack detection [15]. In practical applications, the classification of various types of distress are often required. However, the deep learning models that have been developed have rarely met this requirement sufficiently. Various DCNN have exhibited outstanding performance in the task of image classification. Prominent examples include Visual Geometry Group (VGG) [16], AlexNet [17] and Residual Network (ResNet) [18].

Deep Learning has emerged as a valuable technique for road crack detection. However, its reliance on substantial labeled training data poses challenges. In this study, we propose harnessing the power of pre-trained DCNN models with transfer learning to enable automated detection of pavement distress. By leveraging this approach, we aim to highlight its notable advantages and benefits. The following sections of this paper are structured as follows: a comprehensive description of the proposed methodology is presented, which includes a comprehensive description of the utilized datasets and model performance measures. This is followed by a comprehensive presentation of the experimental results and an in-depth discussion of the findings.

2 Datasets and Method

2.1 Datasets

In this study, we used two databases. The first one is the Mendely Asphalt Crack Dataset [19], which was used for binary classification. The Mendely Asphalt Crack Dataset consists of 400 images of pavement cracks. Each road image has a resolution of 480 × 480 pixels. This dataset was graciously contributed by Jayanth Balaji et al. [19]. Figure 1 provides visual samples from this dataset, illustrating both cracked and intact pavements. These pavement images were labeled into two classes: crack and no crack. The second database [20] consists of various road distress types, including longitudinal cracks, transverse cracks, pothole and alligator cracks. This database comprises 79 road images with dimensions of 2048 × 1635 pixels. Figure 2 visually depicts samples from the dataset, showcasing four distinct classes of distressed pavement. After performing the labeling process, both datasets were divided into two parts, with 67% of the images allocated for model training and the remaining 33% for model validation.

2.2 Method

DCCNs have proven to be incredibly efficient in handling visual information, including images and videos. DCNNs often demand extensive annotated image datasets to attain optimal predictive accuracy. Nevertheless, obtaining such data can be challenging and labeling them incurs substantial costs, particularly in various domains. Given these obstacles, leveraging pre-trained DCNN features from established models like VGG, Resnet and GoogLeNet has proven highly beneficial for addressing multi-domain image classification challenges [21].

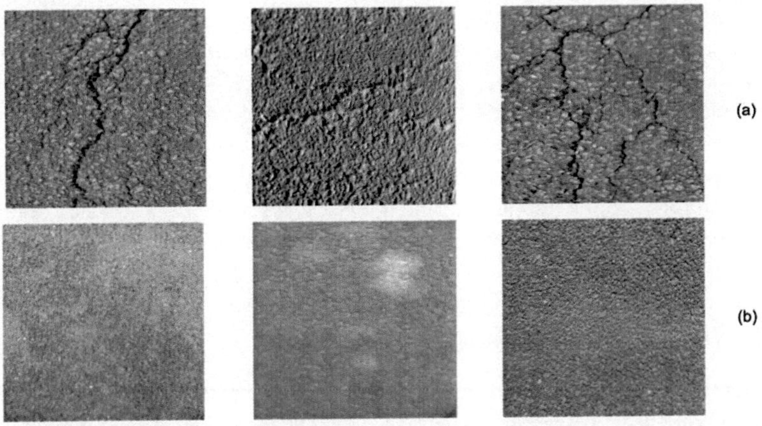

Fig. 1. Samples of cracked and no cracked pavement images. (**a**): crack image, (**b**): no crack image.

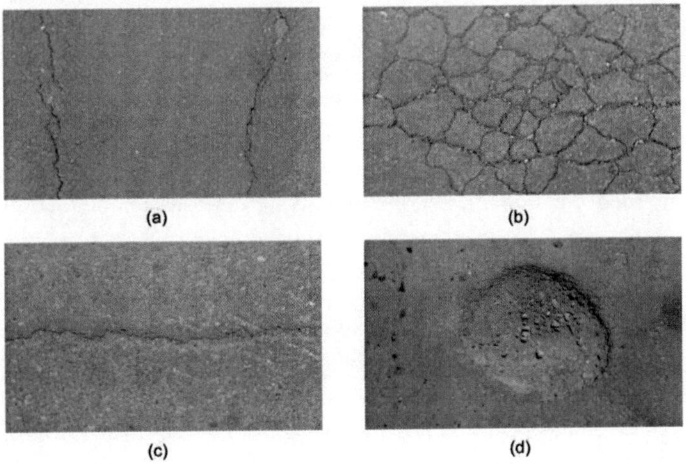

Fig. 2. Samples of distressed pavement images. (**a**): longitudinal crack, (**b**): alligator crack, (**c**): transverse crack and (**d**): pothole.

In this research, a two-stage strategy is implemented. Initially, the pavement images are categorized into two distinct groups: images including cracks and images without any cracks. Subsequently, for the images categorized as crack, the specific types of crack present in the images are detected as such as transverse crack longitudinal crack, alligator crack and pothole. In the initial stage, we used a binary classification model, while in the subsequent stage, we adopted a multi-label classification model. Both VGG-16 and VGG-19 networks are utilized as models for binary classification as well as multi-label classification tasks.

In this work, we used two versions of the VGG DCNN model: VGG-16 and VGG-19. Regarding the VGG-16 DCNN model, it comprises 16 convolutional layers, each having a receptive field size of 3 * 3, totaling 138 million parameters and incorporates five spatial pooling layers, each employing a size of 2 * 2 for max-pooling, and three fully-connected layers. The last layer serves as the softmax layer. The VGG-19 DCNN model consists of 19 convolutional layers, with each layer having a receptive field size of 3 * 3, resulting in a total of 144 million parameters. Additionally, the model incorporates five spatial pooling layers that employ a max-pooling technique with a size of 2 * 2. Furthermore, it includes three fully-connected layers, and the last layer serves as a softmax layer. Table 1 illustrates the employed architecture of VGG 16 and 19 and the number of parameters of each layer.

2.3 Model Performance Measures

There are four widely employed performance metrics: precision, F1 score, recall and accuracy, which are utilized to assess the effectiveness of trained CNNs in image classification tasks. The following section presents a definition to each of these measures.

Accuracy (Acc) is a metric that represents the ratio of correctly identified images, whether they contain distress or not. The calculation of this metric is performed using Eq. (1).

$$\mathrm{Acc} = \frac{\mathrm{TP} + \mathrm{TN}}{\mathrm{TP} + \mathrm{TN} + \mathrm{FN} + \mathrm{FP}} \tag{1}$$

Precision (Pr) is a measure that quantifies the ratio of correctly identified images with distress to the total number of images predicted as distress, including both accurate and inaccurate identifications. It is computed using the equation defined in Eq. (2).

$$\mathrm{Pr} = \frac{\mathrm{TP}}{\mathrm{TP} + \mathrm{FP}} \tag{2}$$

Recall (Rec) is a metric that indicates the ratio of correctly identified images with distress to the total number of images that actually contain distress. It is calculated using the equation defined in Eq. (3).

$$\mathrm{Rec} = \frac{\mathrm{TP}}{\mathrm{TP} + \mathrm{FN}} \tag{3}$$

The F1 score (F1 S) is a composite measure that combines both precision and recall into a single index. It quantifies the balance between these two metrics and is calculated using the equation defined in Eq. (4).

$$\mathrm{F1S} = \frac{2 \times \mathrm{Pre} \times \mathrm{Rec}}{\mathrm{Pre} + \mathrm{Rec}} \tag{4}$$

3 Results and Discussion

3.1 Testing Models for Binary Image Classification

VGG-19 and VGG-16 models are trained to distinguish between cracked pavement and no cracked pavement images. For the training of both models, we utilized a training configuration consisting of 10 epochs and a batch size of 32. The performance of these

Table 1. The architecture of VGG-16 and VGG-19. Conv.L.: Convolution layer, Pool.L.: pooling layer, F.C.L.: Fully connected layer.

Model	Layer	Output shape	Parameters
VGG-16	Input	224 * 224 * 3	–
	Conv.L. 3 * 3	224 * 224 * 64	1792
	Conv. L. 3 * 3	224 * 224 * 64	36,928
	Pool. L. 2 * 2	112 * 112 * 64	–
	Conv. L. 3 * 3	112 * 112 * 128	73,856
	Conv. L. 3 * 3	112 * 112 * 128	147,584
	Pool. L. 2 * 2	56 * 56 * 128	–
	Conv. L. 3 * 3	56 * 56 * 256	295,168
	Conv. L. 3 * 3	56 * 56 * 256	590,080
	Conv. L. 3 * 3	56 * 56 * 256	590,080
	Pool. L. 2 * 2	28 * 28 * 256	–
	Conv. L. 3 * 3	28 * 28 * 512	1,180,160
	Conv. L. 3 * 3	28 * 28 * 512	2,359,808
	Conv. L. 3 * 3	28 * 28 * 512	2,359,808
	Pool. L. 2 * 2	14 * 14 * 512	–
	Conv. L. 3 * 3	14 * 14 * 512	2,359,808
	Conv. L. 3 * 3	14 * 14 * 512	2,359,808
	Conv. L. 3 * 3	14 * 14 * 512	2,359,808
	Pool. L. 2 * 2	7 * 7 * 512	–
	F.C. L	1 * 14096	102,764,544
VGG-19	Input		–
	Same architecture as VGG-16		
	Conv. L. 3 * 3	14 * 14 * 512	2,359,808
	Conv. L. 3 * 3	14 * 14 * 512	2,359,808
	Conv. L. 3 * 3	14 * 14 * 512	2,359,808
	Pool. L. 2 * 2	7 * 7 * 512	–
	F.C. L	1 * 14096	102,764,544

models is evaluated using the Mendely Asphalt Crack Dataset. The summarized results of the aforementioned performance metrics are presented in Table 2. It is evident from Table 2 that VGG 16 achieves the highest accuracy of 98,48% and precision of 98,59%, recall of 98,41% and 98,48% of F1 score.

The confusion matrix provides information about the count and types of incorrect predictions in each classification, as well as the number of correct predictions. Figure 3 displays the confusion matrices VGG-16 and VGG-19. The matrices demonstrate that VGG-16 exhibits the highest accuracy in correctly classifying distressed images.

Based on the findings, it can be concluded that VGG-16 demonstrates a higher level of performance in binary classification compared to VGG-19, despite both models being trained for only 10 epochs. This indicates that VGG-16 exhibits superior learning and generalization capabilities, enabling it to effectively extract and generalize features in a shorter training period. This is advantageous since VGG-16 is less complex and requires less computational time during calculations.

Table 2. Effectiveness of VGG-16 and VGG-19 for binary classification images.

Model	Acc (%)	Pre (%)	Rec (%)	F1 S (%)
VGG-16	98,48	98,59	98,41	98,48
VGG-19	97,73	97,92	97,62	97,72

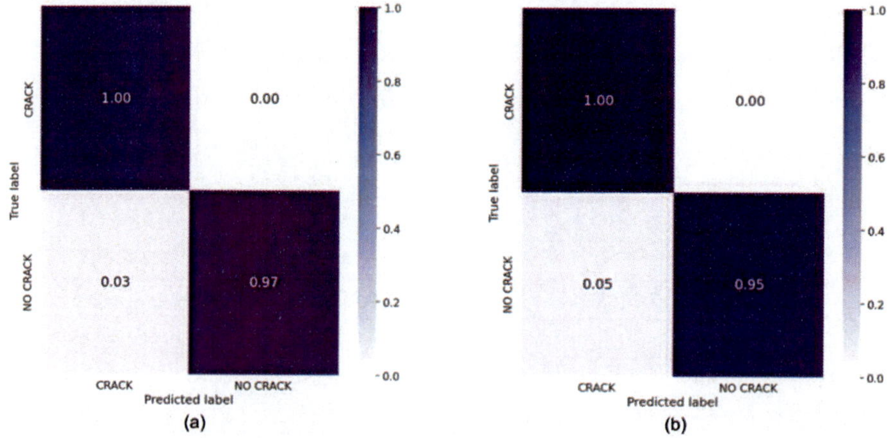

Fig. 3. Confusion matrices of binary classification. (**a**) VGG-16 and (**b**) VGG-19

3.2 Testing of Multi-label Classification Models

For multi-label classification, the dataset of [20] is employed, which consists of four distinct categories: alligator crack, transverse crack, longitudinal crack and potholes. After labeling these various datasets, we proceed to carry out the training process using both VGG16 and VGG19 architectures to assess and compare their performance. The training of this dataset was conducted using 30 epochs and a batch size of 32. Table 3 presents the results of accuracy, precision, recall, and F1 score for both models. It is evident that all performance metrics reveal the power of the VGG-19 model in the multi-class classification of different types of distress pavement. This remarkable performance can be attributed to the depth of the model, which enables it to learn complex and discriminative features from pavement degradation images. It is noteworthy that despite the limited size of the dataset, the model achieved an accuracy of over 94%. This highlights the efficiency and capability of the model to generalize and perform well even with a limited dataset.

To further facilitate a comparison between these two models, we will proceed by presenting the results obtained from the confusion matrix provided in Fig. 4. As we can observe from these results, both models perform well and have nearly similar predictions for the different classes, except for the alligator crack category. According to these results, the model has correctly classified the longitudinal crack category with an accuracy of 79%. For the alligator crack class, the model has achieved an accuracy of 88%. This suggests that the model has demonstrated a satisfactory performance in accurately identifying and classifying images with this both classes. In addition, the model has achieved a flawless accuracy of 100% in classifying images belonging to both the longitudinal crack and pothole classes. This indicates that the model is performing exceptionally well in classifying These two classes.

Table 3. Effectiveness of VGG-16 and VGG-19 for muti-label classification images.

Model	Acc (%)	Pre (%)	Rec (%)	F1 S (%)
VGG-16	85,71	88,64	88,39	86,07
VGG-19	94,28	93,75	96,43	94,50

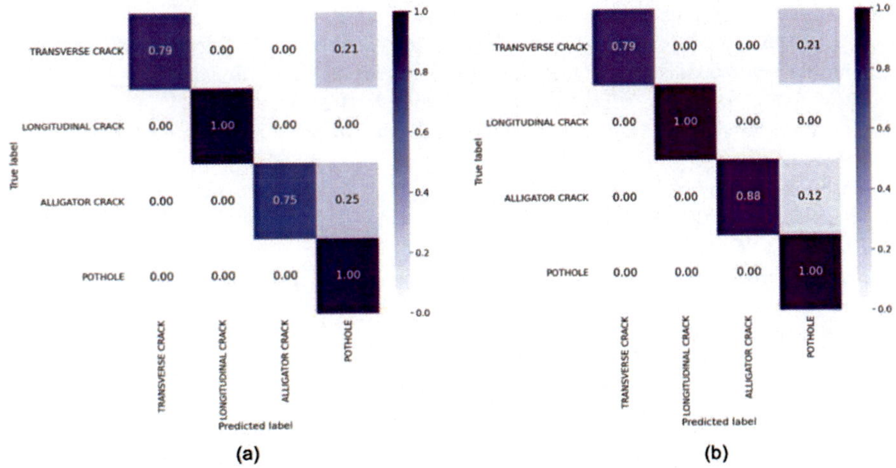

Fig. 4. Confusion matrices of multi-label classification. **(a)** VGG-16 and **(b)** VGG-19.

4 Conclusions

With the advancements in pavement distress inspection technology, pavement degradation detection has moved from manual surveys to automatic methods. Deep learning has demonstrated remarkable effectiveness in classifying pavement degradation. However, it often necessitates extensive annotated image datasets to achieve the highest level of predictive accuracy. In order to address this issue, we suggest employing a Deep Convolutional Neural Network model with transfer learning to identify and classify the distress pavement. This study is conducted in two stages. Firstly, we tested the effectiveness of the VGG-16 and VGG-19 models for binary classification, specifically distinguishing between crack and no crack images. According to the results obtained, VGG-16 demonstrated better performance, even with a smaller number of epochs. It can be concluded that VGG-16 is a powerful model for binary classification, ensuring both computational time optimization and achieving a high accuracy level. Secondly, we proceeded with training the VGG-16 and VGG-19 models for multi-label classification using a small dataset consisting of 79 images. The objective was to classify these images into four distinct classes: alligator crack, transverse crack, longitudinal crack and potholes. According to the results obtained, VGG-19 outperformed VGG-16 in all performance metrics. It can be concluded that VGG-19 is a powerful model for classifying various pavement degradation types while ensuring optimizing computational time, and achieving a higher accuracy level without the need for a large dataset.

References

1. Cheng, H.D., Miyojim, M.: Automatic pavement distress detection system. Inf. Sci. **108**(1), 219–240 (1998). https://doi.org/10.1016/S0020-0255(97)10062-7
2. Li, D., Duan, Z., Hu, X., Zhang, D., Zhang, Y.: Automated classification and detection of multiple pavement distress images based on deep learning. J. Traffic Transp. Eng. (Engl. Edn.) **10**(2), 276–290 (2023). https://doi.org/10.1016/j.jtte.2021.04.008
3. Gopalakrishnan, K.: Advanced pavement health monitoring and management. IGI 771 Global Videos (2016)
4. Cheng, H.D., Chen, J.-R., Glazier, C., Hu, Y.G.: Novel approach to pavement cracking detection based on fuzzy set theory. J. Comput. Civ. Eng. **13**(4), 270–280 (1999). https://doi.org/10.1061/(ASCE)0887-3801(1999)13:4(270)
5. Kirschke, K.R., Velinsky, S.A.: Histogram-based approach for automated pavement-crack sensing. J. Transp. Eng. **118**(5), 700–710 (1992). https://doi.org/10.1061/(ASCE)0733-947X(1992)118:5(700)
6. Oliveira, H., Correia, P.L.: Automatic road crack segmentation using entropy and image dynamic thresholding. In: 2009 17th European Signal Processing Conference, pp. 622–626 (2009)
7. Cheng, H.D., Shi, X.J., Glazier, C.: Real-time image thresholding based on sample space reduction and interpolation approach. J. Comput. Civ. Eng. **17**(4), 264–272 (2003). https://doi.org/10.1061/(ASCE)0887-3801(2003)17:4(264)
8. Ayenu-Prah, A., Attoh-Okine, N.: Evaluating pavement cracks with bidimensional empirical mode decomposition. EURASIP J. Adv. Signal Process. **2008**(1), Art. no. 1 (2008). https://doi.org/10.1155/2008/861701

9. Subirats, P., Dumoulin, J., Legeay, V., Barba, D.: Automation of pavement surface crack detection using the continuous wavelet transform. In: 2006 International Conference on Image Processing, pp. 3037–3040 (2006). https://doi.org/10.1109/ICIP.2006.313007
10. Beamlet Transform-Based Technique for Pavement Crack Detection and Classification - Ying - 2010 - Computer-Aided Civil and Infrastructure Engineering - Wiley Online Library. https://doi.org/10.1111/j.1467-8667.2010.00674.x. Accessed 02 Jun 2023
11. Koch, C., Georgieva, K., Kasireddy, V., Akinci, B., Fieguth, P.: A review on computer vision based defect detection and condition assessment of concrete and asphalt civil infrastructure. Adv. Eng. Inform. **29**(2), 196–210 (2015). https://doi.org/10.1016/j.aei.2015.01.008
12. Lin, J., Liu, Y.: Potholes detection based on SVM in the pavement distress image. In: 2010 Ninth International Symposium on Distributed Computing and Applications to Business, Engineering and Science, Aug. 2010, pp. 544–547 (2010). https://doi.org/10.1109/DCABES.2010.115
13. Kapela, R., et al.: Asphalt surfaced pavement cracks detection based on histograms of oriented gradients. In: 2015 22nd International Conference Mixed Design of Integrated Circuits & Systems (MIXDES), Jun. 2015, pp. 579–584. https://doi.org/10.1109/MIXDES.2015.7208590
14. Zhang, L., Yang, F., Zhang, Y.D., Zhu, Y.J.: Road crack detection using deep convolutional neural network. In: 2016 IEEE International Conference on Image Processing (ICIP), Sep. 2016, pp. 3708–3712. https://doi.org/10.1109/ICIP.2016.7533052
15. Cha, Y.-J., Choi, W., Büyüköztürk, O.: Deep learning-based crack damage detection using convolutional neural networks. Comput. Aided Civ. Infrastruct. Eng. **32**(5), 361–378 (2017). https://doi.org/10.1111/mice.12263
16. K. Simonyan and A. Zisserman, "Very Deep Convolutional Networks for Large-Scale Image Recognition," *arXiv.org*, Sep. 04, 2014. https://arxiv.org/abs/1409.1556v6 (accessed Jun. 03, 2023)
17. Krizhevsky, A., Sutskever, I., Hinton, G.E.: Imagenet classification with deep convolutional neural networks. Commun. ACM **60**(6), 84–90 (2017)
18. He, K., Zhang, X., Ren, S., Sun, J.: Deep residual learning for image recognition. In: Proceedings of the IEEE Conference on Computer Vision and Pattern Recognition, 2016, pp. 770–778. Accessed: Jun. 03, 2023. [Online]. Available: https://openaccess.thecvf.com/content_cvpr_2016/html/He_Deep_Residual_Learning_CVPR_2016_paper.html
19. Jayanth Balaji, A., Thiru Balaji, G., Dinesh, M.S., Nair, Binoy, Harish Ram, D.S.: Asphalt Crack Dataset, vol. 2, (2019). https://doi.org/10.17632/xnzhj3x8v4.2
20. GitHub - summoners29/Asphalt-Pavement-Crack-Data: Asphalt pavement crack image data. https://github.com/summoners29/Asphalt-Pavement-Crack-Data. Accessed 03 Jun 2023
21. Scopus - Document details - Deep Convolutional Neural Networks for Computer-Aided Detection: CNN Architectures, Dataset Characteristics and Transfer Learning | Signed in. https://www.scopus.com/record/display.uri?eid=2-s2.0-84969962996&origin=inward&txGid=521d73da0f0142ca4f9722abc4dd4481. Accessed 29 May 2023

Geomechanical Classification and Observational Method for Deep Urban Excavations in Shale Formations

Ghizlane Boulaid[1](✉), Latifa Ouadif[1], Lahcen Bahi[1], Mohamed Ben Ouakkass[1], Rhita Bennouna[1], and Ahmed Skali Senhaji[2]

[1] Mohammadia School of Engineering, Laboratory of Applied Geophysics, Geotechnics, Engineering Geology and Environment, Mohammed V University, Rabat, Morocco
ghizlane_boulaid@um5.ac.ma

[2] Society for Technical and Economic Studies, Rabat, Morocco

Abstract. Deep urban excavations are crucial for the development of modern cities, but they pose complex challenges due to the specific geological conditions of urban environments. Shale formations, commonly encountered in these contexts, require specific excavation approaches due to their varied geomechanical behavior and propensity for fracturing. To better understand their behavior, this article employs classification systems such as RQD, Rock Mass Rating (RMR), and Geological Strength Index (GSI) to assess the quality and stability of shale formations at different depths. The project site corresponds to a deep excavation in the center of the future Casablanca building, reaching a depth of over 22 m. The excavation involves reinforcing the party wall composed of an R + 15 building and surrounding roads on three sides. Prior to and during excavation works, several measurement devices were installed to monitor the horizontal displacement of the excavation screen and adjacent structures, as well as to measure soil settlement behind the anchored walls using Shape Accel Array inclinometers and cyclops. The application of the observational method allows real-time monitoring of the soil during excavation, enabling the rapid detection of unexpected or abnormal behaviors. By comparing the observed measurements with the expected values, adjustments can be made to the design, excavation sequence, and support systems to ensure stability and safety of the works. The study also explores the correlation between the results of geomechanical classifications and the instrumentation of deep excavation, highlighting the importance of real-time monitoring to detect potential risks associated with excavations in fractured shale formations. The findings of this study provide essential insights for the design and implementation of appropriate support measures in similar projects worldwide, contributing to the safer and more efficient execution of deep urban excavations in geologically complex environments.

Keywords: Deep excavation · Geomechanical classification observational method · Urban Area

K. Baba et al. (Eds.): GeoME 2023, SPEES, pp. 356–367, 2023.
https://doi.org/10.1007/978-3-031-49345-4_34

1 Introduction

Deep urban excavations represent engineering projects of paramount importance in the development of modern cities. Whether for constructing skyscraper foundations, transport tunnels, or underground networks, these works are highly complex due to the specific geological conditions of urban environments [1, 2]. Among the geological formations frequently encountered in these contexts, shales play a significant role. However, their varied geomechanical behavior and propensity for fracturing demand specific excavation approaches to ensure the safety and stability of the projects. To gain a better understanding of their geomechanical behavior, the article utilizes classification systems such as the Rock Quality Designation (RQD) [3], the Rock Mass Rating (RMR) [4], and the Geological Strength Index (GSI) [5] to evaluate the quality and stability of shale formations at different depths [6, 7].

It is in this context that the geomechanical classification of shales and the application of the observational method [8, 9] become highly relevant. The presented article focuses on the application of the observational method in the field of deep urban excavations, offering an adapted approach to deal with the unpredictable challenges of the subsurface in real-time. The observational method allows for meticulous monitoring of soil behavior during the excavation process, providing the opportunity to promptly identify any unexpected or abnormal behavior. By comparing measured displacements to expected values, adjustments can be made to the design, excavation sequence, and support systems used to ensure the stability and safety of the works [10, 11].

By exploring the correlation between the results of geomechanical classifications and deep excavation instrumentation, this article highlights the importance of real-time monitoring to detect potential risks associated with excavations in fractured shale formations.

2 Project Background

The city of Casablanca is a city located on the Atlantic coast in northwest Africa, at a latitude of 33° 35′ North and a longitude of 7° 25′ West. The city is bordered by several geographical elements: to the north, it is bordered by the Atlantic Ocean, to the east by the Benslimane plateau, and to the south and west by the plains of the Chaouia region.

Due to its privileged geographical location, Casablanca has witnessed remarkable development in recent years. The number of construction projects in the city has continued to increase to meet the growing demand for infrastructure and housing.

In this context, the case study presented is part of a construction project aimed at building a 20-story building, consisting of twenty floors above ground, as well as five levels of underground floors.

2.1 Regional Geology

The basement of the Casablanca region, dating back to the Cambrian and Ordovician periods, has been primarily affected by the Hercynian orogeny. Although more recent tectonic phases have occurred, they have only resulted in low-amplitude faults and movements. The Paleozoic basement has undergone folding and uplift, exhibiting numerous faults.

However, the sub-horizontal cover, composed of Permian-Triassic, Cretaceous, Miocene, and Plio-Quaternary formations, is weakly deformed. These sedimentary layers were deposited on the folded and uplifted Paleozoic basement, creating a relatively flat surface. Therefore, the sedimentary cover is generally horizontal, although locally it may be affected by minor faults and folds (see Fig. 1).

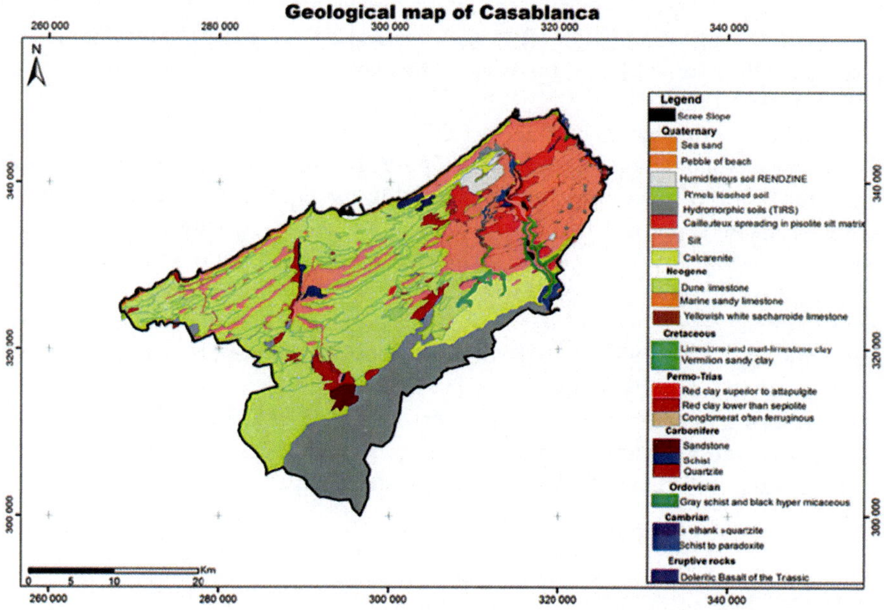

Fig. 1. Map of litho-structural units in the Casablanca region

3 Geotechnical Context of the Study Site

The analysis of the SC1 and SC2 logging profiles revealed that the studied region is mainly composed of homogeneous, monoclinal schist formations, which are affected by multiple sets of discontinuities. The S0 schistosity is the most dominant, with a horizontal or sub-horizontal orientation and an average dip of 44°. The F1 family exhibits a similar orientation and dip as the schistosity, with a spacing ranging from 5 to 40 cm. The F2, F3, F4, and F5 families have specific dips ranging from 39 to 78°, and different filling characteristics. Additionally, there are other minor secondary families with low dips and limited aperture.

3.1 Rock Identification

After synthesizing the various boreholes, core boxes, and logging results, a synthetic cross-section was established for the study area (see Fig. 2). This cross-section is characterized by the following lithological units:

- A thick 4.5 m layer of fill material at the surface.
- A 1.8 m thick layer of calcarenite.
- A basal conglomerate marking the angular unconformity between the Quaternary and the Paleozoic, characterized by the presence of bioclasts such as echinoderms and gastropods with a thickness of 2.00 m.
- A substratum consisting of pelites and graywackes with a thickness of 13.50 m.

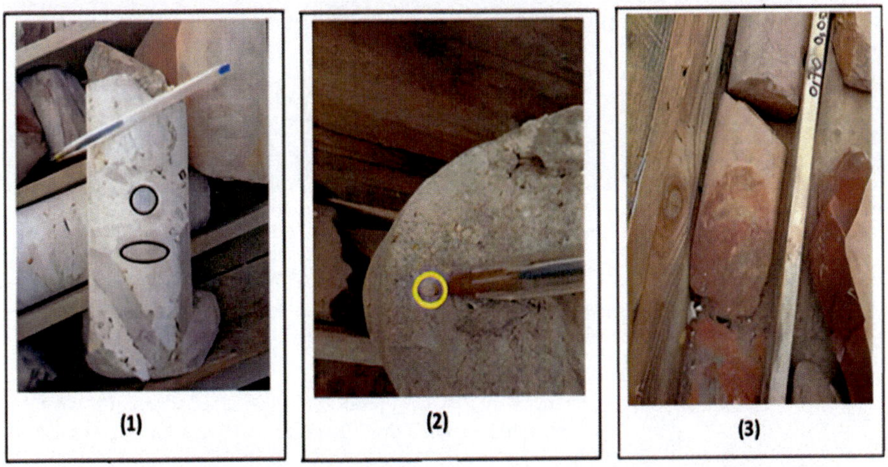

(1) Basal conglomerate (2) Calcarenite (3) Pelites with graywackes

Fig. 2. Sample for borehole SC1–17

4 Geomechanical Classifications

Fractured rock classifications have undergone continuous evolution over more than a century, offering significant value during feasibility studies and preliminary project design. They are particularly useful when hydrological, mechanical and in-situ stress information is lacking. Classification systems take into account several parameters essential for assessing the stability of rock masses, such as the presence of water, roughness, spacing, description of discontinuities, as well as the strength of the rock matrix [7, 12, 13].

We focus on quantitative classifications, also known as geomechanical classifications. Which quantify the quality of a rock mass using an empirical notation, describing a range of terms from very poor rock to very good rock. The main objectives of these classifications are as follows:

- Indirectly assess the large-scale mechanical properties of a fractured rock mass, including estimation of deformation modulus, simple compressive strength, cohesion and angle of internal friction.
- Provide recommendations for supporting structures.

The most commonly used geomechanical classifications for the design of supports in underground excavations are RQD [3] and Rock Mass Rating RMR [4], as well as GSI.

4.1 Assessment of Rock Mass Fracturing by RQD

(Deere et al., 1967) [3] developed the Rock Quality Designation (RQD) to quantitatively assess the influence of fracturing on the behavior of the rock mass. This assessment is based on the analysis of core samples taken during drilling. RQD corresponds to the percentage of intact rock fragments longer than 10 cm, relative to the total length of the borehole.

$$RQD = \frac{\sum \text{Length of pieces} > 10\text{ cm}}{\text{Total drilling length}} \times 100 \quad (1)$$

Table 1 presents the RQD classification according to Deere et al., 1967 [3].

Table 1. Rock quality designation (RQD) classification

Description of fracturing	RQD	Notes
Very bad	0 – 25	1. If the rock quality designation (RQD) is recorded or measured as ≤ 10 (including 0), a default value of 10 is applied to assess Q
Bad	25 – 50	
Fair	50 – 75	
Good	75 – 90	2. RQD intervals of 5, such as 100, 95, 90, etc., are considered adequately accurate
Excellent	90 –100	

The results of RQD (Rock Quality Designation) measurements for cores drilled at various depths have been compiled in Table 2. The values obtained show that rock quality varies considerably with depth, with RQD values ranging from 0 to 80%. It is important to note that the minimum RQD values are shown for each core, allowing for areas of low rock quality that may have a significant impact on the stability of excavations or foundations.

According to the results, three horizons can be distinguished in the study area:

- The first horizon, called "Fractured Altered Shales" (FAS), is located between a depth of 5.40 m and 10 m. The average RQD of this horizon is 47%, indicating that the rock quality is moderate to poor.
- The second horizon, called "Highly Altered Shales" (HAS), is located between 10 m and 17 m depth and has an average RQD of 33%.

Table 2. RQD variation as a function of depth

Profondeur (m)	RQD moy	RQD min	RQD max
5,4	14	0	20
6,70	48	44	50
7,90	51	50	52
8,50	35	30	42
9,30	53	50	56
10,10	45	42	50
11,40	39	36	40
12,90	43	29	53
14,40	30	30	30
15,90	27	0	56
17,00	29	20	41
17,40	46	44	50
18,90	46	44	47
19,40	61	53	80

- The third horizon, called "Moderately Fractured Shales" (MFS), is located between 17 m and the bottom of the borehole and has an average RQD of 51%.

Overall, the RQD values of the traversed shale formations are generally between 25% and 50%, indicating a poor overall quality of the rock mass. These values reflect a high degree of fracturing in the studied shale formations.

4.2 Bieniawski Classification (RMR)

Bieniawski (1973) [4] introduced the Rock Mass Rating (RMR) system, a geomechanical classification method that has undergone continuous improvements to cater to diverse applications, including slopes, deep excavations, mines, and tunnels.

Unlike RQD, it uses multiple parameters to describe the formation, including intact material compressive strength or point load strength, hydrostatic state, as well as characteristics of discontinuities such as orientation, spacing, aperture, continuity, and filling.

Core drilling and logging (SC1, SC2) provide data on the rock formation as well as the fractures and discontinuities present. Rock samples obtained through the drilling data have allowed for compression strength testing. Considering the water table level at – 3.00 m/TN, it is important to take into account the hydrostatic state of the rock formation, as this can influence wall stability and shear strength.

The RMR classification method utilizes five parameters to assess rock masses and consider joint orientation:

- A_1: Rock strength (UCS or Franklin test)

- A_2: Fracturing: RQD
- A_3: Joint spacing (all types of discontinuities: bedding, schistosity, fractures, joints)
- A_4: Nature of joints
- A_5: Groundwater conditions

The sum of the first five indices forms the basic RMR. For underground works, it is necessary to incorporate the correction factor (B) to obtain the final RMR, which is expressed as follows:

$$RMR = A_1 + A_2 + A_3 + A_4 + A_5 + B \quad (2)$$

The overall score is then calculated by adding the scores for each parameter, and it corresponds to a rock quality class IV (altered rock), with a rock mass cohesion of 200 kPa and a friction angle of 35°, suggesting that the study area is composed of altered rocks with low compressive strength and low shear strength capacity.

4.3 Geological Strength Index Classification

The estimation of GSI relies on a direct assessment of the rock mass structure, allowing for an evaluation of rock quality in the field. This parameter varies within a range of 5 to 85. In terms of classification, values close to 5 indicate very poor material quality, while values close to 85 represent excellent material quality [5].

GSI helps estimate the reduction in rock mass strength based on geological conditions. It is a crucial parameter in the criterion developed by Hoek & Brown, 1997 [5]. In our case, with a Hoek and Brown criterion equal to or less than 20, it indicates a very poor quality of the rock mass.

5 The Observational Method

The observational method is used to anticipate and predict the structural and geotechnical behavior of monitored structures. It involves the installation of specific instrumentation to monitor critical parameters in real-time, with predefined and adapted measurement frequencies based on the interpretation of results [10, 11, 14]. The monitoring can be adjusted as the construction progresses to meet the objectives related to excavations and support systems. This approach helps mitigate risks that cannot be eliminated through predictive evaluations.

In our case, the instrumentation approach is used to predict the geotechnical and structural behavior of the anchored walls at the property boundary (front and adjoining walls) during the excavation phase. During the excavation, multiple measurement devices have been installed to monitor the horizontal displacements of the excavation wall and adjacent structures, as well as to measure the settlement of the soil behind the anchored walls (Fig. 3).

Fig. 3. Cross-section of retaining structures (left), installation of Shape Accel Array sensors (right)

5.1 Definition of Alert Thresholds and Precision of Measurements

A detailed approach involves defining thresholds that serve as a reference for triggering alarms and alerts. A calibration period is necessary to validate provisional thresholds and verify the proper functioning of the monitoring system, as well as the adequate implementation of safety instructions.

According to standard NF P 94–270 [15], the predictions of displacements for a wall supported by soil nailing are based on empirical rules that utilize displacement parameters indicated in Table 3.

Table 3. Approximation of displacements at the top of a soil nailed structure

	Semi-rocky soils	Sands	Clays
$\delta_v = \delta_h$	h/1000	2h/1000	4h/1000

With:
h: height of the soil nailed structure
δ_v: vertical displacement of the top of the facing
δ_h: horizontal displacement of the top of the facing

Two thresholds have been established for the construction phase, namely the alert threshold and the alarm threshold. When the first threshold is reached, an expert committee is mobilized to provide information and analyze the situation. If the second threshold, known as the alarm threshold, is exceeded, a siren sounds, and immediate evacuation of the construction site is triggered.

The studied excavation is located near a building with isolated footings. In order to enhance safety, we have set a displacement threshold lower than the one prescribed by the NF P 94–270 standard, set at 5 mm for walls and foundations (L/1000 = 5.0 mm), where L represents the length between supports.

Table 4 presents the displacement thresholds and specific velocities for the R + 15 building and the E1 wall.

Table 4. Predefined alarm thresholds.

Measurement point	Threshold 1	Threshold 2
15-story building	±3 mm	±5 mm
Wall E1	±3 mm	±5 mm

6 Results and Discussion

6.1 Rock Mass Classification

The results allowed for the classification of the schist formations based on their degree of fracturing, with RQD values indicating a fair to poor quality of the rock mass.

Geomechanical classifications, including Rock Quality Designation (RQD), Rock Mass Rating (RMR), and Geological Strength Index (GSI), were used to assess the rock mass quality. The results indicate a poor quality of the rock mass, with low compressive strength and reduced ability to withstand shear forces. These findings highlight the importance of implementing appropriate support and drainage measures to ensure the safety of excavation works in the fractured rock formations of the Casablanca region.

Table 5 summarizes the geomechanical classification of geological formations at different depths obtained from core drilling.

Table 5. Geotechnical characteristics of the layers

Depth	Facies	RQD	RMR	GSI	c	φ
TN à 3.0	Backfill	–	–	–	5	20
3.0 à 5.5	Limestone sandstone	–	–	–	80	35
5.5 à 10	Highly fractured shale	37	23	18	200	35
10.0 à 17.0	Fractured shale	27	25	20	200	35
< 17	Intact shale	40	25	20	200	35

6.2 Field Instrumentation

The observational method allows real-time monitoring of displacement and provides the opportunity to identify and respond to any unexpected or abnormal behavior during the excavation process. By comparing the measured displacements with the considered values, adjustments can be made to the design, excavation sequence, and support systems to ensure stability and safety of the works.

6.3 Vertical Displacements of the Adjacent Building

The measurements taken by the hydrostatic cells reveal no significant vertical movement on the adjacent 15-story building (R + 15). The measurements remain stable, and the only observed movements are attributed to temperature variations, typically between 12 PM and 7 PM during the day, amplified by recent periods of high heat. A graph illustrating the temperature variations has been included to demonstrate the influence of these changes. Thus, the small peaks of vertical movement recorded correspond to an increase in temperature during the day.

Figure 4 presents the settlement measurements of the adjacent building (R + 15) during the excavation works in September and October 2018, conducted using the hydrostatic cells.

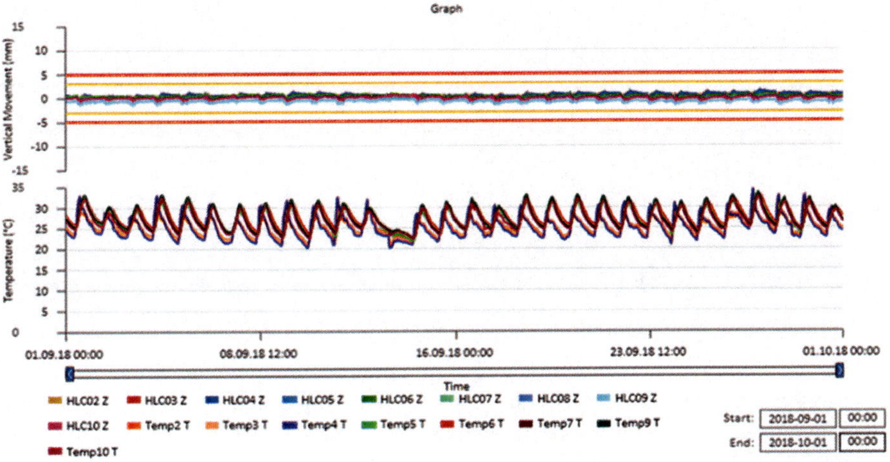

Fig. 4. Vertical displacements measured by hydrostatic cells

6.4 Horizontal Displacements of the Soil Nailed Wall

Two Shape Accel Array (SAA) sensors, named SAA1 and SAA2, were installed at the beginning of the excavation works to monitor the movements of the soil nailed wall on the side of the studied face, adjacent to a 15-story building. Displacement measurements were taken from week 38, after grout injection and system initialization. A small deformation of 2 mm was observed, which remains below the 5 mm displacement threshold. The results obtained by SAA1 and SAA2 sensors for the months of September and December 2018 are presented in Fig. 5.

7 Conclusions

For monitoring the excavation, several measurement devices were installed. The SAA sensors installed at the beginning of the excavation works monitored the horizontal movements of the soil nailed wall adjacent to a 15-story building. The measurements revealed a small deformation of 2 mm, remaining below the 5 mm displacement threshold. Thus, the soil nailed wall demonstrated satisfactory stability. The results obtained from the hydrostatic cells and tiltmeters showed the absence of significant movements or tilting, indicating that the building was not affected by the excavation works.

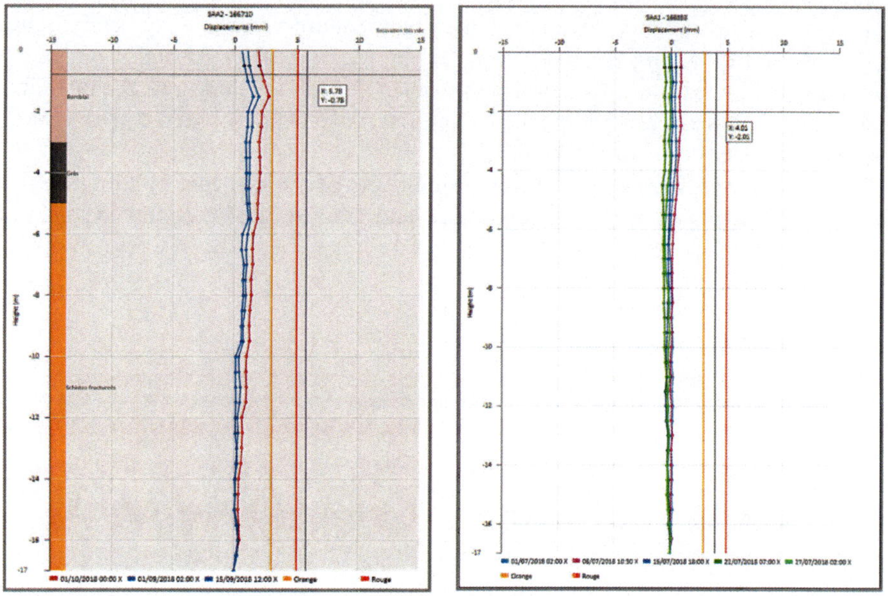

Fig. 5. Horizontal displacements measured by sensors SAA1 and SAA2

The modeling and monitoring of the excavation behavior enabled necessary measures to be taken to ensure the safety of the works.

The conclusions drawn from this study will provide essential insights for the design and implementation of appropriate support measures in similar projects worldwide.

References

1. Boulaid, G., Bahi, L., Ouadif, L.: Risk assessment of excavation work of different types of construction projects using AHP. Int. J. Civ. Eng. Technol. **9**(12), 861–870 (2018)
2. Boulaid, G., Latifa, O., Bahi, L.: Study of the Factors Affecting the Quality and Safety of Deep Excavations in Urban Areas of Casablanca-Settat Province-Morocco. Int. J. Adv. Sci. Eng. Inf. Technol. **12**, 2174 (2022). https://doi.org/10.18517/ijaseit.12.6.16264
3. Deere, D.U., Hendron, A.J., et al.: Design of surface and near surface constructions in rock. In: Proceedings on 8th US Symposium on Rock Mechanics. C Fairhurst N. Y. AIME, pp. 237–302 (1967)
4. Bieniawski, Z.T.: Engineering Classification of Jointed Rock Masses. Trans. South Afr. Inst. Civ. Eng. **15**(12), 335–344 (1973)
5. Hoek, E., Brown, E.T.: Practical estimates of rock mass strength. Int. J. Rock Mech. Min. Sci. **34**(8), 1165–1186 (1997). https://doi.org/10.1016/S1365-1609(97)80069-X
6. Hoek, E., Kaiser, P.K., Bawden, W.F.: Support of Underground Excavations in Hard Rock (1995)
7. Lahmili, A., Ouadif, L., Akhssas, A., Bahi, L.: Rock stability analysis – A case study. MATEC Web Conf. **149**, 02072 (2018). https://doi.org/10.1051/matecconf/201814902072
8. Finno, R.J., Calvello, M.: Supported Excavations: Observational Method and Inverse Modeling. J. Geotech. Geoenviron. Eng. **131**(7), 826–836 (2005). https://doi.org/10.1061/(ASCE)1090-0241(2005)131:7(826)

9. Peck, R.B.: Advantages and limitations of the observational method in applied soil mechanics. Géotechnique **19**(2), 171–187 (1969). https://doi.org/10.1680/geot.1969.19.2.171
10. Chen, Y.: Application of new observational method on deep excavation retaining wall design in London Clay, Thesis. University of Cambridge (2020). https://doi.org/10.17863/CAM.50486
11. Spross, J., Johansson, F.: When is the observational method in geotechnical engineering favourable? Struct. Saf. **66**, 17–26 (2017). https://doi.org/10.1016/j.strusafe.2017.01.006
12. Driouch, A., Latifa, O., Lahmili, A., Amine, B.: Correlation between Q-System and rock mass rating of the serpentine rock mass at the Bou Azzer Mine in the central Anti-Atlas of Morocco. Ann. Romanian Soc. Cell Biol. **25**, 12726–12733 (2021)
13. Zerradi, Y., Lahmili, A., Souissi, M.: Stability of a rock mass using the key block theory: a case study. In: E3S Web of Conferences, vol. 150, p. 03024 (2020). https://doi.org/10.1051/e3sconf/202015003024
14. Nejjar, K.: 'Comportement des parois de soutènement dans un contexte exceptionnel (grande profondeur, formations déformables, environnement sensible)', thèse de doctorat. Université Grenoble Alpes (2019). https://doi.org/10.13140/RG.2.2.31102.33607
15. NF P94–270. (2020). Accessed: May 13, 2023. [Online]. Available: https://www.boutique.afnor.org/fr-fr/norme/nf-p94270/calcul-geotechnique-ouvrages-de-soutenement-remblais-renforces-et-massifs-e/fa198932/260518

Modeling the Fatigue Behavior of Pavement Using the Finite Element Method

Omar Ben Charhi(✉) and Khadija Baba

GCE Laboratory, Mohammadia School of Engineers, Mohammed V University, Rabat, Morocco
bencharhi.omar@gmail.com

Abstract. The primary function of pavement is to withstand traffic stresses while ensuring the comfort and safety of users, guaranteeing adequate ride quality, and providing good vehicle grip. The stresses applied to the pavement depend on road traffic and thermal stresses, which often occur simultaneously. A moving vehicle generates not only vertical stresses but also tangential stresses (longitudinal and lateral), exerting traction at the base of the layers. To analyze and predict the fatigue behavior of pavement, various principles and methods are employed. These approaches can be broadly categorized into two groups. The first category involves modeling the pavement as a thin plate using a linear elastic monolayer model. This approach aims to predict the fatigue characteristics of asphalt mixes, taking into account the response of a single layer under loading conditions. By considering the mechanical properties of the asphalt mixture, this model attempts to assess the fatigue life of the pavement. The second category of methods adopts an empirical modeling approach to understand the behavior and performance of the pavement. In this approach, an elastic multilayer model is employed to predict the pavement's response to loading and evaluate its performance. This model takes into account the distinct layers of the pavement, each with its own material properties, such as thickness, modulus, and strength. By considering the interaction between these layers, the model provides insights into the fatigue life of the pavement and its ability to withstand traffic-induced stresses. The above methods have shown limitations in modeling the fatigue behavior of pavement. These limitations can hinder the accuracy and reliability of fatigue predictions, potentially leading to premature pavement deterioration or inefficient maintenance strategies. Therefore, in this article, we propose a new approach to address these limitations. Our proposed approach incorporates the dynamic effects of vehicles on the pavement by utilizing the finite element method (FEM).

Keywords: Modeling the pavement · Asphalts mixes · The fatigue of pavement · Multilayer model · The finite element method

1 Introduction

Pavement, as a vital element of transportation infrastructure, serves as a fundamental pillar in facilitating the efficient and safe mobility of vehicles. It plays a pivotal role in accommodating the myriad of stresses exerted by traffic, while simultaneously prioritizing user comfort, ensuring superior ride quality, and maintaining optimal vehicle

K. Baba et al. (Eds.): GeoME 2023, SPEES, pp. 368–379, 2023.
https://doi.org/10.1007/978-3-031-49345-4_35

grip. The attainment of these essential attributes is integral to enhancing road user satisfaction and overall transportation efficiency [1]. The stresses applied to the pavement depend on the combined effects of road traffic and thermal stresses, which often occur concurrently. When a vehicle is in motion, it not only induces vertical stresses due to its weight but also generates tangential stresses in both longitudinal and lateral directions. These tangential stresses exert traction at the base of the pavement layers, contributing to the overall loading conditions experienced by the pavement structure [2].

To analyze and predict the fatigue behavior of pavement, a wide range of principles and methods are employed, each with its unique characteristics and objectives. These approaches can be broadly categorized into two groups, providing distinct perspectives and insights into pavement fatigue analysis. The first category involves modeling the pavement as a thin plate using a linear elastic monolayer model. This approach focuses on predicting the fatigue characteristics of asphalt mixes by considering the response of a single layer under various loading conditions. By assuming linear elastic behavior, the model takes into account the material properties of the asphalt mixture, to estimate the fatigue life of the pavement [3]. The pavement, represented by a thin plate with Young's modulus E1 and Poisson's ratio ν1, is placed on an infinite soil with Young's modulus Es and Poisson's ratio νs. Under the assumption that the pavement slides perfectly on its support. Bending problems lend themselves to a number of simplifying assumptions (Navier's assumptions for thin plates) [4]. This leads to the assumption that in the pavement layer:

The mid-plane coincides with the neutral axis.

The cross-sectional planes remain flat during deformation.

The normal stresses in the transverse direction can be neglected. Under these assumptions, the vertical displacements w of the neutral axis of the plate satisfy the Lagrange equation for thin plates as shown in Eq. 1 [5].

$$\mathrm{D}.\Delta^2\mathrm{w} = \mathrm{p} \tag{1}$$

With

$$\mathrm{D} = \frac{\mathrm{E}_1\mathrm{H}^2}{12\left(1 - \nu_1^2\right)} \tag{2}$$

where H is the thickness of the plate, E1 and ν1 are the Young's modulus and Poisson's ratio of the plate material. In this equation, D represents the stiffness characteristic of the plate, w represents the vertical displacement of the neutral axis, and p denotes the sum of vertical pressures. This model does not take into account discontinuities and is only applicable in cases where the pavement structure is not significantly different from the natural ground. It also cannot model multi-layered structures.

In contrast, the second category employs an empirical modeling approach to predict the behavior and performance of pavement by applying an elastic multilayer model. This method considers the pavement as a multi-layered structure with distinct layers possessing their own material properties, such as thickness, modulus, and strength. By capturing the interaction between these layers, the model provides a comprehensive understanding of the pavement's response to fatigue-inducing loads [6].

The pavement is modeled as an infinite linear elastic multilayer system and is supported by an infinite soil. The load is assumed to be circular, which greatly simplifies the problem by making it axisymmetric. Solving an elasticity problem in cylindrical coordinates reduces to finding a stress function φ (r, z) with a double Laplacian equal to zero, as indicated in the Eq. 3:

$$\Delta^2\varphi(r, z) = 0 \tag{3}$$

In the case of a pavement structure composed of multiple layers, often referred to as an n-layer system, the analysis involves seeking n distinct functions, each specific to layer i, which satisfy the equation $\Delta^2\varphi$ (r, z) = 0 and are subject to the appropriate boundary conditions. To accurately capture the behavior of the pavement structure, all layers are considered as individual elastic solids rather than assuming a plate-like behavior as seen in alternative models. This approach allows for a more realistic representation of the complex response of each layer to external loading and environmental factors. One key aspect in the analysis of multi-layered pavements is the consideration of the interfaces between adjacent layers. These interfaces can either be bonded, where the layers are tightly interconnected, or unbonded, allowing for relative movement and potential separation between layers. The choice of bonded or unbonded interfaces depends on the materials used and the design requirements. This consideration ensures that the model accounts for the potential effects of interface conditions on the overall performance and behavior of the pavement structure. However, in reality, the loads applied to the pavement are not static but dynamic in nature, continually changing as vehicles move across the surface. This dynamic nature of the loads includes vehicle speeds, acceleration, deceleration, and the effects of vehicle interactions with the pavement. These dynamic loads impose additional challenges for accurately modeling and analyzing the behavior of the pavement structure [7].

Previous methods employed to model the fatigue behavior of pavements have demonstrated certain limitations, hindering their ability to accurately capture the complex nature of pavement fatigue. Recognizing these shortcomings, this article introduces a novel approach aimed at addressing these limitations and improving our understanding of pavement fatigue behavior.

In this study, we propose the development of a new methodology that takes into account the influence of dynamic loads on asphalt mixes when evaluating the fatigue performance of pavements. By incorporating the finite element method (FEM), a powerful numerical technique widely used in engineering analyses, we aim to provide a more comprehensive and accurate assessment of the pavement's fatigue life.

Unlike the previous methods that relied on simplified assumptions and linear models, our approach acknowledges the dynamic nature of the loads that pavements are subjected to in real-world conditions. This includes considering vehicle speed, axle loads, and tire characteristics that contribute to the dynamic loadings. By incorporating these dynamic load effects into the modeling process, we can better simulate and analyze the actual response of the pavement under realistic traffic conditions.

The finite element method allows us to divide the pavement structure into smaller elements, enabling a detailed analysis of stress distribution, strain accumulation, and potential failure mechanisms within the layers. Through this approach, we can investigate the cumulative effects of repeated loading cycles and predict the fatigue life of the pavement more accurately [8].

By employing the proposed approach, we aim to enhance our understanding of the fatigue behavior of pavements and provide valuable insights for pavement design, maintenance, and management. This method has the potential to improve the accuracy of fatigue life predictions, leading to more effective and informed decision-making processes related to pavement infrastructure.

2 Materials and Methods

The general approach methodology encompasses several crucial steps: data collection, pavement behavior analysis, action selection, determination of thicknesses, and dynamic modeling of the pavement structure. These aspects have been thoroughly elaborated and refined when compared to alternative methods.

The primary objective of data collection is to amass comprehensive information on various aspects, including traffic, geotechnical factors, and general data such as climate, environment, and regional geology. Traffic classes, as defined in the Moroccan catalog of new pavement structures are considered [7], and some of these classes have been further subdivided into subclasses to widen the options for reinforcement and surface layer materials. The geotechnical and general data play a vital role in determining the appropriate pavement thickness.

The general data encompasses the history of the pavement and the geological, hydrogeological, and climatic conditions that characterize the pavement's environment. This data enables the identification of homogeneous zones with similar characteristics, on which subsequent actions are defined.

The process of selecting the structures to be implemented involves a preliminary decision-making stage where actions to rehabilitate the pavement are chosen, essentially forming a pivotal aspect of the study and a form of pavement diagnosis.

Designing the pavement involves the meticulous determination of various factors, including the appropriate thicknesses and materials to be implemented. This multifaceted endeavor requires a comprehensive understanding of the pavement's intended use, the expected traffic loads, and the prevailing environmental conditions. One of the key aspects of pavement design is the determination of the suitable thicknesses for each layer of the pavement structure. This involves a rigorous analysis of the anticipated traffic loads, including type of vehicles that will be using the pavement, as well as the frequency of heavy trucks or other heavy vehicles. By considering these aspects, we can calculate the required thicknesses for the various layers to ensure the pavement's durability and longevity.

After that, the dynamic modeling of pavement structure using the finite element method (FEM). In this process, the pavement structure is discretized into a mesh of small elements, with each element representing a portion of the pavement. These elements are interconnected at nodes, and mathematical equations are used to describe the behavior of each element based on its material properties and response to dynamic loading.

The finite element method enables us to simulate the dynamic interactions between different layers and components of the pavement. This includes the effects of dynamic loads from traffic, such as the impact forces and vibrations generated by vehicles moving over the pavement surface. By accurately capturing these dynamic forces, engineers can obtain valuable insights into how the pavement responds to various traffic loads and environmental factors.

Notably, the study has introduced various types of unconventional materials in Moroccan road engineering to maximize the utilization of locally available resources. The flowchart below (see Fig. 1) provides an overview of the adopted methodology, summarizing the steps involved in the process.

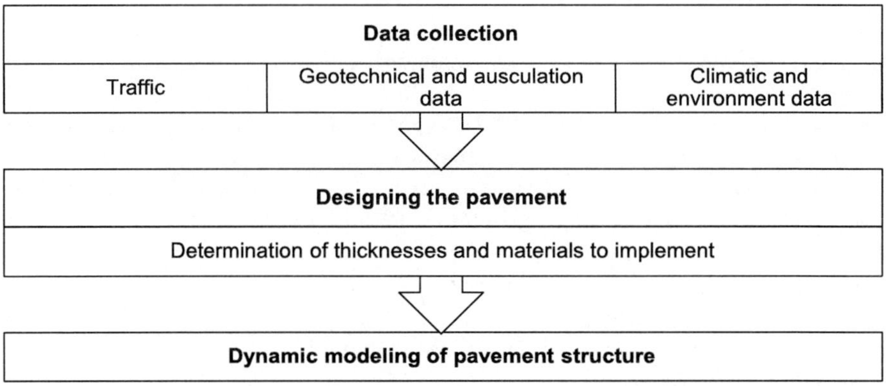

Fig. 1. The flowchart of the adopted methodology

Experimentation The pavement structure to be defined is situated in the bustling industrial area of Had Souelem, which is located within the Province of Berrechid, Morocco. The area is known for its significant industrial activities, with various manufacturing and commercial establishments contributing to a substantial flow of vehicular traffic. As a critical transportation route within this industrial zone, the pavement faces unique challenges due to the heavy loads and frequent movement of trucks, delivery vehicles, and other industrial traffic.

The climate and environmental conditions in this region also play a significant role in influencing the pavement's performance and durability. Berrechid, being part of Morocco, experiences a Mediterranean climate with hot and dry summers and mild winters. These weather conditions, along with the potential for temperature variations, can impact the pavement's materials and structural integrity over time.

To support our proposed methodology, a comprehensive approach comprising field investigations, laboratory testing, and analysis will be implemented. Field investigations will entail sample collection of road materials for testing, conducting geotechnical surveys, and measuring the road's structural response. In parallel, laboratory testing will be conducted to assess the properties of road materials, including soil and aggregates, to determine their suitability for the proposed reinforcement techniques. By combining these methods, we aim to gain valuable insights into the pavement's condition and material characteristics, essential for developing effective reinforcement strategies and ensuring the long-term durability of the road infrastructure.

The pavement structure exhibits flexibility, and selecting the appropriate structure is contingent on the overall framework and unique characteristics of the specific project (see Table 1).

Table 1. Type of pavement structure

Surface course	Base course	Sub base course
Asphaltic concrete	Bituminous concrete gravel (GBB)	Untreated gravel

In a pavement, the sub base course is composed of untreated granular materials with a relatively low modulus of rigidity. When subjected to rolling loads from traffic, flexible pavements experience deformation, and the materials may either rebound to their original position or retain a residual deformation. As a result of the low stiffness of the granular materials in the pavement structure, the vertical forces generated by traffic loads are transmitted to the subgrade course.

The climatic environment Four distinct zones are established in accordance with the average annual precipitation, measured in millimeters (mm), and calculated over an extended period of about 30 years. The study region presents an average annual precipitation of 449 mm and an average annual temperature of 18 °C, as demonstrated (see Fig. 2).

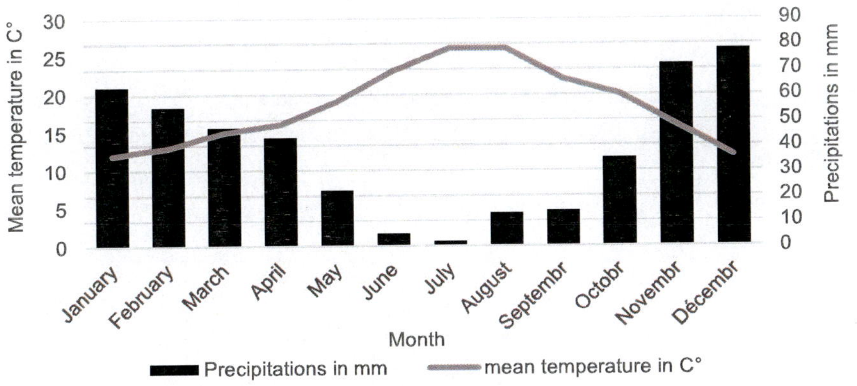

Fig. 2. Precipitation and mean temperature data collected for each month in the study area

Pavement design The design of pavements holds significant significance in establishing durable, secure, and high-performance road infrastructure. Developing a resilient and sustainable pavement structure requires comprehensive analysis of numerous factors. Among the various parameters that influence pavement design, three stand out as particularly critical: traffic, climatic conditions, and the subgrade course. A thorough understanding and appropriate consideration of these key elements are essential to attaining optimal pavement performance and ensuring its long-lasting functionality.

To determine the heavy vehicle traffic class, we will utilize the formula in Eq. 4, as shown below:

$$\mathrm{NEE13T} = \mathrm{C1} \times \mathrm{C2} \times \mathrm{C3} \times \mathrm{C4} \times \mathrm{N4} \times \mathrm{N2} \times \mathrm{F1} \quad (4)$$

where:

NEE13T represents the cumulative traffic expressed in the number of standard 13-ton vehicles.

C1 denotes the width coefficient.

C2 signifies the average aggressiveness coefficient.

C3 accounts for the increase in heavy vehicles.

C4 represents the growth of heavy vehicles.

N4 indicates the cumulative traffic.

N2 represents the updated traffic of heavy vehicles weighing more than 8 tons.

F represents the traffic distribution coefficient.

3 Results

3.1 Results of Reconnaissance Work

A field survey was carried out to collect essential information regarding the study sections. The survey involved conducting manual assessments that were strategically distributed across the designated areas. The primary goal was to gather precise data to facilitate the evaluation and analysis of the project. The manual surveys yielded the following findings (see Table 2).

Table 2. Type of pavement structure

Sounding depth	Type of material
[0 à – 0,1 m]	Embankment
[– 0,1 m à – 0,6 m]	Topsoil
[– 0,6 m à – 2,3 m]	Turf loam
[– 2,3 m à – 3 m]	Loamy tuff

The land's average configuration, based on the survey results, is described as follows:

The embankment has an average thickness of 10 cm.

This is followed by topsoil with an average thickness of 50 cm.

Next, there is turf loam with an average thickness of 1.7 cm.

In the coastal region, there is a presence of loamy tuff, which extends until the end of the soundings.

3.2 Proctor Test Results

The Proctor test, also known as the Proctor compaction test, is a crucial laboratory procedure extensively used in pavement design. It plays a vital role in determining the optimal compaction characteristics of the soil or aggregate materials that form the foundation layers of a pavement structure.

The test involves compacting a representative sample of the soil or aggregate at various moisture content levels and measuring its resulting dry density and corresponding compaction effort. The objective is to identify the moisture content at which the material achieves maximum dry density and optimum compaction, ensuring its ability to withstand the anticipated traffic loads and environmental conditions (see Table 3).

Table 3. Proctor test results

Density in T/m^3	Water content in %
1,89	7,9
1,93	10,7
1,96	12,6
1,9	14,5
1,84	16,2

3.3 Laboratory Tests Results

The subgrade soils encountered during the borehole investigation were subjected to representative sampling, and these samples were subsequently analyzed in the laboratory. The laboratory analyses yielded valuable information about the soil characteristics, enabling the identification of a particular soil family with properties (see Table 4).

Table 4. Laboratory tests results

Liquid limit (V.B)	Maximum dry density (γdmax) in T/m^3	Optimal water content (Wopt) in %	The California Bearing Ratio (CBR)
1,09	1,89 to 1,96	7,9 à 16,2%	6

These soils have a fines content of 17%, while the liquid limit (V.B) has a value of 1.09. According to the Moroccan guide for road terraces (GMTR) [8], this soil falls under the classification B5.

To conduct further evaluations of the soil's characteristics, California Bearing Ratio (CBR) tests were performed [9]. The testing process involved subjecting molded specimens to Modified Proctor optimal water content and compaction energy, with either 25 or 50 Proctor blows. The results obtained from the Modified Proctor Compaction Test are as follows:

Maximum dry density (γdmax) ranging from 1.89 to 1.96 T/m^3.

Optimal water content (Wopt) ranging from 7.9% to 16.2%.

Additionally, the California Bearing Ratio (CBR) after saturation, corresponding to a compaction level of 95% Modified Proctor Optimal (MPO), was found to be approximately 6 on average. This classification indicates a lift class of type P1 for the platform.

The results of the adopted pavement structure

The design and dimensioning of pavements are vital steps in ensuring the longevity and stability of road infrastructure. In this study, the dimensioning process was carefully carried out, following the guidelines outlined in the adopted methodology. The findings regarding the adopted pavement structure are presented (see Fig. 3).

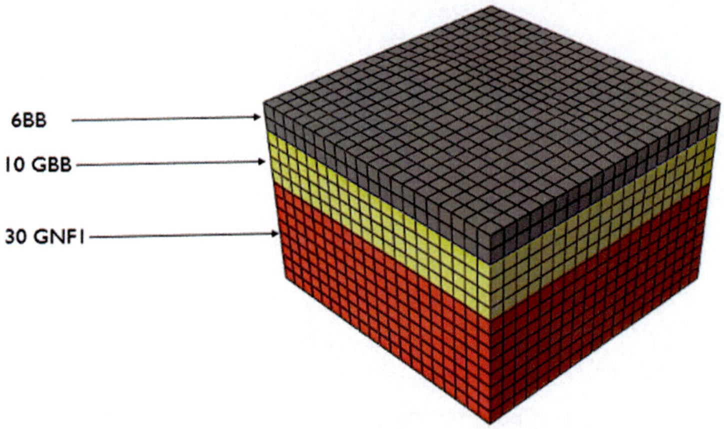

Fig. 3. The adopted pavement structure

The adopted pavement structure is: **6BB + 10GBB + 30GNF1**.

With:

BB: Asphalt Concrete 0/10.

GBB: Grave Bituminous Concrete 0/14.

GNF1: Untreated Gravel 0/40.

Dynamic modeling of pavement structure

To validate the presented pavement structure, a comprehensive series of assessments will be carried out. This validation process involves modeling the pavement structure to evaluate the stresses and deformations resulting from a unit type load [10]. The approach focuses on critical aspects, including the selection of the structure type and materials, determination of permissible stresses in various materials, and calculation of appropriate thickness for each layer.

For the standard 13-ton axle, the following assumptions will be considered: twin-wheel axles supporting a load of 13 tons, vertical pressure of 0.6620 MPa, contact radius of 0.125 m, and twin center distance of 0.375 m. The pavement structure to be modeled is detailed (see Table 5).

Table 5. The pavement structure to be modeled

Pavement material	Thickness of the pavement layer in cm	Young's modulus in Mpa	Poisson's ratio ν
BB	6	4320 Mpa	0,35
GBB	10	7500 Mpa	0,35
GNF1	30	600 Mpa	0,35
PF2 of the supporting floor	Infinite	50 Mpa	0,35

To ensure the reliability and effectiveness of the pavement structure, a comprehensive validation process will be undertaken. This process involves conducting a series of checks and dynamic modeling of the pavement structure using the finite element method (FEM). By employing FEM, a powerful numerical technique, we can accurately simulate and analyze the pavement's behavior under varying traffic loads and environmental conditions (see Figs. 4 and 5).

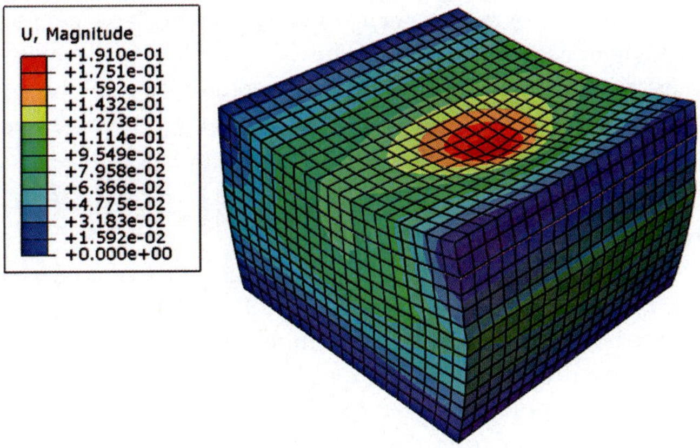

Fig. 4. The depth of road rut based on the position.

The finalized pavement structure has undergone thorough verification, confirming that the calculated stresses within the pavement are well below the permissible limits. This ensures the pavement's ability to endure anticipated loads and traffic conditions while staying within safety thresholds.

Based on these verified results, the ultimate pavement structure has been determined and is outlined (see Table 6).

Table 6 provides comprehensive details, including layer thicknesses and material types, for each component within the pavement system.

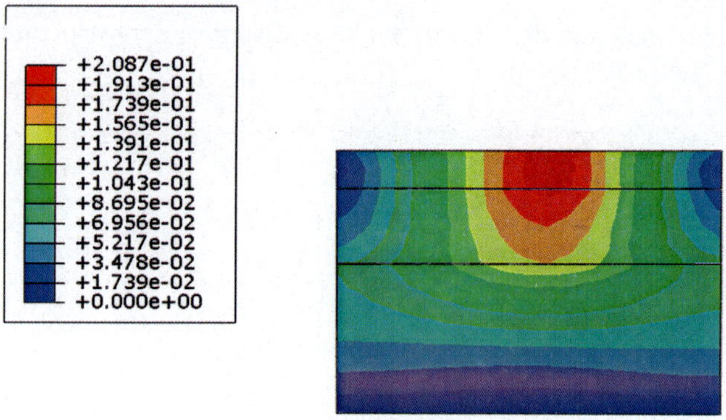

Fig. 5. The maximum vertical deformation as a function of position.

Table 6. The final pavement structure

Type of pavement	Pavement structure
Flexible pavement	6 BB + 10GBB + 30 GNF1

The presented final pavement structure is the outcome of meticulous analysis and consideration of traffic volume, material properties, desired durability and performance. This design ensures the pavement's capacity to withstand predicted stresses and maintain its integrity throughout its intended service life.

By adhering to the established pavement structure, optimal performance of the road infrastructure is anticipated, providing efficient support for vehicular traffic and contributing to a safe and reliable transportation network.

4 Conclusions

In conclusion, this article introduces an innovative and cutting-edge approach for modeling pavements through the application of the finite element method. The seamless integration of experimental investigations and advanced modeling techniques has proven the robustness and efficacy of the proposed pavement structure in withstanding a diverse range of stresses.The findings from this comprehensive study are highly encouraging, as the calculated stresses within the pavement consistently fall well below the allowable stress thresholds. This crucial validation confirms the structural integrity and reliability of the pavement, instilling confidence in its ability to withstand real-world traffic loads and environmental conditions.

References

1. Zhang, J., Xiong, C., Wang, D., Wei, D.: Investigation of the pavement ride quality of urban roads based on a three-level evaluation system. J. Traffic Transp. Eng. (Engl. Edn) **7**(2), 228–240 (2020)

2. Taheri, A., Fwa, T.F., Chang, M.: T: Analysis of pavement responses under moving surface loads: investigation of different dynamic tire–pavement contact models. J. Transp. Eng. Part B Pavements **146**(1), 04019054 (2020)
3. Ambassa, Z., Allou, F., Petit, C., Eko, R.M.: Fatigue life prediction of an asphalt pavement subjected to multiple axle loadings with viscoelastic FEM. Constr. Build. Mater. **43**, 443–452 (2013)
4. Dynnikova, G.Y.: The Lagrangian approach to solving the time-dependent Navier-Stokes equations. Doklady Phys. **49**, 648–652 (2004)
5. Mohammadi, S., Valizadeh, N., Ghorashi, S.S., Shojaee, S., Ghasemzadeh, H.: Analysis of thin plates by a combination of isogeometric analysis and the Lagrange multiplier approach. Comput. Methods Civ. Eng. **3**(2), 47–66 (2012)
6. Luo, X., Gu, F., Zhang, Y., Lytton, R.L., Zollinger, D.: Mechanistic-empirical models for better consideration of subgrade and unbound layers influence on pavement performance. Transp. Geotech. **13**, 52–68 (2017)
7. Morovatdar, A., Ashtiani, R.S.: Influence of acceleration and deceleration of Super Heavy Loads (SHLs) on the service life of pavement structures. In: Eleventh International Conference on the Bearing Capacity of Roads, Railways and Airfields, pp. 335–345(2021)
8. Asim, M., Ahmad, M., Alam, M., Ullah, S., Iqbal, M.J., Ali, S.: Prediction of rutting in flexible pavements using finite element method. Civ. Eng. J. **7**(8), 1310–1326 (2021)
9. Direction Routiere.: Catalog Types of New Pavement Structures. Direction Routiere, Rabat (1995)
10. Direction Routiere.: Moroccan Guide for Road Terraces. Direction Routiere, Rabat (2001)
11. Wimalasena, K., Gallage, C.: Predicting california bearing ratio (CBR) value of a selected subgrade material. In Road Airfield Pavement Technol. **193**, 547–558 (2022)
12. Pérez-Acebo, H., Linares-Unamunzaga, A., Rojí, E., Gonzalo-Orden, H.: IRI performance models for flexible pavements in two-lane roads until first maintenance and/or rehabilitation work. Coatings **97** (2020)

Modeling Post-Covid 19 Variability of Potholes According to Visual Inspection Results: GIS Approach

Mohammed Amine Mehdi[1](✉), Toufik Cherradi[1], Imane Mehdi[2], Chaimae Merimi[3], Said El Karkouri[4], and Ahmed Qachar[5]

[1] Civil Engineering and Construction Laboratory, Mohammadia School of Engineers, Rabat, Morocco
aminemehdi@research.emi.ac.ma

[2] Quality Safety and Maintenance Laboratory, Mohammadia School of Engineers, Rabat, Morocco
aminemehdi@research.emi.ac.ma

[3] Faculty of Science, Laboratory of Applied Chemistry & Environment, Mohammed I University, Oujda, Morocco

[4] National Center for Road Studies and Research, Rabat, Morocco

[5] Moroccan Road Directorate, Ministry of Equipment, Transport, Logistics and Water, Rabat, Morocco

Abstract. The 2020s and 2021s have been marked by a viral pandemic (Covid-19) which has had a major impact on all vital sectors (economy, society, etc.) and has imposed a kind of slowed-down life throughout the world. This paper proposes a modelling of pothole variation, before and after the covid-19 period based on visual inspections. The purpose of the study is to geolocalize on thematic maps the pavement's state of potholes by assigning the deterioration level. This pavement has undergone several maintenance interventions, in this respect, using a deterioration matrix prepared by the Moroccan National Center for Road Studies and Research, we initially prepared a reduced visual inspection data carried out between 2018 and the close of 2020. These will be implemented in a Microsoft-Excel-GIS Model proposed, to facilitate the reading of the data to display information in a thematic map view. This study is a first step to help decision-makers to keep the pavement level.

Keywords: GIS modeling · Potholes · Pavement deterioration · Moroccan road · Road inspection · Road infrastructures

1 Introduction

GIS has not been developed for road management in every country in the world, even though this application appeared fairly quickly [1]. In France, GIS suppliers are pushing this type of application forward, presenting GIS as an alternative to the road database. However, in a GIS, road data form an information layer, like others (relief, hydrology,

K. Baba et al. (Eds.): GeoME 2023, SPEES, pp. 380–389, 2023.
https://doi.org/10.1007/978-3-031-49345-4_36

urbanization zones, parcels, settlement network...) [2]. It is therefore possible to model interactions between roads and other environmental or contextual elements. In addition, GIS offers geographic cross-referencing functions that facilitate and accelerate this type of operation. This article presents a spatial evolution study aimed at correlating spatial entities generated by the reduction of qualitative data relating to the results of visual inspections realized by the *CNER* [3, 4]. This method is based firstly on an assignment of deterioration state matrices, and secondly on a validation of the model in Excel MS, in order to implement the data prepared in a prototype-GIS model to project the variability of potholes before and after the COVID-19 period on topographical maps [5].

Pull-outs is a cavity in the pavement with cut edges, created when the road surface crumbles and the pavement materials (asphalt mixes) disperse. The term comes from a time when road surfaces didn't exist. Gallinaceans used to nestle in holes in the middle of roads and lay their eggs there from time to time. Feathering is the tearing away of gravel from the coating layer, due to poor dosage or lack of cleanliness of the gravel, or to a compaction problem [6] (Fig. 1).

Fig. 1. Phothole

2 History of GIS Applications in Road Pavement Management

2.1 International Context

Aerospace applications produce a simplified perspective of a complex system based on the geometric branch of mathematics known as topology, which deals with the spatial relationships that correlate spatial entities. A useful asset is conceived to facilitate the acquisition, the handling, analysis, modeling and visual display of spatially referenced data through hardware, software, personnel, business processes and organizations. It is mainly used to solve global management and planning problems [7]. GIS has been used in several pavement management projects. It facilitates the analysis of various operations and planning projects related to pavement management, notably scale, time and format, while improving the measurement, cartography, supervision and simulation of various geographical phenomena [8].

The appropriate selection of spatial tools, the creation of an adequate base map and the linking of these attributes in spatial systems and information management systems, and their correlation with spatial and cartographic information, are all crucial concerns for developing and implementing space- based pavement management systems (PMS) [9]. (Elhadi, 2009) quoted (Meyers et al., 2003), who reported that testing complements a parameter package comprising the distance, surface deformation, road surface temperature and applied load. These parameters are recorded at each test site [10].

In Spain, a GIS-PMS model was developed in 2008 using the GIS capabilities to simplify determining the present pavement condition, based on forecasts pavement behavior models, the development of proposals and the selection of solutions. The models of evolution and prediction of pavement behavior are used at network level for the selection of maintenance strategies, at project level in pavement design, life-cycle cost analysis, or for the selection of affordable optimal maintenance concept and planning work. Models support evaluation of maintenance strategies or approaches for the entire pavement network, and the prioritization of projects with specific maintenance work [11].

In Saudi Arabia, Jeddah Municipality's GIS technology is a leading reference in road surfacing for display on scree data visualization for key indicators and appropriate visualization, terrain survey, model creation, mapping and management and rehabilitation recognition using single numbers [12].

2.2 Local Context

The National Center for Roads Studies and Research (*CNER*) began integrating GIS and cartography into road pavement management in 1995, in the following sequence:

- 1995–2000: The effective implementation of SYGER (with Oracle as DBMS and Micro- station as operating software and a network of PCs), the addition of other data in the RDB and the development of cartographic applications;
- 2000–2004: This period is mainly marked by the continued development of the GIS and its availability to other users for exploitation, particularly at regional level, by providing them with the necessary licenses for Oracle and Micro-station;
- 2004–2009: a period marked by CNER's move towards the use of new technologies, in particular Geographic Information Systems/GIS for the development of SYGER (MAPINFO, GEOMAPGIS, ORACLE).

According to a literature review, a pavement management model based on GIS has been proposed to effectively manage pavement maintenance. This tool includes the factors identified in the literature review phase.

3 Research Significance and Methods

In Morocco, there is no mapping system for the management of civil engineering road infrastructures in particular, or at least the initiators of this project do not have such a tool. Furthermore, none of the GIS systems that have been developed is dedicated to road infrastructure management and the quantification of condition indicators depending on monitoring and road inspection results [3]. It is with this in mind that we are carrying out this study, the general aim of which is to simplify the reading of the state of evolution of potholes on a 50-kilometer-long stretch of soft carriageway on Moroccan National Road number 06 [13].

3.1 Moroccan Method: Visual Inspection

The visual survey consists of covering the entire paved road network, over a period of 1 to 3 months, in order to record deterioration in sections of around ten meters every 200m. The visual inspection focuses mainly on cracking, pull-outs and potholes to assess pavement surface quality [14]. The concept is based on monitoring of regular 200m intervals (Fig. 2), indicating at each kilometer point (KP) the existence, absence, and severity of every deterioration condition.

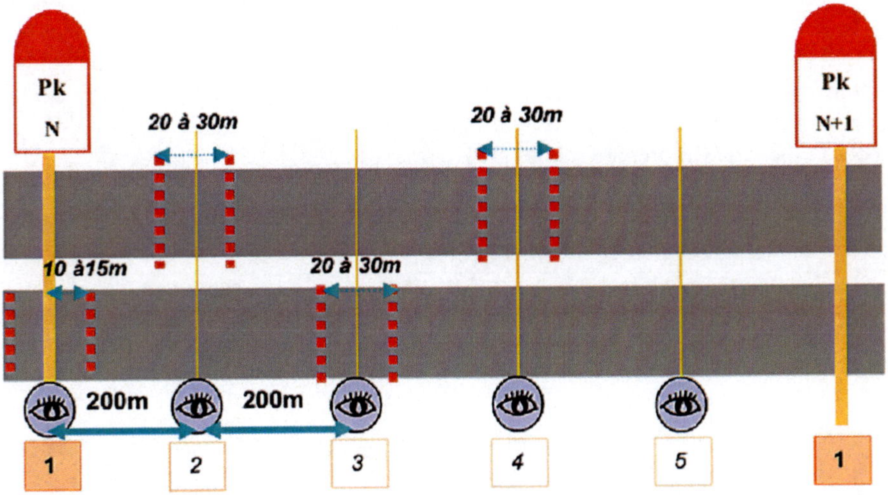

Fig. 2. Principle of visual pavement inspection applied in Morocco.

3.2 Deformation Matrix

The number of potholes sampled is represented qualitatively for each 1 km of road length by the following matrix:

Table 1. Potholes condition index

Deterioration level	Potholes
A	0
B	1
C	2–4
D	5–10

3.3 GIS Approach

For a GIS study, pavement condition data is a critical input for decision making, constituting the second step in the pavement management process. These data are used to identify current maintenance and rehabilitation needs, to forecast future requirements and to assess the overall impact on the network. Consequently, the type of data and detail required depends on the pavement management process used. This is irrespective of whether such strategies, and the associated investment decisions, are derived by expert systems approaches [15]. Collecting pavement condition data can be a complex process. Selecting an appropriate method is therefore an important step [16].

In our approach, the basis and design methods for the road database, and an Excel sheet to simplify data input and to automate the calculations required to determine the value of the pothole condition and SUI and to arrange the data in customized tables tailored GIS connection capability and methods of use. The creation of the PMS database for the study area was achieved by transforming the qualitative data (Table 1) into quantitative values on ArcGIS software (Table 2), in order to implement the MS-Excel model in the ArcGIS prepare model [17].

Table 2. Transforming qualitative results in ArcGIS

Potholes deterioration level	Reading ArcGIS
A	1
B	2
C	3
D	4

The study methodology adopted is summarized in Fig. 3.

3.4 Study Area

The study section is part of National Road 06, linking the town of Khemisset and Meknes. It extends along a 50-km stretch from kilometer point KP 0 + 081 and ending at KP 0 + 130. The section is defined by the high traffic levels estimated annually in 2017 at 3639207 vehicle km/day, and on the other hand on the interesting geographical position. Figure 4 shows the location of the study road section on the map of Morocco.

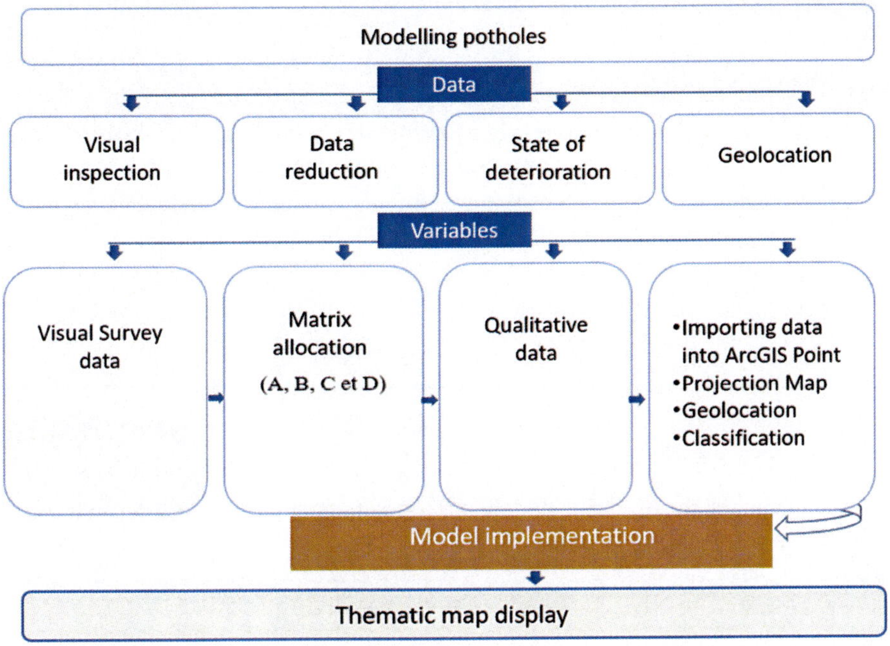

Fig. 3. Study methodology

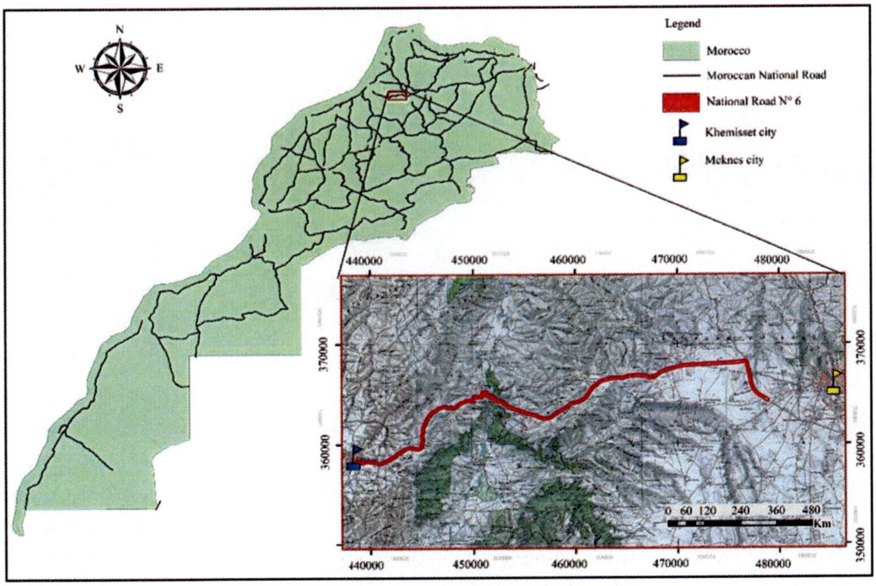

Fig. 4. Study area location

4 Results

4.1 Model Validation

The model developed is verified by kriging before it is compiled. In geostatistics, kriging is the linear estimating method that ensures minimum variance. It is the best unbiased linear estimator, based on an objective method [17]. After verifying the kringing, the process of the GIS model is shown as follows (Fig. 5).

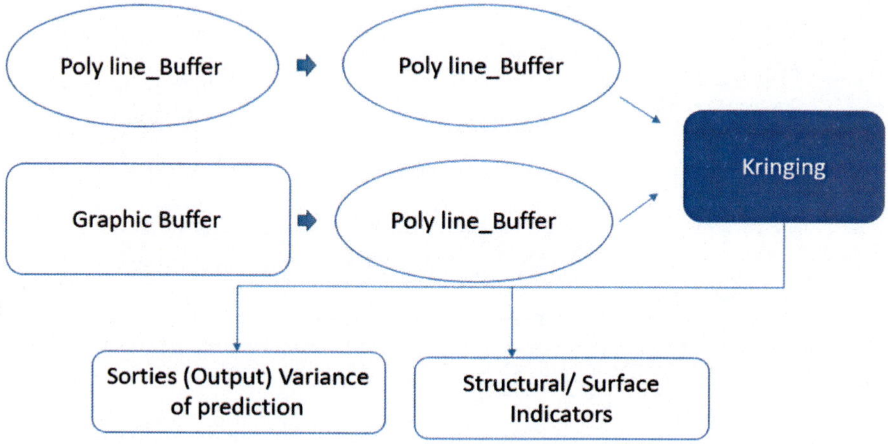

Fig. 5. Kringing validation model

4.2 Pothole Modeling During 2018 (Before the COVID-19 Period)

See Fig. 6.

4.3 Modelling Potholes After Deconfinement ('2020)

See Fig. 7.

5 Discussion

This geospatial study showed the variation in pothole condition showed some stability in condition between 2018 and 2020, with degradation condition residing in level A (Acceptable). These results confirmed that the decrease in traffic due to the blockage during containment, and the absence of maintenance ensured stability, aggravated the situation on this stretch of road. This confirms the impact of traffic on the development of this type of surface degradation.

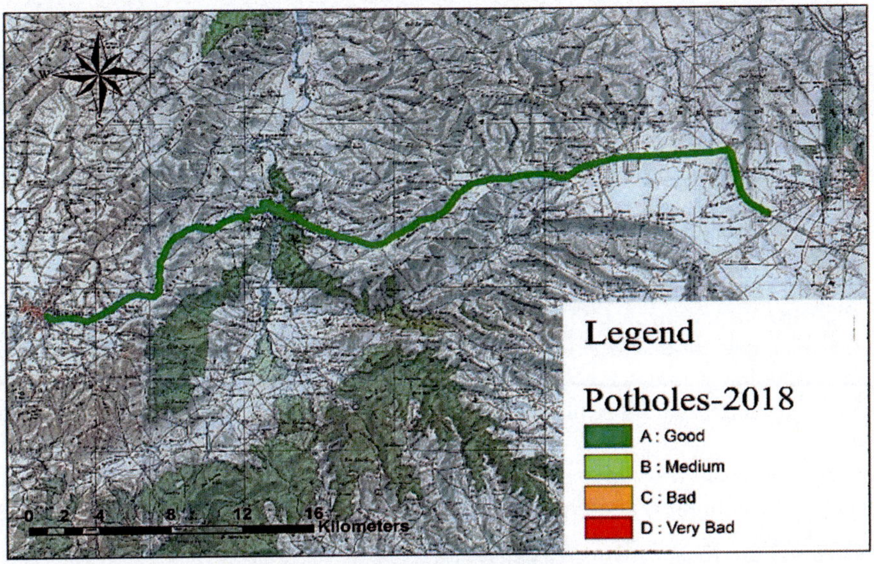

Fig. 6. Phothole mapping (2018)

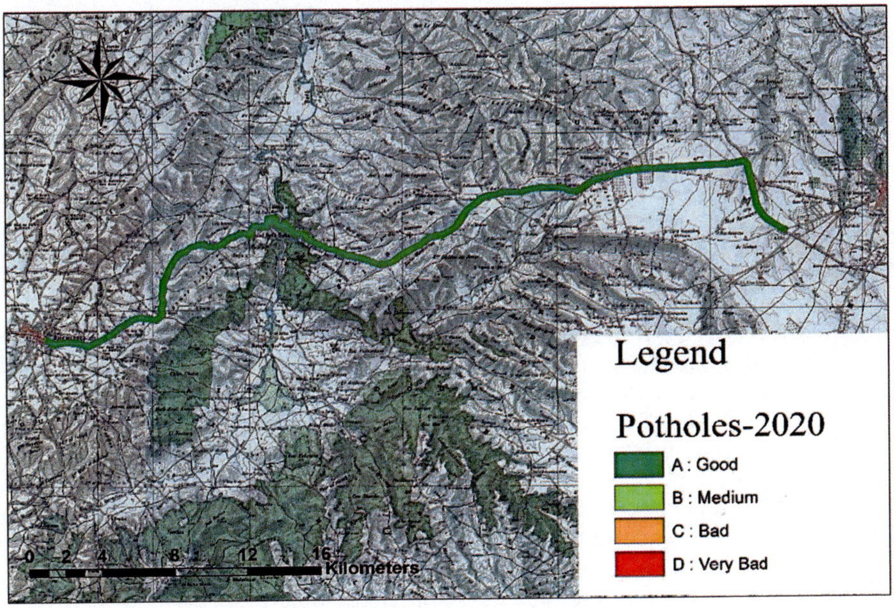

Fig. 7. Phothole mapping (2020)

6 Conclusions

This study was carried out solely on the basis of the results of environmental inspections (in the field) during 2018 and 2020 (during Covid-19). The classification of degradations by deformation matrices, the analysis of results and GIS modeling led us to propose corrective measures. These results take into account the diagnostic findings and the type of traffic the road was expected to support between 2018 and 2020. Essentially, this study has shown that:

- GIS functions as a visualization tool that enables road managers to track maintenance performance and inspection priorities throughout the life of each road section.
- The management of this section would be more specifically dedicated to the orange and red colored area, which represents the most degraded sections. In this way, road decision- makers could be alerted and therefore react quickly to severely deteriorated sections.
- Each road section in this zone is linked to its geographical location, making it an essential key to programming possible actions.

In the future perspectives, we expect to:

- Complete the present study, by analyzing the results of surface degradations such as stripping and cracking. As well as auscultation by Deflectograph and APL, to deduce the values of deflection and evenness on the one hand, and to study the evolution of the structural indicator of this pavement on the other. And to compare it with the pavement studied in this paper: Study the futuristic forecast of the surface and structural indicators evolution this pavement.
- Deduce possible future scenarios and optimize the budgets allocated for the repair, maintenance and monitoring of this pavement.

References

1. Ngene, B.U., Bassey, D.E., Busari, A.A., Bamigboye, G.O., Nworgu, A.T.: Influence of GIS on sustainable pavement maintenance: a comparative review. In: IOP Conference Series: Materials Science and Engineering, vol. 1036, no. 1, p. 012039. IOP Publishing (2021)
2. Mauduit, C., Robert, C.: Un SIG sur les indices de gel de référence: Quelle plus-value pour le dimensionnement au gel des infrastructures françaises? Revue générale des routes (872) (2008)
3. Mehdi, M.A., Cherradi, T., Bouyahyaoui, A., El Karkouri, S., Qachar, A.: Evolution of a flexible pavement deterioration, analyzing the road inspections results. Mater. Today Proc. **58**, 1222–1228 (2022)
4. Mehdi, M.A., Cherradi T., Qachar M., Chigr A.: Analysis of structural and surface deteriorations of a flexible pavementbased on inspection results using the MCA method. Mater. Today Proc. **45**(Part 8), pp. 7538–7546 (2021)
5. Mehdi, M.A., Cherradi, T., Elkarkouri, S., Qachar, A.: The Macroscopic Effect of COVID 19 on flexible pavement condition indicators based on analysis of road inspection results. In: Advanced Technologies for Humanity: Proceedings of International Conference on Advanced Technologies for Humanity (ICATH'2021) (pp. 441–452). Springer International Publishing, Cham (2022)

6. Koch, C., Brilakis, I.: Pothole detection in asphalt pavement images. Adv. Eng. Inform. **25**(3), 507–515 (2011)
7. Lewis, S., Sutton, J.: Demonstration Project No. 85 : GIS/Video Imagery Applications. Washington DC: Federal Highway Administration (1993)
8. Folger, P.: Geospatial information and geographic information systems (GIS) : an overview for congress (2011)
9. Transportation Officials, & AASHTO Joint Task Force on Pavements. (2001). Pavement management guide. AASHTO
10. Elhadi, H.M.A.: GIS, a tool for pavement management (2009)
11. Frias, V.G., Chaparro, T.S.: Managing the national road network maintenance in Spain. In: European Transport Conference 2008. Seminar Roads Policy, Delivery & Operation (2008)
12. Al-Mansour, A.I., Al-Qaili, A.H.: An application of android sensors and google earth in pavement maintenance management systems for developing countries. Appl. Sci. **12**(11), 5636 (2022)
13. Mehdi, M.A., Cherradi, T., Bouyahyaoui, A., El Karkouri, S., Qachar, A.: GIS approach for the pavement surface index modeling using a road survey data. **17**(20) (2022). ISSN/1819–6608
14. Mehdi, M.A., Cherradi, T., Bouyahyaoui, A., EL Karkouri, S.A.I.D., Qachar, M.: Applying Moroccan method for evolutionary study of the flexible pavement deflection based on the Lacroix deflectograph results. Jilin Daxue Xuebao (Gongxueban)/J. Jilin Univ. (Eng. Technol. Edn.). **41**(08–2022). ISSN: 1671-5497 E-Publication Online Open Access
15. Kaseko, M.S., Ritchie, S.G.: A neural network-based methodology for pavement crack detection and classification. Transp. Res. Part C Emerg. Technol. **1**(4), 275–291 (1993)
16. Mehdi, M.A., Cherradi, T., El Karkouri, S. and Qachar, A.: Applying geographic information systems (GIS) for surface condition indicators modeling of a flexible pavement. In: E3S Web of Conferences, vol. 298, p. 04001. EDP Sciences (2021)
17. Mehdi, M.A., Cherradi, T., Bouyahyaoui, A., El Karkouri, S., Qachar, A.: Evolution study of the pavement structural indicator based on evenness and deflection results using a GIS tool mapping. Geomate J. **23**(95), 144–153 (2022)

The Traditional Building Materials in the Ksours of Rissani

Sana El Malhi[1](✉), Latifa Ouadif[1], and Driss El Hachmi[2]

[1] L3GIE Mohammadia Engineering School Mohammed V University, Rabat, Morocco
Sanaaelmalhi@gmail.com

[2] EMM - MSME Faculty of Science Mohammed V University, Rabat, Morocco

Abstract. The preservation and restoration of historical monuments are essential for cultural heritage and identity. In Morocco, traditional construction materials have been used for centuries in the construction of historical monuments. This article explores the traditional construction materials used in the historic ksour of Rissani, located in the southeastern region of Morocco. Focusing on materials such as rammed earth, stone, and wood, the study examines their advantages, limitations, and relevance in the contemporary context. The study was conducted through a documentary analysis of scientific literature and field data collection in the Tafilalet ksours. Data was gathered through interviews with local experts, surveys among the local populations, and direct observations. The article delves into the significance of these materials in terms of durability, aesthetic appeal, energy efficiency, and cultural heritage preservation. By understanding their characteristics and considering their integration into modern architectural practices, the article underscores the importance of preserving these traditional construction techniques. Through this exploration, it aims to shed light on the unique essence of Rissani's ksour and their invaluable contribution to Morocco's cultural and heritage wealth.

Keywords: Ksour · Rammed earth · Mud brick · Stone · Palm wood · Durability

1 Introduction

The ksour of Tafilalet, nestled in the enchanting region of Sijilmassa in southwestern Morocco, are captivating historical treasures. These ancient fortified villages, characterized by their distinctive architectural style and traditional building techniques, offer a captivating glimpse into the rich cultural heritage of the area. In this article, we embark on a journey to explore the extraordinary world of earthen construction materials that have been utilized in the construction of the ksour of Tafilalet.

For centuries, the local inhabitants have harnessed the abundant resources of the surrounding landscape to create remarkable structures that harmoniously blend with their natural surroundings. Earthen construction materials, such as rammed earth, adobe, and clay, form the very foundation of these awe-inspiring ksour, embodying the essence of sustainable and environmentally-friendly building practices (Gil-Piqueras and Rodríguez-Navarro [8].

K. Baba et al. (Eds.): GeoME 2023, SPEES, pp. 390–399, 2023.
https://doi.org/10.1007/978-3-031-49345-4_37

The unique characteristics of these earthen materials have contributed to the enduring charm and resilience of the ksour of Tafilalet. The use of rammed earth, for instance, provides exceptional thermal insulation, keeping the interiors cool during scorching summers and warm during chilly winters. The inherent thermal mass of these materials regulates temperature fluctuations, reducing the reliance on artificial heating and cooling systems. Furthermore, the local availability and affordability of these materials have made them an integral part of the vernacular architecture of the region.

The construction techniques employed in the ksour of Tafilalet reflect a deep understanding of the local climate, culture, and available resources. The meticulous craftsmanship exhibited in the formation of earthen walls, intricately designed decorative elements, and sturdy structural components is a testament to the ingenuity and expertise of the local builders.

While these traditional construction practices have stood the test of time, the preservation and revival of earthen construction face contemporary challenges. Rapid urbanization, changing lifestyles, and modern building materials have influenced the way buildings are designed and constructed. However, by recognizing the intrinsic value and sustainable advantages of earthen construction, there is a growing appreciation for the revival and integration of these techniques in contemporary architecture.

In this article, we delve deep into the world of earthen construction in the ksour of Tafilalet, unraveling the secrets of their unique materials, exploring their environmental benefits, and highlighting their cultural significance. By understanding the historical context, properties, and potentials of these earthen materials, we hope to inspire a renewed appreciation for the architectural wonders of Tafilalet's ksour and foster a sustainable approach to construction that embraces both the past and the future.

2 Materials and Methods

2.1 Description of the Ksours and Their Cultural and Architectural Importance

Rissani's ksours are traditional fortified structures that were once used as fortified villages. They were designed to provide protection from tribal conflict and invasion, while serving as economic and social centers for local communities. The ksours are generally built using local materials such as raw earth and stone, thus reflecting the adaptation to the natural resources available in the region and their adaptation to the semi-arid climate of the south eastern region of Morocco (Fig. 1).

These ksours have a characteristic architecture with their thick walls, their defense towers, their narrow passages, their interior courtyards and their large giant arched door decorated with carved motifs of the culture and aesthetics of the region. They are often arranged in a compact way, thus forming fascinating architectural ensembles. Additionally, the ksours feature unique ornamental details, such as carved patterns and decorative wooden elements, which add an artistic touch to their construction (Fig. 2).

2.2 The Traditional Materials and Techniques Used in the Ksours of Rissani

In this research we proceeded to a field visit to some ksours of rissani (ksar abbar, ksar oulad abdelhalim), in order to make a survey with the local expert (lmaalem) to know the

Fig. 1. Aerial view ksar oulad abdelhalim in Rissani

Fig. 2. The main entrance to ksar abbar in rissani

adapted techniques and the materials in the construction of these ksours, with the taking of photos for the exterior architecture and the main elements of the structure (walls, post, roof, etc.).

The key element in the traditional construction method of ksours is the use of earth material, which is employed in various situations due to its ideal suitability in effectively adapting to the warm and dry climate of the region.

Among the building techniques adapted using earth as the main material, we have rammed earth is called "allouh", and we also have mud bricks called "toub", these two techniques are used separately in the different parts of the construction.

– **Rammed earth**

The technique of rammed earth involves compacting layers of moist soil within a wooden formwork. This method enables the creation of robust continuous load-bearing walls with considerable thickness (40–100 cm). The construction process begins with horizontal masonry until the entire perimeter is completed. After allowing sufficient drying time (usually not less than one week, depending on the climate) to prevent wall deformation or collapse, the construction proceeds to higher levels (Fig. 3).

Fig. 3. Rammed earth wall

The building is lifted to add subsequent layers using shifting formwork, block by block. This approach maintains a relatively uniform wall thickness along the perimeter and throughout the height of the structure. When constructing masonry buildings, careful attention is required to ensure proper interlocking between blocks, walls, and partitions. The dimensions of the formwork can vary, and in the ksours of tafilalet, the average size is about 200 cm in length, 80 cm in height, and a width ranging from 60 to 100 cm, depending on the building's height. The height of the plans is determined by the number of rammed earth blocks used [5].

– **Mud bricks**

Mud bricks are one of the traditional building materials used in ksours, these bricks are made from raw clay, sometimes mixed with aggregates such as sand then shaped and dried in the sun [15].

the size of the bricks varies according to the needs and the tradition, generally the width of brick equal to a half-length and the height equal to a half width.

the brick earth must be carefully selected rich in clay (at least 40%), and water until you obtain a magnificent and plastic paste, you can add vegetable fibers such as straw to fight against cracking of the bricks during drying due to the high percentage of clay in the mix.

the bricks are formed by hand, molded and left to dry in the sun, once dried can be stored and used later [3] (Fig. 4).

the drying time varies according to the climate and the season (2 to 3 days in summer and 10 to 15 days in winter).

Fig. 4. Mud brick sun dried. Credit by Baglioni [5]

the bricks are then used in the construction of walls, arcades, vaults and other structural elements of buildings.

– **Stone**

Stone is another traditional building material used in ksours. Local craftsmen use locally quarried stones, usually limestone or sedimentary rock.

Stone construction involves cutting and shaping the stones as needed and then assembling them using traditional lime-based mortars. The craftsmen use specific bonding techniques, such as the "boutisse" or stone on stone", to guarantee the stability and solidity of stone structures. Stone is often used either in foundations or as a base for walls, and also in the construction of posts in mixed masonry with earthen bricks [9].

Typically, the base of the ksar consists of a variety of stones arranged in layers of 0.50 to 0.80 m in height. These stones are positioned on a trench measuring 0.50 to 1.00-m-wide and are held together by a layer of mud. This construction method ensures the wall's stability [10].

All the buildings are generally built on a stone base of various hight. The presence of a good stone basement limits the water capillary rise and protect the wall from water splashes ans human actions.

– **Wood**

The wood used in its ksours is palm wood and cedar wood to utilize the sturdy and elaborate timber frames. Often made of interlocking beams and joists to provide structural stability.

Wooden doors are an essential element of ksour architecture. They are often decorated with carved patterns and traditional ironwork. Wooden windows are commonly used in ksours. They usually have wooden frames and shutters which make it possible to regulate the ventilation and the luminosity inside the buildings. The wooden windows also add a touch of charm and authenticity to the architecture of the ksours [3].

In addition to structural elements, wood is also used to create decorative elements such as carved beams, relief patterns and wooden friezes. These elements contribute to the overall aesthetics of buildings and showcase local craftsmanship (Figs. 5 and 6).

Fig. 5. Wooden roof

3 Results and Discussion

3.1 The Advantages of Using Traditional Building Materials in the Ksours of Rissani

The use of these traditional earth-based building materials in this region brings many benefits. For example, rammed earth, wood and stone have natural insulating properties. They have a low thermal mass, which means that they can absorb, store and release heat more slowly than modern materials such as concrete. This helps maintain a more stable interior temperature and reduces heat fluctuations during hot periods, also creating thick rammed earth walls with low thermal conductivity helps to reduce the transmission of heat through the walls of buildings, thus contributing to better thermal insulation. These materials also have the ability to regulate humidity inside buildings. El Azhary [6].

Fig. 6. Wood frame

For example, rammed earth has a good water vapor permeability, which helps to limit the accumulation of moisture and prevent problems of excessive humidity, it helps to maintain a comfortable and healthy indoor environment, especially in dry climates in which the air dehydration can be a problem [11].

these materials are characterized by their durability, rammed earth and stone can withstand bad weather and extreme climatic conditions, intense heat, temperature variations and strong winds, while wood used properly can last for decades or even centuries.

local and economic availability is another advantage of these traditional building materials used in the ksours of tafilalet, they are often available locally, which reduces transport costs and makes them more economical compared to imported materials, for example earth Raw material is used in abundance due to the presence of clay and sand in the region. The availability of these materials has facilitated their use in construction and has helped to strengthen the local economy and preserve natural resources [13].

When it comes to environmental impact, these traditional materials are often considered more environmentally friendly than modern materials [12]. They are natural, renewable and require less energy to produce and maintain. They also reduce the carbon footprint associated with modern building materials [7].

Traditional materials give the ksours a unique aesthetic character and a strong cultural identity. Rammed earth, stone and wood are appreciated for their natural appearance, their texture, their carved patterns and their harmonious integration into the local environment. They contribute to preserve the cultural identity of the region and to maintain the link with the architectural heritage.

3.2 The Limitations of Using Traditional Building Materials in the Ksours of Rissani (Arial 10)

While earthen construction seems to be in perfect harmony with Tafilalet environment, it does exhibit some problems and limitations. Generally speaking, earthen construction deficiencies are mainly caused by three factors: atmospheric (wind and water), building design and the construction system [4].

Evidently, wind and water can affect any building. Whether in earthen or modern construction, cracks and other pathologies are caused by the infiltration of water. Earthen construction is very sensitive to water infiltration because rammed earth is not waterproof. Water can easily infiltrate the mansion and lead in the short-term to serious pathologies. In addition, rain attacks the surface plaster, which serves as the guarding shield of the house. The removal of the surface plaster accelerates the deterioration of earthen construction. Hence, a proper and frequent maintenance of plaster is essential to preserve earthen construction [4].

The design of earthen construction in Tafilalet is another cause of pathologies. Rammed walls are highly sensitive to water. A good way to enforce their water' resistance is the presence of a solid stone basement [14]. However, due to the scarcity of stones in the region, stone basement is either unavailable or insufficiently high [3]. Obviously, stone basement can protect walls against water-infiltration and enforce their durability.

Pathologies are also related to building techniques [14]. Walls are built separately without joints or connections in the corner. For economic factors, people tend to minimize the use of rammed earth. In the long term, walls start to separate from each other causing vertical cracks.

Some traditional materials such as wood may be more susceptible to attacks from pests and pests such as termites or fungi. This can lead to structural damage and require appropriate prevention and treatment measures.

Another disadvantage is that these materials may sometimes not meet modern building standards for seismic resistance or other specific criteria. Adjustments and additions may be necessary to comply with current regulations, which may result in additional costs or loss of authenticity of the structures.

Earthen constructions in this region require regular maintenance and periodic repairs to preserve their original condition. This may involve restoration, consolidation or replacement of damaged elements. The availability of qualified professionals to carry out this work can sometimes be a challenge.

4 Conclusions

In conclusion, the exploration of traditional construction materials in the ksour of Rissani reveals their significant advantages and limitations. These materials, including rammed earth, stone, and wood, have played a crucial role in shaping the unique architectural heritage of the region.

One of the key advantages of these traditional materials is their inherent durability. The use of rammed earth walls, for example, has demonstrated remarkable resilience against the harsh climate conditions of Rissani, including extreme heat and aridity. The thermal mass properties of these materials allow for a more stable indoor temperature, providing natural insulation and reducing the need for excessive energy consumption.

Moreover, the authentic aesthetics of the traditional materials contribute to the cultural identity and sense of place in the ksour. The earthen tones, intricate stonework, and wooden elements blend harmoniously with the natural surroundings, showcasing the craftsmanship and artistry of the local communities. The preservation of these architectural traditions not only conserves the tangible heritage but also honors the intangible cultural values and knowledge embedded in their construction techniques.

However, it is important to acknowledge the limitations and challenges associated with traditional construction materials. Their susceptibility to weathering and erosion requires regular maintenance and periodic repairs. The availability of skilled craftsmen well-versed in these traditional techniques may also be limited, posing obstacles to the preservation and restoration efforts. Additionally, the compliance of these materials with modern building codes and regulations may necessitate adaptations and compromises to ensure structural integrity and safety.

The integration of traditional materials in contemporary architectural practices offers an opportunity to strike a balance between heritage conservation and sustainable development. By harnessing the advantages of traditional materials while addressing their limitations, innovative approaches can be employed to enhance their performance, such as reinforcing techniques, protective coatings, and complementary use of modern materials where necessary.

In summary, the use of traditional construction materials in the ksour of Rissani represents an invaluable cultural heritage that needs to be safeguarded. The sustainable and aesthetic qualities of rammed earth, stone, and wood contribute to the preservation of the region's architectural identity and environmental harmony. Recognizing the importance of these materials, further research, documentation, and collaboration between heritage conservation experts, local communities, and policymakers are essential to ensure their continued preservation and transmission to future generations. Embracing a holistic approach that combines tradition and innovation will enable the ksour of Rissani to thrive as living cultural landmarks while embracing the challenges of a rapidly changing world.

References

1. Baglioni, E., et al.: Cultural landscape of the Draa Valley, Morocco. In: Digital proceeding of XI Conferencia Internacional sobre el Extudio y conservacion de la Arquitechtura patrimonial de Tierra (2012)
2. Baglioni, E., Mecca, S.: As tipologias habitacionais no vale do draa (marrocos) a ca-patio. In: Terra em seminario: confrence proceedings of 6° Seminario de Arquitectura de Terra en Portugal (2010)
3. Baglioni, E.: Sustainable vernacular architecture: the case of the Drâa Valley Ksur (Morocco). In: Proceedings: Sustainable Architecture and Urban Development, pp. 12–14 (2010)
4. Baglioni, E.: Draa valley Earthen architecture: construction techniques, pathology and intervention criteria. J. Mater. Environ (2016)
5. Baglioni, E.: Tecniche construtive in terra cruda nella valle del draa, morocco. In: unipublished graduation thesis Architecture faculty of Florence University, Italy, pp. 1–13 (2009)
6. El Azhary, K.C.: Energy efficiency and thermal properties of the composite material clay-straw. Energy Procedia. 160–164 (2017)

7. El Azhary, K.O.: Energy efficiency of a vernacular building design and materials in hot arid climate: experimental and numerical approach. Int. J. Renew. Energy Dev. **10**(3), 481 (2021)
8. Gil-Piqueras, T., Rodríguez-Navarro, P.: Tradition and sustainability in vernacular architecture of Southeast Morocco. Sustainability **13**(2), 684 (2021)
9. Khaloufi, B.: The use of natural resources for housing in rural areas of Morocco (1956)
10. Kolbl, O., Boussalh, M., Fadli, B., Larbi, B., Naji, M.: Syntèse de l'inventaire du patrimoine Architectural de la Vallée du Draa. Ouarzazate (2010)
11. Lamrani Alaoui, A.M.: Optimization of thermal efficiency in traditional clay-based buildings in hot dry locations, case study: the South Estern Region of Morocco (2022)
12. Lawrence, M.: Reducing the environmental impact of construction by using renewable materials. J. Renew. Mater. 163–174 (2015)
13. Mellaikhafi, A.O.: Characterization and thermal performance assessment of earthen adobes and walls additive with different date palm fibers (2021)
14. VV.AA.: Inventaire du patrimoine architectural de la Vallée du Draa. Ministère de la culture du Maroc (CERKAS), Bureau d'Architecture et d4Urbanisme H. Hostettler (2005)
15. Zaghez, I., Attoui, R., Saou-Dufrêne, B.: Preservation and sustainability of earthen architecture. In: The Vernacular Ksar of Khanguet Sidi Nadji in Algerian Sahara as a Case Study (2023)

Spectral Responses and Building Elevations: A Case Study of the Mediterranean Earthquake in Morocco

Mohamed Ahatri(✉), Yassine Razzouk, and Khadija Baba

Civil Engineering and Environment Laboratory (LGCE), Mohammadia School of Engineering, Mohammed V University Rabat, Rabat, Morocco
mohamed.ahatri@gmail.com

Abstract. Morocco is located in the "SEISMIC BELT"; the most seismically active region where the majority of the world's earthquakes occur. More precisely, the region is known as the Ibero-Maghreb Mediterranean Region, which includes high-risk locations due to seismic instability. It should be mentioned that the National Institute of Geophysics' seismic monitoring stations record hundreds of earthquakes of varying magnitudes each year. The objective of this article is to examine the effects of Spectral Responses on architectural elements and to juxtapose these effects with the Moroccan seismic construction regulations (RPS 2011). This analysis is conducted within the context of the Mediterranean earthquake. The earthquake was recorded at a station situated in the city of Zeghanghane, located in Nador, Morocco.

Keywords: Seismic event · Spectral behavior · Soil dynamics · Mediterranean seismic activity

1 Introduction

Seismic Action refers to the dynamic forces and vibrations generated by an earthquake that affect structures and the surrounding environment. These actions are caused by the release of accumulated stress along geological faults, resulting in the propagation of seismic waves through the Earth's crust. Sources of seismic action primarily include tectonic activities, such as the movement of tectonic plates, fault slip, and volcanic activity [1]. Anthropogenic sources like mining and reservoir-induced seismicity can also contribute to seismic actions. Geotechnical conditions play a crucial role in how seismic waves interact with the Earth's subsurface [2]. The properties of the soil and rock layers, as well as their density, stiffness, and damping characteristics, influence the transmission and amplification of seismic waves. Soft soils tend to amplify ground motions, whereas stiff soils tend to diminish their impact [3].

Seismic waves exhibit distinct types, each playing a unique role in earthquake dynamics. Primary Waves (P-waves) propagate through solids, liquids, and gases with rapid speed, causing particles to move in the direction of the wave. Conversely, Secondary

K. Baba et al. (Eds.): GeoME 2023, SPEES, pp. 400–413, 2023.
https://doi.org/10.1007/978-3-031-49345-4_38

Waves (S-waves) travel solely through solids, inducing particle movement perpendicular to their path and potentially causing significant structural damage. Surface Waves, operating along the Earth's exterior, generate the majority of ground tremors during earthquakes. Their complex motion encompasses both longitudinal and transverse components. Notably, Love waves and Rayleigh waves, both forms of surface waves, can trigger severe ground shaking and structural resonance [4]. In seismic analysis within civil engineering, velocity emerges as a vital parameter denoted as "v." Often expressed as "shear wave velocity" (Vs) or "compression wave velocity" (Vp), shear wave velocity, particularly pertinent for comprehending wave propagation through diverse soil and rock layers, influences ground motions, with lower velocities in softer soils amplifying the impact [5].

The primary objective of this research is twofold: firstly, to develop and assess the ground motion's spectral response at ZGH station [6], and concurrently, to investigate the structural stability through spectral methodology. This method seeks to understand how seismic characteristics affect structures of varying elevations. Additionally, the paper aims to establish a comparative analysis between seismic evaluations conforming to the Moroccan seismic regulations RPS 2011 [7, 8] and spectral assessments for a Nador-based structure positioned at different heights, subjected to a 6.3 Mw earthquake. In pursuit of these objectives, the initial phase involves the generation of response spectra utilizing a Matlab Code [6, 8].

2 Seismic Structure Analysis (Moroccan Seismic Construction Rules RPS 2011)

The "Paraseismic Construction Regulations (RPS 2011)" have two main goals: protecting material assets and ensuring public safety during earthquakes [7]. The complete structure, along with its entirety of structural elements, should demonstrate a significantly elevated likelihood concerning the design seismic forces. This prerequisite ensures the attainment of an adequate safety standard for human lives prior to, during, and following a substantial seismic event. Additionally, it is demanded that the building as a whole, along with all of its structural and non-structural components, be safeguarded in a reasonable manner against both appearance damage and the restriction of the building's intended use in the event of an earthquake [9, 10].

According to the 2011 update to the Moroccan seismic rules RPS 2000 [11], the following factors are used to assess an earthquake's effect on a structure:

- The maximum soil acceleration Amax (Table 1):

- The maximum soil velocity Vmax (Table 2):

- Site influence (Table 3):

Maximum acceleration and response spectrum are regarded sufficient for the Moroccan regulation's application, even though other characteristics are involved.

Table 1. The maximum soil acceleration for each seismic zone [7].

Seismic zone acceleration (Za)	A max (%g)
0	4
1	7
2	10
3	14
4	18

Table 2. The maximum soil velocity for each seismic zone [7].

Seismic zone acceleration (Zv)	V max (m/s)
0	0,05
1	0,07
2	0,10
3	0,13
4	0,17

Table 3. Site coefficient [7].

Site class	Soil profil name	Site coefficient
S1	Rock	1
S2	Very dense soil and soft rock	1,2
S3	Stiff soil	1,4
S4	Soft soil	1,8
S5	Special conditions	Determined by an expert

3 Reminder of the Duhamel Integration Method

The representation of earthquake motion through response spectra adds a dimension of frequency-based movement content analysis, encompassing displacement, velocity, and acceleration. The primary aim of this approach is to capture earthquake characteristics by evaluating the reactions of simple structures. Notably, an acceleration response spectrum characterizes the curve depicting the maximum acceleration experienced by oscillators with varying degrees of freedom and natural frequencies, encompassing a mass (m), stiffness (k), and damping (C).

When represented in accelerogram form, the equation governing forced oscillations post-earthquake can be expressed as:

$$ma(t) + Cv(t) + kD(t) = -ma_g(t) \tag{1}$$

To ascertain the maximum acceleration sustained by mass (M), the solution to the differential equation is determined. Utilizing the damping definition ζ, the equation is transformed as follows [12]:

$$a(t) + 2\omega v\zeta(t) + \omega^2 d(t) = -a_g(t) \tag{2}$$

The equation for relative displacement (d(t)) is given by [13]:

$$d(t) = -\frac{1}{\omega}\int_0^t a_g(\tau)e^{-\zeta\omega(1-\tau)}\sin\omega(t-\tau)d\tau \tag{3}$$

For relative velocity (v(t)), the equation stands as [14]:

$$v(t) = -\int_0^t a_g(\tau)e^{-\zeta\omega(1-\tau)}\cos\omega(t-\tau)d\tau \tag{4}$$

The pseudo-acceleration equation (a(t)) is defined as [15]:

$$a(t) = +\omega\int_0^t a_g(\tau)e^{-\zeta\omega(1-\tau)}\sin\omega(t-\tau)d\tau \tag{5}$$

Utilizing the earthquake registration (accelerogram) formulas, the maximum response values for displacement $[d(t)]_{max}$, velocity $[v(t)]_{max}$, and acceleration $[a(t)]_{max}$ can be systematically determined for various oscillators, incorporating a wide range of periods and possible damping. These values contribute to the creation of response spectra graphs. The displacement response spectrum S_d, velocity response spectrum S_v, and acceleration response spectrum S_a can be calculated as follows [12, 16]:

$$S_d = [d(t)]_{max} = S_v/\omega = D_{max} \tag{6}$$

$$S_v = [v(t)]_{max} = \omega S_d \tag{7}$$

$$S_a = [a(t)]_{max} = -\omega^2 d(t) = \omega S_v = \omega^2 S_d \tag{8}$$

4 Case Study

4.1 Earthquake Data

On January 25, 2016, at 4:22 a.m., a severe earthquake struck the Mediterranean between Spain and Morocco [17, 18], affecting the beaches of Nador and El Hoceima. On the Richter scale, it had a magnitude of 6.3. The epicenter was found at (35.586, – 3.690) at a depth of 26 km.

In its preliminary evaluation of the earthquake's consequences. USGS stated that there was "a low probability of loss and damage (Figs. 1 and 2)."

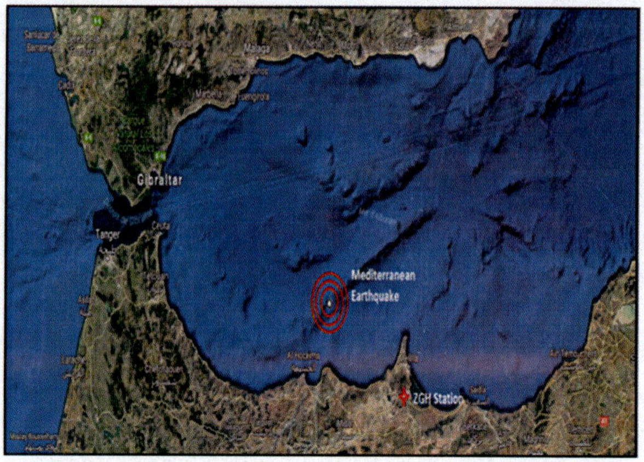

Fig. 1. The location of the earthquake and the ZGH station

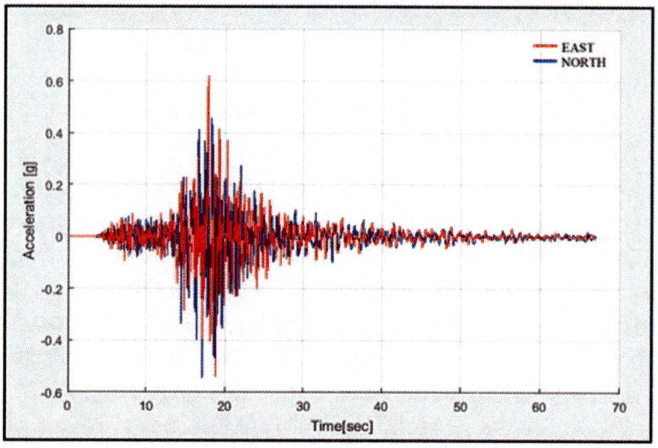

Fig. 2. Acceleration

4.2 Basic Data of the Structures Studied

In accordance with the Earthquake regulation 2000 revised 2011, We studied four identical types of structure at various heights (Basement + GF + 3, Basement + GF + 5, Basement + GF + 7 and Basement + GF + 9). Furthermore, we performed a spectral assessment utilizing the response spectra derived from the seismic activity in the Mediterranean region, specifically focusing on the ZGH station seismic recordings. These analyses, distinct yet interrelated, were executed through the utilization of the Robot Structural Analysis software. This software, known for its application of the finite element method, served as the platform for conducting both investigations [19].

5 Results and Discussion

5.1 The Response Spectra

Derived from the time-sequenced documentation of the Mediterranean seismic event, the response spectra are acquired through the utilization of a MatLab software application [6, 8].

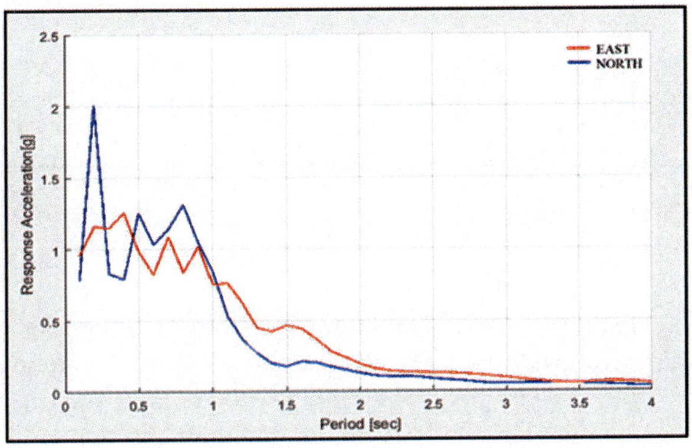

Fig. 3. Spectrum of acceleration response (Sa)

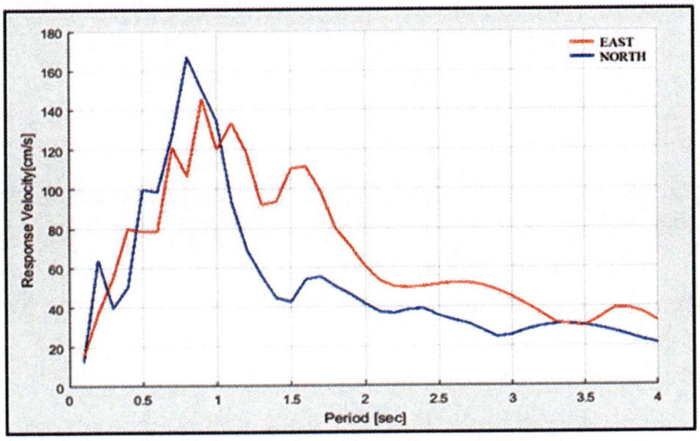

Fig. 4. Velocity response sspectrum (Sv)

Figures 3, 4, and 5 illustrate the response spectra under 5% damping conditions.

The spectrum of the accelerogram displays an array of non-uniform peaks.

Meanwhile, the displacement spectra exhibit a rise corresponding to the period, whereas the velocity and acceleration spectra depict a trend shift with a decrease across different periods.

Within the realm of civil engineering, the focus of structural designers centers on the frequency band spanning 1 to 10 Hz [20]. In this frequency range, the amplification characteristics of ground motion play a significant role in shaping the impact of seismic events.

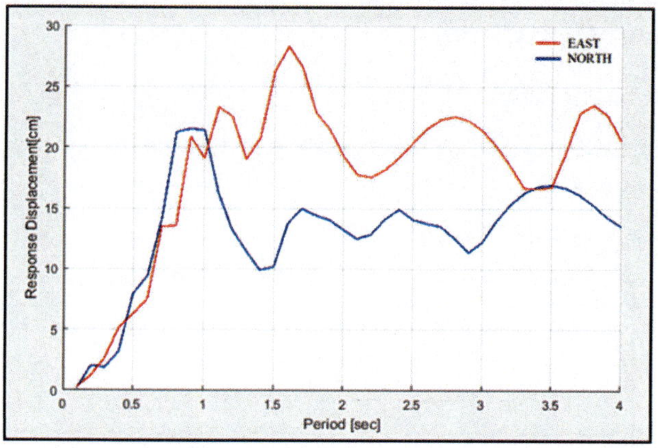

Fig. 5. Spectrum of displacement response (Sd)

Conclusively, referring to the formulated equations (expressions: (10, 11, and 12) [21], he differential response spectra for various time intervals can be derived.

5.2 The Impact of Response Spectra on Buildings of Various Heights

In order to explore the influence of seismic characteristics on structures, considering their diverse elevations, we conducted an examination of structural stability utilizing the spectral methodology. Subsequently, a comparative analysis was carried out between seismic assessments executed in adherence to the Moroccan seismic code RPS 2011, and spectral evaluations for a structure situated at varying elevations within Nador, exposed to a 6.3 Mw earthquake. The Robot Structural Analysis software facilitated the determination of vectors, forces, reactions, and structural displacements (Tables 4, 5 and 6).

As demonstrated in the presented tables and figures, both analyses indicate a proportional growth in both displacements and stresses within the structure as the number of floors increases (Figs. 6, 7 and 8).

By comparing the results of the two analyzes, we have found that the seismic analysis had higher values than the spectral analysis.

The findings demonstrate that spectral analysis has a greater impact than seismic analysis, which may allow the coefficients of regulatory requirements to be reviewed in order to alleviate them.

Table 4. The values of the extreme deflection according to seismic and spectral analysis.

Building	Type	Max/Min	Value (cm)
Basement + GF + 3	Seismic analysis	Max	0.4
		Min	– 0.4
	Spectral analysis	Max	0.1
		Min	– 0.4
Basement + GF + 5	Seismic analysis	Max	0,6
		Min	– 0,6
	Spectral analysis	Max	0.2
		Min	– 0.5
Basement + GF + 7	Seismic analysis	Max	0,7
		Min	– 0,7
	Spectral analysis	Max	0.4
		Min	– 0.7
Basement + GF + 9	Seismic analysis	Max	0,8
		Min	– 0,9
	Spectral analysis	Max	0.4
		Min	– 0.9

Table 5. The values of efforts according to Seismic and Spectral Analysis.

Building	Type	Max/Min	Effort/x [kN]	Effort/y [kN]	Effort/z [kN]	Moment/x [kNm]	Moment/y [kNm]	Moment/z [kNm]
Basement + GF + 3	Seismic analysis	Max	1169,51	17,73	125,37	5,53	26,74	27,53
		Min	− 24,92	− 9,32	− 136,13	− 5,56	− 123,58	− 23,77
	Spectral analysis	Max	1169,51	14,05	125,37	5,53	21,32	23,87
		Min	− 19,46	− 7,46	− 136,13	− 5,56	− 123,58	− 23,46
Basement + GF + 5	Seismic analysis	Max	1439,07	24,83	124,88	5,45	41,36	38,59
		Min	− 50,95	− 12,8	− 136,61	− 5,48	− 124,66	− 33,85
	Spectral analysis	Max	1439,07	20,3	124,88	5,45	32,06	31,56
		Min	− 40,77	− 10,34	− 136,61	− 5,48	− 124,66	− 29,74
Basement + GF + 7	Seismic analysis	Max	1671,69	30,59	124,56	6,72	53,25	47,58
		Min	− 79,64	− 15,65	− 136,89	− 5,39	− 125,34	− 42,87
	Spectral analysis	Max	1671,69	31,38	124,56	6,22	59,11	48,81
		Min	− 89,21	− 16,9	− 136,89	− 5,39	− 125,34	− 44,4
Basement + GF + 9	Seismic analysis	Max	1872,72	35,83	124,37	8,29	65,86	55,68
		Min	− 110,93	− 19,36	− 137,05	− 5,83	− 139,32	− 51,18
	Spectral analysis	Max	1872,72	28,88	124,37	7,12	48,61	44,91
		Min	− 81,72	− 14,51	− 137,05	− 5,83	− 125,76	− 44,34

Table 6. Reactions according to seismic and spectral analysis

Building	Type	Max/Min	Reaction/x [kN]	Reaction/y [kN]	Reaction/z [kN]	Moment/x [kNm]	Moment/y [kNm]	Moment/z [kNm]
Basement + GF + 3	Seismic analysis	Max	89,16	135,09	1169,51	59,62	23,59	9,38
		Min	− 52,49	− 122,66	− 316,84	− 34,97	− 16,29	− 8,44
	Spectral analysis	Max	43,3	76,64	1169,51	38,49	16,57	7,63
		Min	− 44,06	− 122,66	− 234,75	− 32,82	− 16,29	− 8,44
Basement + GF + 5	Seismic analysis	Max	109	167,23	1439,07	76,55	31,28	12,03
		Min	− 73,01	− 163,54	− 400,76	− 45,02	− 21,85	− 10,78
	Spectral analysis	Max	82,22	122,65	1439,07	61,53	26,28	10,14
		Min	− 59,16	− 143,98	− 313,32	− 39,09	− 21,85	− 10,78
Basement + GF + 7	Seismic analysis	Max	127,3	188,15	1671,69	88,86	37,66	14,43
		Min	− 89,13	− 200,15	− 458,67	− 52,97	− 27,49	− 13,27
	Spectral analysis	Max	131,27	193,8	1671,69	88,82	36,51	14,01
		Min	− 97,4	− 211,8	− 508,88	− 54,91	− 27,49	− 13,27
Basement + GF + 9	Seismic analysis	Max	146,71	207,4	1872,72	100,45	43,84	16,91
		Min	− 105,44	− 232,79	− 517,59	− 60,97	− 33,46	− 15,9
	Spectral analysis	Max	115,17	162,5	1872,72	86,52	37,52	14,13
		Min	− 83,66	− 204,25	− 376,92	− 54,86	− 33,46	− 15,9

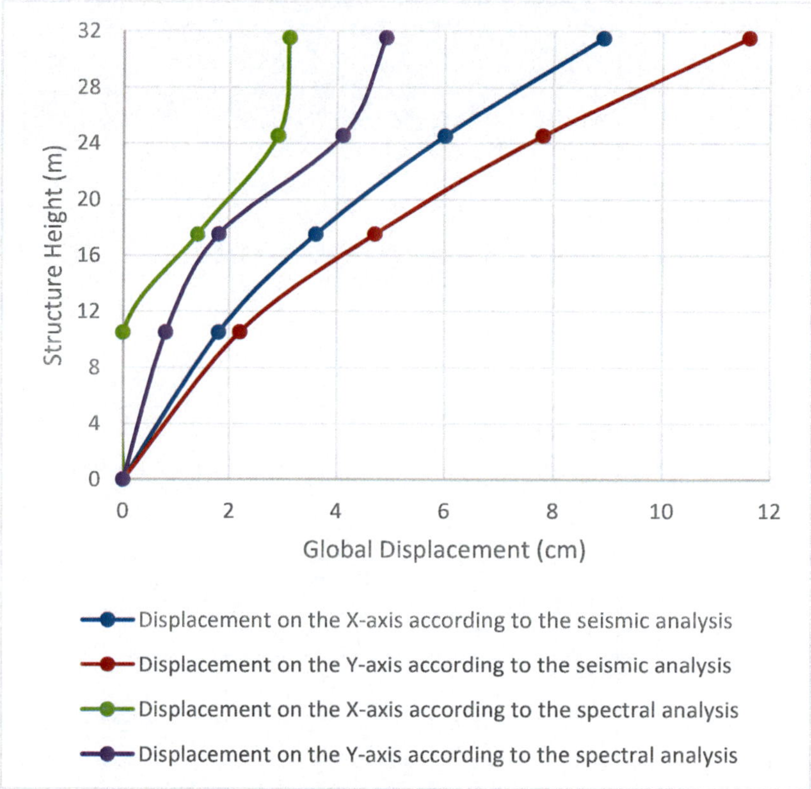

Fig. 6. Global Displacements according to both "seismic and spectral" analysis of buildings.

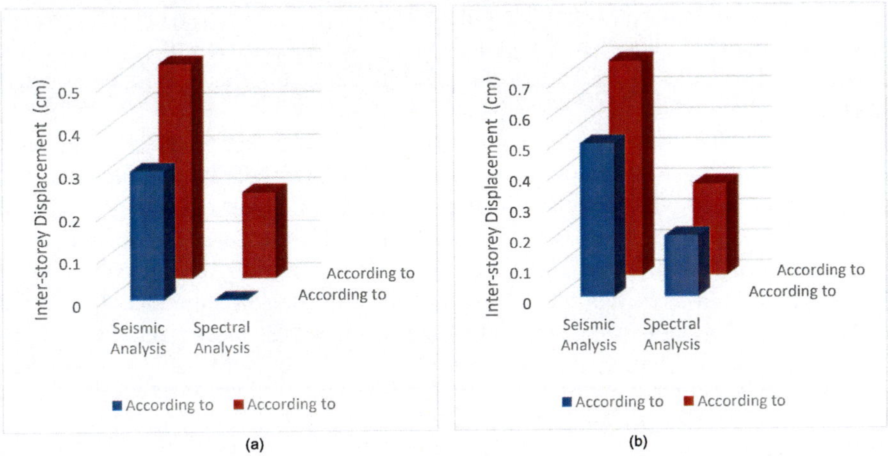

Fig. 7. Global inter-storey displacements according to both "seismic and spectral" analysis of (**a**) 4-storey buildings (Basement + GF + 3) and (**b**) 6-storey buildings (Basement + GF + 5).

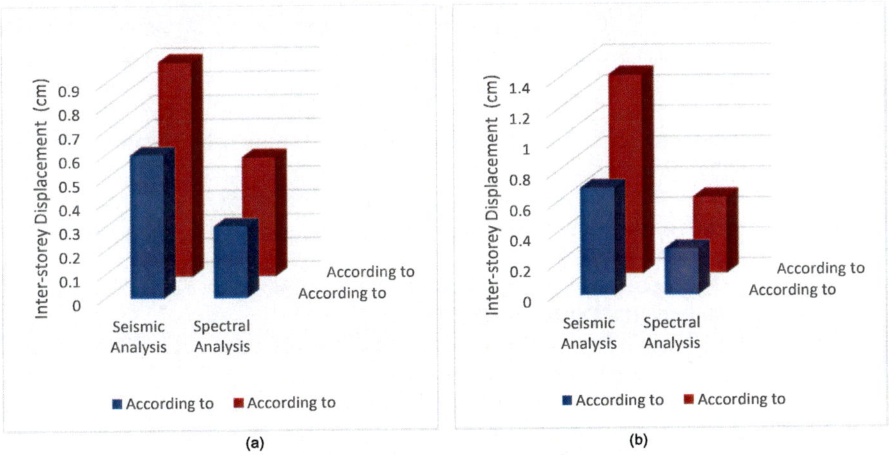

Fig. 8. Global inter-storey displacements according to both "seismic and spectral" analysis of (**a**) 8-storey buildings (Basement + GF + 7) and (**b**) 10-storey buildings (Basement + GF + 9).

6 Conclusion

In the realm of civil engineering, the utilization of Robot Structural Analysis software, which operates through the finite element method, has been meticulously developed to cater to structural computations. Specifically designed to encompass a 5% absorption factor for both soil and constructions, this software has proven invaluable for comprehending the dynamic behavior of constructions subjected to seismic forces.

Our investigation unveiled a noteworthy contrast in the realm of global and inter-floor displacement, deflections, and reactions within the context of our case study. This discrepancy came to light through a comparison between two distinct analytical approaches:

one adhering to seismic construction regulations and the other employing response spectra analysis. Notably, this disparity underscores that the inclusion of a response spectrum derived from real-world scenarios aligns with the requirements dictated by seismic regulations.

Conversely, this study accentuates the imperative of a comprehensive assessment that integrates response spectra acquired from accelerograms. This complementary approach, when coupled with seismic construction mandates, grants a more comprehensive understanding of a structure's potential behavior under the duress of seismic events. These analytical insights hold particular significance for vital infrastructure elements such as bridges and enduring institutions like hospitals, which are anticipated to withstand the sway of seismic forces. This relevance becomes even more pronounced in cases where structures are positioned in close proximity to the epicenter of an earthquake.

References

1. Razzouk, Y., Baba, K., Ahatri, M.: The influence of spectral responses on the structures heights: case of the rhiss river earthquake in morocco (6.3 mw)-seismogenic source 4 (rif oriental-al hoceima-alboran). ARPN J. Eng. Appl. Sci. **17**(6), 645–651 (2022)
2. El Majid, A., Cherradi, C., Baba, K., Razzouk, Y.: Laboratory investigations on the behavior of CBR in two expanding soils reinforced with plant fibers of varying lengths and content. Mater. Today Proc. p. S2214785323037884 (2023). https://doi.org/10.1016/j.matpr.2023.06.395
3. Razzouk, Y., Ahatri, M., Baba, K., El Majid, A.: The impact of bracing type on seismic response of the structure on soft soil. CEA **11**(5), 2706–2718 (2023). https://doi.org/10.13189/cea.2023.110534
4. Majid, A.E., Baba, K.: Assessing the impact of plant fibers on swelling parameters of two varieties of expansive soil. Case Stud. Chem. Environ. Eng. **8**, 100408 (2023). https://doi.org/10.1016/j.cscee.2023.100408
5. Cornell, C.A.: Engineering seismic risk analysis. Bull. Seismol. Soc. Am. **58**(5), 1583–1606 (1968). https://doi.org/10.1785/BSSA0580051583
6. Ahatri, M., Baba, K., Touijrate, S., Bahi, L.: Characteristics of spectral responses for a ground motion from Mediterranean earthquake—ZEGHANGHANE Station (6.3Mw) in Morocco, and its Influence on the Structures. MATEC Web Conf. **149**, 02041 (2018). https://doi.org/10.1051/matecconf/201814902041
7. M. of Housing and U. Policy, Seismic Building Code RPS 2000 - Version 2011 (2011)
8. Ahatri, M., Baba, K., Touijrate, S., Bahi, L.: The Seismic motion parameters effects on response spectra: comparison between El Centro 1940 and Imperial Valley 1979 earthquakes. Int. J. Civ. Eng. Technol. (IJCIET)-Scopus Indexed. **9**(10), 1610–1618 (2018)
9. Simou, S., Baba, K., Akkouri, N., Lamrani, M., Tajayout, M., Nounah, A.: Mechanical characterization of the adobe material of the archaeological site of Chellah. In: Ameen, H., Jamiolkowski, M., Manassero, M., Shehata, H. (eds.) Recent Thoughts in Geoenvironmental Engineering, in Sustainable Civil Infrastructures. Springer International Publishing, Cham (2020), pp. 118–130. https://doi.org/10.1007/978-3-030-34199-2_8
10. Simou, S., Baba, K., Akkouri, N., Lamrani, M., Tajayout, M., Nounah, A.: Mechanical characterization and reinforcement of the adobe material of the Chellah archaeological site. In: E3S Web Conference, vol. 150, p. 03022 (2020). https://doi.org/10.1051/e3sconf/202015003022
11. M. of Housing and U. Policy, Seismic Building Code RPS 2000 (2000)
12. Kramer, S.L.: Performance-based earthquake engineering: opportunities and implications for geotechnical engineering practice. In: Geotechnical Earthquake Engineering and Soil

Dynamics IV, Sacramento, California, United States: American Society of Civil Engineers, pp. 1–32 (2008). https://doi.org/10.1061/40975(318)213

13. Hudson, D.E.: Dynamics of structures: theory and applications to earthquake engineering, by Anil K. Chopra, Prentice-Hall, Englewood Cliffs, NJ, 1995. No. of pages: xxviii + 761. ISBN 0-13-855214-2. Earthquake Engng. Struct. Dyn. **24**(8), 1173 (1995). https://doi.org/10.1002/eqe.4290240809
14. Yamada, M., Yamada, M., Hada, K., Ohmi, S., Nagao, T.: Spatially dense velocity structure exploration in the source region of the Iwate-Miyagi Nairiku earthquake. Seismol. Res. Lett. **81**(4), 597–604 (2010). https://doi.org/10.1785/gssrl.81.4.597
15. Mouzzoun, M.: Seismic performance assessment of reinforced concrete buildings using pushover analysis. IOSR-JMCE **5**(1), 44–49 (2013). https://doi.org/10.9790/1684-0514449
16. Razzouk, Y., Ahatri, M., Baba, K., El Majid, A.: Optimal bracing type of reinforced concrete buildings with soil-structure interaction taken into consideration. Civ. Eng. J. **9**(6), 1371–1388 (2023). https://doi.org/10.28991/CEJ-2023-09-06-06
17. Kariche, J., Meghraoui, M., Timoulali, Y., Cetin, E., Toussaint, R.: The Al Hoceima earthquake sequence of 1994, 2004 and 2016: Stress transfer and poroelasticity in the Rif and Alboran Sea region. Geophys. J. Int. **212**(1), 42–53 (2018). https://doi.org/10.1093/gji/ggx385
18. Cherkaoui, T.-E.: Sismicité de la région Taza-Al Hoceima-Taounate, cas du séisme d'Al Hoceima du 24 février 2004 (2004)
19. Analysis and design of multi-storeyed building using Autodesk robot structural analysis professional 2016. In: Int. J. Mod. Trends Eng. Res. **3**(8), 13–19 (2016). https://doi.org/10.21884/IJMTER.2016.3003.3N6QF
20. Abd El-Aziz Khairy Abd El-Aal: Ground motion prediction from nearest seismogenic zones in and around Greater Cairo Area, Egypt. Nat. Hazards Earth Syst. Sci. **10**(7), 1495–1511 (2010). https://doi.org/10.5194/nhess-10-1495-2010
21. Reyes, J.C., Ardila-Bothia, L., Smith-Pardo, J.P., Villamizar-Gonzalez, J.N., Ardila-Giraldo, O.A.: Evaluation of the effect of containers on the seismic response of pile-supported storage structures. Eng. Struct. **122**, 267–278 (2016). https://doi.org/10.1016/j.engstruct.2016.04.051

An Examination of the Effects of Overestimating Mechanical Properties on Instabilities: The Tangier-Kenitra High-Speed Line Case Study

Ghizlane Ardouz(✉) and Khadija Baba

GCE Laboratory, Mohammed V University, Rabat, Morocco
Ghizlane.ardouz@gmail.com

Abstract. The deformation and failure mechanisms observed in slopes, particularly within excavated areas, are intricate and influenced by a multitude of factors. Overestimating the mechanical characteristics of the soils composing these slopes, primarily because of their inherent unpredictability, is the main cause of these instabilities. The intricacy of this situation makes it a daunting task to accurately characterize these soils, especially when dealing with coarse-grained varieties. Conducting tests on a representative volume of these soils using conventional testing devices proves challenging. Hence, granulometric reconstitution methods are employed. The capping method involves removing larger elements, which results in an underestimation of geotechnical properties. On the contrary, the substitution method replaces elements exceeding the maximum diameter permissible for laboratory tests with smaller elements, often leading to an exaggeration of the mechanical characteristics of the soils. The specific excavation under examination is situated in the rural locality of Hjar Nhal in the northernmost region of Morocco, along the high-speed train route between Kenitra and Tangier. This excavation is 1170 m long and reaches a height of more than 60 m along its axis. The site's soil is predominantly composed of pelitic formations and has a large number of water sources present at the excavation's bottom. Upon opening the excavation, a notable heterogeneity within the massif became evident, with intermittent sandstone layers appearing sporadically. Particularly noteworthy is the distinct dissimilarity between the Western slope, largely comprised of sandstone and exhibiting stability, and the Eastern slope, mainly consisting of pelitic material and showcasing multiple slides and extensive fracturing. The Western slope displays two major failures where sandstone horizons interface with pelitic ones. Inclinometers installed in the area also detected slickensides, which could contribute to the initiation of landslide mechanisms. The excavation has been subject to numerous analyses and expert evaluations, shedding light on its complexities due to multiple instances of sliding. These failures can be attributed to various factors, including the unfavorable slope dip (approximately 3 horizontal to 1 vertical) and the heightened vulnerability of the pelitic material to massif decompression. Furthermore, the presence of fractures within the massif enables water infiltration, and the existence of permeable but disjointed sandstone horizons leads to interstitial over-pressures. These water movements likely play a pivotal role in triggering landslides. In summary, the diagnosis and analysis highlight that these instabilities primarily arise from an overestimation of the mechanical properties of the soils forming the excavation.

K. Baba et al. (Eds.): GeoME 2023, SPEES, pp. 414–423, 2023.
https://doi.org/10.1007/978-3-031-49345-4_39

Keywords: Coarse-grained soils · Granulometric reconstitution methods · Capping · Substitution · Mechanical properties

1 Introduction

In nature, granular soils are widespread, particularly in mountainous regions. These soils comprise a mixture of various elements with various types and dimensions, ranging from tiny fractions of millimetres (like clay particles) to relatively large pieces of several tenths of centimeters (e.g., boulders). A thorough understanding of these materials is crucial when it comes to building on this sort of soil or using them for embankments or excavations. However, the presence of these bigger components in coarse soils makes it difficult to characterise these soils mechanically. Particularly, there are currently no reliable methods (e.g., deformability characteristics, failure criteria, etc.) to rationally include these soils in projects for calculating and dimensioning structures [1].

The conventional geotechnical methods for characterizing these soils are often highly complicated, if not impossible, due to the heterogeneity present (e.g., stones, gravels, blocks) at decimetric to metric sizes, for a minimum of two explanations [2]. Initially, traditional in-situ assessments like dynamic penetrometry, static testing, and pressiometric analysis demonstrate a lack of effectiveness in comprehensively describing the extensively varied composition of these coarse materials [2]; Furthermore, due to the dimensions of the largest elements present in these soils, obtaining a representative soil volume for testing becomes a challenging task beyond the capabilities of standard soil testing equipment. In order to tackle this concern, it is common practice to omit sizeable particles or blocks over fifty or one hundred milimiters during the sampling procedure. By doing this, the quantity of soil sample material collected and delivered to the lab for later geotechnical testing can be regulated. Unfortunately, this practise frequently results in an underestimation of the material's mechanical properties [2].

Diverse characteristics of the coarse sediments found in the digs played a crucial role as the primary contributing factor to these instabilities. The encountered soils display discontinuities that, although helping water infiltration, also pose challenges in their accurate characterization, potentially leading to an overestimation of their mechanical properties.

2 Materials and Methods

2.1 Geotechnical Model

The focus of this study is on an excavation situated in the rural locality of Hjar Nhal in the northernmost region of Morocco, along the high-speed train route between Kenitra and Tangier. This excavation is 1170 m long and reaches a height of more than 60 m along its axis [3].

The Hjar Nhal hill serves as the northwest limit of the Numidian sandstone aquifer. This geological structure, which currently covers the clay pelites of the Tangier Lower Unit, was moved during the Rifan tectonic processes. The aquifer is characterized by alternating layers of sandstone and pelites in its upper portion, while the lower part of this formation consists of more clay-rich pelites.Unfortunately, piezometric measurements could not be conducted on the superposed aquifers due to the absence of selective piezometers. However, it is evident that the sandstone exhibits relatively high porosity (n = 12%) and thus has the potential to serve as an aquifer, with the alternating clayey levels of pelites likely forming the confining boundaries of these aquifers [3]. This observation gains further support from the presence of numerous springs at the excavation's base, aligning precisely with the interface between the Numidian sandstones and the Tangier nappe's pelites.

Geotechnical examinations carried out in the proximity of the structure encompassed a range of in-situ and laboratory tests.

However, distinguishing between the various facies of pelites posed a significant challenge. To address this issue, a comparison of various reconnaissance methods was carried out, leading to the following observations:

The drilling parameters recorded during destructive drilling did not provide a clear identification of sandstone horizons from purely pelitic horizons. The advancement speeds were similar and generally low for both horizons, and the injection pressures showed significant variability within the same horizon (where analysis at a suitable scale was possible). These variations were likely related to fracturing rather than the nature of the materials crossed.

- The velocities recorded in the seismic surveys also did not allow a straightforward identification of the materials. The scale of alternations between sandstone and pelitic banks was too small for this method to be effective. Instead, the variations in seismic velocities appeared to be more related to the intensity of material alteration. In horizons identified as intact, the recorded velocities were high and characteristic of rocky materials, regardless of whether sandstone horizons were dominant or not.
- Examination of the cores (via photographs) seemed to be more useful in differentiating sandstone banks, which often appeared more massive and had a beige or white color, from pelitic banks, which were generally darker and recognizable by the presence of shrinkage cracks. However, there were locally thick levels with an intermediate light gray appearance, making their nature difficult to determine.
- The gamma ray diagraphs showed significant contrasts between purely sandstone horizons and pelitic horizons. They proved valuable in characterizing the alternating pelitic and sandstone layers, which were typically 10 cm to 1 m thick.

As a result, the identification of the thickness of alteration relied on descriptions of facies variations in borehole sections and seismic surveys, supplemented by information from gamma ray logs.

The following material categories are defined in the study:

- E/C: Colluvial cover soils with a dominant clayey composition.
- E5a: Alteration products of pelites, typically found at depths between 0 and 10 m.
- E5: Pure pelites with a rocky nature.
- E5g: Alternation of pelites and sandstone, where sandstone banks represent between 15% and 40% of the composition.
- E5g/G: Alternation of pelites and sandstone in approximately equal proportions.

- G2: Sandstone, possibly with minor pelitic banks (constituting less than 20% of the composition).
- Ga: Alteration of sandstone.

The identification of the E5g formation primarily relies on the thickness of the sandstone beds. We established a criterion where sandstone banks exceeding 2 to 3 m in thickness are considered suitable for separate extraction. These horizons are then categorized as G2. Below this thickness, avoiding mixing with the pelites becomes challenging, and the designation E5g or E5g/G is used, depending on the percentage of sandstone present. Similarly, when the pelite banks surpass 2 to 3 m in thickness, we utilize the classification E5 or E5a, depending on their compactness (Fig. 1).

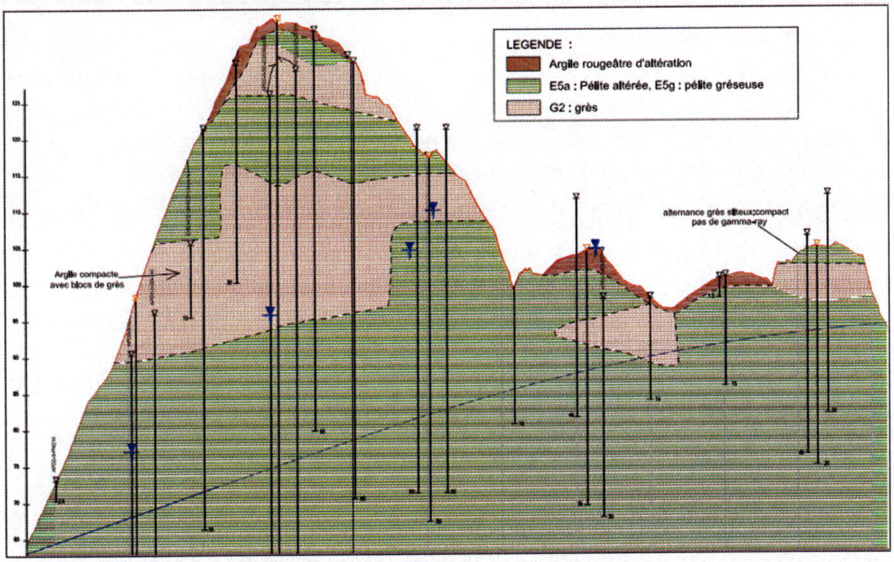

Fig. 1. Geotechnical longitudinal profile of the excavation with the location of the tests applied

Table 1 outlines the characteristics attributed to each facies.

Table 1. Soil properties utilized in the analyses.

Soil layer	Cohesion (kPa)	Friction angle (°)	Bulk density (kN/m^3)
Clays	5	25°	18.7
Altered pelites	28	17°	19
Sandstone	Substratum		

3 Results: Examination of the Generated Events

Upon the commencement of the excavation work, a multitude of observations were recorded:

- The massif exhibited a substantial heterogeneity, with sandstones being in the minority and occurring as discontinuous banks limited to 1 or 2 m in thickness. Only a few isolated pockets or certain banks on the western slope deviated from this pattern.
- A noticeable asymmetry between the western slope, which mainly consisted of sandier materials and showed no evidence of instability, and the eastern slope, which predominantly comprised pelitic materials and displayed multiple landslides.
- The massif showed significant fracturing, particularly evident on the western slope, where two major failures brought sandstone horizons into contact with pelitic horizons.

Additionally, the boreholes conducted for inclinometer installation unveiled the presence of a smooth interface, which could potentially contribute to the initiation of sliding mechanism. As a result, the excavation site witnessed a series of landslides.

The landslide occurred in several stages:

- The first sliding event (G1) took place when the excavation slopes were at a 2 h/1 v gradient. As a response to this, a decision was taken to adjust the slopes to a 3 h/1 v ratio.
- Subsequently, a second sliding event (G2) transpired, swiftly followed by a thirdevent (G3), which reactivated despite the slopes being adjusted to 3 h/1 v.
- A fourth sliding event (G4) materialized when a failure oriented at N240 emerged at the slope's peak, where the gradient was configured at 2.5 h/1 v.

These four sliding incidents are visually depicted in Fig. 2.

The following observations about the visual qualities can be made:

- The different landslides generally have a direction of N350 (approximately) indicated by arrows, which is oriented at an angle of about 20° relative to the slope.
- There is a clearly visible upper detachment fissure that roughly parallels the slope's entry into the ground and runs parallel to N240 (which is perpendicular to the direction of sliding).
- A cutting (flaw) is observed with orientations spanning from N270 to N330, possibly correlating with failures akin to those visibly present on the western slope. These cutting features might intersect different segments of the sliding.
- In relation to the sliding occurrence (G2), the implanted inclinometer was disrupted down to a depth of 5 m, underscoring the magnitude of movement at that specific point.

4 Discussions and Analysis of the Facts

Based on the studies conducted on the cuttings, the failure mechanism of the landslides can be attributed to several contributing factors:

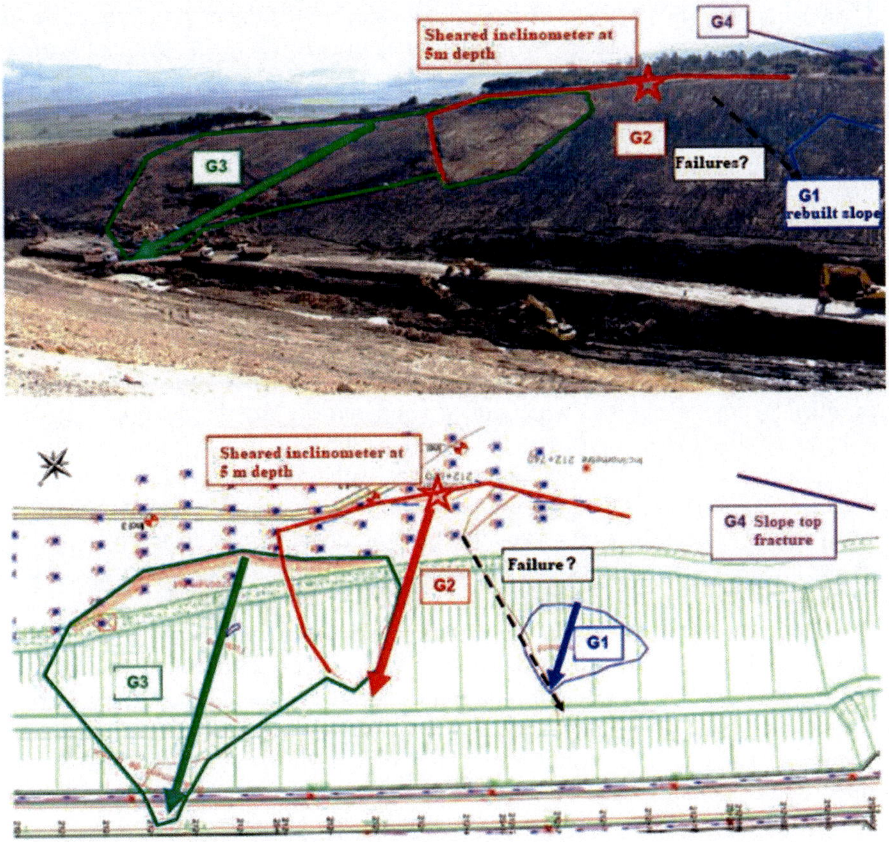

Fig. 2. The multiple landslides transpired on the eastern slope of the excavation

- Unfavorable Inclination: The excavation slopes were laid at about 3 h/1 v, which might have created an unfavorable slope angle, potentially leading to instability.
- High Pelites Susceptibility to Decompression: The pelites in the massif displayed high sensitivity to the influence of decompression. This decompression may result in moves greater than the maximum stress resistance of the pelites, allowing a change to residual properties and raising the risk of landslides.
- Presence of Discontinuities: The existence of discontinuities within the massif provided pathways for water infiltration, potentially increasing pore water pressures and reducing the overall stability of the slopes.
- Sandstone Banks and Permeable Horizons: The existence of sandstone formations, which could exhibit both permeability and discontinuity, facilitated interstitial overpressures due to water flow. These water flows likely played a significant role in triggering the landslides.

- Overestimation of Mechanical Characteristics: The heterogeneity of the soils composing the excavation made it challenging to accurately characterize them. As a result, there might have been an overestimation of the mechanical properties of the materials, leading to unexpected instability.

These findings highlight the complex interaction of various geotechnical factors that contributed to the occurrence of the landslides. Understanding these factors is crucial for implementing effective mitigation measures and ensuring the stability and safety of the excavation site in the future.

The stability analysis of the excavation through the application of the finite element approach has confirmed our hypothesis regarding the overestimating of the mechanical characteristics of the soils. The results obtained from the analysis closely matched the actual events that occurred. This overestimation can be attributed to the fact that the soils are grained, characterized by a widely distributed particle size distribution (Fig. 3).

Characterizing these coarse soils in the laboratory or in situ presents significant challenges, primarily due to the high cost, lack of appropriate equipment, and the time-consuming nature of testing large soil samples. Conventional testing equipment often struggles to analyze a representative volume of material for the investigation of coarse-grained soils. To address this, grain size reconstruction methods, such as capping, substitution, and parallel grading, are employed.

The substitution method involves replacing elements with dimensions exceeding the largest diameter permitted by lab tests with smaller-sized constituents (Fig. 4). While maintaining the same volume proportion (fv) of gravels, this process reduces their maximum diameter (Dmax) and particle spread (dmin/dmax). Our previous study, "The Influence of the Fundamental Parameters on the Mechanical Behavior of Coarse-Grained Soils," demonstrated that a more uniform (tight) grading leads to increased shear strength. Therefore, the substitution method has a tendency to overestimate the geotechnical properties of gravelly soils.

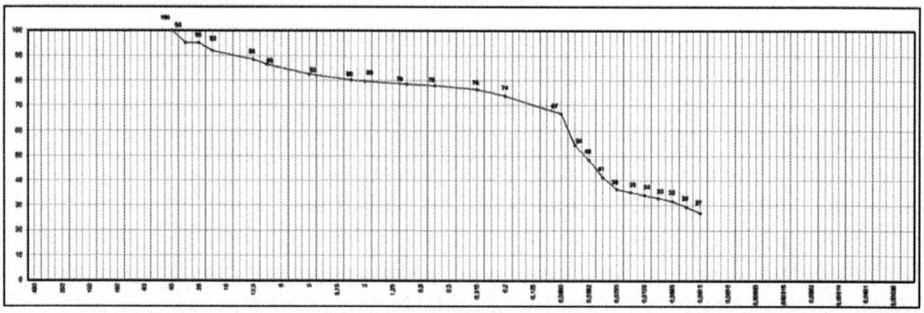

Fig. 3. Size distribution curve of densely packed surface clays containing sandstone blocks (E/C)

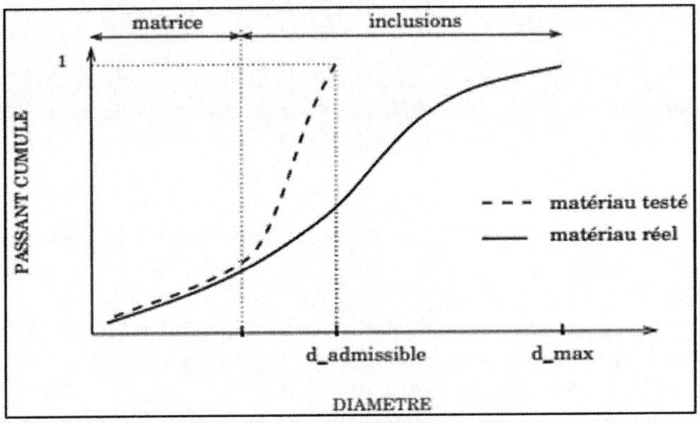

Fig. 4. Illustrative representation of the substitution technique

5 Conclusions

The excavation discussed in this study underwent several calculations to assess its stability using a slope ratio of 3 h/1 v. However, after excavation, landslides occurred, potentially resulting from a range of factors:

- Existence of Porous Sandstone Formations: The presence of porous sandstone banks, which can be discontinuous, facilitated water infiltration, potentially contributing to the instability of the slopes.
- Exaggeration of Mechanical Properties: The soils forming the excavation displayed heterogeneity, making it difficult to accurately characterize them. This led to an overestimation of their mechanical properties, possibly influencing the occurrence of landslides.

Coarse matrix soils represent a specific category of coarse soils, comprising large grains (inclusions) of varying sizes surrounded by fine sandy, silty, or clayey matrices. Comprehending the geomechanical response of these soils presents notable difficulties, primarily attributed to the presence of substantial elements that obstruct or complicate the effective acquisition of samples for analysis.

Representative samples typically require extracting large masses of materials, which becomes impractical when large-sized elements are present (a minimum required mass of 125 kg of soil when dmax = 100 mm). Consequently, substitution of material is often necessary during sample collection for laboratory or in-situ analysis. However, this method can have implications on the mechanical properties of the material.

The equilibrium analysis of the dig, conducted using the finite element approach, supported our hypothesis of overestimating the mechanical characteristics. The results obtained were close to the actual events that occurred, validating the impact of the overestimation on the stability of the excavation slopes.

References

1. Kučová, E., Kuvik, M.: Evaluation of the mechanical properties of fluvial coarse-grained soils from dynamic penetration test. IOP Conf. Ser. Mater. Sci. Eng. **1209**(1), 012081 (2021). https://doi.org/10.1088/1757-899x/1209/1/012081
2. Chang, W.J., Phantachang, T.: Effects of gravel content on shear resistance of gravelly soils. Eng. Geol. **207**, 78–90 (2016). https://doi.org/10.1016/j.enggeo.2016.04.015
3. Xun, X., Liang, H., Xiaodong, H.: Study on the relationship between energy evolution and strength parameters of coarse-grained soils in direct shear test. J. Eng. Geol. **29**(5), 1331–1341. https://doi.org/10.13544/j.cnki.jeg.2020-522. (In Chinese)
4. Dorador, L.: Experimental investigation of the effect of broken ore properties on secondary fragmentation during block caving. Ph.D. Thesis, University of British Columbia, Vancouver, Canada (2016). https://doi.org/10.14288/1.0319061
5. Dorador, L.: Density estimation of broken materials in sinker mining (block/panel caving). In: 50th Chilean Geotechnical Congress, 3–5 December, 2018, Valparaiso, Chile. (In Spanish) (2018)
6. Stacho, J., Sulovska, M.: Shear strength properties of coarse-grained soils determined using large-size direct shear test. Civ. Environ. Eng. **18**(1), 244–257 (2022). https://doi.org/10.2478/cee-2022-0023
7. Dorador, L.: A review of the parallel granulometry methodology or homothetic curve scaling applied to the geotechnical characterization of coarse grained materials. In: 50th Chilean Geotechnical Congress, 3–5 December, 2018, Valparaiso, Chile. (In Spanish) (2018)
8. Dorador, L., Hoz, K., Salazar, F.: Considerations in the geotechnical characterization of coarse grained materials. In: 50th Chilean Geotechnical Congress, 3–5 December, 2018, Valparaiso, Chile. (In Spanish) (2018)
9. Humire Guarachi, F.A.: Liquefaction and post-liquefaction behavior of coarse-grained soils. Ph.D. Thesis, University of California, Davis, United States (2022)
10. Dorador L., Poblete, M., Foretic, I.: Minimum and maximum densities in coarse-grained materials-preliminary results of an ongoing test program. In: 50th Chilean Geotechnical Congress, 3–5 December, Valparaíso, Chile. (In Spanish) (2018)
11. Dorador, L., Urrutia, J.: Geotechnical characterisation of coarse-grained soils containing weak and strong particles mixtures. In: 70th Canadian Geotechnical Conference (GeoOttawa 2017), 1–4 October, 2017, Ottawa, Canada (2017)
12. Steger, S., Mair, V., Kofler, C., Pittore, M., Zebisch, M., Schneiderbauer, S.: Correlation does not imply geomorphic causation in data-driven landslide susceptibility modelling – Benefits of exploring landslide data collection effects. Sci. Total Environ. **776**, 145935 (2021). https://doi.org/10.1016/j.scitotenv.2021.145935
13. Ravindran, S., Gratchev, I.: Effect of water content on apparent cohesion of soils from landslide sites. Geotechnics **2**(2), 385–394 (2022). https://doi.org/10.3390/geotechnics2020017
14. Ardouz, G., Baba, K.: Numerical analysis of instabilities affecting an excavation on the high speed line in Northern Morocco. In: Advanced Numerical Methods in Foundation Engineering, GeoMEast 2019, Sustainable Civil Infrastructures, Springer, Cham, Switzerland (2020). https://doi.org/10.1007/978-3-030-34193-0_9
15. Ardouz, G., Baba, K., Ouadif, L.: Modeling landslides by the finite element method: application to an embankment on a railway in the Moroccan Rif. In: Contemporary Issues in Soil Mechanics, GeoMEast 2018, Sustainable Civil Infrastructures, Springer, Cham, Switzerland (2019). https://doi.org/10.1007/978-3-030-01941-9_11
16. Ardouz, G., Baba, K., Bahi, L., Cherradi, C.: Comparison of analytical and numerical methods in the analysis of the stability of an excavation of the high-speed line in northern Morocco. In: 7 th International Congress Water, Waste and Environment (EDE7–2019), E3S Web of Conferences, vol. 150, 1–7 (2020). https://doi.org/10.1051/e3sconf/202015003006

17. Moulay Smaîne, G., Gueye, B., Zoubir, B.: Influence of grain size coarse soil on shear strength. In: International Congress on Materials & Structural Stability, MATEC Web of Conferences (2014). https://doi.org/10.1051/matecconf/20141103009
18. Honkanadavar, N.P., Sharma, K.G.: Modeling the triaxial behavior of riverbed and blasted quarried rock fill materials using hardening soil model. J. Rock Mech. Geotech. Eng. **8**(3), 350–365 (2016). https://doi.org/10.1016/j.jrmge.2015.09.007
19. Ovalle, C., Frossard, E., Dano, C., Hu, W., Maiolino, S., Hicher, P.Y.: The effect of size on the strength of coarse rock aggregates and large rock fill samples through experimental data. Acta Mech. **225**(8), 2199–2216 (2014). https://doi.org/10.1007/s00707-014-1127-z
20. Abozraig, M., Ok, B., Yildiz, A.: Determination of shear strength of coarse-grained soils based on their index properties: a comparison between different statistical approaches. Arab. J. Geosci. **15**(7), 1–17 (2022). https://doi.org/10.1007/s12517-022-09875-w
21. Rashidi, M., Rasouli, H.: Initial hypotheses for modeling and numerical analysis of rock fill and earth dams and their effects on the results of the analysis. Adv. Civ. Eng. **2018** (2018). https://doi.org/10.1155/2018/3974675
22. Sepehr, A., Hassanzadeh, M., Rodriguez-Caballero, E.: The protective role of cyanobacteria on soil stability in two Aridisols in northeastern Iran. Geoderma Reg. **16** (2019). https://doi.org/10.1016/j.geodrs.2018.e00201
23. Wen, R., Tan, C., Wu, Y., Wang, C.: Grain size effect on the mechanical behavior of cohesionless coarse-grained soils with the discrete element method. Adv. Civ. Eng. **2018** (2018). https://doi.org/10.1155/2018/4608930
24. Zhang, Z.L., Xu, W.J., Xia, W., Zhang, H.Y.: Large-scale in-situ test for mechanical characterization of soil–rock mixture used in an embankment dam. Int. J. Rock Mech. Min. Sci. **86**, 317–322 (2016). https://doi.org/10.1016/j.ijrmms.2015.04.001

Expansive Soils: Is It a Hidden Problem in Civil Engineering?

Mounir Bouassida[1](✉), Sergio Andrew Manigniavy[1], Nabil Kazi Tani[3,4], and Yosra Bouassida[1,2]

[1] National Engineering School of Tunis, ENIT, LR14ES03 Geotechnical Engineering and Georisk Research Laboratory, University of Tunis El Manar, Tunis, Tunisia
mounir.bouassida@enit.utm.tn

[2] University of Tunis, Ecole Nationale Supérieure d'Ingénieurs de Tunis—ENSIT, Tunis, Tunisia

[3] University Abou Bekr Belkaid, RISAM Laboratory, Tlemcen, Algeria

[4] Higher School in Applied Sciences of Tlemcen—ESSAT, Tlemcen, Algeria

Abstract. Expansive soils currently pose serious pathologies and significant damages to lightweight constructions. Geologically, expansive soils exist in most regions of the world. In geotechnical engineering, this soil category deserves more attention from geotechnical engineering researchers. The constructions are threatened because the volume variation in expansive soils is tributary of the variation of the water content. Furthermore, repairing solutions of affected constructions is of a high cost. So far, attention made by investigators, focused first, on the occurrence of the swelling phenomenon, and, second, the efficient methods to quantify the swelling pressure. The key issue, in practice, is to mitigate the swelling phenomenon of foundations built on expansive soils. Among the proposed solutions, chemical treatments and improvement solutions, like the use of granular materials revealed successful. This study aims to develop a method to identify if a clayey soil is expansive or non-expansive through analyzing consolidation test results. The proposed method considers the swelling index (C_s) and the compression index (C_c). The ratio "C_c/C_s" appears a good indicator, by identified thresholds values to make the difference between expansive and non-expansive clay ones. After, the proposed method of identification, when the C_c/C_s ratio exceeds 15, the swelling pressure vanishes. In turn, when C_c/C_s ratio is lower than 10, the swelling pressure increases as this ratio decreases.

Keywords: Characterization · Disaster · Expansive soils · Oedometer test · Swelling

1 Introduction

Everyone is aware that temperatures are subject to huge fluctuations. As a result, the water content of the soil varies seasonally.. Some categories of soil are very sensitive to this variation, especially clayey soils. Swelling soils are clayey soils that shrink when dried and swell when wetted. Nowadays, the phenomenon of shrink-swell of this soil is

K. Baba et al. (Eds.): GeoME 2023, SPEES, pp. 424–431, 2023.
https://doi.org/10.1007/978-3-031-49345-4_40

not limited to semi-arid (and/or arid) zones, but can be found almost everywhere in the world. Several African countries are affected by the shrink-swell behavior (Fig. 1), and there have been several reports of studies from North Africa in the last decades.

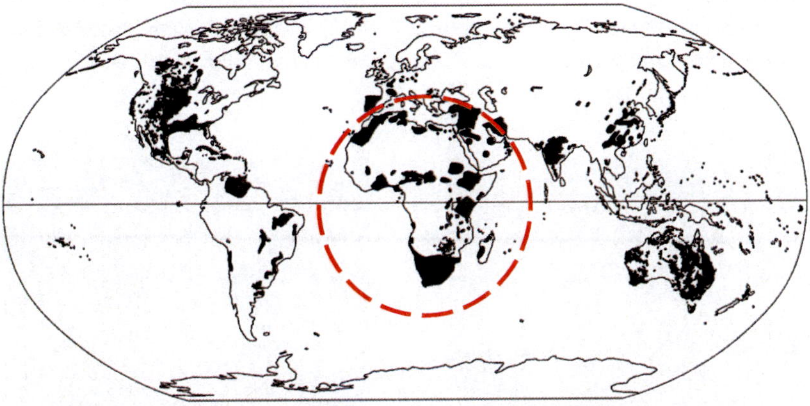

Fig. 1. Update of Nelson et al. 2015 [1]: reported expansive soils sites

The volume of swelling soils changes depending on the moisture content. The process of soil swelling occurs gradually and is influenced by the degree of saturation and by the structure of the soil particles. During this process, the movement of the water molecules and other cations in between clay particles and between the lattice layers of clay particles increase the swelling potential [17].

It often causes serious damage to buildings, especially lightweight buildings. Expansive soils generally consist of clays, predominantly of the smectite group (montmorillonite). With regard to the damage caused by swelling soil, the study of the behavior of foundations on this type of soil is a topic of great interest.

A number of researchers are working on the shrink-swell behavior, to find appropriate techniques for mitigating the effect of this phenomenon on infrastructure [2]. The principal challenge is to find ways to reduce the effect of swelling phenomenon when building structures in contact with swelling soils. Mixture or chemical treatments are the most popular research topics for the improvement of this problematic soil.

2 Damages and Costs

Swelling soils are very dangerous since it counts among the top six natural hazards. In the United States of America, this hazard is ranked the second of the most destructive natural disasters [3].

2.1 Structural Damages Due to Swelling Soils

As the seasons change, the position of water table changes the foundations on swelling soils move upwards. Heave and settlements are currently observed for constructions built on swelling soils. The most common deformation observed on buildings is the uplift causing cracks in the structure: wall, slab, etc. (Fig. 2). The triggering of cracks in lightly loaded structures results from an easier vertical displacement due to the pressure induced by the underlined expansive layers [4]. Heavy loaded buildings and moderately rigid constructions are less affected by the shrink-swell phenomenon due to their self-weight, a high-applied load, high embedment of the foundation's and moderate stiffness. For buildings constructed on expansive soils, the fissures' size increases over time until it leads to the soil failure [4].

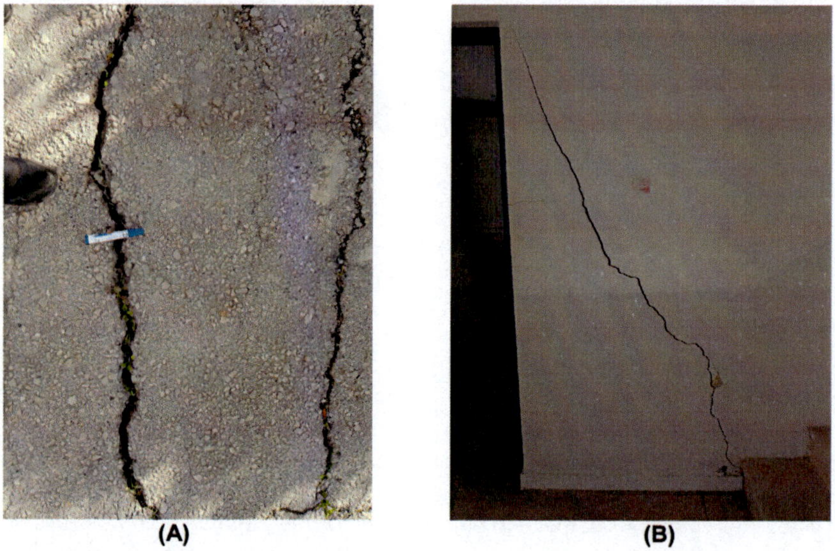

Fig. 2. Damages caused by the shrink-swell behavior of expansive soils soils in Tunisia (National Engineering School of Tunis): **A.** Fissures on a pavement and **B.** Cracks in a wall

The existence of swelling soils may be the trigger of slope instabilities, heave buckling of pavement, tunnels collapse, destruction of buried or hydraulic structures (e.g. pipelines or irrigation systems), etc.…

2.2 Financial Costs Associated with Expansive Soils

All over the world, there is a big lot of money due to considerable damages caused to buildings by the effect of shrink-swell behavior of expansive soils. The estimated annual global cost of damages due to expansive soils is over 23 billion US dollars [5]. For instance, in Sudan, the cost of damages caused by the swell-shrink behavior of soil exceeds 6 million dollars [6]. In United States, classified as the most affected country by swelling soil problems, annual losses are more than 15,000 million dollars in damages to pavements, pipelines and other infrastructures [7]. The assessment of financial losses due to the problems of swelling soils has not yet been carried out in several countries. Due to the reduction in the structure lifetime, there is a huge loss of investments.

Traffic roads are the blood vessels of development. In its design and construction, the choice of a route is the first and most important decision to take. The latter is often influenced by the social-economical, environmental and political factors that are given a high priority in the dimensioning phase. In addition, there are several factors to take into account, such as the existence of swelling soils on the road, which can be very dangerous, when ignored.

The swelling and shrinkage behaviors of clay is an alarming and extremely devastating hazard, with enormous repair costs. This is the reason why Jones and Holtz [8] classified expansive soils as a hidden disaster. Therefore, it is important to develop the suitable ways to characterize swelling soils and to tackle their effects.

2.3 Mitigation Methods

The improvement of expansive soils is a current solution to mitigate the swelling effect. Among the current techniques, the use of granular materials and the chemical treatments by soil admixture were practiced in several regions. Granular piles' technique, of current use at any season, greatly decreases the heave of expansive soils [9]. Indeed, the intercalation of a granular layer between the foundation and swelling soils also reduces the swelling potential [2]. The treatment of expansive soils with a sulfonated oil reduce the shrinkage and swelling potentials after several wetting-drying cycles, [10]. The mixture of an expansive clay with 10% of cement and lime reduces the swelling potential by 41%, [10]. In fact, the lime treatment, by hydration, reduces the soil sensitivity to water because the moisture content is decreased [12]. Noted that the cement content is of about 60% of lime, the mix of expansive soils by cement is similar to that by added lime. Moreover, the mixture of an expansive soil with a dune sand provides a significant reduction in the swelling phenomenon [13]. The mix of an expansive soil with 45% sand reduces the swelling potential by 65%. One concludes three commonly used techniques to mitigate the swelling effect: replacing the expansive soil, building sufficient heavy structures and separating the structure from the expansive clay layer.

3 Methods to Characterize Expansive Soils

3.1 Existing Approaches

Many investigators suggested different approaches to characterize expansive soils. After Sridharan and Prakash [14] two categories of methods apply. The first category comprises the identification of the soil mineralogy such as the X-ray diffraction analysis, differential thermal analysis, dye adsorption, and scanning electron microscopy. The second category of characterization methods included the inferential tests. Meanwhile, other methods considered the clay fraction, and current parameters as the liquid limit, plasticity index and shrinkage limit. Meanwhile, performing swell test ther using the oedometer apparatus is the more considered.

3.2 Proposed Method: The Use of C_C/C_S Ratio

For characterizing expansive soils, the proposed method uses the ratio between the compression index (C_C) and swelling index (C_S) which are defined from the results of an oedometer test usually performed on a saturated soil specimen subjected to increased applied loads. This method compiled 48 oedometer tests results on clayey soils from three countries: Tunisia (36 data), Algeria (08 data) and Morocco (04 data). Table 1 presents the collected data from Morocco, the remaining data were published in Bouassida et al. 2022, [2]. The performed oedometer tests were accompanied by the measurement of the swelling pressure.

Table 1. Consolidation test results on clayey soils from Morocco [15]

Site	Sample	C_C	C_S	C_C/C_S	Swelling pressure (MPa)	Expansive
Settat plateau	BH28	0.320	0.060	5.33	0.681	Yes
	BH31	0.240	0.050	4.80	0.452	Yes
	BH36–2	0.180	0.070	2.57	0.221	Yes
	BH36–3	0.430	0.130	3.31	0.152	Yes

Table 1 shows that the C_C/C_S ratio depends on the swelling pressure (σ_S). From the data in Table 1 and recently published ones [16], Fig. 3 shows the variation of C_C/C_S ratio versus the swelling pressure. The interpretation of those results is discussed in the next paragraph.

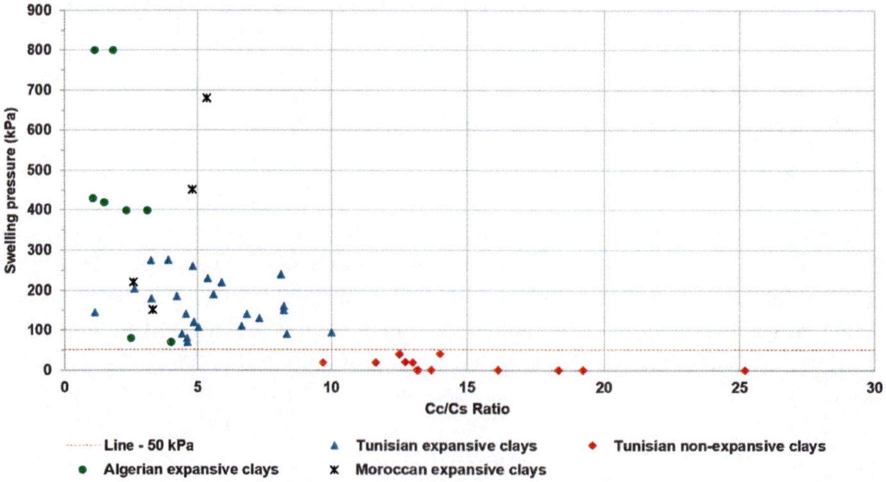

Fig. 3. C_C/C_S ratio versus swelling pressure of clayey soils from Tunisia, Morocco and Algeria.

4 Discussion of Results

The chart in Fig. 3 indicates that a non-expansive soil is characterized by a Cc/Cs ratio higher than 10 and a swelling pressure lower than 50 kPa. In turn, the soil is expansive when the Cc/Cs ratio is lower than 10 and the swelling pressure is higher than 50 kPa. Earlier, Chen (1975) approved this approach, [3], by stating: the swell potential is marginal when the swelling pressure does not exceed 50 kPa. Elsewhere, Table 2 shows that the swelling pressure is zero when the C_C/C_S ratio exceeds 15. Therefore, the suggested classification combines the swelling and non-swelling soils. Moreover, using the results in Table 2, Fig. 4 shows that a compression index equal 10 times the swelling index, corresponds to a marginal swelling index.

Table 2. Swelling soil classification, update of Bouassida et al. 2022 [2]

Soil	C_c/C_s	Swelling pressure (kPa)
Swelling	< 10	> 50
Non-swelling	> 10	< 50
Highly non-swelling	≥ 15	≈ 0

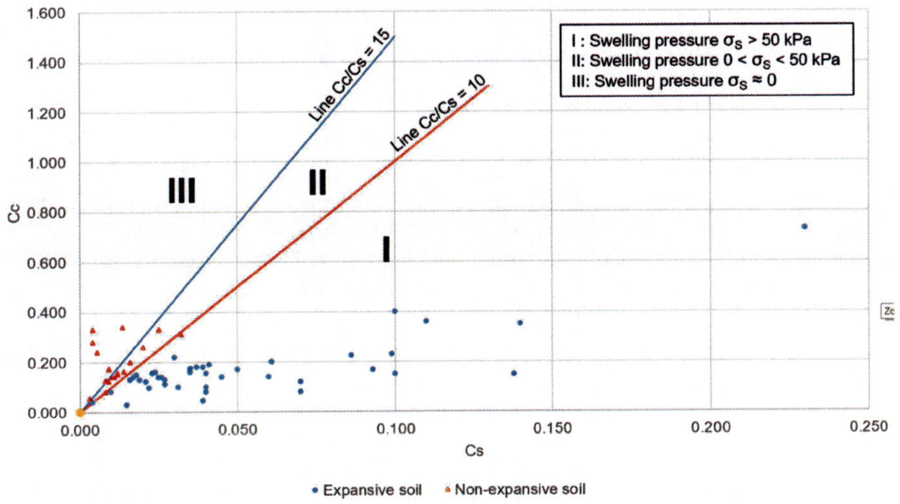

Fig. 4. Compression index (C_C) versus swelling index (C_S) of 48 clayey soils from Tunisia, Algeria and Morocco.

5 Conclusions

This paper reviewed, in brief, the behavior of expansive soils, the occasioned damages of constructions in contact with expansive soils and the proposed methods to mitigate the swelling phenomenon.

Then, the carried out research focused on the characterization of expansive soils by suggesting a new method based on the ratio between the compression and swelling indices and the swelling pressure σ_S. Those parameters are determined from the oedometer test.

The data collected from forty-eight (48) oedometer tests carried out in Tunisia, Morocco and Algeria showed that the C_C/C_S ratio depends on the swelling pressure.

The proposed classification that resulted from the collected data plotted on the chart C_C/C_S versus σ_S, indicate three possibilities. First, when C_c/C_s ratio exceeds the soil is non-expansive with a swelling pressure close to zero. Second, an expansive soil corresponds to a C_c/C_s ratio lower than 10 and swelling greater than 50 kPa. Third, the soil is moderately expansive with a swelling pressure lower than 50 kPa and the C_c/C_s ratio is in the range 10 to 15.

The results presented herein confirm the proposed method of classification after compiling a different set of data from Tunisia, Algeria and other countries, Manigniavy et al., 2023 [16].

Therefore, dependent less of the method of measurement of the swelling pressure, when the ratio C_c/C_s is lower than 10, the soil is the more likely expansive, and for which the determination of the swelling pressure should be determined with care. In addition, the relationship between the shrink–swell behavior and the measured swelling pressure needs more highlights.

The confirmation of the suggested method of characterization is tributary of a larger number of data to collect from several regions where the presence of expansive soils revealed a serious problem in terms of identification and design of foundations.

References

1. Nelson, J.D., Chao, K.C., Overton, D.D., Nelson E.J.: Foundation Engineering for Expansive Soils. John Wiley & Sons. 978-1-118-41799-7, Hoboken, New Jersey (2015)
2. Bouassida, M., Manigniavy, S.A., Azaiez, D., Bouassida, Y.: New approach for characterization and mitigation of the swelling phenomenon. Front. Built. Environ. **8**, 836277 (2022). https://doi.org/10.3389/fbuil.2022.836277(2022)
3. Chen, F.H.: Foundations on Expansive Soils. Elsevier Science Publishing Company Inc., New York (1975)
4. Elarabi, H.: Damage mechanism of expansive soils. In: Proceeding of the 2nd International Conference on Geotechnical Engineering, ICGE'10, pp. 125–131, Tunisia (2010)
5. Amakye, S.Y., Abbey, S.J.: Understanding the performance of expansive subgrade materials treated with non-traditional stabilisers: a review. Clean. Eng. Technol. **4**, 100159 (2021). https://doi.org/10.1016/j.clet.2021.100159
6. Osman, M.A., Charlie, W.A.: Expansive soil in Sudan. BBRI current papers. No. CP.3/83. Building and Road Research Institute. University of Khartoum, Sudan (1983)
7. Jones, L.D., Jefferson, I.: Institution of Civil Engineers Manuals Series (2019). http://nora.nerc.ac.uk/id/eprint/17002/1/C5_expansive_soils_Oct.pdf. Accessed 30 Mar 2020
8. Jones Jr, D.E., Holtz, W.G.: Expansive soils-The hidden disaster. American Society of Civil Engineers. **43**(8) (1973)
9. Raghuram, A.S.S., Phanikumar, B.R., Purnanandam, K., Srirama Rao, A.: Heave and pullout studies of geogrid-encased granular pile-anchors in expansive clay beds. Proc. Inst. Civ. Eng. Ground Improv. **175**(4), 236–246 (2022). https://doi.org/10.1680/jgrim.19.00081

10. Soltani, A., Raeesi, R., O'Kelly, B.C.: Cyclic swell-shrink behavior of an expansive soil treated with a sulfonated oil. Proc. Inst. Civ. Eng. Ground Improv. **175**(3), 166–179 (2020). https://doi.org/10.1680/jgrim.19.00084
11. Mahamedi, A., Khemissa, M.: Cement and lime stabilization of compacted expansive clay. In: Proceedings of the 3rd International Conference on Geotechnical Engineering ICGE'13, pp. 369–377 Hammamet, Tunisia (2013)
12. Kechouane, Z., Nechnech, A.: Characterization of an expansive clay treated with lime: effect of compaction on the swelling pressure. In: 4th International Congress in Advances in Applied Physics and Materials Science (APMAS 2014) AIP Conf. Proc. 1653, 020057–1–020057–8 (2015). https://doi.org/10.1063/1.4914248
13. Gueddouda, M.K., Goual, I., Lamara, M., Goual, S.: Amélioration des propriétés physico-mécaniques des argiles gonflantes stabilisées par ajout de sable de dune. In: 3ème Conférence Maghrébine en Ingénierie Géotechnique CMIG'13, pp. 221–226, Algier, Algeria (2013)
14. Sridharan, A., Prakash, K.: Expansive soil characterization: an appraisal. INAE Lett. **1**(1), 29–33 (2016). https://doi.org/10.1007/s41403-016-0001-9
15. Ouakkass, B.M., Ouadif, L., Bahi, L., Akhssas A.: Predictive risk mapping related to the shrink-swell of clays under the railway track track in the Settat Plateau (Morocco). ARPN J. Eng. Appl. Sci. **14**(issue 13) (2019)
16. Manigniavy, S.A., Bouassida, Y., Azaiez, D., Bouassida, M.: Using compression and swelling indices to characterize expansive soils. In: Atalar, C., Çinicioğlu, F. (eds) 5th International Conference on New Developments in Soil Mechanics and Geotechnical Engineering. ZM 2022. Lecture Notes in Civil Engineering, Vol. 305. Springer, Cham (2023). https://doi.org/10.1007/978-3-031-20172-1_10
17. Li, M., Wei, Y., Liu, Y., Jin, J.: A framework for interpreting lateral swelling pressure in unsaturated expansive soils. Article ID 6626835l (2021). https://doi.org/10.1155/2021/6626835

Water Management

Accumulation of Potentially Hazardous Elements in Water of a Significant Reservoir in Marmara Region of Türkiye

Cem Tokatlı[1](✉) and Esengül Köse[2]

[1] İpsala Vocational School, Trakya University, İpsala, Edirne, Türkiye
tokatlicem@gmail.com

[2] Eskisehir Vocational School, Eskişehir Osmangazi University, Eskişehir, Türkiye

Abstract. Reservoirs are significant lacustrine freshwater habitats built for various purposes such as drinking water supply, irrigation and flood protection. However, various factors such as contamination, sedimentation and excessive use of waters may cause them to be useless. Çokal Dam Lake is one of the most significant reservoirs located in the Marmara Region of Türkiye. It was built on the Kocadere Stream for irrigation and drinking water supply in the Tekirdağ Province. In the current research, we aimed to determine the accumulation levels of potentially hazardous elements (PHEs) in water of this significant dam lake watershed. Water samples were collected from the Çokal Reservoir (ÇDL1, ÇDL2), Aksakal (AS) and Çayırlar (ÇS) Streams that are feeding the reservoir and the Kocadere (KS) Stream before it falls into the Saros Bay in the rainy season of 2020. According to the data obtained as a result of our study, the average levels of investigated PHEs were recorded as follows: 900 ppb for B, 0.84 ppb for Cr, 52 ppb for Mn, 6.16 ppb for Ni, 1.46 ppb for Cu, 7.72 ppb for Zn, 4.34 ppb for As, 1.88 ppb for Se, 0.19 ppb for Cd and 0.55 ppb for Pb. As a result of applied Cluster Analysis (CA), three clusters were composed named as "relatively more contaminated zone—C1 (KS)", "relatively moderate contaminated zone—C2 (ÇS)" and "relatively less contaminated zone—C3 (ÇDL1, ÇDL2 and AS)".

Keywords: Çokal Reservoir · Water quality · Potentially hazardous elements · Cluster Analysis

K. Baba et al. (Eds.): GeoME 2023, SPEES, pp. 435–442, 2023.
https://doi.org/10.1007/978-3-031-49345-4_41

1 Introduction

Environmental pollution that is known as the unnatural degradation of the environment by human hands can also be defined as the events that adversely affect the life activities of the living elements of foreign substances that mix heavily with the air, water and soil, and cause structural damage on the inanimate elements and deteriorate their qualities. Many organizations are working to raise awareness about environmental pollution and take precautions. In addition to these studies, there are important steps we can take individually to protect the environment, such as sorting waste, promoting recycling, and saving energy. Lacustrine habitats such as natural lakes and artificial dam lakes are among the most significant freshwater bodies because of their rich biodiversity and providing irrigation and drinking water to local people [1–3].

Many regions of Türkiye are located in the semi-arid and arid zone of climate. Due to the insufficient rainfall in many regions in terms of plant production, irrigation is required to obtain a high level of yield and quality products. The more effective and efficient use of water resources in Türkiye has a great importance to meet the need for quality drinking and potable water as a result of the country's population exceeding 80 million, to meet the intense water need of the rapidly developing industry in recent years and to provide the amount required for agricultural irrigation. The need for quality water for all these uses can only be met with the right soil and water management [4, 5].

Wastewaters can generally contain harmful pathogenic microorganisms, organic and inorganic compounds, phenols, oxidizers, sulfates, hydrocarbons, oils and various heavy metals. The discharge of these wastewaters to streams, rivers, natural lakes or artificial stagnant water sources such as dams affects the quality of freshwater resources and causes ecosystems to be described as contaminated. For this reason, monitoring the quality parameters of both surface and groundwater resources and preventing pollution have a very important place in the environmental policies of countries. In order to prevent pollution, the pollution status of water resources and the changes in their physical, chemical and biological properties should be followed and evaluated frequently. This becomes even more important when water sources are contaminated by heavy metals, especially when toxic elements such as As, Ni, Cr, Pb and Cd are toxic even at very low concentrations. Because when heavy metals contaminate the biological system, they cause great environmental problems since they do not deteriorate or disappear [6, 7].

Reservoirs are significant lacustrine freshwater habitats built for various purposes such as drinking water supply, irrigation and flood protection. They are built to safely store the water in large quantities. When this aim is achieved, it is tried to provide objectives such as irrigation, hydroelectricity, recreation, water supply and flood prevention. Dam lakes also play a significant role in the supply of water for domestic and industrial use. In addition, they are of great importance for agricultural life. However, various factors such as contamination, sedimentation and excessive use of waters may cause them to be useless. They are being constructed by State Hydraulic Works (DSI) in Türkiye and there are approximately 900 dam lakes in operation in Türkiye [8, 9].

Çokal Dam Lake is one of the most significant reservoirs in the Marmara Region of Türkiye. It was built on the Kocadere Stream in the Tekirdağ Province of Thrace Region between 1997 and 2002 for irrigation and drinking water supply [10, 11].

The body volume of the reservoir is about 4.065,000 m^3 and its height from the stream bed is 81 m. At the normal water level, the lake volume is about 204 hm^3 and the lake area is 10 km^2. Çokal Dam Lake provides 14 hm^3 of drinking water and irrigation services to an area of 10,660 ha [10, 11].

Potentially hazardous elements (PHEs) may cause significant health problems. Although PHEs occur naturally on the earth's crust, they may enter in large quantities into the water causing serious damage to living things and the environment as a result of mainly human-induced activities. Drinking water is known as one of the main routes of their entry into the human body. Numbers of studies have been performed to examine the effects of PHEs in surface and groundwater ecosystems [12–16].

The main purpose of this scientific research was to investigate the accumulations of PHEs in water of Çokal Dam Lake watershed and to evaluate the water quality by comparing our data with some national and international limit values of PHEs.

2 Materials and Methods

2.1 Collecting Surface Water Samples from Reservoir

Aksakal and Çayırlar Streams are the main streams feeding the Çokal Dam Lake. The consisted fluvial ecosystem at the output of the reservoir is named as Kocadere Stream and it flows into the Gulf of Saros. Water samples were collected from the Çokal Reservoir (ÇDL1, ÇDL2), Aksakal (AS), Çayırlar (ÇS) and Kocadere (KS) Streams in the rainy season (autumn) of 2020. The Çokal Reservoir basin map and the sampling points with the coordinate information are given in Fig. 1.

2.2 PTEs Analysis and Statistical Application

pH values of surface water samples (sws) (one liter) were set to 2 (with adding HNO_3). Sws were filtered (with a filter of cellulose nitrate—0.45 μm). Capacities of the sws were made up to 50 ml (with ultrapure water). Then the levels of PTEs were read with an Agilent 7700 xx ICP—MS (TS EN/ISO IEC 17025) [17].

Cluster Analysis (CA) was applied to data by using PAST 4.07.

3 Results and Discussion

The concentrations of investigated PHEs in water of Çokal Dam Lake, Aksakal, Çayırlar and Kocadere Streams are given in Fig. 2. The order of elements detected in water is generally as follows: B > Mn > Zn > Ni > As > Se > Cu > Cr > Pb > Cd. Also the average levels of investigated PHEs were recorded as 900 ppb for B (90 ppb–3799 ppb), 0.84 ppb for Cr (0.164 ppb–1.90 ppb), 52 ppb for Mn (0.674 ppb–249 ppb), 6.16 ppb for Ni (2.76 ppb–12.56 ppb), 1.46 ppb for Cu (0.90 ppb–2.83 ppb), 7.72 ppb for Zn (4.8 ppb–16.6 ppb), 4.34 ppb for As (1.31 ppb–9.57 ppb), 1.88 ppb for Se (0.52 ppb–6.18 ppb), 0.19 ppb for Cd (0.041 ppb–0.351 ppb) and 0.55 ppb for Pb (0.461 ppb–0.676 ppb).

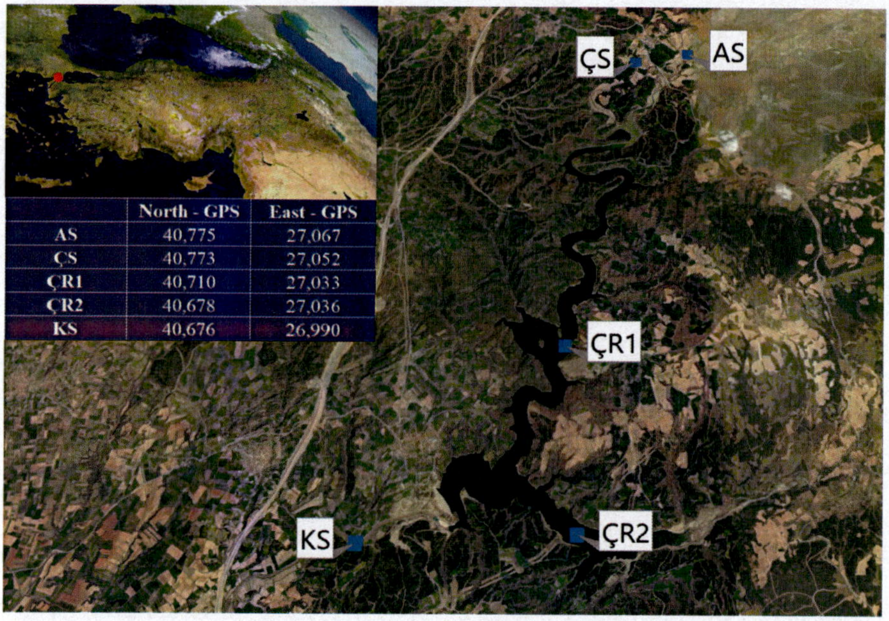

Fig. 1. Çokal Dam Lake watershed and selected stations

Values of PHEs recorded in water of Çokal Dam Lake Basin components with some standard values are given in Table 1. While the Kocadere Stream has IV. Class water quality (wq) in terms of boron accumulations and the Çayırlar Stream has II class wq in terms of manganese accumulations, all the other locations have I. class wq in terms of all the determined PHEs [18]. In addition, all the basin components, except boron values in water of Kocadere Stream and manganese values in water Çayırlar Stream, were below the drinking water standards [19–21] in terms of all the investigated PHEs.

Cluster Analysis (CA), which is being widely used to classify the investigated habitats, is an effective statistical tool [22–27]. The detected data of applied CA in the current study with the similarity coefficients of investigated locations are given in Fig. 3. According to the results of current CA application, 3 clusters were recorded. Cluster 1 (C1—Relatively less contaminated zone) is consists of the following stations: ÇR1 (Çokal Dam Lake), ÇR2 (Çokal Dam Lake) and AS (Aksakal Stream). Cluster 2 (C2—Relatively moderate contaminated zone) is consists of the following station: ÇS (Çayırlar Stream). Cluster 3 (C3—Relatively more contaminated zone) is consists of the following station: KS (Kocadere Stream).

In a similar CA application performed in China, CA was used to assess the quality of stagnant and fluvial ecosystems of a Xiangjiang watershed. As a result of this research, CA grouped thirty-four locations into three clusters [28]. In another study performed in Türkiye, water quality of Ergene River Basin was assessed by using some statistical techniques. As a result of this research, three clusters were formed similarly as the current research [29].

Table 1 Surface and drinking water standards and values of PHEs in the basin

	Surface Water Standards				Drinking Water Standards			Cokal Dam Lake Basin*				
	Water Quality Classes [18]											
	I clean	II less polluted	III polluted	IV high polluted	TS266 [19]	EC [20]	WHO [21]	AS	ÇS	ÇR1	ÇR2	KS
As (μg/L	20	50	100	> 100	10	10	10	3.33 I. Class	9.56 I. Class	2.31 I. Class	1.31 I. Class	5.19 I. Class
Cu (μg/L	20	50	200	> 200	2000	2000	2000	1.33 I. Class	0.90 I. Class	0.95 I. Class	1.29 I. Class	2.83 I. Class
B (μg/L	1000	1000	1000	> 1000	1000	1000	500	127.51 I. Class	377.51 I. Class	107.29 I. Class	89.82 I. Class	3799 **IV. Class**
Zn (μg/L	200	500	2000	> 2000	-	-	-	4.81 I. Class	16.56 I. Class	5.25 I. Class	4.95 I. Class	7.04 I. Class
Cd (μg/L	2	5	7	> 7	5	5	3	0.05 I. Class	0.04 I. Class	0.08 I. Class	0.06 I. Class	0.35 I. Class
Cr (μg/L	20	50	200	> 200	50	50	50	0.21 I. Class	1.44 I. Class	0.16 I. Class	0.50 I. Class	1.90 I. Class
Pb (μg/L	10	20	50	> 50	10	10	10	0.46 I. Class	0.68 I. Class	0.52 I. Class	0.54 I. Class	0.56 I. Class
Ni (μg/L	20	50	200	> 200	20	20	70	6.93 I. Class	12.56 I. Class	2.76 I. Class	3.57 I. Class	4.95 I. Class
Se (μg/L	10	10	20	> 20	10	10	10	0.57 I. Class	0.85 I. Class	1.04 I. Class	0.74 I. Class	6.18 I. Class
Mn (μg/L	100	500	3000	> 3000	100	100	100	9.33 I. Class	248.57 **II. Class**	1.35 I. Class	1.83 I. Class	0.67 I. Class

*II – III. – IV. Class water qualities are given in bold

4 Conclusions

In this research, water quality of Çokal Dam Lake Basin components were investigated in terms of inorganic contaminant accumulations of water. All the investigated fluvial and lacustrine habitats were found as quite high-water quality and they have I. class wq, in general. It was also recorded that the lentic habitats of the watershed (mid and output of Çokal Dam Lake) were found as relatively less contaminated components of the basin, while the lotic habitats (Aksakal, Çayırlar and Kocadere Streams) were found as relatively more contaminated components. Population growth and industrialization are increasing day by day, and this situation causes pollution of the soil, vegetation, surface and groundwaters and atmosphere. It is clear that serious problems may arise in the quality and amount of water that can be used for irrigation, with the pollution of rivers, lakes and other water resources. Therefore, in order to prevent possible environmental and health risks, it is necessary to constantly monitoring the water quality of Çokal Dam Lake, which is an important irrigation source in Tekirdağ.

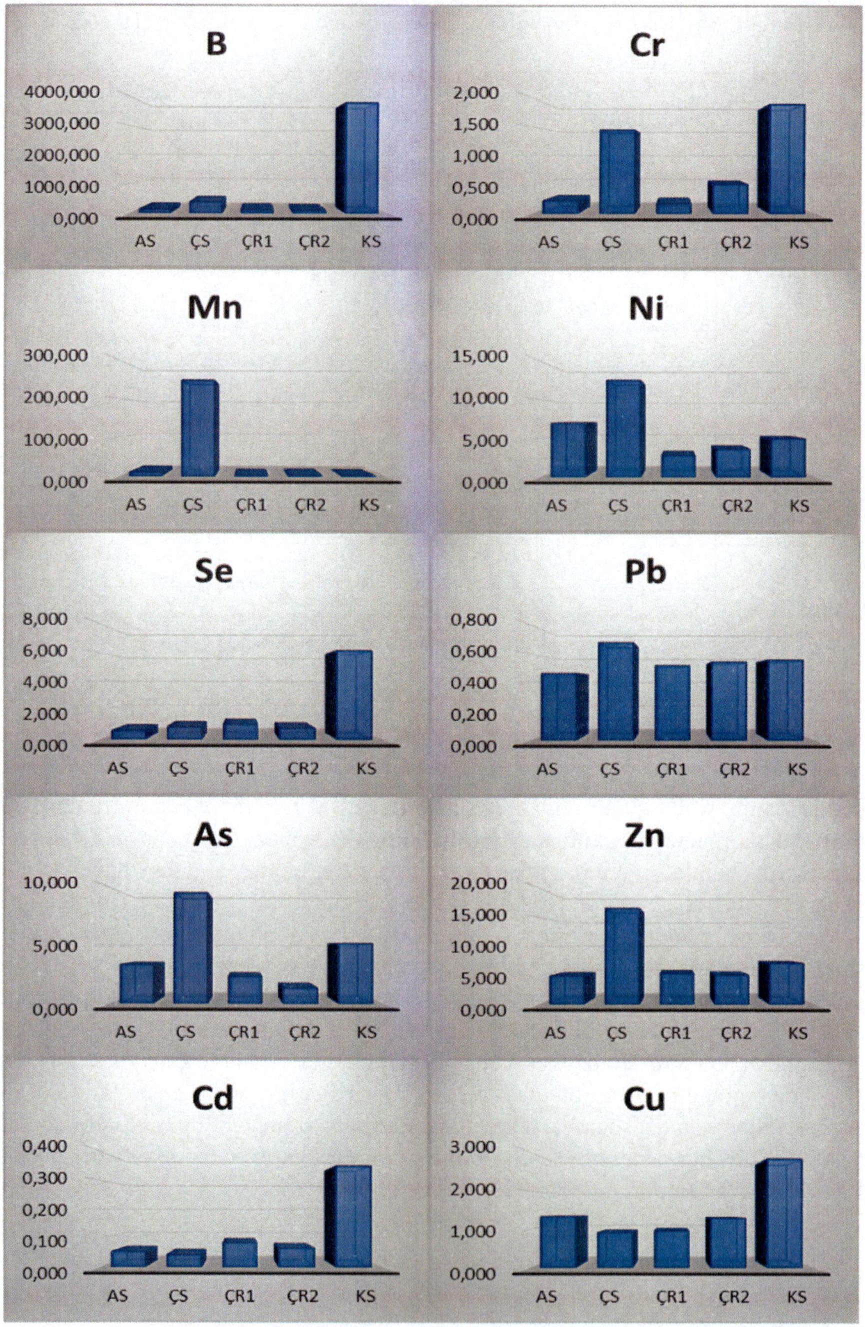

Fig. 2. Levels of PHEs in water (μg/L)

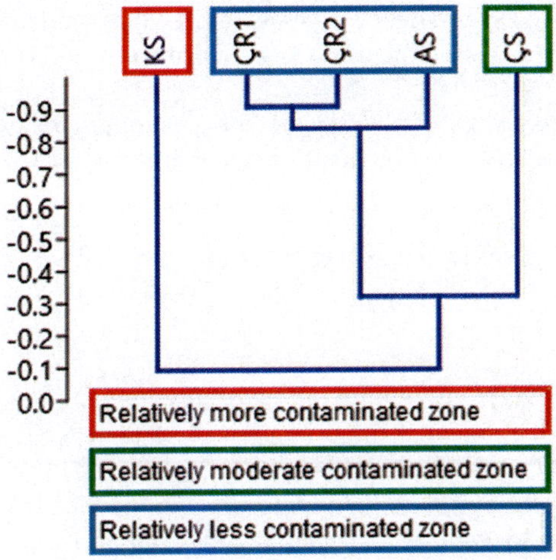

	AS	ÇS	ÇR1	ÇR2	KS
AS	1,00				
ÇS	0,37	1,00			
ÇR1	0,87	0,31	1,00		
ÇR2	0,80	0,27	0,91	1,00	
KS	0,07	0,18	0,06	0,05	1,00

Fig. 3. CA diagram

References

1. Jannat, J.N., et al.: Hydrochemical assessment of fluoride and nitrate in groundwater from east and west coasts of Bangladesh and India. J. Clean. Prod. (2022). https://doi.org/10.1016/j.jclepro.2022.133675
2. Onur, Ş.G., Tokatlı, C.: Comparison of fluoride contents in terms of teeth health and water quality in drinking water at the northern and southern regions of Meriç River Basin (Edirne/Turkey). Int. J. Agric. Environ. Food Sci. **4**(2), 173–180 (2020)
3. Tokatlı, C., Köse, E., Çiçek, A., Emiroğlu, Ö.: Pesticide accumulation in Turkey's Meriç River basinwater and sediment. Pol. J. Environ. Stud. **29**(1), 1–6 (2020)
4. Tokatlı, C.: Use of the potential ecological risk index for sediment quality assessment: a case study of dam lakes in the Thrace part of the Marmara Region. Aquatic Sci. Eng. **34**(3), 90–95 (2019)
5. Mutlu, E., Tokatlı, C., Islam, A.R.T., İslam, S., Muhammad, S.: Water quality assessment of Şehriban Stream (Kastamonu, Türkiye) from a multi-statistical perspective. Int. J. Environ. Anal. Chem. (2023). https://doi.org/10.1080/03067319.2023.2197114
6. Tokatlı, C., Baştatlı, Y., Elipek, B.: Water quality assessment of dam lakes located in Edirne province (Turkey). Sigma J. Eng. Nat. Sci. **35**(4), 743–750 (2017)
7. Köse, E., et al.: Assessment of ecologic quality in terms of heavy metal concentrations in sediment and fish on Sakarya River and dam lakes, Turkey. Soil Sediment Contamination: Int. J. **29**(3), 292–303 (2020)
8. Tokatlı, C.: Pesticide accumulations in water and sediment of dam lakes located in Thrace part of Marmara Region (Turkey). Aquatic Res. **3**(3), 124–134 (2020)
9. Varol, M., Ustaoğlu, F., Tokatlı, C.: Ecological risks and controlling factors of trace elements in sediments of dam lakes in the black sea region (Turkey). Environ. Res. **205**, 112478 (2022)
10. Anonymous: Çanakkale Province 2020 Environmental Status Report. Republic of Türkiye, Canakkale Governorship, Provincial Directorate of Environment and Urban Management (2020)
11. Anonymous: Tekirdağ Province 2020 Environmental Status Report. Republic of Türkiye, Tekirdağ Governorship, Provincial Directorate of Environment and Urban Management (2020)

12. Arslan, N., Tokatlı, C., Çiçek, A., Köse, E.: Determination of some metal concentrations in water and sediment samples in Yedigöller Region (Kütahya). Rev. Hydrobiol. **4**(1), 17–28 (2011)
13. Arslan, N., Köse, E., Tokatlı, C., Emiroğlu, Ö., Çiçek, A.: Ecotoxicological effects of solid waste storage areas on aquatic systems: example of Yedigöller, Kütahya. Karaelmas Sci. Eng. J. **2**(1), 20–26 (2012)
14. Varol, M., Tokatlı, C.: Impact of paddy fields on water quality of Gala Lake (Turkey): an important migratory bird stopover habitat. Environ. Pollut. **287**, 117640 (2021)
15. Tokatli, C.: An investigation on heavy metal accumulations in water, sediment and fishes of Emet Stream. Ph. D. Thesis, Dumlupınar University, Institution of Science, Department of Biology (2012)
16. Tokatlı, C., Emiroğlu, Ö., Çiçek, A, Köse, E., Başkurt, S., Aksu, S., Uğurluoğlu, A., Şahin, M., Baştatlı, Y.: Investigation of toxic metal bioaccumulations in fishes of Meriç River Delta (Edirne). Anadolu Univ. J. Sci. Technol.—C Life Sci. Biotechnol. **5**(1), 1–11 (2016)
17. Environmental Protection Agency (EPA) METHOD 200.7.: Determination of Metals and Trace Elements in Water and Wastes by Inductively Coupled Plasma-Atomic Emission Spectrometry (2001)
18. Turkish Regulations: Yüzeysel Su Kalitesi Yönetimi Yönetmeliği, 15 Nisan 2015 tarihli Resmi Gazete, Sayı: 29327 (2015). http://suyonetimiormansu.gov.tr
19. TS 266 (Turkish Standards): Sular-İnsani tüketim amaçlı sular. Türk Standartları Enstitüsü, ICS 13.060.20 (2005)
20. EC (European Communities): European Communities (drinking water) (no. 2), Regulatıons 2007, S.I. No. 278 of 2007 (2007)
21. WHO (World Health Organization): Guidelines for Drinking-water Quality. World Health Organization Library Cataloguing-in-Publication Data, NLM classification: WA 675 (2011)
22. Tabachnick, B.G., Fidell, L.S.: Using Multivariate Statistics, 3rd edn. Harper Collins College Publishers, New York (1996)
23. Tokatlı, C., Çiçek, A., Köse, E.: Groundwater quality of Türkmen Mountain (Turkey). Pol. J. Environ. Stud. **22**(4), 1197–1208 (2013)
24. Tokatlı, C.: Application of water quality index for drinking purposes in dam lakes: a case study of Thrace Region. Sigma J. Eng. Nat. Sci. **38**(1), 393–402 (2020)
25. Tokatlı, C., Köse, E., Çiçek, A., Emiroğlu, E., Baştatlı, Y.: Use of cluster analysis to evaluate surface water quality: an application from downstream of Meriç River Basin (Edirne, Turkey). Int. J. Adv. Sci. Eng. Technol. Spl. (3) (2015)
26. Çiçek, A., Köse, E., Emiroğlu, Ö., Tokatlı, C., Başkurt, S., Sülün, Ş: Boron and arsenic levels in water, sediment and tissues of Carassius gibelio (Bloch, 1782) in a dam lake. Pol. J. Environ. Stud. **23**(5), 1843–1848 (2014)
27. Ustaoğlu, F., İslam, S., Tokatlı, C.: Ecological and probabilistic human health hazard assessment of heavy metal in sera lake nature park sediments (Trabzon, Turkey). Arab. J. Geosci. **15**, 597 (2022)
28. Zhang, Q., et al.: Assessment of surface water quality using multivariate statistical techniques in red soil hilly region: a case study of Xiangjiang watershed, China. Environ. Monit. Assess **152**, 123–131 (2009)
29. Tokatlı, C.: Water quality assessment of Ergene River Basin using multivariate statistical analysis. J. Limnol. Freshwater Fisheries Res. **6**(1), 38–46 (2020)

Hydraulic Study of Fan Spillway Using Computational Fluid Dynamics (CFD) and Experimental Approaches

Hamza Souli(✉), Jihane Ahattab, and Ali Agoumi

Hassania School of Public Works, Hydraulic, Environmental, Marine, Soil and Structural Systems, Casablanca, Morocco
Souli.hamza.cedoc@ehtp.ac.ma

Abstract. The optimization of spillways in dams is very important, especially in the current conditions of climate change and the occurrence of extreme events. Our objective is to study the flow behavior in fanned spillway, which is complicated because shock waves are present, vibrations and high order of turbulence. The hydraulic structure can be damaged by these transverse waves, increase the cost and time of construction. The purpose of this paper is to examine the water surface distribution, the flow patterns in these structures to have a better understanding of the hydraulic difficulties and to test several alternatives for a better performance of the spillway.

Theoretical, empirical and numerical approaches were used. RNG k- ε model of turbulence based on the Volume of fluid method was chosen. Also, in order.

The numerical results obtained with Flow 3D hydro were validated for reliability and accuracy through theoretical calculations. Furthermore, a comparison was made among the measured data, theoretical calculations, and numerical outcomes.

At last, an optimal bottom slope and an optimized arc crested weir were chosen.

The outcomes of the physical measurements and numerical methods show a reliable consistency and robust correspondence; therefore, the numerical model assumes a vital role in facilitating the study, optimization, and support of decision-makers in making dependable and accurate choices.

Keywords: Computational fluid dynamics CFD · Turbulence · VOF · Fluid mechanics · Hydraulic structures · Spillways

1 Introduction

The critical element of a dam is its spillway, which serves to manage overflow during floods. Designing and sizing this structure present significant challenges for engineers, considering factors such as geotechnical, geological, and hydraulic considerations. Selecting the appropriate spillway type is a complex decision that necessitates the collaboration of multiple experts in their respective fields to identify the most effective and efficient variant [1].

K. Baba et al. (Eds.): GeoME 2023, SPEES, pp. 443–453, 2023.
https://doi.org/10.1007/978-3-031-49345-4_42

Fan spillways are hydraulic structures designed to collect water in a frontal direction. The water is then directed to a collecting trough, where its hydraulic energy is dissipated before entering the spillway chute, and eventually, it reaches the structure of dissipation, also known as the ski jump. See Fig. 1 [2].

The unique aspect of this design lies in its fan-shaped contraction, which means we cannot assume that the frontal inflow after passing the arc-crested weir will not produce cross-waves [3–5].

This type of structure tends to experience flow concentration at the center, leading to the formation of hazardous transversal waves, while the sidewalls may have zero flow due to the influence of centrifugal forces. As a result, extensive research has been conducted to study the flow dynamics of fan channel spillways [3].

Hartung and Knaus [6, 7] have developed a formulation to estimate the maximum wave height h_M in the collecting trough as a function of the Froude number. The Froude number is defined as a parameter used to characterize the flow regime in this context.

$$F_1 = \frac{V}{\sqrt{gh_1}}$$

$$\left\{ h_M = 3.66\gamma\left(2 + F_1^2\right)^{-0.5} if 1 < F_1 < 3.5 \right. \quad (1)$$

where:

$$F_1 = \frac{V}{\sqrt{gh_1}}$$

With V: is the velocity (m/s).
h_1: is the water depth (m).

The US army corps of engineers adopted this type of spillway for the Genegantslet dam [8].

In this paper, we try to model the fan shape spillway using 3D numerical and physical approaches. The aim is to have a better understanding of the behavior of the flow in these types of structures and to estimate the water level, velocity and pressure fields [8–10].

Despite the interest of this conception, there are few studies on this subject. In this case, we will use the experimental data collected from our physical model to validate our numerical model. Then we will try to compare physical and numerical methods in terms of the velocity and pressure at the spillway chute and the aggressiveness of shock waves by testing several bottom slope and the scour parameters downstream.

For our case study, Table 1 summarizes the parameters taken into account for the design of our structure.

2 Experimental Setup and Numerical Approaches

2.1 Physical Experiments

The scaled model [9–12] of this case study a 3D representation physical model with a scale of 1/50 of the nature size. The experimental facility is divided into two sections:

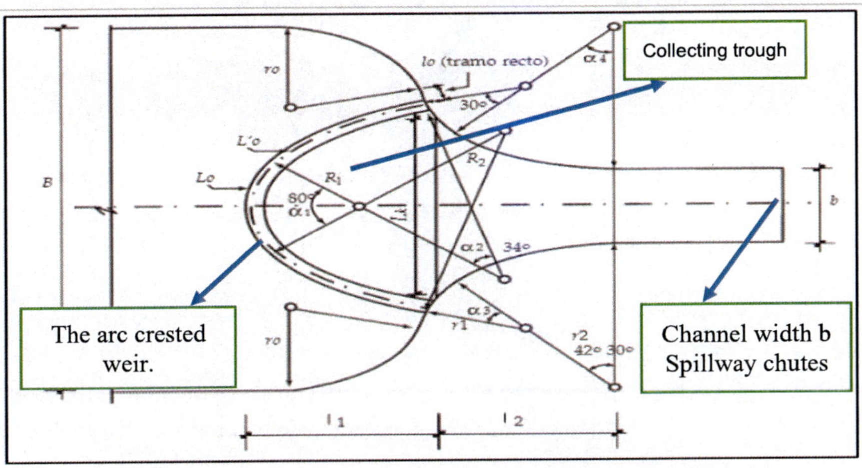

Fig. 1. Geometric parameters for sizing a fan shaped spillway [2]

Table 1. Geometric parameters for our case study

b	B	L	L_0	R1	$\frac{B}{L_0}$	$\frac{R_1}{L_0}$	$\frac{L_k}{L_0}$	$\frac{l_0}{L_0}$	$\frac{b}{L_0}$	$\frac{r_1}{L_0}$	$\frac{r_2}{L_0}$	α_1	α_2	α_3
20	70	70.8	80.87	70	0.86	0.685	0.675	–	0.86	0.268	0.0071	76	19	44

- The part before the main flow comprises a rectangular reservoir with specified dimensions of 10 m x 12 m.
- Additionally, there is a downstream section with a rectangular shape, specifically intended to investigate issues related to scour. However, it is important to note that the findings from this downstream section are not within the scope of the current article.

Instrumentation

Rolling point instrumentation with high accuracy used to measure the flow height above the weir; (±1 mm)
Magnetic flow meter to measure the flow rate;
Current meter for velocity;

The spillway consists of:

- A weir, 80 m long, developed in a duckbill shape.
- A convergent weir (or duckbill) reducing the width to 20 m;
- A chute spillway with a constant width of 20 m and a slope of 22%.
- A cylindrical spoon (ski jump) for the restitution of the water (Fig. 2).

The flow behavior of fan shaped spillways differs significantly from classic smooth spillways by: the collecting trough and the arc crested weir, which generates rooster tails waves.

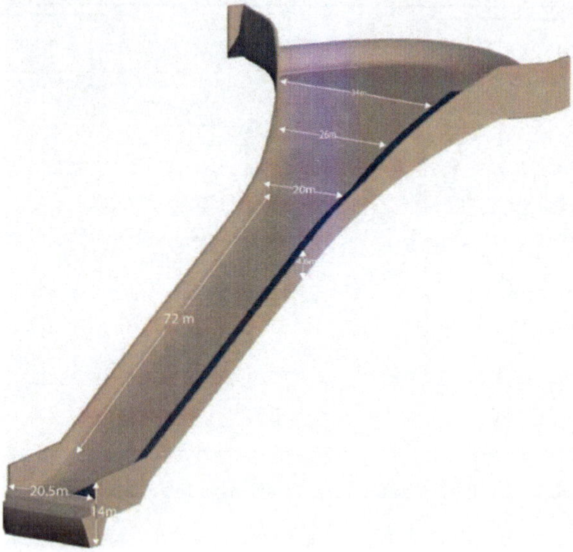

Fig. 2. 3D model of the studied spillway

- Different head discharges were tested (starting from 1 m to 3.85m).
- The laboratory data samples will be presented alongside the numerical approaches for each aspect of the current study to facilitate comparison.

2.2 Numerical Model and Flow Simulation

To explore the flow patterns, water surface profile, and flow behavior in each segment of our spillway, a series of steps were taken to construct the numerical model. For this purpose, we employed the software Flow 3D Hydro, known for its excellence in simulating free surface flows. With this software, we modeled the aggressiveness of cross-waves in the collecting trough and computed velocity and pressure fields. Subsequently, we compared these results with our experimental data to validate the accuracy of our numerical model.

In our simulation, we employed the Reynolds-averaged Navier-Stokes (RANS) equations and the RNG k-ε model of turbulence. The hydraulic parameters were calculated for each cell using a staggered grid System.

2.3 Grid Analysis Study

To design the geometry, we utilized a CAD software [16] to create the arc-crested weir and the entire spillway shape. A mesh independence analysis was conducted, involving meshes of different starting from coarse to fin mesh sizes to determine the optimal cell size.

The grid convergence study revealed that, for this specific study, the most suitable cell size is 0.1 m (refer to Table 2). To find this correct value, we compared the computed values of hydraulic parameters with those obtained from physical modeling.

For the first part of the analysis, we experimented with three grid resolutions: 0.1 m, 0.2 m, and 0.5 m. We discovered that the most appropriate cell size is 0.1 m. Subsequently, for the entire spillway, we tested two cell sizes: 0.3 m and 0.8 m, and determined that the proper cell size is 0.3 m.

The total number of meshes utilized ranged from 500,000 cells to 15,256,894 cells. Table 2 provides a summary of the cases tested, indicating the cell size and the number of cells for each specific case.

Table 2. Grid sensitivity analysis

Head discharges (m)	part	Cell size (m)	Total number of cells
1 to 3.85	Collecting trough	0.5 m 0.2 m 0.1 m	500000 4 000 000 8 000 003
1 to 3.85	Whole spillway	0.8 m 0.3 m	5 245 236 **15 256 894**

2.4 RANS Governing Equations

The RNG k-ε turbulence model are defined by the following equations [17]:

The governing equations of the RNG k-ε (Eqs. 2–11) model yield the velocity and pressure fields, providing essential insights into the flow characteristics and behavior of the fluid in the simulated system (Table 3).

$$\frac{\partial k}{\partial t} + \frac{\partial k u_i}{\partial x_i} - \frac{\partial}{\partial x_i}\left(\mu_{keff}\frac{\partial k}{\partial x_i}\right) = G - \varepsilon \tag{2}$$

$$\frac{\partial \varepsilon}{\partial t} + \frac{\partial \varepsilon u_i}{\partial x_i} - \frac{\partial}{\partial x_i}\left(\mu_{\varepsilon eff}\frac{\partial \varepsilon}{\partial x_i}\right) = (C_{1\varepsilon} - R_\varepsilon)\frac{\varepsilon}{k}G - C_{2\varepsilon}\frac{\varepsilon^2}{k} \tag{3}$$

$$D_{keff} = \nu_t + \nu \tag{4}$$

$$\mu_{\omega eff} = \alpha_{\omega i}\nu_t + \nu \tag{5}$$

$$\mu_{\varepsilon eff} = \frac{\nu_t}{\sigma_\varepsilon} + \nu \tag{6}$$

$$\nu_t = C_\mu\frac{k^2}{\varepsilon} \tag{7}$$

$$S_{ij} = \frac{1}{2}\left(\frac{\partial u_j}{\partial x_i} + \frac{\partial u_i}{\partial x_j}\right) \tag{8}$$

$$S_2 = 2S_{ij}S_{ij}$$

$$G = 2\nu_t S_{ij} S_{ij} \tag{9}$$

$$R_\varepsilon = \frac{\eta\left(1 - \frac{\eta}{\eta_0}\right)}{1 + \beta^3} \tag{10}$$

$$\eta = \sqrt{S_2}\frac{k}{\varepsilon} \tag{11}$$

where:

Table 3. Turbulence model constants

σ_ε	$C_{1\varepsilon}$	$C_{2\varepsilon}$	σ_k	η_0	β	C_μ
0.71942	1.42	1.68	0.71942	4.38	0.012	0.0845

With:

V	Velocity [m/s]
g	Gravity constant [m/s^2]
h	Flow depth [m]
ρ	Water density [kg/m^3]
σ	Water surface tension [N/m]
ν	Water kinematic viscosity [m^2/s]
u_i	Velocity component of x_i [m/s]
$\mu_{keff}, \mu_{\varepsilon eff}$	The diffusivity for k, ε respectively
ν_t	The turbulent kinetic viscosity [m^2/s]
G	The production of turbulence due to shear
K	Turbulent kinetic energy
ε	Dissipation rate

2.5 Free Surface Modelling

The Volume of Fluid (VOF) method is used to estimate the water surface variations. It employs the fraction function α as an estimator to determine the proportion of the cell occupied by water, air, or both. The software utilizes the Tru-VOF method [18, 19] to capture the interface between water and air, and the VOF equation is used for interface tracking.

$$\frac{\partial \alpha}{\partial t} + u_i \frac{\partial \alpha}{\partial x_i} = 0 \tag{12}$$

3 Results

3.1 Shock Waves in the Collecting Trough

To emphasize the hydraulic issues linked to the other tested variants in the collecting trough, we introduced slope variations (3%, 4%, 5%). The simulated models revealed the presence of shock waves, which intensified with steeper slopes. These shock waves would submerge the flow in the collecting trough, inevitably leading to a reduction in the spillway's capacity.

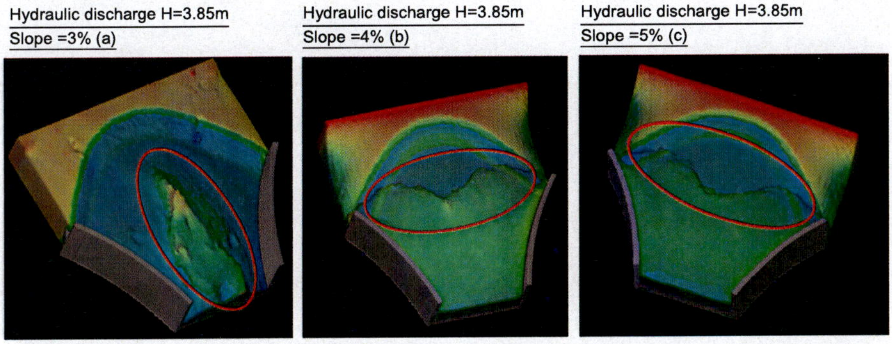

Fig. 3. Shock waves in the collection trough for different slopes found using flow 3D

The observations become evident as slopes surpass 4%, wherein the formation of two distinct wave fronts becomes apparent in cases (b) and (c). This occurrence is a result of the interference principle and poses a significant threat to the integrity of the hydraulic structure, potentially leading to damage.

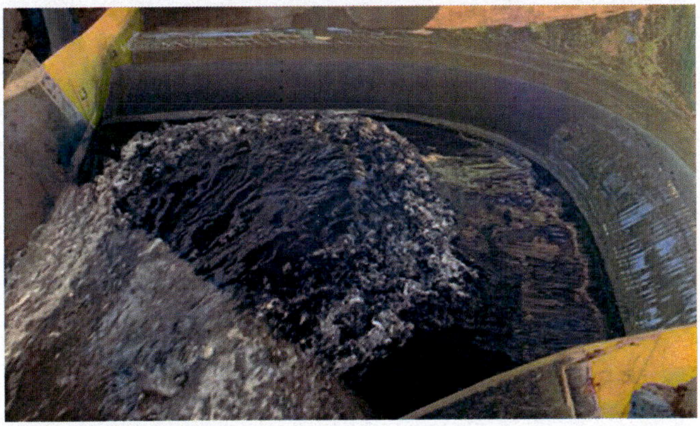

Fig. 4. Example of Shock waves in the collection trough in the physical model for bottom slope 3%

The flow behavior shows the same aspect in terms of the flow patterns (see Figs. 3 and 4), the aggressive waves were formed in the receiving trough.

3.2 Velocity and Pressure Fields in the Whole Structure

In order to comprehensively investigate the behavior of the runner, we carefully established measuring points to accurately calculate the velocity and pressure values. This approach allowed us to gain insight into the magnitude of these fields, assess the flow conditions, and proactively evaluate the likelihood of cavitation-related risks. The data obtained from these measurements served as a crucial factor in understanding and optimizing the performance of the runner under various flow scenarios (Fig. 5).

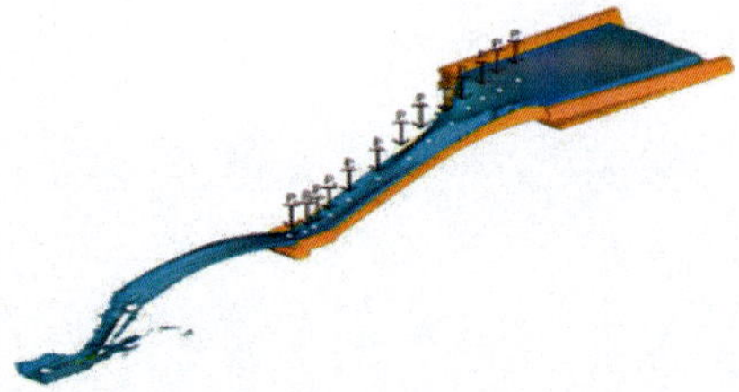

Fig. 5. Points for velocity and static pressure measurements and calculations

The figure presented offers a comparison between numerical values and those obtained from the physical model. The results indicate that the average error remains below 5%. This demonstrates the viability of numerical methods in estimating hydraulic parameter values, thereby enabling the optimization of structures and proactive identification of potential hydraulic malfunctions (Fig. 6).

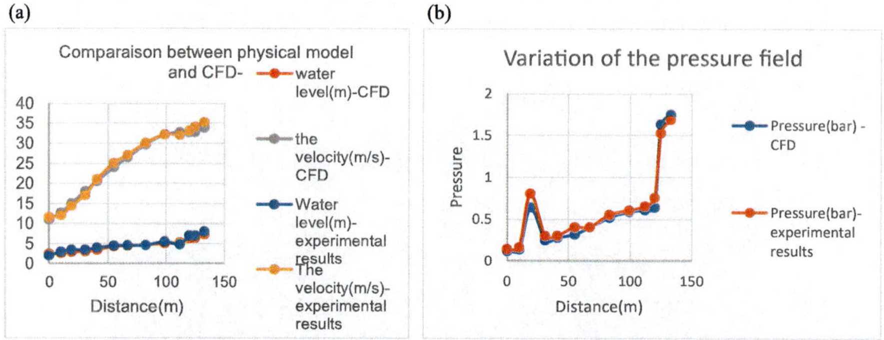

Fig. 6. Comparison between physical model and CFD—(a) Velocity and water level in the spillway chute— (b) static pressure

3.3 The Length of the Jet Downstream of the Dam

The aim of this part is to numerically calculate the length of the jet at the spoon and estimate the scouring depth. Knowledge of these data plays a vital role in spillway operation, as it is essential to have a clear idea of the damage that can be caused by a flowing jet downstream of a dam. These data will help the design engineer to anticipate the protection measures needed to minimize disastrous damage [20].

With regard to the water jet, we compared theoretical and numerical values of the impact distance of the water jet from the ski jump (see Fig. 7).

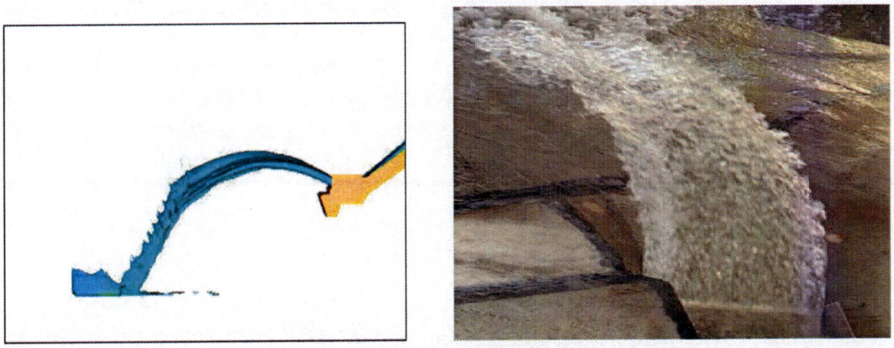

Fig. 7. Jet detachment in ski jumping in the numerical model (Left), physical model (right)

Table 4. Theoretical and numerical values for the maximum impact distance of the water jet

Theoretical jet impact distance (m)	Numerical jet impact distance (m)
90	89.07

The table values demonstrate that both the theoretically and numerically calculated impact distances of the water jet are similar and considerably distant from the toe of the dam.

Table 5. Theoretical and numerical values for the maximum impact distance of the water jet water jet

Theoretical scouring (m)	Numerical scouring (m)
30	27.05

We conducted a comparison between the theoretical and numerical values of the scouring distance concerning the depth of the erosion pit (Tables 4 and 5).

4 Conclusions

In conclusion, this paper highlights the significance of the bottom slope in the collecting trough to mitigate cross-waves in this critical area. Our objective was to reduce the impact of transversal waves, spray formation, water wings, and wave interferences, thus enhancing the future conception of this crucial zone for hydraulic designers. To achieve this, we modified the bottom slope.

To validate our numerical approach regarding flow patterns and jet behavior, we utilized a physical model, considered the gold standard in hydraulic analysis. The numerical results exhibited a reasonable agreement with the physical model's outcomes. Both velocity and pressure fields followed similar paths and displayed nearly equivalent values, with minor fluctuations attributed to numerical limitations like solver constraints and turbulence models.

Furthermore, we calculated jet length and scouring depth for both physical and numerical models, revealing a difference of no more than 5%. This implies that we can efficiently compute and optimize our design using CFD methods, then corroborate the final design using a physical model to minimize time and costs while achieving high accuracy.

References

1. Duncan, W., Huntley, C., Hokenstrom, J., Cudworth, A., McDaniel, T.: Design of small dams. A water resources technical publication. Final report, Bureau of Reclamation, Denver, CO (United States). Engineering and Research (1987)
2. Alegret-Brena, E., Martínez-González, Y.: An integrated study of the fan spillway. Tecnol. Cienc. Agua, vol. 1, n° 2, pp. 37–57 (2010)
3. Souli, H., Ahattab, J., Agoumi, A.: Investigating supercritical bended flow using physical model and CFD. Model. Simul. Eng. e5542589 (2023). https://doi.org/10.1155/2023/5542589
4. I. (International C. on L. Dams), Spillways, shockwaves and air entrainment: review and recommendations: Bulletin 81. ICOLD Paris (1992)
5. Mousavimehr, S.M., Yamini, O.A., Kavianpour, M.R.: Performance assessment of shock-waves of chute spillways in large dams. Shock Vib. (2021)
6. Chanson, H.: Self-aerated flows on chutes and spillways. J. Hydraul. Eng. **119**(2), 220–243 (1993)
7. Falvey, H.T.: Cavitation in chutes and spillways. US Department of the Interior, Bureau of Reclamation Denver, CO, USA (1990)
8. U. Spillways: In Design of small dams. US Department of the Interior, Bureau of Reclamation: Washington, DC, USA (1987)
9. Erpicum, S., Tullis, B.P., Lodomez, M., Archambeau, P., Dewals, B.J., Pirotton, M.: Scale effects in physical piano key weirs models. J. Hydraul. Res. **54**(6), 692–698 (2016)
10. Heller, V.: Scale effects in physical hydraulic engineering models. J. Hydraul. Res. **49**(3), 293–306 (2011)
11. Chanson, H., Pfister, M.: Scale effects in modelling two-phase air-water flows. In: Proceedings of the 35th IAHR World Congress, TPU, pp. 1–10 (2013)
12. Pfister, M., Chanson, H.: Two-phase air-water flows: scale effects in physical modeling. J. Hydrodyn. Ser. B, **26**(2), 291–298 (2014)
13. Chanson, H.: Hydraulic design of stepped spillways and downstream energy dissipators. Dam Eng. **11**(4), 205–242 (2001)
14. Baylar, A., Emiroglu, M.E., Bagatur, T.: An experimental investigation of aeration performance in stepped spillways. Water Environ. J. **20**(1), 35–42 (2006)
15. Boes, R.M., Hager, W.H.: Two-phase flow characteristics of stepped spillways. J. Hydraul. Eng. **129**(9), 661–670 (2003)

16. Falck, B., Falck, D., Collette, B.: Freecad [How-To]. Packt Publishing Ltd (2012)
17. Wilcox, D.C.: Turbulence modeling for CFD, vol. 2. DCW industries La Canada, CA (1998)
18. Akbar, M., Huali, P., Guoqiang, O.: Numerical investigation of pressure profiles and energy dissipation across the stepped spillway having curved treads using FLOW 3D. Arab. J. Geosci. **15**(15), 1–19 (2022)
19. Mansoori, A., Erfanian, S., Moghadam, F.K.: A study of the conditions of energy dissipation in stepped spillways with Λ-shaped step using FLOW-3D. Civ. Eng. J. **3**(10), 856 (2017)
20. Ghaderi, A., Daneshfaraz, R., Torabi, M., Abraham, J., Azamathulla, H.M.: Experimental investigation on effective scouring parameters downstream from stepped spillways. Water Supply **20**(5), 1988–1998 (2020)

Piano Key Weirs: A Bibliometric Study of Patterns and Trends in Scientific Literature

Yousra Marghoub(✉), Driss Khomsi, Naoual Semlali Aouragh Hassani, and Amal Aboulhassane

Hydraulic Systems Analysis Team (EASH), Mohammadia School of Engineers, Mohammed V University in Rabat, Rabat, Morocco
marghoub@emi.ac.ma

Abstract. Piano Key Weir (PKW) is a non-linear weir structure that has proven its high efficiency in increasing the discharge capacity compared to regular weirs. Experimental researches conducted in the late 1900s presented the PKW as an upgrade of the Labyrinth weir that offers additional discharge capacity with a reduced footprint. Since then, several experimental and numerical studies were performed to understand the different aspects of this hydraulic structure. Regardless of how promising this field is becoming, there is a lack of a quantitative study.

A bibliometric analysis, which is a method used to study the patterns and trends in scientific literature, was conducted using the SCOPUS database and VOSviewer software. The purpose of this study is to provide a comprehensive quantitative view of the researches performed on Piano Key Weirs, to decipher the intellectual structure of the field as well as to identify collaboration patterns.

The results show that the number of publications on Piano Key Weirs has been increasing in recent years. In addition, thematic scientific conferences contributed notably to the advancement of this field. Researchers from various countries have investigated PKWs, with France and Algeria being the initiators in scientific research. Notably, Iran has recently showed interest in studying PKWs, positioning itself as the second highest contributor in terms of publication count. The study concludes that piano key weirs are an active area of research and provides an insightful overview of the current state of the field for researchers, engineers and practitioners. Therefore, it unravels collaboration patterns within the field.

Keywords: Piano Key Weirs · Bibliometric Analysis · VOSviewer

1 Introduction

Bibliometric analysis, a statistical and mathematical method, examines patterns found in available published literature [1]. This approach offers valuable insights into the historical evolution of scientific production, the research trends and knowledge gaps [2].

The use of bibliometrics to evaluate scientific production began in the 1960s. Eugene Garfield, founder of the Institute for Scientific Information (ISI), developed the Science Citation Index (SCI), which tracked the citation of scientific articles across a wide range

K. Baba et al. (Eds.): GeoME 2023, SPEES, pp. 454–463, 2023.
https://doi.org/10.1007/978-3-031-49345-4_43

of journals [3]. The SCI became a valuable resource for scholars seeking to assess the impact of their work and track the development of scientific fields [4].

This method essentially relies on citation and co-citation analysis. Citation analysis is founded on the premise that authors cite documents they deem significant for advancing their research. Consequently, highly cited papers are often to have the greater impact on the discipline [5].

Co-citation analysis on the other hand, requires identifying the articles that are frequently cited together by other articles. It relies on the assumption that a pair of articles are co-cited frequently by many authors, they must have an intellectual or thematic connection and they start to form a cluster of research [6].

The development of databases, such as Scopus and Web Of Science, as well as the internet have revolutionized the collection and analysis of citation data, allowing researchers to conduct more comprehensive and efficient bibliometric studies [7].

However, it is important to note that while bibliometric analysis is a powerful tool, it should be used in conjunction with qualitative reviews. Qualitative analysis, including extensive reading and critical assessment, remains primordial for a comprehensive understanding of a research field [8].

In fact, Piano key weirs have become an interesting research field. This innovative non-linear weir design, introduced by Hydrocoop [9, 10], has gained significant interest in recent years due to their efficient water management and flood control capabilities. This inventive design increases the allowable weir length within a given spillway channel width and improves upon the labyrinth weir by integrating upstream and/or downstream overhangs. This design increases the discharge capacity and reduces the structural footprint, which makes the piano key weir a compelling solution for dam rehabilitation [11].

Following the successful experience of the piano key weir at the Goulours dam in France [12], there was an increasing interest in PKWs. This led to a surge in research in many laboratories around the world to better understand the hydraulic behavior of PKWs [13–16] and study the influence of geometrical parameters on their performance [17, 18], as well as the evaluation of its cost effectiveness as a viable rehabilitation solution [19]. As a result of this work, over 30 PKWs were constructed for dam rehabilitation [20–23], and new dams [15, 24] in 12 years.

By analyzing the bibliographic data related to PKWs, this study aims to identify the key authors, publications and countries that have contributed to the development of this technology. In addition, it will allow researchers to visualize complex relationships within the scientific literature.

2 Materials and Methods

This study aims to quantitively analyze the bibliography on piano key weir using VosViewer. In recent years, bibliometric reviews have grown in popularity due to the availability and accessibility of scientific database like SCOPUS and Web of Science, as well as the bibliometric software such as Gephi, Leximancer and VOSviewer [25].

The power of bibliometric analysis lies in its ability to rigorously analyze a large amount of unstructured data, enabling scholars to comprehend the intellectual structure of their field, identify research gaps, generate innovative research ideas and identify collaboration patterns.

VOSViewer is a powerful software for generating and displaying bibliometric networks. It analyzes and visualizes the relationships between articles, authors, keywords, and other entities within a given field of study. VosViewer can create various network maps, such as co-authorship, co-citation, and keyword co-occurrence networks. It is widely used in bibliometric research and information science for visualizing complex relationships within scientific literature.

While other bibliometric software tools such as CiteSpace, SciMAT, and BibExcel are available, researchers may prefer VOSviewer for many reasons. For starters, it is free and open-source software, making it available to researchers without financial constraints. Secondly, VOSviewer is simple to use, with a variety of functions and visualization options that can be tailored to the user's preferences. Lastly, VOSviewer is compatible with a number of bibliographic databases, including Scopus, Web of Science, and PubMed, making data retrieval and importation a seamless process. In addition, VOSviewer features an assortment of analytical tools, such as clustering and density visualization, to enable in-depth analysis and interpretation of bibliometric data.

SCOPUS is the bibliographic database of peer reviewed literature selected to provide our dataset. It was created in 2004 by the publishing house Elsevier to help researchers, students and professionals explore and analyze the scientific literature. It is worth noting that Scopus offers access to over 75 million records for more than 24000 journals, which empowers our study with an important coverage [26].

Our dataset includes all articles, conference papers, book chapters and reviews published in the SCOPUS database up to March 8th, 2023. For the primary search, we used the keyword "Piano Key Weir" in the article title, abstract, and keywords. We then refined the search by excluding keywords from aberrant subjects' areas and limiting the results to articles, conference papers, reviews, and conference reviews. This refinement reduced our sample from 230 to 139 results. The dataset was then downloaded in plaintext format for further processing by the software. Considering the importance of this step in the accuracy of our results, a meticulous manual review of the sample was conducted by the authors to eliminate any aberrant data.

The final sample was then subjected to a performance analysis based on publications and citation metrics, along with a bibliometric analysis of authors and countries.

3 Results and Discussion

3.1 Performance Analysis

Our dataset consists of a total of 139 papers. Articles are the most prevalent type of document with a count of 72, followed by conference papers, with a count of 33. In addition, the dataset includes 30 book chapters and 4 review papers [10, 27–29]. These figures highlight the variety of document types in the dataset (See Fig. 1).

The 139 papers that form our dataset were published between 2003 and 8th march 2023, the date of extraction of this data. Within 18 active years of publication, the publications gathered a total citation of 1464, with an average of 10.5 citations per

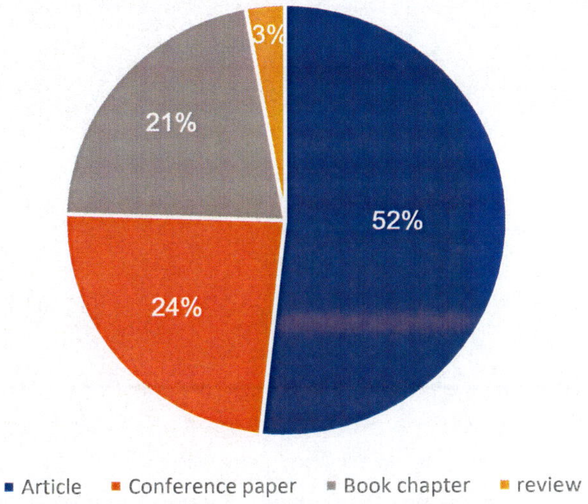

Fig. 1. The distribution of documents by type

publication on PKWs. 105 publications have been cited, which accounts for 75% of all publications with an average citations per cited publication of 13.9.

Our dataset was authored by a total of 243 contributing authors. Among them, 11 papers were solely authored. The remaining documents were the result of collaboration between multiple authors. This provides a collaborative index of 92%, which indicates a significant level of collaboration among researchers.

3.2 Publication Trend

The first paper involving piano key weirs was published in 2003 by Lempérière and Ouamane [10] (See Fig. 2). Up until 2009, only 5 papers were published in SCOPUS with a relatively high citation rate. In 2011, 19 papers were published. This year's papers mark the highest number of citations with 374 citations.

The highest number of publications in a single year was 31 in 2013. Yet, the citation count does not necessarily follow the same trend as the number of publications. There is no clear trend in the citation count over the years, with some years having high citation counts and others having low counts.

The publications number during the years 2020 and 2021 is relatively high, but the citation count does is not as high as in recent years. In summary, publications number has increased over time, but the citation count does not necessarily follow the same pace.

The significant surge of publication during 2011 and 2013 can be attributed to the hosting of an international conference on PKW [30, 31]. These scientific gatherings have a profound impact on research output in the field, acting as catalysts for increased productivity. The fact that two conferences were organized specifically on PKW reflects the importance and interest generated by this innovative design. It is worth mentioning that a third international workshop on PKW was held in 2017, but the publications from this event are not indexed in Scopus, the database utilized for this particular study.

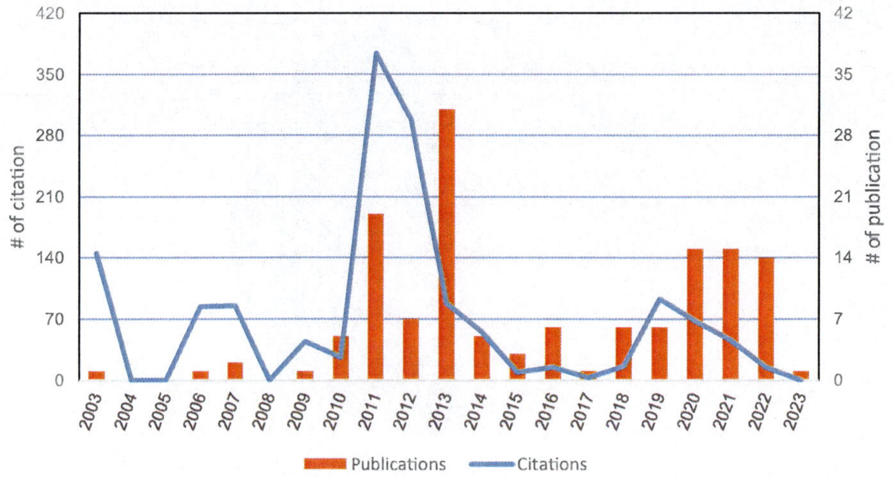

Fig. 2. PKW's publications and citations evolution

3.3 Country Distribution

All articles came from 23 different countries. The country with the highest number of documents published about piano key weirs is France with 32 documents followed by Iran with 26 documents and India with 21 documents. However, in term of citations, France leads the list with 634 citations. Switzerland has the second-highest number of citations (348), despite having only 13 documents, which indicate a high citation impact. Algeria follows with 280 citations for only 7 documents published demonstrating the impact of its research.

Japan is among the countries with the lowest number of documents, only 3 papers, but with relatively high number of citations (82), which indicates the significant impact of its research papers (See Fig. 3).

The results of bibliographic coupling among countries shows that Algeria and France are the first countries to discuss Piano key weirs as they were the first to study this innovative design [10] (See Fig. 4). Despite being ranked second in terms of publication counts, Iran has recently expressed interest in studying PKWs. This suggests that Piano key weir is an active field of study that is gaining interest within the international scientific community. In addition, this analysis will allow researchers to identify organizations around the globe interested to collaborate in the field of PKWs.

3.4 Journal of Publication

All documents about piano key weirs were published in a total of 59 journals. Almost 59% of documents and 78% of total citations were published in the top 10 contributing journals. "Labyrinth and piano key weirs II" is the book with the highest number of documents published on piano key weirs. In term of citations, the 6 documents published by the "Journal of Hydraulic Research" gained the highest number of citations (319 citations), which reflects the significant impact and influence of the research on

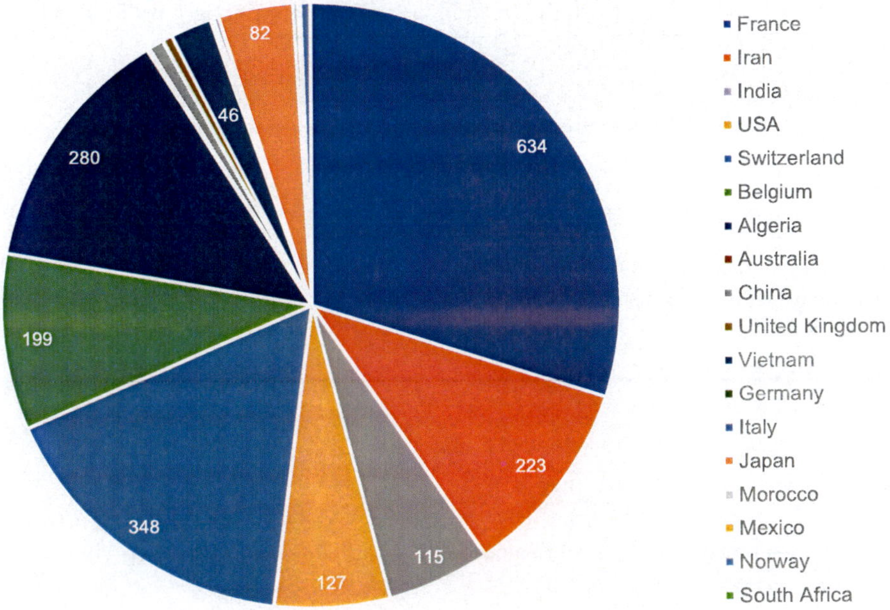

Fig. 3. Country distribution of citations

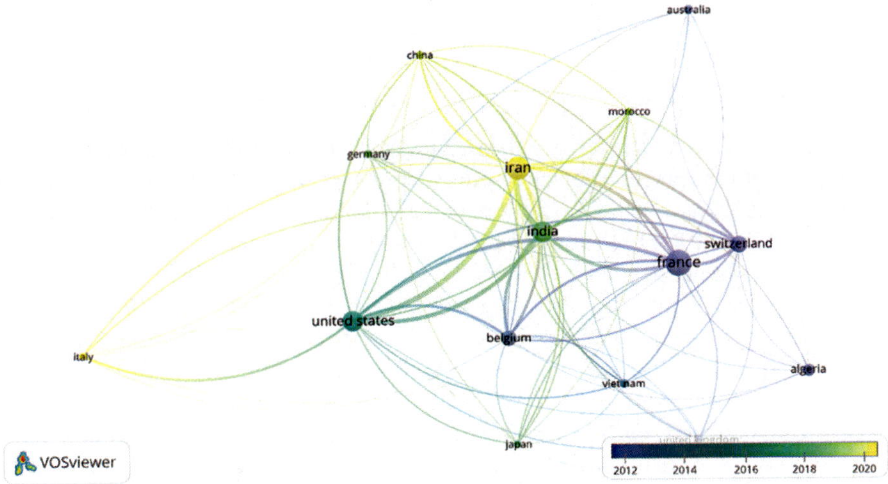

Fig. 4. Overlay visualization of country-based bibliographic coupling

PKWs published in the journal. In addition, the "Journal of Hydraulic Research" has an impressive H-index of 89, a testament to its exceptional contribution in the scientific community. The "International Journal on Hydropower and Dams" have published a modest number of articles (4 papers) on the subject of PKWs, however, those articles received a significant citations count (279 citations). This suggests that this journal may focus on a specific subtopic within the field of PKWs and attract a substantial attention within the specialized communities (See Table 1).

Table 1. Most prolific journals based on the number of publications

Source title	# articles	# citations	H-index [32]
Labyrinth and Piano Key Weirs II	30	88	6
Labyrinth and Piano Key Weirs—Proceedings of the International Conference on Labyrinth and Piano Key Weirs, PKW 2011	15	221	11
Journal of Hydraulic Research	6	319	89
Journal of Hydraulic Engineering	5	157	132
International Journal on Hydropower and Dams	4	279	17
Flow Measurement and Instrumentation	4	35	67
Iranian Journal of Science and Technology—Transactions of Civil Engineering	4	16	22
Water (Switzerland)	4	13	85
Water Supply	4	6	46
ISH Journal of Hydraulic Engineering	3	7	23

3.5 Most Contributing Authors

Using the citation analysis of our dataset on VosViewer, we can visualize the most contributing authors to the subject of piano key weirs. From a total of 243 contributing authors, Laugier f. is the most contributing author considering publications count. Laugier f. published 13 publications, with 5 of them being first-authored. Erpicum S. has produced 10 documents, with 2 of them being first authored. Schleiss a.j, on the other hand, co-authored 9 publications with a cumulative citation count of 296. While Ouamane, A only published 7 documents on PKWs, he has a significant number of citation count indicating the importance of his work. Tullis B.P., Sharma N., Crookston B.M., Boillat J.-L., and Pirotton M. have also made notable contributions with their respective publication records and citation count. These results exhibit information about the productivity of contributing authors to the research domain of piano key weir and the impact of their work (see Table 2 and Fig. 5).

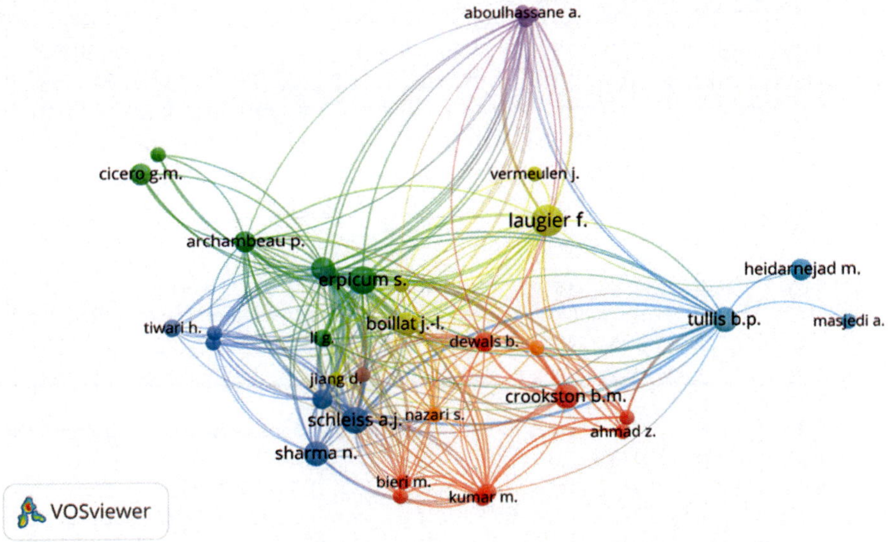

Fig. 5. Network visualization of citation analysis by authors

Table 2. Top 10 contributing authors based on publication number

Author	# Publications	# first authored publication	Citations	Mean citation
Laugier f	13	5	250	19.2
Erpicum s	10	2	207	20.7
Schleiss a.j	9	0	296	32.9
Tullis b.p	8	1	109	13.6
Sharma n	8	2	80	10.0
Crookston b.m	8	4	67	8.4
Ouamane a	7	2	288	41.1
Boillat j.-l	7	0	249	35.6
Pirotton m	7	0	163	23.3

4 Conclusion

The present bibliometric analysis conducted on the literature of Piano key weirs has unraveled PKW as an innovative spillway design, showcasing its superior hydraulic efficiency and cost effectiveness. The study not only revealed the pioneering and influential researches in the field but also identified contributing authors.

Furthermore, the expansion of PKWs have benefited from the international workshops dedicated to exploring the latest studies on labyrinth and PKWs. These seminars provide a platform for scholars and practitioners to exchange ideas, share findings and foster collaborations, resulting in advancement in PKW research.

Overall, this bibliometric review highlights the interest of PKWs and emphasizes the importance of ongoing research and collaboration to advance its development and application in the field of hydraulic engineering.

As for future work, this bibliometric analysis can provide deeper insights by conducting keyword co-occurrence analysis or even co-citation analysis.

References

1. Diodato, V., Gellatly, P.: Dictionary of Bibliometrics, Routledge (1994)
2. White, H., McCain, K.: Visualizing a discipline: an author co-citation analysis of information science. J. Am. Soc. Inf. Sci. **49**, 327–355 (1998)
3. Danesh, F., Mardani-Nejad, A.: A historical overview of bibliometrics. In: Handbook Bibliometrics, pp. 7–18 (2021)
4. Garfield, E., Sher, I.: New factors in the evaluation of scientific literature through citation indexing. Am. Doc. **14**, 195–201 (1963)
5. Garfield, E.: Citation indexes for science. Science **122**, 108–111 (1955)
6. Small, H.: Co-citation in the scientific literature: a new measure of the relationship between two documents. J. Am. Soc. Inf. Sci. **24**(4), 265–269 (1973)
7. Mongeon, P., Paul-Hus: The journal coverage of Web of Science and Scopus: a comparative analysis. Scientometrics **106**, 213–228 (2016)
8. White, H.D., Griffith, B.C.: Author cocitation: a literature measure of intellectual structure. J. Am. Soc. Inf. Sci. **32**, 163–171 (1981)
9. Blanc, P., Lempérière, F.: Labyrinth spillways have a promising future. Hydropower Dams **8**(4), 129–131 (2001)
10. Lempérière, F., Ouamane, A.: The Piano Keys Weir: a new cost-effective solution for spillways. Int. J. Hydropower Dams **10**(5), 144–149 (2003)
11. Barcouda, M., Cazaillet, O., Cochet, P., Jones, B., Lacroix, S., Laugier, F., Odeyer, C., Vingny, J.: Cost-effective increase in storage and safety of most dams using fuse gates or P.K. Weirs. In: Proceedings of the 22nd Congress. ICOLD, Barcelona, Spain (2006)
12. Laugier, F.: Design and construction of the first Piano Key Weir (PKW) spillway at the Goulours dam. Int. J. Hydropower Dams **14**(5), 94–101 (2007)
13. Machiels, O., Erpicum, S., Dewals, B., Archambeau, P., Pirotton, M.: Experimental observation of flow characteristics over a Piano Key Weir. J. Hydraul. Res. **49**(3), 359–366 (2011)
14. Machiels, O., Erpicum, S., Archambeau, F., Dewals, B., Pirotton, M.: Method for the preliminary design of Piano Key Weir. la Houille Blanche **4**(5), 14–18 (2012)
15. Khanh, M.H.T., Hien, T.C., Hai, N.T.: Main results of the PK weir model tests in Vietnam (2004 to 2010). In: Labyrinth and Piano Key Weirs, p. 191 (2011)
16. Da Singhal, G., Sharma, N.: Rehabilitation of Sawara Kuddu Hydroelectric Project–Model studies of Piano Key Weir in India. In: International Workshop on Labyrinths and Piano Key Weirs PKW 2011, pp. 241–250 (2011)
17. Hien, T., Son, H., Khanh, M.: Results of some piano keys weir hydraulic model tests in Vietnam. In: Proceedings of 22e congrès des grands barrages, CIGB/ICOLD, Barcelona (2006)
18. Ouamane, A., Lempérière, F.: Study of various alternatives of shape of piano key weirs. In: HYDRO 2010–Meeting Demands in a Changing World (2010)
19. Aboulhassane, A., Rhouzlane, S., Ouazar, D., Sounni Sliten, H.: Assessment of Piano key weirs cost-effectiveness: a Moroccan case study. Int. Rev. Civil Eng. (IRECE) **8**(5), 212–220 (2017)

20. Leite Ribeiro, M., Albalat, C., Boillat, J.-L., Schleiss, A.J.: Rehabilitation of St-Marc dam: experimental optimization of a piano key weir. In: 32nd IAHR Congress, Venise, Italy (2007)
21. Laugier, F., Lochu, A., Gille, C., Ribiero, M.L., Boillat, J.L.: Design and construction of a labyrinth PKW spillway at Saint-Marc dam, France. Int. J. Hydropower Dams **16**(5), 100–107 (2009)
22. Pinchard, T., Boutet, J.M., Cicero, G.M.: Spillway capacity upgrade at Malarce dam: design of an additional Piano Key Weir spillway. In: Proceedings of International Workshop on Labyrinths and Piano Key Weirs PKW 2011, pp. 233–240 (2011)
23. Erpicum, S., Machiels, O., Dewals, B., Archambeau, P., Pirotton, M.: 1D numerical modeling of the flow over a Piano Key Weir. In: Labyrinth and Piano Key Weirs-PKW, pp. 151–158 (2011)
24. Lempérière, F., Vigny, J.P., Ouamane, A.: General comments on Piano key weirs: the past and present. In: International Conference on Labyrinth and Piano Key Weirs, Liege, Belgium (2011)
25. Donthu, N., Kumar, S., Mukherjee, D., Pandey, N.: How to conduct a bibliometric analysis: an overview and guidelines. J. Bus. Res. **133**, 285–296 (2022)
26. "Elsevier" [Online]. Available: https://www.elsevier.com/solutions/scopus/how-scopus-works/content. Accessed 21 Mar 2023
27. Singh, D., Kumar, M.: Hydraulic design and analysis of piano key weirs: a review. Arab. J. Sci. Environ. **47**(4), 5093–5107 (2022)
28. Tuan, L.A., Hiramatsu, K.: Hydraulic investigation of piano key weir. Rev. Agric. Sci. **8**, 310–322 (2020)
29. Crookston, B.M., Erpicum, S., Tullis, B., Laugier, F.: Hydraulics of Labyrinth and Piano Key Weirs: 100 years of prototype structures, advancements and future research needs. J. Hydraul. Eng. **145**(12), 1–7 (2019)
30. Erpicum, S., Laugier, F., Boillat, J.L., Pirotton, M., Reverchon, B., Schleiss, A.J.: Labyrinth and Piano Key Weirs. CRC Press, Liège (2011)
31. Erpicum, S., Laugier, F., Pfister, M., Pirotton, M., Cicero, G.M., Schleiss, A.J.: Labyrinth and Piano Key Weirs II. CRC Press, Paris (2013)
32. SCImago Journal Rank, SCImago, [Online]. Available: https://www.scimagojr.com/journalrank.php. Accessed 31 May 2023

Evolution of Desalination in Morocco

Zineb Chari(✉), Essediya Cherkaoui, Mohamed Khamar, and Abderrahman Nounah

Civil and Environmental Engineering Laboratory (LGCE), High School of Technology Salé, Mohammed V University in Rabat, Rabat, Morocco
zinebchari@gmail.com

Abstract. The situation of Morocco in a semi-arid zone highly exposed to climatic changes has made it vulnerable to the impacts of climate change. As a result of these changes, Morocco has witnessed an increase in the frequency and intensity of extreme weather events such as droughts and floods.

With this worsening water scarcity situation, the Moroccan government has implemented a project called the National Water Plan, which has set the following major objectives to address climate variability and its impact on water resources:

- Governance and water resource management.
- Interconnection of hydraulic basins.
- Construction of dams.
- Reuse of wastewater for irrigation of green spaces.
- Mobilization of non-conventional resources.

Indeed, the national water plan has given great importance to seawater desalination. This technique has been used in Morocco since 1977 and has been developed through the various stations that have been established in the Saharan provinces as well as other regions of the kingdom.

Overall, desalination has become a crucial solution for Morocco to ensure water security and meet the growing demand for water. It provide drinking water to urban areas, irrigation water to agriculture, and industrial water to various sectors.

Keywords: Climate change · Desalination · Drought · Development

1 Introduction

Morocco, positioned in a semi-arid zone to the south of the Mediterranean, is exceptionally susceptible to climate change. This vulnerability is evident through the rising temperatures and episodes of heavy rainfall. Specifically, between 1981 and 2020, the country witnessed a substantial rise in the national average temperature exceeding 1.4 °C [1]. Furthermore, there was a noteworthy decline in overall precipitation, estimated at 20% [2] from 1960 to 2018.

K. Baba et al. (Eds.): GeoME 2023, SPEES, pp. 464–471, 2023.
https://doi.org/10.1007/978-3-031-49345-4_44

The progression of droughts, worsened by the effects of climate change, is leading to persistent water shortages in numerous hydraulic basins. These shortages have negative consequences across all sectors that rely on water [3]. The prolonged absence of rainfall in the region is identified as the primary cause of both edaphic (related to soil) and hydrological droughts. The Ministry for Water [4] has conducted assessments that reveal a projected deficit in all basins by 2050, amounting to 4 million cubic meters (MMm^3) if climate change is not considered. However, when the impacts of climate change are taken into account, this deficit is estimated to increase to 7 MMm^3. In an effort to mitigate this shortfall, the government has embraced a strategy outlined in the National Water Plan (NWP) Project, which revolves around three complementary approaches [5]:

1. Pursuing the policy of dams to increase water storage capacity…
2. Developing interconnections between dams, to contribute to better water management.
3. Promoting of the mobilization of unconventional water resources.

Indeed, the utilization of seawater desalination has become imperative [3], not only to tackle a portion of the existing and anticipated water deficit but also to guarantee water security for meeting the demands of drinking water supply and industrial water requirements.

Starting from 1977 [8], when the first seawater desalination plant was constructed in Boujdour, Morocco has actively pursued the development of desalination by implementing the desalination plants in the Saharan provinces, which are characterized by scarce precipitation (with an annual precipitation rate of less than 100 mm [1]. With the spread of drought in our country, other regions in north of Morocco, such as the El Hoceima, have also benefited from desalinated water. In fact, in 2020, a desalination plant with a capacity of 200 L [7] per second was commissioned.

With the worsening water scarcity situation in Morocco, the Moroccan government has implemented a project called the National Water Plan [3], which has set the following major objectives to address climate variability and its impact on water resources:

- Governance and water resource management.
- Interconnection of hydraulic basins.
- Construction of dams.
- Reuse of wastewater for irrigation of green spaces.
- Mobilization of non-conventional resources.

2 Government's Strategy Towards the Drought Situation.

The National Water Plan 2050 project aims to bridge the gap between water demand and resources by 2050 [6]. This gap is estimated to be around 7 billion cubic meters of water. This will be achieved through the adoption of a series of technical solutions, primarily including:

- Reducing water losses in the transportation and distribution of drinking water to 0.4 billion cubic meters.
- Saving 1.8 billion cubic meters of water per year in the agricultural sector through irrigation modernization.

- Constructing dams and interconnections to provide an additional 3 billion cubic meters.
- Implementing desalination methods to produce 1 billion cubic meters of water.
- Utilizing wastewater treatment for an additional 0.3 billion cubic meters.
- Collecting rainwater for an additional 0.3 billion cubic meters.

3 Evolution of Desalination in Morocco:

3.1 Desalination/demineralization in Morocco Between 1975 and 1995

This period, between 1975 and 1995, is characterized by the testing of technologies for seawater desalination and/or brackish water demineralization [8].

The following graph (Fig. 1) shows the evolution of flow rates produced by desalination and/or demineralization stations using various technologies, namely electrodialysis, distillation, and reverse osmosis.

This period is characterized by the construction of the demineralization plants in Tarfaya and Smara cities and the desalination plant in Boujdour city, using multiple techniques and processes such as electrodialysis, distillation, and reverse osmosis. It is a period of experience feedback to assess the most suitable desalination process for the national context.

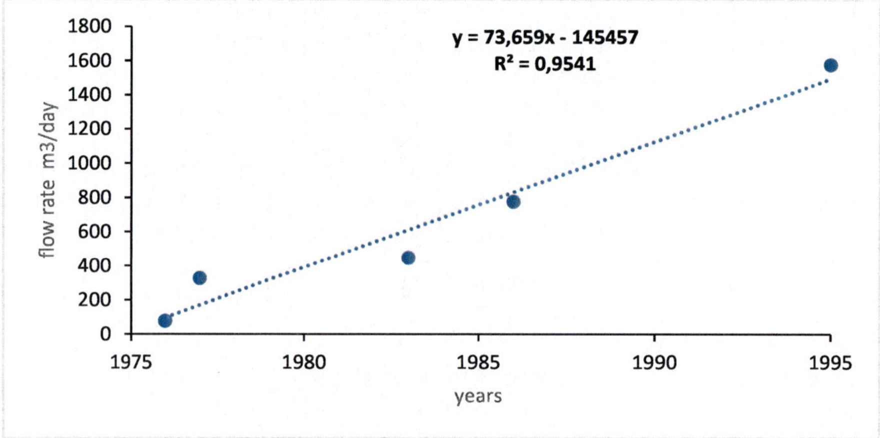

Fig. 1. Evolution of desalination/demineralization production flow rate between 1975 and 1995

3.2 Desalination in Morocco Between 1995 and 2022 [7]

Starting from the year 1995, we will present the evolution of desalination in Morocco without considering the demineralization of brackish waters.

The period between 1995 and 2022 is characterized by the commissioning of twelve desalination plants, with the largest being the one in Agadir, with a capacity of 400,000 m^3/day (275,000 m^3/day in the first phase). The technology adopted for all these plants is reverse osmosis (Fig. 2).

Between the years 1995 and 2012, seawater desalination has been focused on the Saharan provinces, particularly in the regions of Laayoune-Sakia Elhamra and Dakhla-Oued Dahab, characterized by low rainfall levels, with an average observed for the past decade of around 60 mm [12]. The desalination projects implemented in these regions primarily aim to meet the population's drinking water needs.

Starting in 2016, with recurring droughts, desalination has been adopted by other regions in addition to the Saharan provinces. One example is the desalination project at the industrial complex of Jorf Lasfar in the Casablanca-Settat region, with an annual capacity of 25 million cubic meters. This project was carried out by the Office Chérifien des Phosphates (OCP) to meet its industrial water needs.

Another project was implemented by the National Office of Electricity and Drinking Water (ONEE) in the Tanger-Tetouan-El Hoceima region in northern Morocco, with an annual capacity of 6.2 million cubic meters to meet the drinking water needs of the city of El Hoceima.

The year 2022 was marked by the commissioning of a large desalination plant in Chtouka, in the Souss Massa region, with an annual capacity of 100.4 million cubic meters. This project was a collaboration between the Ministry of Agriculture and the National Office of Electricity and Drinking Water.

The same year also saw the commissioning of the new desalination plant in Laayoune with an annual capacity of 9.5 million cubic meters. This is the third desalination plant in the city, which has experienced significant socio-economic development [13].

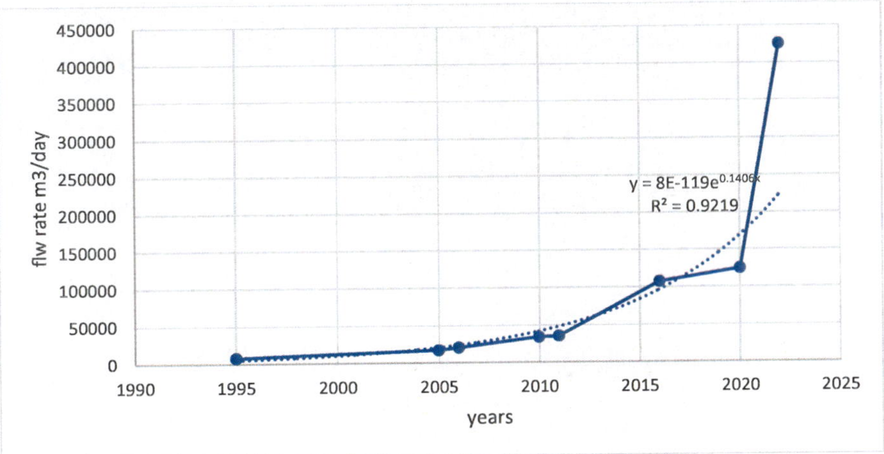

Fig. 2. Evolution of desalination production flow rate between 1995 and 2022

3.3 Desalination in Morocco Taking into Account Ongoing Construction Projects [9]

In addition to the operational plants, several other stations are under construction, such as the one in Dakhla with a capacity of 68,000 m^3/day, the Sidi Ifni station with a capacity of 8,650 m^3/day, and the smaller desalination plant in Tarfaya with a capacity of 860 m^3/day.

To meet the demand for both drinking water and irrigation water, several desalination plants have been put into operation, with the largest being the Chtouka plant with a capacity of 275,000 m^3/day (150,000 m^3/day for drinking water and 125,000 m^3/day for irrigation), and the ElHoceima plant with a capacity of 17,000 m^3/day (drinking water).

As for the industrial water needs of the Cherifian Office of Phosphate (OCP), they have been met through the El Jorf desalination plant with a capacity of 68,000 m^3/day, and the Boukraa plant with a capacity of 7,000 m^3/day.

The following map (Fig. 3) displays all the desalination plants aimed at meeting the demands for drinking water, irrigation, and industrial needs.

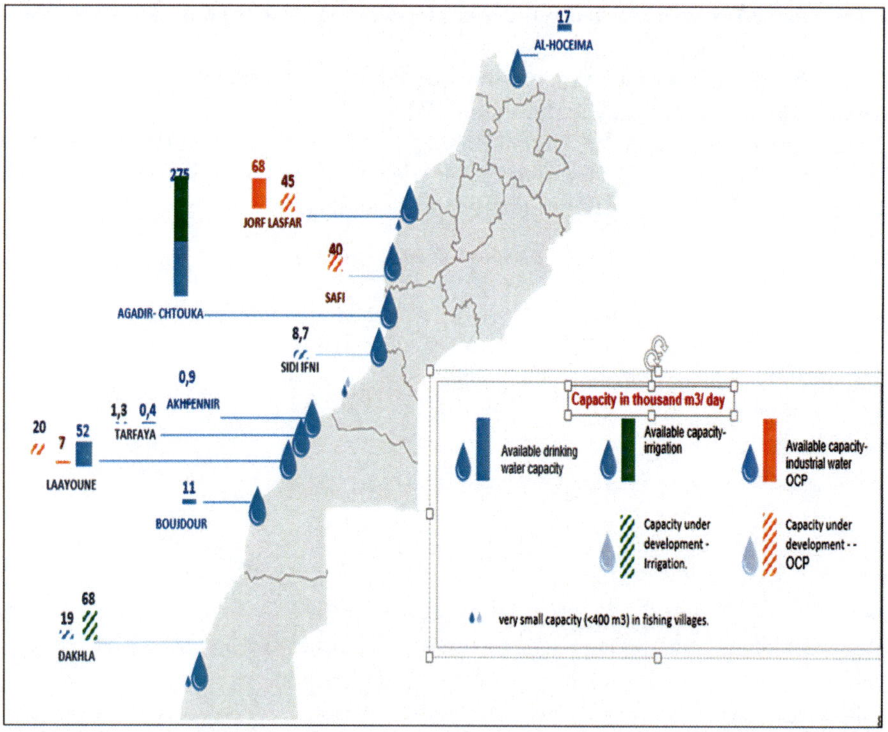

Fig. 3. Desalination plants in Morocco, taking into account ongoing construction projects

3.4 Desalination Plants Projected to Be Commissioned Starting from 2027 and Currently Under Development

The National Water Plan 2050 project, in line with the orientations of the New Development Model [5], envisions desalination projects with a capacity of nearly 1 billion m^3. These projects will be implemented gradually, taking into account the intensity of climate change.

The following map presents ongoing development projects, including the major desalination plant in Casablanca regions, which will have a capacity of 300 million m^3 per year upon completion [10] (Fig. 4).

The Casablanca desalination project aims to establish a desalination plant with a projected capacity of 300 million cubic meters per year to supply and secure the water demand of the Casablanca-Settat region, as well as meet the irrigation needs of approximately 50 hectares of agricultural land in the vicinity of Sidi Rahhal.

This project will be carried out in two phases. The first phase, with a capacity of 200 million cubic meters per year, is scheduled to be operational by 2027. The second phase, with a capacity of 100 million cubic meters per year, is planned to be operational by 2035 [11].

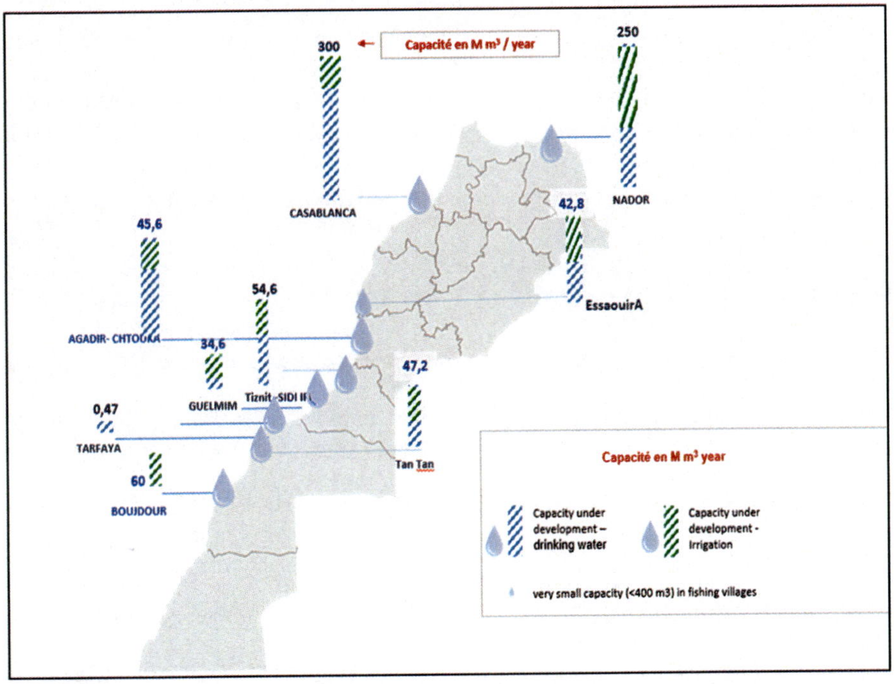

Fig. 4. Desalination plants projected to be starting from 2027 and currently under development

4 Conclusion

The main strategic orientations of the national water plan project are based on the following three important axes [3]:

- Pursuit and strengthening the development of supply policy
- Water demand management
- Preserving the management of extreme climate events.

The use of desalination in addressing these three axes will allow for [8]:

- Securing drinking water supply in all coastal regions
- Reducing water stress in coastal regions
- Meeting industrial water needs in coastal areas
- Preserving existing irrigated areas or developing new ones (such as Dakhla, currently being launched)
- Enhancing the resilience of water supply systems to climate change.

Key recommendations [4]

- Optimize the program for constructing new dams by targeting watersheds that have proven efficiency, economic profitability, and minimal negative social and environmental impacts. Promote seawater desalination primarily for drinking water supply and, to a lesser extent, for irrigation of high-value crops in areas where farmers have the capacity to bear the cost of desalinated water. Modernize irrigation systems while ensuring adequate support for farmers in terms of production, marketing, and value-added activities. Promote the reuse of treated wastewater, particularly for watering green spaces and crops, ensuring compatibility with required treatment processes.

References

1. Report from the Ministry of Equipment, Transport, Logistics, and Water (DRPE), February 2021 - Morocco, State of the climate in 2020. 3
2. DEPF: Morocco facing climate change: situation, impacts, and response policies in the Climate and Development Report in Morocco. Sectors of water and agriculture. Policy Brief - DEPF No.18 (2020)
3. Climate and Development Report in Morocco Middle East and North Africa Oct 2022 page 3. 23
4. White Paper on Water Resources in Morocco - for sustainable management ensuring the country's water security Oct 2022
5. The New Development Model, Kingdom of Morocco, April 2021, General and thematic report
6. National Water Plan-Kingdom of Morocco (2021)
7. Webinar about the opportunities in seawater desalination in Morocco, Dutch Embassy in Rabat, June 2022
8. Formation of ONEE-BO- Desalination in Morocco June 2005
9. The second international symposium on nanomaterials and membrane science for water, energy and environment desalination and reuse face to water scacity June 2022

10. Budget Engagement Program 2023, ONEE (National Office of Electricity and Drinking Water)
11. Study on the Implementation of a Sea Water Desalination Project in the Casablanca-Settat Region - Water Research and Planning Department (2021)
12. Monograph of the Laayoune Boujdour-Sakia Lhamra region, 2013 - Regional Directorate of Laayoune, High Commission for Planning. Page 5
13. Review of Geographical Space and Moroccan Society, Issue 70, April 2023, page 154

Linear Hydrography Mapping Using Airborne Lidar

Kawtar Chaari(✉) and Latifa Ouadif

Mohammadia School of Engineers, Laboratory of Applied Geophysics, Geotechnics, Engineering Geology and the Environment (L3GIE), Rabat, Morocco
kawtar_chaari@um5.ac.ma

Abstract. Mapping and hydrological modeling work together to prevent climate hazards, namely floods. The mapping of the hydrological network is generally based on aerial or satellite photos. Admittedly, this method does not reflect the reality on the surface. Unlike the lidar as an active sensor it is able to pass through the gaps and reach the surface and the objects lying below. It is in this context that we have launched a campaign of topographic surveys of upper ZIZ watershed covering approximately 400 km^2 using state-of-the-art technology, airborne lidar. The objective of this work is to present the difference in the level of precision between a hydrographic network resulting from lidar technology and the one resulting from an aerial imagery. We apply a methodology based on, first, the production of DEM from Lidar data then extracts hydrological network as well as the delimitation of sub watershed and finally compare the result with photogrammetry DEM wand our reference network using stereographic mode. The high resolution and precision of LiDAR data can't replace definitively the work in the field. Certainly, the combination of this technology with field visits and aerial photography helps with the organization of in site work and logistics much easier for the researchers and scientists.

Keywords: Lidar · Hydrography · Remote sensing · Aerial photos · Cartography

1 Introduction

Field observations are an essential phase and the starting point for any study and analysis of landscape change and its evolution. In this phase, specialists use geospatial data to organize field observations.

The evolution of geospatial data over the years has made it possible to detect morphology, topography and changes in the terrain without having to travel to the site. The hydrographic network being a large-scale study, is generated from digital models coming from different sources, namely: satellite images, aerial photography (photogrammetry) and manual surveys (Farr and Kobrick, 2000; Kinsey-Henderson and Wilkinson, 2013; Han et al., 2012). Digitals models derived from these methods now have coarseness, as they failed to easily identify topography mainly in areas of high-density plantations or forest.

K. Baba et al. (Eds.): GeoME 2023, SPEES, pp. 472–480, 2023.
https://doi.org/10.1007/978-3-031-49345-4_45

The development of topographic surveying tools (total stations and GNSS receivers) has improved the resolution of the details surveyed, which leads to a better analysis of the morphology of the terrain. Admittedly, the acquisition rate is limited by the duration and accessibility to the field.

In the 1990s, specialists and decision-makers turned their interest to airborne lidar (detection and distance estimation using light) technology. A new source of data, the point cloud, which stands out for its precision and speed. The nature of the derivatives of this technology (Slatton et al., 2007) has made it possible to increase and facilitate large-scale research related to climatic hazards.

2 Materials and Methods

2.1 State of Art

Understanding the connection between the different processes and mechanisms that form a landscape is one of the main motivations for geomorphological studies. Several models and algorithms have been created and developed to test these mechanisms and phenomena like landslide, flow calculation, hydrological network detection, ect. At first scientists used contour maps with an accuracy ranging from 10 to 100 m as a basis for this research. With the digitization and development of digital terrain models, a simple shift of a detail on the map of a few centimeters compared to traditional maps can cause a shift of hundreds of meters on the ground. Over time, several countries around the world have decided to move from traditional mapping to lidar technology as a basis for mapping geological and natural hazards [16] (Madin and Frankel, 2010). France with its Lidar HD program from IGN which managed to cover 202,312 km^2 of its territory with an accuracy of 10 points/m^2 (mid-March 2023). United States public funds have been set up for the acquisition of lidar data (e.g. Oregon, Kentucky, Pennsylvania, North Carolina, South Carolina). The success of these programs provides the opportunity for professionals and institutions to implement and study phenomena related to landscape change, flood prevention, urban and forest management, etc.

2.2 Study Area

The Ziz basin occupies part of the eastern High Atlas. This dominates, to the south, the domain of the hamadas by the great southern Atlas massif (south Atlas furrow) and, to the north, the Moulouya and the high eastern plateaus. The outlet of the upper Ziz basin is controlled by the Sise station at Foum Zaabel. The volume of average annual inflows from the watercourse at this site is 104 Mm^3, which can vary from 6 Mm^3 (1983–84) to 321 Mm^3 (1995–96). The perimeters of the upper basin are irrigated by small and medium hydraulic facilities (PMH) for an area of 4970 ha and a mobilized volume of 35 Mm^3. Apart from prolonged droughts, the needs of this area are met by surface water. The Ziz wadi has its source in the eastern High Atlas and ends, after 282 km, in the Sahara desert. It is considered one of the longest and most powerful wadis in North Africa (Fig. 1).

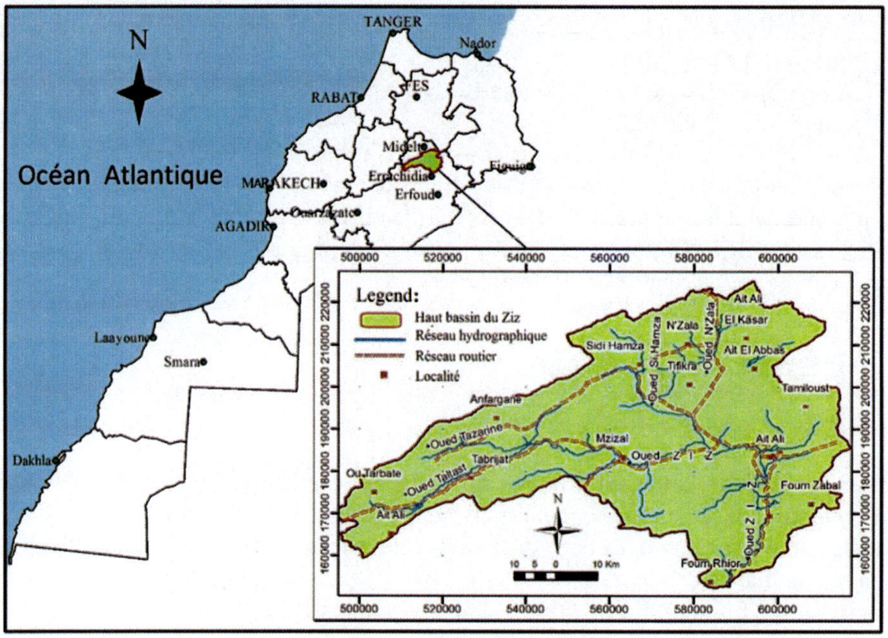

Fig. 1. Location of study area: Ziz basin

2.3 Collection and Processing of LiDAR Data

The LiDAR data was acquired as part of a project to create an early warning platform, commissioned by the Guir, Ziz and Ghris Watershed Agency. The LiDAR readings were carried out by S.E.P.R.E.T-Sarl. The LiDAR used in this study is the ALS70 HP and UltraCam camera in a Cessna 404 aircraft from October 5, 2021 to December 15, 2021. The data was delivered in LAS 1.2 format with a point spacing of 2 points/m^2 simultaneously with an aerial imagery with a resolution of 20 cm. The reference system for this dataset is EPSG26191 (Lambert Nord Maroc).

3 Results and Discussion

The lidar technology can obtain a cloud of points which is characterized by return number (varies between 0 and 6), intensity, three-dimensional coordinates (x, y, z). We will use the raw Lidar data to first filter data into two classes ground and on ground. The availability of lidar data makes digital models more geometrically strong (Jos. M, Estefanía. A, Miguel. E, Juan. G 2023).

In order to extract the hydrographic network, we first started by extracting the digital terrain model from the Lidar data which will then allow us to extract the hydrographic network, delimit the sub-watersheds and a basis for the geometric creation of hydraulic modelling.

3.1 Production of the Digital Elvation Model

The creation of the digital terrain model which is mainly based on the elimination of vegetation and buildings. We have filtered the ground class automatically by combining the reflection, the intensity of the echoes returned and the geometry defined by the neighboring points [15].

It has been observed that the quality of the digital terrain model mainly depends on the nature of the topography of the study area and the density of points per square meter [21].

Figure 2 presents the final result of the production of the numerical model derived from the LiDAR data. It is a GIS file in raster format and continuous type with a spatial resolution of 1 m.

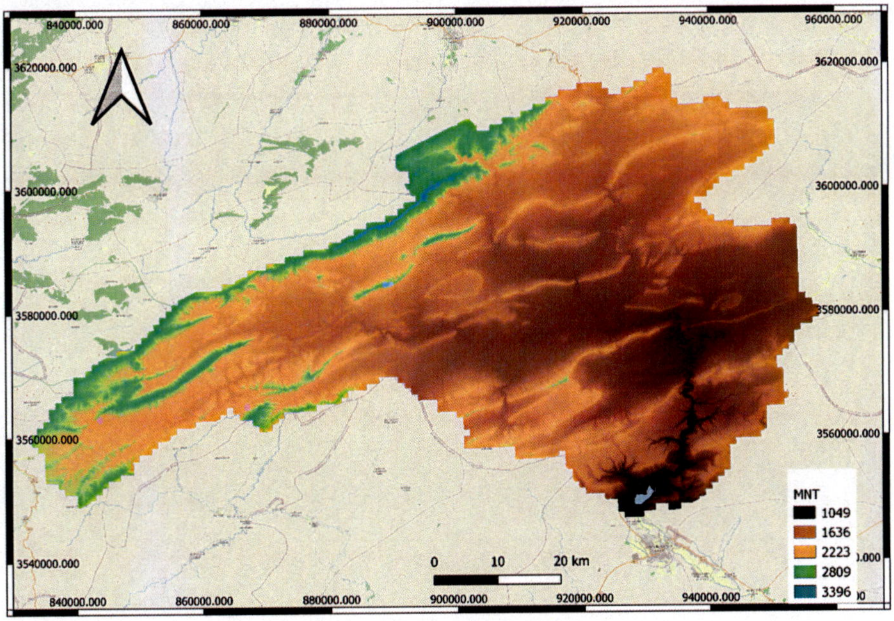

Fig. 2. Digital elevation model

3.2 Extraction of the Hydrographic Network and the Subwatershed

The extraction of the hydrographic network from a high-resolution digital terrain model plays a very important role in hydrological studies (Moore et al., 1991; Garbrecht and Martz, 2000; Wilson, 2012). Generally, there are two extraction approaches. The first is based on the recognition of river valleys or ridge lines from topographic and geomorphological data [5, 23, 29] (Tribe, 1992; Rulli, 2010; Matsuura and Aniya, 2012; Peltier, 2013). The second approach is based on the determination of surface flow paths from the digital terrain model. (O'Callaghan and Mark, 1984; Jenson and Domingue, 1988;

Tarboton, 1997; Martz and Garbrecht, 1999; Turcotte et al., 2001; Nobre et al., 2011; Choi, 2012).

This approach is adopted in software such as QGIS and ESRI ArcHydro Tools (Moore et al., 1993; Yang et al., 2005), TOPAZ (Martz and Jong, 1988; Martz and Garbrecht, 1999), GRASS (Metz et al., 2011) and TecDEM (Shahzad and Gloaguen, 2011). To make sure that the model does not include any sink, the "Fill Sinks (wang and liu)" tool from the SAGA library on QGIS is iteratively exploited until the DEM keeps the same histogram. The extraction of the hydrographic network, we exploited the utilities of the software QGIS (Channel networks and drainage basins, Upslope area) (Fig. 3).

The following map presents the result of the extraction of the hydrographic network as well as the black points identified at the level of this basin, and which will be the essential part in the delimitation of the sub-watershed, and the hydraulic models. We have cut the upper Ziz into seven sub-basins, and each black dot presents the outlet of a sub-watershed (Tables 1 and 2).

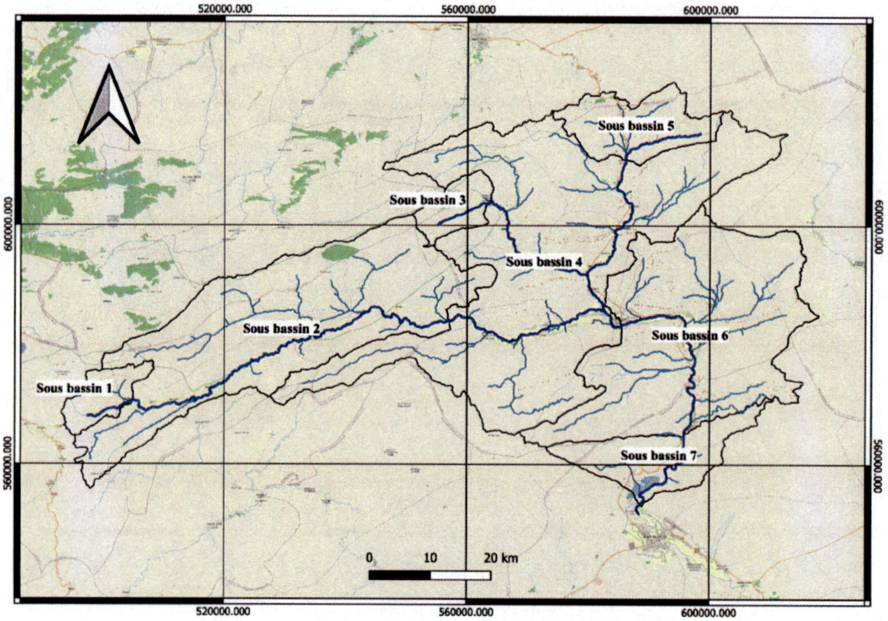

Fig. 3. Hydrogical network and subwatershed

3.3 Discussion

The results obtained from this study are raw data. In order to validate the accuracy of these data, we used a reference hydrological network.

We divided the project data into three categories: high-density vegetation areas, medium-density vegetation areas, and bare ground area. With the stereographic interpretation, we can compare the hydrological network producing lidar data with that of

Table 1. Geomorphological data of the subwaterhed

Watershed	Area (km^2)	Perimeter (km)	Length (km)	Zmax (m)	Zmin (m)	Difference in height (m)
BV 1	92.22	65.12	9.72	3038.56	2103.21	935.35
BV 2	1100.10	294.63	89.97	3331.14	1498.19	1832.95
BV 3	91.95	58.01	10.81	3428.08	1644.36	1783.72
BV 4	1641.55	458.17	69.87	3679.92	1308.18	2371.74
BV 5	216.84	108.97	14.96	2806.56	1621.03	1185.53
BV 6	1049.89	241.35	55.83	2311.94	1215.13	1096.81
BV 7	235.12	107.69	25.69	2247.72	1056.22	1191.50

Table 2. Geomorphological data of the subwatershed

Watershed	Slope (m/m)	Equivalent length (km)	Equivalent width (km)	Gravelius index	Horton index
BV 1	0.10	29.70	3.11	1.91	4.74
BV 2	0.02	140.59	7.82	2.51	6.11
BV 3	0.17	25.64	3.59	1.71	4.25
BV 4	0.03	223.45	7.35	3.19	11.75
BV 5	0.08	50.61	4.28	2.09	7.25
BV 6	0.02	112.22	9.36	2.10	9.40
BV 7	0.05	49.50	4.75	1.98	4.58

the aerial photo and our reference network. The first category, the offset is 50 cm (photo) and 7cm (reference network). The second category, the offset is 35 cm (photo) and 5 cm (reference network). The third category the offset is 5 cm (photo) and 2 cm (reference network).

Comparing the hydrographic network derived from Lidar data (2points/m^2) and that of aerial photographs (GSD 20 cm) with the reference network, we have observed that the difference between the two networks in the open areas is less than 5cm, on the other hand, in areas of high and medium density plantations, the DTM of aerial photography is less suitable for the reference network.

The high precision of LiDAR data brings several advantages to geomorphological studies and research. We manage to detect the fine details of the relief, including: wrinkles, invoices which are generally neglected in the other types of data. This will allow to better obtain and analyze the studied landscape in combination with other data sources.

LiDAR remains a sophisticated medium-fast instrument in terms of acquisition, we manage to capture and cover large areas. Certainly, the process of processing and producing point clouds is long and requires a precise system architecture, without neglecting the manual control of these data which is necessary after the automatic classification.

4 Conclusion

In large scale surveys it is nearly impossible to visit all sites to interpret landforms. The use of the remote sensing technology as LiDAR offers many advantages, efficiency and cost over traditional technologies. It provides high resolution and accurate data over large areas. Additionally, this powerful technology improves our understanding of hydrological network and reduces the risks associated with flooding.

The comparison of the two DEMs LiDAR and photogrammetry with our reference network in stereographic mode to show that the added value that lidar brings to photogrammetry at the level of precision is seen in the vegetation areas. The precision of LiDAR data cannot definitively replace the work in the field. Certainly, the combination of this technology with field visits and aerial photography will facilitate the logistics of research for geoscientists.

References

1. Abdelkarim, H.M. Alogayell, I.I. Alkadi, I., Youssef, I.: Spatial–temporal prediction model for land cover of the rural–urban continuum axis between Ar-Riyadh and AlKharj cities in KSA in the year of 2030 using the integration of CA–Markov model, GIS-MCA, and AHP. Appl. Geom. **14**(3), 501–525 (2022). https://doi.org/10.1007/s12518-022-00448-w. Last accessed 02 Feb 2023
2. Ahnert, F.: Brief description of a comprehensive three-dimensional processresponse model of landform development. Z. Geomorphol. **25**, 29–49 (1976)
3. Amouch, S., Akhssas, A.: Drought variability in Agadir's Region (Southern Morocco)—recent and future trends. Ecol. Eng. Environ. Technol. **24**(2), 241–250 (2023). https://doi.org/10.12912/27197050/157169
4. Amouch, S., Akhssas, A., Bahi, L., Bennouna, R.: Characterisation of recent and future climatic trends in the region of guelmim (Morocco). In: E3s Web of Conferences 150, 0 (2020). https://doi.org/10.1051/e3sconf/20201500
5. Band, L., Wood, E.F.: Computer graphics for distributed hydrologic modeling. Paper No. H22B-03. Am. Geophys. Union, Spring Meet., Baltimore, Md (1986)
6. Bowles, C., Cowgill, E.: Discovering marine terraces using airborne LiDAR along the Mendocino-Sonoma coast, northern California. Geosphere **8**, 386–402 (2012)
7. Burns, W.J., Coe, J.A., Kaya, B.S., Ma, L.: Analysis of elevation changes detected from multi-temporal LiDAR surveys in forested landslide terrain in western Oregon. Environ. Eng. Geol. **16**, 315–341 (2010)
8. Webster, C., Essery, R., Mazzotti, G., Jonas, T.: Using just a canopy height model to obtain lidar-level accuracy in 3D forest canopy shortwave transmissivity estimates. Agric. Forest Meteorol. **338**, 109429 (2023). ISSN 0168-1923. https://doi.org/10.1016/j.agrformet.2023.109429
9. Carter, W., Shrestha, R., Tuell, G., Bloomquist, D., Sartori, M.: Airborne laser swath mapping shines new light on Earth's topography. EOS Trans. Am. Geophys. Union **82**, 549–555 (2001)

10. Challis, K.: Airborne laser altimetry in alluviated landscapes. Archaeol. Prospect. **13**, 103–127 (2006)
11. Church, M.: Refocusing geomorphology: field work in four acts. Geomorphology 1–9 (2013). https://doi.org/10.1016/j.geomorph.2013.01.014
12. Dietrich, W.E., Bellugi, D., Real de Asua, R., Iordache, I., Allen, D., Napolitano, M., Trso, M.: The use of airborne laser swath mapping data in watershed analysis to guide restoration priorities: the Napa River Watershed Study. AGU Fall Meeting (2004)
13. Duffy, B., et al.: Fault kinematics and surface deformation across a releasing bend during the 2010 MW 7.1 Darfield, New Zealand, earthquake revealed by differential LiDAR and cadastral surveying. Geol. Soc. Am. Bull. **125**, 420–431 (2012)
14. Fatiha, A., Latifa, O., Ahmed, A.: Monitoring land use and land cover changes using remote sensing techniques and the precipitation-vegetation indexes in Morocco. Ecol. Eng. Environ. Technol. **24**(1), 272–286 (2023). https://doi.org/10.12912/27197050/154937
15. Vosselman, G.: Slope based filtering of laser altimetry data. Int. Arch. Photogrammetry Remote Sens. **33**(Part B3), 935–942 (2000)
16. House, P.K.: My love affair with lidar: confessions of an inveterate surficial mapper. Geol. Soc. Am. Abstr. Programs **42**, 39 (2010)
17. Jasiewicz, J., Metz, M.: A new GRASS GIS toolkit for Hortonian analysis of drainage networks. Comput. Geosci. **37**(8), 1162–1173 (2011). https://doi.org/10.1016/j.cageo.2011.03.003
18. Jennifer, R., Bernhard, H., Stefaan, D., Jos, O., Jan, D.: Drainage ditch extraction from airborne LiDAR point clouds. ISPRS J. Photogramm. Remote. Sens. **146**, 409–420 (2018). https://doi.org/10.1016/j.isprsjprs.2018.10.014
19. Bodoque, J.M., Aroca-Jiménez, E., Eguibar, M.Á., García, J.A.: Developing reliable urban flood hazard mapping from LiDAR data. J. Hydrol. **617**, 128975 (2023). https://doi.org/10.1016/j.jhydrol.2022.128975. Last accessed 13 May 2023
20. Roering, J.J., Mackey, B.H., Marshall, J.A., Sweeney, K.E., Deligne, N.I., Booth, A.M., Cerovski-Darriau, C.: 'You are HERE': connecting the dots with airborne lidar for geomorphic fieldwork. Geomorphology **200**, 172–183 (2013). https://doi.org/10.1016/j.geomorph.2013.04.009
21. Liu, X., Zhang, Z., Peterson, J., Chandra, S.: LiDAR-derived high quality ground control information and DEM for image orthorectification. GeoInformatica **11**(1), 37 (2007)
22. del Val, M., Iriarte, E., Arriolabengoa, M., Aranburu, A.: An automated method to extract fluvial terraces from LIDAR based high resolution digital elevation models: the Oiartzun valley, a case study in the Cantabrian Margin. Quat. Int. **364** (2015). https://doi.org/10.1016/j.quaint.2014.10.030
23. Passalacqua, P., Do Trung, T., Foufoula-Georgiou, E., Sapiro, G., Dietrich, W.E.: A geometric framework for channel network extraction fromlidar: nonlinear diffusion andgeodesic paths. J. Geophys. (2010)
24. Yang, P., Ames, D.P., Fonseca, A., Anderson, D., Shrestha, R., Glenn, N.F., Cao, Y.: What is the effect of LiDAR-derived DEM resolution on large-scale watershed model results? Environ. Modell. Softw. **58**, 48e57 (2014). https://doi.org/10.1016/j.envsoft.2014.04.005
25. Rui, B., Tiejian, L., Yuefei, H., Jiaye, L., Guangqian, W.: An efficient and comprehensivemethod for drainage network extraction from DEM with billions of pixels using a size-balanced binary search tree. Geomorphology **238**, 56–67 (2015)
26. Semlali, I., Ouadif, L., Bahi, L.: Landslide susceptibility mapping using the analytical hierarchy process and GIS. Curr. Sci. J. **116**(5), 773–779 (2019)
27. Souidi, H., Ouadif, L., Bahi, L., Elhachmi, D., Edderkaoui, R.: Application of stochastic process & cellular automata integrated to GIS for land monitoring: Coastal Chaouia, Morocco. Int. J. Adv. Res. Eng. Technol. (IJARET) **11**(5), 245–252 (2020)

28. Vosselman, G.: Slope based filtering of laser altimetry data. IAPRS XXXIII, 935–942. Research **115**(F1) (2000). https://doi.org/10.1029/2009JF001254F01002
29. Yoeli, P.: Computer-assisted determination of the valley and ridge lines of digital terrain models. Int. Yearb. Cartogr. **24**, 197–205 (1984)

Author Index

K. Baba et al. (Eds.): GeoME 2023, SPEES, pp. 481–482, 2023.
https://doi.org/10.1007/978-3-031-49345-4